SELECTED PAPERS OF DEMETRIOS G. MAGIROS

DEMETRIOS G. MAGIROS

# Selected Papers of Demetrios G. Magiros

## *Applied Mathematics, Nonlinear Mechanics, and Dynamical Systems Analysis*

Edited by

S. G. Tzafestas

*Electrical Engineering Department, National Technical University, Athens, Greece*

SPRINGER-SCIENCE+BUSINESS MEDIA, B.V.

**Library of Congress Cataloging in Publication Data**

CIP

Magiros, Demetrios G.
Selected papers of Demetrios G. Magiros.

"Published on behalf of the Greek Mathematical Society."
Bibliography: p.
1. Mathematics–Collected works. 2. Nonlinear mechanics–Collected works. 3. System analysis–Collected works. I. Tzafestas, S. G., 1939- . II. Hellénikê Mathêmatikê Hetaireia (Greece)
QA3.M277 1985 510 85-2204

ISBN 978-94-010-8869-5 ISBN 978-94-009-5368-0 (eBook)
DOI 10.1007/978-94-009-5368-0


Originally published by D. Reidel Publishing Company, Dordrecht, Holland 1985
Softcover reprint of the hardcover 1st edition 1985

# FOREWORD

*The theory of nonlinear oscillations and stability of motion is a fundamental part of the study of numerous real world phenomena. These phenomena, particularly auto-oscillations of the first and second kind, capture, parametric, subharmonic and ultraharmonic resonance, asymptotic behavior and orbits' stability, constitute the core of problems treated in "Nonlinear Mechanics", and their study is connected with the names of H. Poincaré, A.M. Lyapunov, N.M. Krylov and N.N. Bogolyubov.*

*Professor Demetrios Magiros, a widely known scientist in the theories of oscillations and nonlinear differential equations, has devoted his numerous works to this significant part of modern physical science. His scientific results can be classified in the following way:*

*1) creation of methods of analysis of subharmonic resonances under the nonlinear effect,*
*2) determination and analysis of the main modes of nonlinear oscillations on the basis of infinite determinants,*
*3) analysis of problems of celestial mechanics,*
*4) classification of stability of solutions of dynamic systems concepts,*
*5) mathematical analogs of physical and social systems.*

*He has developed new methods and solutions for a great number of difficult problems of nonlinear mechanics making a significant contribution to the theory and applications of the field.*

*Urgency, depth of perception of the considered phenomena, and practical directness are characteristics of his work.*

*D.G. Magiros' scientific findings have been utilized in the research of many scientists and can be applied to a variety of branches of mechanics, physics and engineering, in which nonlinear oscillating processes and problems of stability are involved.*

*Dr Magiros participated in several International Conferences, among which the International Conferences on Nonlinear Oscillations held in Kiev (1969, 1981).*

*We are sure that the publication of Professor D.G. Magiros' Selected Papers will be accepted by numerous specialists, working on the theory and application of nonlinear oscillations, with great interest.*

*Yu. A. Mitropolsky, Academician*
*A.A. Martynyuk*
*Ukrainian Academy of Sciences*
*Kiev, USSR*

# IN MEMORY OF DIMITRIS MAGIROS

*The Greek Mathematical Society is pleased to sponsor the publication of the "Selected Papers" of a prominent member Dimitris G. Magiros, honoring his memory and his work.*

*The distinguished mathematician graduated from the National Capodistrian University of Athens and continued his studies and work in the USA, where he dedicated his time principally to research in applied mathematics and mechanics.*

*Our colleague Dimitris, in spite of the fact that he had spent most of his scientific life abroad, was an untiring collaborator of our Society and has particularly shared our efforts for the advancement of mathematics in Greece.*

*We believe that this volume, where most of his publications -scattered in various scientific sources- are made collectively accessible, will facilitate the propagation of the thoughts of an inspiring researcher for the benefit of the international mathematical and systems science community.*

*The Governing Board of the*
*Greek Mathematical Society*

# PREFACE

*It is a distinct honor and privilege to be the editor of the Selected Papers of Dr Demetrios G. Magiros.*

*Magiros' scientific work continued for about forty years bearing important contributions of theoretical and practical significance. His enormous interest in the applications of mathematics, and the versatility of problems which he has solved in the fields of his applications are reflected in these papers.*

*Magiros' thoughts have left an impact on a variety of fields such as nonlinear differential equations, subharmonic oscillations, principal modes of nonlinear systems, celestial and orbital mechanics, stability analysis of nonlinear dynamical systems, precessional phenomena, and other.*

*In his investigation, Magiros starts from a philosophical basis and follows an axiomatic principle aiming always at the mathematical description and interpretation of the reality with emphasis on the applications.*

*Dr Magiros was an enthusiastic teacher with the talent to transmit both the depth and width of his knowledge. I personally remember his excellent lectures at NRC "Demokritos" in Athens (1973) on mathematical modeling and stability of dynamical systems. His impressive set of lecture notes, listed at the end of this book, covers a variety of topics from special functions and transforms to numerical methods for solution of nonlinear differential equations and optimal control problems.*

*We have decided with Mrs Helen Magiros to publish this volume hoping that it would serve both as an important reference in the field of nonlinear mechanics and dynamical systems, and as a source of inspiration to all young mathematicians and systems scientists who are challenged to solve real problems of our space age.*

*Many thanks are expressed to the Greek Mathematical Society for encouraging and supporting this publication.*

*Spyros G. Tzafestas*

# ACKNOWLEDGEMENTS

*The publishers, the Greek Mathematical Society, the Editor, and Mrs Helen Magiros wish to thank the following publishers for granting permission to reprint the papers which appear in this volume (the identification of papers refers to the List of Contents):*

*Athens Academy of Sciences (Practika-Proceedings of Athens Academy of Sciences) for papers 2, 3, 4, 5, 8, 9, 10, 12, 13, 14, 15, 17, 21, 22, 23, 24, 29, 33, 34, 39 and 40.*

*Ukrainian Academy of Sciences (Ukrainian Mathematical Journal and Matematicheskaya Fizika) for papers 41, 42 and 43.*

*John Wiley and Sons, Inc. (Communications of Pure and Applied Mathematics) for paper 7.*

*Academic Press, Inc. (Information and Control) for papers 16 and 31.*

*The American Institute of Physics (Journal of Mathematical Physics) for paper 20.*

*The Franklin Institute (Journal of the Franklin Institute) for papers 25, 26, 27 and 32.*

*It is also acknowledged that papers 1, 18, 19 and 30 are reprinted from Proceedings of the National Academy of Sciences (U.S.A.), papers 28, 37 and 38 from the Proceedings of International Astronautical Congresses held in Madrid (1966), Buenos Aires (1969) and Brussels (1971), respectively, and paper 36 from the Proceedings of the VIIIth Symposium on Space Technology and Science (Tokyo, 1969).*

*The work described in the papers of the volume was done by D.G. Magiros while he was with General Electric Co. (ReEntry Systems Operation, Space Systems Division, Philadelphia, PA., U.S.A.).*

*Finally, the publishers, The Greek Mathematical Society, the Editor and Mrs Magiros would like to thank A.J. Dennison, Joseph B. Keller and George Reehl, co-authors of D.G. Magiros, for granting permission to reprint the respective papers (7, 28, 29, 32, 37).*

# CONTENTS

## PART I
## APPLIED MATHEMATICS AND MODELLING



* *Numbers in brackets refer to the Complete Chronological List of Magiros' publications.*

## PART II
## NONLINEAR MECHANICS

### 1. Subharmonic Oscillations and Principal Modes

**2. Celestial and Orbital Mechanics**



## PART III
## DYNAMICAL SYSTEMS ANALYSIS

**1. Stability Analysis**

## APPENDIX: PAPERS IN RUSSIAN

# PART I

# APPLIED MATHEMATICS AND MODELLING

In this Part, papers dealing with differential equations (DE), mathematical modelling, and linearization of nonlinear models are presented.

In paper 1 the reduction of a class of nonautonomous DE to another class of autonomous DE is studied, via an exact method which uses an appropriate transformation of the independent variable. A necessary and sufficient condition for this reduction is derived, and the equivalence of the above classes of DE as well as the concept of "irreducibility strips" of the reducible DE are discussed. In paper 2 (and paper 41 published in Russian) the existence and construction of several general solutions of nonlinear ordinary DE is investigated. To this end, appropriate restrictions of quantities of the DE, or the factorization of the DE, are employed. In paper 3, the way in which the arbitrary constants enter the general solutions of nonlinear DE is studied. It is found that the arbitrary constants appear either "linearly" in a few cases, or "rationally linearly" in some other cases, or in a "complicated nonlinear manner" in the majority of the situations. The singularities, particularly the essential singularities, make the general solutions as well as the introduction of the arbitrary constants in them, very complicated.

Paper 4 presents a "classical procedure" for determining "actual" mathematical solutions of well-posed (real) systems, either physical or social. Three basic steps in the procedure can be distinguished, namely: contruction of models of the system, selection of a well-posed model, and solution of this model. Paper 5 provides a selection of applications from various fields in an attempt to illustrate the procedure outlined in paper 4. Paper 6 gives some more material along the lines of the two previous papers.

In paper 7, the diffraction of a cylindrical wave by a perfectly conducting, semi-infinite thin screen with a cylindrical tip is analyzed by three different methods. These methods are: (i) the geometrical theory of diffraction, (ii) expansion in radial eigenfunctions, and (iii) the Watson transformation of the angular eigenfunction expression. Using the third method it is shown that the geometrical theory of diffraction is correct for the problem at hand, giving the asymptotic form of the exact solution. The result determines how the field in the shadow depends upon the wavelength and the curvature of the shadow forming object.

In papers 8 and 9 linearization methods for nonlinear models of physical phenomena are presented. Paper 8 refers to linearization by exact methods (exact transformations of variables, restriction of variables and parameters) which lead to general solutions of the nonlinear systems in closed form. Paper 9 refers to linearization by approximate methods and conditions are imposed for the linearization to be accepted. In paper 10, various properties which characterize linear and nonlinear physical systems are discussed.

In the final paper of this part (Paper 11) the creation of mathematical models of physical, technological and real-life systems is discussed. A "classical method" of creating mathematical models is proposed (see Papers 4 and 5) which is clarified by a variety of remarks and illustrated applications (of "complete" and "incomplete" cycle) from many fields.

# A CLASS OF NONAUTONOMOUS DIFFERENTIAL EQUATIONS REDUCIBLE TO AUTONOMOUS ONES BY AN EXACT METHOD

By Demetrios G. Magiros

GENERAL ELECTRIC COMPANY, VALLEY FORGE SPACE TECHNOLOGY CENTER, PHILADELPHIA

*Communicated by Leon Brillouin, June 23, 1967*

1. *Introduction.*—This paper is concerned with the reduction of a class of nonautonomous differential equations to another class of autonomous ones by using an exact method based on an appropriate transformation of the independent variable. A condition necessary and sufficient for the reducibility is found and proved.

By autonomous differential equations we mean differential equations where there is no term containing explicitly the independent variable, and by nonautonomous ones those where at least one term contains explicitly the independent variable.

The reduction of nonautonomous differential equations into autonomous ones is important in theory and very desirable in practice. Some nonautonomous equations may be reduced into autonomous ones by either analytical exact methods or approximate ones, but the first are complicated and the second are based on assumptions rarely found in reality. The ideal methods for this reducibility are those which are short and exact and of practical usefulness, if such methods exist.

In this paper we discuss the conditions under which a nonautonomous equation of the form

$$\ddot{x} + p_1(t)\dot{x} + p_2(t)x = 0 \tag{1}$$

can be reduced to an autonomous one of the form

$$x'' + a_1x' + a_2x = 0 \tag{2}$$

by using an exact method. The coefficients $p_1, p_2$ in equation (1) are time-dependent, and the coefficients $a_1$ and $a_2$ in equation (2) are constants. The "dots" and "primes" in these equations denote differentiation with respect to the original time $t$ and to the transformed time $\tau$, respectively; $p_1$ and $p_2$ are single-valued, real-valued, and differentiable functions of $t$ in an interval $I$: $t_0 < t < t_1$, where equation (1) is valid; $t_0$ may be zero and $t_1$ infinite.

The following theorem gives a condition under which equation (1) is reducible to equation (2).

2. Theorem. *A necessary and sufficient condition for the equation* (1) *to be reducible to equation* (2) *is that the coefficients $p_1$ and $p_2$ of equation* (1) *must satisfy the identity*

$$\frac{1}{p_2}\left\{2p_1 + \frac{\dot{p}_2}{p_2}\right\}^2 \equiv \gamma^2, \tag{3}$$

*where $\gamma^2$ is a nonnegative fixed constant, $p_1$ and $p_2$ have the properties indicated above, and in addition, $p_2$ is positive.*

*Proof:* In equation (1), we change the independent variable $t$ into another independent variable $\tau$ according to a given transformation:

$$\tau = \varphi(t). \tag{4}$$



Reprinted from the *Proceedings of the National Academy of Sciences* **58**, No. 2 (1967), 412–419.

By using this transformation, the derivatives $\dot{x}$ and $\ddot{x}$ of equation (1) are given by

$$\dot{x} = x'\dot{\varphi}, \qquad \ddot{x} = x'\ddot{\varphi} + x''(\dot{\varphi})^2 \tag{5}$$

when, inserting (5) into (1), we can get the equation

$$x'' + \frac{p_1\dot{\varphi} + \ddot{\varphi}}{(\dot{\varphi})^2} x' + \frac{p_2}{(\dot{\varphi})^2} x = 0, \tag{6}$$

where the independent variable is $\tau$, its coefficients are dependent on $p_1$, $p_2$, $\varphi$, and then they are functions of $t$, and $\varphi$ is nonconstant.

If equation (1) is reducible to equation (2), then the functions $p_1$, $p_2$, $\varphi$ must be restricted in such a way that the coefficients of equation (6) are constants, that is:

$$\frac{p_1\dot{\varphi} + \ddot{\varphi}}{(\dot{\varphi})^2} = a_1, \qquad \frac{p_2}{(\dot{\varphi})^2} = a_2. \tag{7i), (7ii}$$

From (7*ii*), we have

$$\dot{p}_2 = 2a_2\dot{\varphi}\ddot{\varphi}, \tag{8}$$

and the elimination of $\dot{\varphi}$ and $\ddot{\varphi}$ between (7) and (8) gives the condition of the reducibility of equation (1) into equation (2).

From (7*ii*), we have

$$(\dot{\varphi})^2 = \frac{p_2}{a_2}, \qquad \dot{\varphi} = \pm\sqrt{\frac{p_2}{a_2}}, \tag{9}$$

and inserting (9) into (8) we have

$$\ddot{\varphi} = \pm \frac{\dot{p}_2}{2\sqrt{a_2p_2}}. \tag{10}$$

Finally, inserting (9) and (10) into (7*i*) we have

$$\frac{1}{p_2}\left\{2p_1 + \frac{\dot{p}_2}{p_2}\right\}^2 \equiv \gamma^2, \tag{11}$$

where

$$\gamma^2 = \frac{4a_1^2}{a_2}. \tag{11a}$$

The quantity of the left-hand-member of (11) is the "reducibility quantity," the nonnegative constant $\gamma^2$ is the "reducibility constant," and the condition (11) the "reducibility condition" of equation (1).

By the above we prove that if equation (1) is reducible to equation (2), then $p_1$ and $p_2$ must satisfy condition (11), which then is a necessary condition for the reducibility. Now we can see that the converse is true; that is, if $p_1$ and $p_2$ satisfy the condition (11), then equation (1) is reducible to equation (2).

We select the transformation $\varphi$ in such a way that $(\dot{\varphi})^2$ is a constant multiple of $p_2$, as formula (7*ii*) shows, $(\ddot{\varphi})^2 = p^2/a^2$, then we will have

$$\dot{\varphi} = \pm\frac{1}{\sqrt{a_2}}\sqrt{p_2}, \qquad \ddot{\varphi} = \pm\frac{1}{2\sqrt{a_2}}\frac{1}{\sqrt{p_2}}, \qquad \varphi = \pm\frac{2}{2\sqrt{a_2}}p_2^{3/2} + K, \tag{12}$$

$K$ is an arbitrary constant. Inserting the first two of (12) into formula (7*i*), we can have the coefficient $a_1$ in terms of the functions $p_1$, $p_2$, the transformation $\varphi$, and the arbitrary coefficient $a_2$, when equation (2), corresponding to equation (1), is constructed.

If $\rho_1$ and $\rho_2$ are the eigenvalues of equation (2), its general solution is given by

$$x(\tau) = K_1 e^{\rho_1 \tau} + K_2 e^{\rho_2 \tau}. \tag{13}$$

where $K_1$, $K_2$ are arbitrary constants, and inserting $\tau = \varphi(t)$ in (13), we have the general solution of equation (1) in the form

$$x(t) = K_1 e^{\rho_1 \varphi(t)} + K_2 e^{\rho_2 \varphi(t)}. \tag{14}$$

*Remark 1:* The reducibility condition (11) of equation (1) into equation (2) is independent of the transformation $\varphi$ used.

*Remark 2:* The above procedure shows that $p_2$ and $a_2$ must be positive.

*Remark 3:* If the reducibility constant $\gamma^2$ exists, then, by using (11*a*), we can determine either $a_1$ in terms of $a_2$ and $\gamma^2$, or $a_2$ in terms of $a_1$ and $\gamma^2$, without using the transformation function $\varphi$. Then the constant coefficients of equation (2) will be either $a_1$, $a_2 = 4a_1^2/\gamma^2$, or $a_1 = \pm(\gamma/2)\sqrt{a_2}$, $a_2$, if $\gamma \pm 0$; and $a_1 = 0$, $a_2 =$ arbitrary, if $\gamma = 0$; or $a_1 =$ arbitrary, $a_2 = 0$, if $\gamma = \infty$.

3. COROLLARY. *From the reducibility condition* (11) *we have*

$$p_1 = \frac{1}{2}\left\{\pm\gamma\sqrt{p_2} - \frac{\dot{p}_2}{p_2}\right\} \tag{15i}$$

$$\dot{p}_2 + 2p_1p_2 = \pm\gamma p_2^{3/2} \tag{15ii}$$

*Then, if the coefficient $p_2$ of equation* (1) *is given, we can calculate the other coefficient $p_1$ of equation* (1) *in such a way that equation* (1) *is reducible to equation* (2). *The coefficient $p_1$ is given by* (15*i*). *$\gamma^2$ is the nonnegative fixed reducibility constant. If the coefficient $p_1$ of equation* (1) *is given, we can calculate the coefficient $p_2$ of equation* (1) *for its reducibility. The coefficient $p_2$ is a solution of the specific Bernoulli equation* (15*ii*).

*Remark 4:* Formula (15*i*) implies that if $p_2$ is a positive constant, then, for the reducibility, $p_1$ must be a real constant; consequently, in case $p_1$ is variable and $p_2$ a constant, equation (1) is irreducible to equation (2). The Hermite equation: $\ddot{x} - 2t\dot{x} + cx = 0$, is an example of the last case. But, if $p_1$ is a constant, then there are variable coefficients $p_2$ of equation (1) for its reducibility, and these coefficients are solutions of the Bernoulli equation (15*ii*).

4. *Illustrated Problems.*—The above discussion is clarified by the following problems.

PROBLEM 1. *Given the equation*

$$t^2\ddot{x} + 2t\dot{x} + \mu x = 0, \qquad t \neq 0, \tag{16}$$

*where the coefficients are variable and $\mu$ is a fixed positive constant:*

(i) *check the possibility for its reduction to an equation* (2) *with constant coefficients;*

(ii) *find the appropriate transformation of the independent variables of these equations for the reducibility of equation* (16) *to equation* (2);

(iii) *determine the constants $a_1$ and $a_2$ of the reduced equation* (2), *and*

(iv) *give the form of the general solution of equation* (16), *using the general solution of the reduced equation* (2).

*Solution:* (*i*) Equation (16) can be written in the form

$$\ddot{x} + \frac{2}{t}\dot{x} + \frac{\mu}{t^2}x = 0, \tag{16a}$$

and we have

$$p_1 = \frac{2}{t}, \qquad p_2 = \frac{\mu}{t^2}$$

when condition (11) gives $\gamma^2 = (4/\mu)$, and then equation (16) is reducible to equation (2).

(*ii*) The transformation $\tau = \varphi(t)$, which reduces equation (16) to equation (2), can be found by using (7*ii*) when

$$(\dot{\varphi})^2 = \frac{\mu}{a_2}\frac{1}{t^2}, \qquad \dot{\varphi} = \pm\sqrt{\frac{\mu}{a^2}}\cdot\frac{1}{t}, \qquad \ddot{\varphi} = \mp\sqrt{\frac{\mu}{a_2}}\cdot\frac{1}{t^2} \tag{16b}$$

$$\tau = \varphi(t) = \pm\sqrt{\frac{\mu}{a_2}}\log t + K. \tag{16c}$$

(*iii*) Inserting (16*b*) into (7*i*), we can find

$$a_1 = \pm\sqrt{\frac{a_2}{\mu}}, \qquad a_2 = \text{arbitrary constant}, \tag{16d}$$

then equation (2) is one of the following equations:

$$\left.\begin{aligned} x'' \pm \sqrt{\frac{a_2}{\mu}}\,x' + a_2x &= 0 \\ x'' + a_1x' + \mu a_1{}^2x &= 0 \end{aligned}\right\}, \tag{16e}$$

$a_2$ and $a_1$ are arbitrary constants in the first and second of equations (16*e*), respectively.

(*iv*) The characteristic roots of, say, the second of (16*e*) are

$$\rho_{1,2} = a_1\lambda_{1,2}, \qquad \lambda_{1,2} = \frac{1}{2}\left\{-1 \pm \sqrt{1 - 4\mu}\right\}, \tag{16f}$$

when its general solution is

$$x(\tau) = K_1e^{a_1\lambda_1\tau} + K_2e^{a_1\lambda_2\tau}. \tag{16g}$$

By using transformation (16*c*), function (16*g*) becomes

$$x(t) = \bar{K}_1e^{\lambda_1\log t} + \bar{K}_2e^{\lambda_2\log t}, \tag{16h}$$

which is the general solution of equation (16).

*Remark 5:* Equation (16) is of Euler's type and, as we know, this class of equations accepts solutions of the form $y = t^\lambda$, where $\lambda$ is a root of the characteristic equation, which in the case of equation (16) is $\lambda = \frac{1}{2}\{-1 \pm \sqrt{1 - 4\mu}\}$, and then the general solution of equation (16), by using this method, which characterizes the Euler's equation, is

$$x(t) = c_1 t^{\lambda_1} + c_2 t^{\lambda_2}, \qquad (16i)$$

where $c_1$ and $c_2$ are arbitrary constants. Since $t^\lambda = e^{\lambda \log t}$, solutions (16*h*) and (16*i*) are the same.

*Remark 6:* We can check that the reducibility quantity of the following equations is time-dependent, then these equations are irreducible to equations with constant coefficients:

(*i*) $t^2\ddot{x} + t\dot{x} + (t^2 - n^2)x = 0, \quad t \neq 0$ (Bessel).

(*ii*) $t\ddot{x} + (1 - t)\dot{x} + tx = 0, \quad t \neq 0$ (Laguerre).

(*iii*) $(1 - t^2)\ddot{x} - 2t\dot{x} + n(n - 1)x = 0, \quad t \neq \pm 1$ (Legendre).

(*iv*) $(1 - t^2)\ddot{x} - t\dot{x} + n^2x = 0, \quad t \neq \pm 1$ (Tschebycheff).

PROBLEM 2. *In the equation*

$$\ddot{x} + \frac{2}{t}\dot{x} + p_2(t)x = 0, \qquad t \neq 0, \qquad (17)$$

*calculate the coefficient $p_2$ in such a way that equation* (17) *is reducible to an equation* (2) *with constant coefficients and find the general solution of equation* (17).

*Solution:* The Bernoulli equation (15*ii*), of which the solution $p_2$ gives the coefficient of equation (17) for its reducibility to equation (2), becomes

$$\dot{p}_2 + \frac{4}{t}p_2 = \pm\gamma p_2^{3/2}, \qquad (17a)$$

where $\gamma^2$ is the fixed reducibility constant.

Equation (17*a*), by using the transformation

$$Z = \frac{1}{\sqrt{p_2}} \qquad (17b)$$

can be reduced to the linear equation:

$$\dot{Z} - \frac{2}{t}Z = \pm\frac{1}{2}\gamma \qquad (17c)$$

of which the general solution can be found to be

$$Z = t(c_3 t \pm \tfrac{1}{2}\gamma), \qquad (17d)$$

where $c_3$ is an arbitrary constant.

By using transformation (17*b*), function (17*d*) becomes

$$p_2 = t^{-2}(c_3 t \pm {}^1/_2\gamma)^{-2}, \qquad (17e)$$

which is the coefficient of equation (17) for its reducibility.

For $c_3 = 0$, $\frac{1}{2}\gamma = (1/\sqrt{\mu})$, (17*e*) gives $p_2 = \mu t^{-2}$, the special case of Problem 1.

To calculate the constant coefficients of the reduced equation (2), we determine the appropriate transformation $\varphi(t)$.

Combining (7*ii*) and (17*e*), we have

$$(\dot{\varphi})^2 = \frac{1}{a_2\{t^2(c_3 t \pm \frac{1}{2}\gamma)^2\}} = \frac{1}{t^2\{c_3\sqrt{a_2}\,t \pm \frac{1}{2}\gamma\sqrt{a_2}\}^2},$$

where $a_2$, $c_3$ are arbitrary constants.

If we take the arbitrary constants $a_2$ and $c_3$ such that $c_3 = (1/\sqrt{a_2})$, and put $c = \pm \frac{1}{2}\gamma\sqrt{a_2}$, we have:

$$(\dot{\varphi})^2 = \frac{1}{t^2(t+c)^2}, \qquad \dot{\varphi} = \pm \frac{1}{t(t+c)}, \qquad \ddot{\varphi} = \mp \frac{2t+c}{t^2(t+c)^2}. \tag{17f}$$

$$\varphi(t) = \pm \int \frac{dt}{t(t+c)} = \pm \frac{1}{c}\log\frac{t}{t+c} + K. \tag{17g}$$

We can check that $a_1 = \pm c = \pm \frac{1}{2}\gamma\sqrt{a_2}$, then the corresponding differential equation with constant coefficients is either:

$$x'' \pm \tfrac{1}{2}\gamma\sqrt{a_2}\,x' + a_2 x = 0, \qquad \text{or:}\quad x'' + a_1 x' + \frac{4a_1^2}{\gamma^2}x = 0. \tag{17h}$$

The characteristic roots of (17*h*) are

$$\rho_{1,2} = \frac{a_1}{2}\left\{-1 \pm \sqrt{1 - \left(\frac{4}{\gamma}\right)^2}\right\}, \tag{17i}$$

then

$$x(\tau) = K_1 e^{\rho_1\tau} + K_2 e^{\rho_2\tau} \tag{17j}$$

is the general solution of (17*h*), when

$$x(t) = \bar{K}_1 e^{\rho_1\varphi(t)} + \bar{K}_2 e^{\rho_2\varphi(t)} \tag{17k}$$

is the general solution of equation (17), with $\varphi(t)$ given by (17*g*).

PROBLEM 3. *Calculate the coefficient* $p_2$ *in the equation*

$$\ddot{x} + p_2(t)x = 0 \tag{18}$$

*in such a way that this equation is reducible to an equation* (2) *with constant coefficients and find the general solution of equation* (18).

*Solution:* The reducibility condition (11) in the case of this problem becomes $\dot{p}_2 = \pm\gamma p_2^{3/2}$ where $\gamma^2$ is the reducibility fixed constant, then the function $p_2(t)$ for the reducibility of equation (18) is a positive nonoscillatory function of $t$ given by the formula

$$p_2(t) = (\pm \tfrac{1}{2}\gamma t + K_1)^{-2} \tag{18a}$$

where $K_1$ is an arbitrary constant. The appropriate reducibility transformation $\tau = \varphi(t)$ can be found from

$$(\dot{\varphi})^2 = \frac{1}{a_2}(\pm \tfrac{1}{2}\gamma t + K_1)^{-2}, \qquad \text{or}\quad \dot{\varphi} = \pm \frac{1}{\sqrt{a_2}}(\pm \tfrac{1}{2}\gamma t + K_1)^{-1},$$

then:

$$\tau = \varphi(t) = \frac{2}{\gamma\sqrt{a_2}} \log (\pm \tfrac{1}{2}\gamma t + K_1). \tag{18b}$$

Calculating $a_1$, we find $a_1 = \pm \frac{1}{2}\gamma\sqrt{a_2}$, then $a_2 = (4a_1^2/\gamma^2)$, when the corresponding equation with constant coefficients is

$$x'' \pm \tfrac{1}{2}\gamma\sqrt{a_2}\, x' + a_2 x = 0, \qquad \text{or} \qquad x'' + a_1 x' + \frac{4a_1^2}{\gamma^2}\, x = 0. \tag{18c}$$

The characteristic roots of (18c) are

$$\rho_{1,2} = \tfrac{1}{2}a_1\left\{-1 \pm \sqrt{1 - \left(\frac{4}{\gamma}\right)^2}\right\}, \tag{18d}$$

the general solution of (18) is given by

$$x(t) = K_1 e^{\rho_1\varphi(t)} + K_2 e^{\rho_2\varphi(t)}, \tag{18e}$$

where $K_1$ and $K_2$ are arbitrary constants, $\rho_1$ and $\rho_2$ are given by (18d), and $\varphi(t)$ by (18b).

*Remark 7:* By the above, it is found that if the function $p_2(t)$ of equation (18) is given by (18a), the equation (18) is reducible to equation (18c), when its general solution is given by (18e). If $p_2(t)$ of equation (18) has a form different than (18a), equation (18) is irreducible to an equation with constant coefficients, and in case of irreducibility of equation (18), its general solution, if it is possible to be found, cannot have the form (18e).

This happens when, say, $p_2(t)$ is oscillatory or periodic in $t$, as, e.g., in the case of the Mathieu equation, when $p_2 = k_1 + k_2 \cos 2t$. The solution of a Mathieu equation, as we know, is given by

$$x(t) = K_1(t)e^{r_1 t} + K_2(t)e^{r_2 t}, \tag{18f}$$

where $K_1(t)$ and $K_2(t)$ are purely periodic functions of $t$ of period $\pi$, and the calculation of the constants $r_1$ and $r_2$ necessitates no simple methods. The solution (18f) is not of the form (18e).

5. *Equivalence of the Classes.—Irreducibility strips of the reducible equation* (1): (*a*) The solutions of equation (1) with variable coefficients lead in general to special functions expressed in terms of infinite series or continued fractions or integrals, as, e.g., the solutions of the special differential equations referred to in *Remark* 6. The reducible class of equation (1) has the important property that it can be fully and explicitly treated as the class of equation (2) with constant coefficients, then its solution can be expressed in terms of elementary functions.

Stability properties of the solutions of these two classes of differential equations are preserved when we go from one class to the other through the suitable transformation, and then these classes have a "stability equivalence."

(*b*) We supposed in the preceding that the coefficients $p_1(t)$ and $p_2(t)$ of equation (1) are of ordinary behavior throughout the interval $I$ of the time $t$ where equation (1) is valid, and then all values of this interval are "ordinary points" of the reducible equation (1).

There may be values of the interval $I$ where at least one of these coefficients has a nonordinary behavior, and these values are "singular points" of the equation (1), and equation (1) is irreducible at these points.

If at least one of the coefficients have the form of a ratio, and at a point $t = \bar{t}$ the denominator of the ratio has a zero value but the numerator has a nonzero value, this point $\bar{t}$ is a singular point of the coefficient, then it is a singular point of the solution $x(t)$ of equation (1), either "regular" or "irregular" singular point.

A small neighborhood around any singular point of the solution $x(t)$ corresponds to a narrow strip in the "$t,x$-plane," which is an "irreducibility strip" of equation (1). The solution of a reducible equation (1), outside the irreducibility strips, can be expressed by elementary functions according to formula (14).

Inside these strips, the solution may be expressed by a special series convergent or not,[1, 3(a), 4(a)] or by formulas similar to "Kramer's connection formulas," known from quantum mechanics (cf. ref. 4($b$)).

The subject of the present paper was inspired by ($b$) of reference 3, and was especially prepared to cover practical needs by giving short criteria useful in many fields. It must be supplemented by an extension to some other classes of equations (cf. refs. 2, 5).

[1] Dennery, P., and A. Krzywicki, *Mathematics for Physicists* (New York: Harper and Row, 1967), pp. 298–301.

[2] Erugin, N., *Linear Systems of Ordinary Differential Equations* (New York: Academic Press, 1966), chap. 19.

[3] Ince, E., *Ordinary Differential Equations* (New York: Dover Publications, 1944): (*a*) pp. 160–161, 168–169; (*b*) p. 131.

[4] Kemble, E., *The Fundamental Principles of Quantum Mechanics* (New York: McGraw-Hill Book Co., 1937): (*a*) pp. 140–142; (*b*) pp. 93–103.

[5] Roseau, M., *Vibrations Nonlinéaires et Théorie de la stabilit é* (New York: Springer-Verlag, 1966), pp. 27–34.

ΕΦΗΡΜΟΣΜΕΝΑ ΜΑΘΗΜΑΤΙΚΑ.— **Nonlinear differential equations with several general solutions,** *by Demetrios G. Magiros* *.
Ἀνεκοινώθη ὑπὸ τοῦ Ἀκαδημαϊκοῦ κ. Ἰω. Ξανθάκη.

## I. INTRODUCTION

Nonlinear differential equations constitute today a field of scientific knowledge basic for the investigation of the majority of physical phenomena, technological systems, social problems.

The general solutions of these DE are the most desirable, especially in applications, but only for some nonlinear DE the general solutions are known to exist, and only in a few classes of these DE the general solutions can be determined in a closed form.

While any linear ordinary DE has one general solution, a nonlinear ODE may have either one or several general solutions.

It is necessary here to make clear some concepts related to that of the general solution. We consider a nonlinear ODE (F) valid in a region R of the space of its variables, and a function (Φ) of the variables of (F) containing a number of arbitrary constants independent to each other, which can take the values in a region C of the space of these arbitrary constants. A function coming from (Φ) by a specification of all its arbitrary constants, if it satisfies the DE (F), is called a «particular solution» of (F). The function (Φ), a totality of all the particular solutions of (F) coming from (Φ), is called a «general solution» of (F).

Any function, coming from the general solution (Φ) of (F) by a specification of some only arbitrary constants of (Φ), which, then, contains the unspecified constants of (Φ) as arbitrary parameters, is called a «part of the general solution» of (F).

If there are functions $(\Phi_i)$, $i = 1, 2, \ldots, n$, which have properties similar to that of (Φ) in connection to the DE (F), and, in addition, particular solutions of one of the $(\Phi_i)$ are not identical to particular solutions of any of the others $(\Phi_i)$, these functions must be considered as «distinct general solutions» of (F), and, in this case, we say that the DE

* Δ. Γ. ΜΑΓΕΙΡΟΥ, **Μὴ γραμμικαὶ διαφορικαὶ ἐξισώσεις μὲ πολλὰς γενικὰς λύσεις.**

Reprinted from the *Proceedings of the Athens Academy of Sciences* **52** (1977), 221–229.

(F) has «several general solutions». By this definition, is not excluded that a «common part» among some of $(\Phi_i)$ may exist.

In this paper remarks are given concerning the existence and determination of several general solutions of some nonlinear ODE.

The «restriction of the solutions of the DE», and the «factorization of the DE» can be used as methods for investigation. These methods are explained and illustrated by examples from which we draw certain conclusions.

## II. THE METHOD OF RESTRICTING THE SOLUTIONS

By this method, if it is applicable, one can determine several general solutions of a given nonlinear DE in a closed form. We give an example.

Example 1: $$y'''(1+y'^2)-3y'y''^2=0 \tag{1.1}$$

One can have two appropriate restrictions of the unknown solution of (1.1), and to each restriction a general solution of (1.1) corresponds.

(a): The solution y of (1.1) is restricted by $y''=0$, which implies $y'''=0$. In this case, (1.1) is satisfied simultaneously by:

$$y''=0, \quad y'''=0 \tag{1.2}$$

when the function

$$y=\alpha_1 x+\alpha_2 \tag{1.3}$$

which satisfies (1.2) is a general solution of (1.1). The $\alpha_1$ and $\alpha_2$ are independent to each other and the two-parameter family (1.3) represents «all straight lines» in the x, y - plane.

(b): The solution y of (1.1) is restricted by $y''\neq 0$, and, in this case, one can determine a general solution of (1.1) which is different than (1.3). The DE (1.1) in this case can be written in the form: [1]

$$\frac{y'''}{y''}=\frac{3y'y''}{1+y'^2}=\frac{3}{2}\cdot\frac{d(1+y'^2)}{1+y'^2}. \tag{1.4}$$

An integration of this equation gives:

$$\frac{y''}{(1+y'^2)^{3/2}}=c_1.$$

A new integration leads to:

$$y' = \frac{c_1x + c_2}{\sqrt{1-(c_1x+c_2)^2}} \tag{1.5}$$

and a third integration to:

$$c_1y + c_3 = -\sqrt{1-(c_1x+c_2)^2} \tag{1.6}$$

which can get the form:

$$x^2 + y^2 + C_1x + C_2y + C_3 = 0. \tag{1.7}$$

The $C_1$, $C_2$, $C_3$, functions of $c_1$, $c_2$, $c_3$, are arbitrary constants independent to each other, and the three-parameter family (1.7), which represents «all circles» in x, y-plane, is a general solution of (1.1). There is no specification of the arbitrary constants of (1.3) and (1.7) by which a member of (1.3) becomes identical with a member of (1.7), then (1.3) and (1.7) are two distinct general solutions of (1.1).

## III. THE METHOD OF FACTORIZING THE DE

A possible application of this method can give several general solutions of a DE. We give examples.

Example 2: $x^3y''y''' + x^2y''^2 - 2xy'y'' + 2yy'' = 0.$ (2.1)

The factorization gives:

$$y''(x^3y''' + x^2y'' - 2xy' + 2y) = 0 \tag{2.2}$$

then:

$$\left.\begin{array}{ll}(a): & y'' = 0 \\ (b): & x^3y''' + x^2y'' - 2xy' + 2y = 0, \quad x \neq 0\end{array}\right\} \tag{2.3}$$

the general solutions of which are:

$$\left.\begin{array}{ll}(a): & y = a_1x + a_2 \\ (b): & y = c_1x + c_2x^2 + c_3x^{-1}\end{array}\right\} \tag{2.4}$$

and these functions are the general solutions of (2.1).

The family of straight lines through the origin is a «common part» of the general solutions (2.4) of the DE (2.1).

Example 3: $y'^2 + y(x^2y - xy) - x^3y^2 = 0$ (3.1)

By factorizing one has :

$$(y' - xy) \cdot (y' + x^2y) = 0 \tag{3.2}$$

then :

$$\left.\begin{array}{ll} (a): & y' = xy \\ (b): & y' = -x^2y \end{array}\right\} \tag{3.3}$$

from which :

$$\left.\begin{array}{ll} (a): & y = c_1 e^{x^2/2} \\ (b): & y = c_2 e^{-x^3/3} \end{array}\right\} \tag{3.4}$$

that is the general solutions of the DE (3. 1) :

We remark that from each point of the x, y - plane two, in general, integral curves of (3. 1) pass with different slopes, one belonging to the family (3. 4. a) corresponding to a specific value of $c_1$, and one belonging to the family (3. 4. b) for a specific value of $c_2$, and these values of $c_1$ and $c_2$ are different.

The equation $(xy = -x^2y)$ gives the singular lines: $x = 0$, $x = -1$, $y = 0$, at the points of which the integral curves have just one slope, the slope $-y$ at $x = -1$, and the zero-slope at the coordinate axes, which are the «singular solutions» of the DE (3. 1).

Example 4:

«The general nonlinear DE of the first order :

$$F(x, y, y') = 0 \tag{4.1}$$

in case F is a polynomial for $y'$ of degree m".

In this case, (4. 1) can have the form :

$$F(x, y, y') \equiv y'^m + P_1 y'^{m-1} + \ldots + P_{m-1} y' + P_m = 0 \tag{4.2}$$

where $P_1, \ldots, P_m$, functions of x and y, are continuously differentiable in the region of validity of (4. 2).

$$F_{y'}(x, y, y') \neq 0.$$

The DE (4. 2) if $F_{y'}(x, y, y') \neq 0$, solved for $y'$, gives m simple roots : $y'_1, \ldots, y'_m$, when one can write :

$$F(x, y, y') \equiv (y'_1 - p_1) \cdot (y'_2 - p_2) \ldots (y'_m - p_m) = c \tag{4.3}$$

where $p_1, \ldots, p_m$ are functions of x and y.

The DE (4. 3) is equivalent to the m DE :

$$y'_1 = p_1(x, y), \quad y'_2 = p_2(x, y), \ldots, \quad y'_m = p_m(x, y) \tag{4.4}$$

from which, by integration, one can get the functions :

$$\varphi_1(x, y, c_1) = 0, \quad \varphi_2(x, y, c_2) = 0, \ldots \quad \varphi_m = (x, y, c_m = 0 \tag{4.5}$$

which are the m general solutions of the DE (4. 2).

We remark that m is the number of the factors of (4. 3), but we restrict ourselves to those factors which lead to real functions of real variables, when the number of the general solutions of (4. 2) is either equal or smaller than m. Simple example for that is the DE :

$$F = y'^4 - 1 = 0 \tag{4.6}$$

for which one can have:

$$F \equiv (y'^2 - 1)(y'^2 + 1) = (y' - 1)(y' + 1)(y'^2 + 1) = 0$$

and since $y'^2 + 1 \neq 0$, only the first two factors count, to which the DE: $y' = 1$, $y' = -1$ correspond, and then the general solutions of (4. 6) are :

$$y = x + c_1, \quad y = -x + c_2 \tag{4.7}$$

E x a m p l e 5: $\quad y'^2 y''' + y^2 y''' + y''' = 0 \qquad (5.1)$

The factorization gives :

$$y'''(y'^2 + y^2 + 1) = 0 \tag{5.2}$$

and, since $y'^2 + y^2 + 1 \neq 0$, (5. 2) is equivalent to :

$$y''' = 0 \tag{5.3}$$

of which the general solution is :

$$y = c_1 x^2 + c_2 x + c_3 \tag{5.4}$$

For any set of values of the arbitrary constants for which $c_1 \neq 0$, (5. 4) represents parabolas in the x, y -plane. For $c_1 = 0$, (5. 4) reduces to :

$$y = c_2 x + c_3 \tag{5.5}$$

which represents the straight lines in the x, y -plane.

The functions (5.4) and (5. 5), which contain arbitrary constants and satisfy (5. 1), are not two distinct general solutions of (5. 1), but (5. 5) is «part of the general solution (5. 4)».

Examle 6: «The Emden's Equation:

$$y'' + \frac{2}{x} y' + y^n = 0 \text{ ».} \tag{6.1}$$

We use this basic in Astrophysics DE in order to give appropriate remarks in connection with the general solutions of a DE and the physical reality from which the DE comes.

Emden (1907) [2] examined the thermal behavior of a spherical cloud of gas acting under the mutual attraction of its molecules and subject to the classical laws of Thermodynamics, and found the DE (6.1).

The solutions of (6.1), in a closed form, are:

$$\left.\begin{array}{lll} (a): & n = 0: & y = \alpha + \frac{b}{x} - \frac{x^2}{6} \\ (b): & n = 1: & y = \alpha \frac{\sin x}{x} + b \frac{\cos x}{x} \\ (c): & n = 5: & y = \left(\frac{3\alpha}{x^2 + 3a^2}\right)^{1/2} \end{array}\right\} \tag{6.2}$$

The first and second of the functions (6.2) are general solutions of the corresponding DE, while the third one, which contains $\alpha$ as an arbitrary constant, is a part of the unknown general solution of (6.1).

The physical situation considered by Emden lead him to the boundary conditions:

$$x = 0, \quad y = 1, \quad y' = 0 \tag{6.3}$$

to which the following particular solutions correspond:

$$\left.\begin{array}{lll} (a): & n = 0: & y = 1 - \frac{x^2}{6} \\ (b): & n = 1: & y = \frac{\sin x}{x} \\ (c): & n = 5: & y = \left(1 + \frac{x^2}{3}\right)^{-1/2} \end{array}\right\} \tag{6.4}$$

Comparing the theoretical results (6.4) with the physical situation of his problem, Emden found that these results are not acceptable physically, and that, from a physical point of view, n of (6.1) must have values only between 0 and 5.

The general solution of (6.1) must contain two arbitrary constants, but one of these it that called a «constant of homology», which does not help in finding the general solution of (6.1).

For the solution of (6.1) needed, if n is in (0,5), and x takes the values of an adequate interval, the investigation was turned to use a Taylor's expansion about $x = 0$ and, then, an analytic continuation of the series. Tables have been computed and graphical representations.

Astrophysicists use today the tables of the Emden's functions, appropriately modified, in order to estimate the density and the interior temperature of stars and some other things, by taking into account new physical data.

R e m a r k. If one follows the converse prosedure, that is, if a function with arbitrary constants in a variety of forms is given, and, then, one formulates, in the known way, the correspondent DE, one may get more rich results in connection with the subject of the paper.

## Π Ε Ρ Ι Λ Η Ψ Ι Σ

Αἱ μὴ γραμμικαὶ διοφορικαὶ ἐξισώσεις ἀποτελοῦν σήμερον τὸ πλέον βασικὸν πεδίον ἐπιστημονικῆς γνώσεως διὰ τὴν ἔρευναν φυσικῶν φαινομένων, τεχνολογικῶν συστημάτων, κοινωνικῶν προβλημάτων.

Αἱ γενικαὶ λύσεις τῶν ἐξισώσεων αὐτῶν εἶναι αἱ περισσότερον ἐπιθυμηταὶ λύσεις των, ἀλλὰ μόνον εἰς μερικὰς κλάσεις τῶν ἐξισώσεων αὐτῶν αἱ γενικαὶ λύσεις εἶναι δυνατὸν νὰ προσδιορισθοῦν. Ἐνῷ εἰς τὰς γραμμικὰς διαφορικὰς ἐξισώσεις ὑπάρχει μόνον μία γενικὴ λύσις, εἰς τὰς μὴ γραμμικὰς δύναται κανεὶς νὰ εὕρῃ μίαν ἢ καὶ περισσοτέρας γενικὰς λύσεις.

Ἡ παροῦσα ἐργασία σχετίζεται μὲ τὴν ὕπαρξιν καὶ τὸν προσδιορισμὸν πολλῶν γενικῶν λύσεων εἰς συνήθεις μὴ γραμμικὰς διαφορικὰς ἐξισώσεις. Ὁ «κατάλληλος περιορισμὸς τῶν λύσεων τῆς ἐξισώσεως», ὅπως καὶ ἡ «ἀναγωγὴ τῆς ἐξισώσεως εἰς γινόμενον παραγόντων» χρησιμοποιοῦνται εἰς τὴν ἐργασίαν αὐτὴν ὡς μέθοδοι ἐρεύνης, ποὺ ἐφαρμόζονται εἰς εἰδικὰ παραδείγματα.

Αἱ γενικαὶ λύσεις, παρὰ τὴν χρησιμότητά των, δὲν δύνανται ἐνίοτε νὰ ἐκφράσουν τὴν φυσικὴν πραγματικότητα, καὶ ὁ προσδιορισμός των, ἀπὸ ἀπόψεως ἐφαρμογῶν, εἶναι ἐνίοτε ἄσκοπος. Τοῦτο ὑποδεικνύεται ἀπὸ τὸ τελευταῖον παράδειγμα.

REFERENCES

1. E. Goursat, «Differential Equations», Ginn and Company, Boston (U. S. A.), 1917, p. 5.
2. H. Davis, «Introduction to Nonlinear Differential and Integral Equations», Dover Publications Inc. N. Y. (1962) pp. 371 - 377.

★

Ὁ Ἀκαδημαϊκὸς κ. **Ἰω. Ξανθάκης,** παρουσιάζων τὴν ἀνωτέρω ἀνακοίνωσιν, εἶπε τὰ ἑξῆς:

Ἔχω τὴν τιμὴν νὰ παρουσιάσω τὴν ἐργασίαν τοῦ κ. Δημητρίου Μαγείρου ὑπὸ τὸν τίτλον: «Μὴ γραμμικαὶ διαφορικαὶ Ἐξισώσεις μὲ πολλὰς γενικὰς λύσεις». Αἱ μὴ γραμμικαὶ διαφορικαὶ ἐξισώσεις διαδραματίζουν σπουδαῖον ρόλον εἰς τὴν ἔρευναν τόσον φυσικῶν φαινομένων ὅσον καὶ τεχνολογικῶν συστημάτων καὶ κοινωνικῶν προβλημάτων. Ὑπάρχει ὅμως μία σημαντικὴ διαφορὰ μεταξὺ τῶν γραμμικῶν καὶ μὴ γραμμικῶν διαφορικῶν ἐξισώσεων. Πράγματι ἐνῷ διὰ τὰς γραμμικὰς διαφορικὰς ἐξισώσεις ὑπάρχει πάντοτε μόνον μία γενικὴ λύσις, εἰς τὰς μὴ γραμμικὰς εἶναι δυνατὸν νὰ ὑφίστανται περισσότεραι τῆς μιᾶς γενικαὶ λύσεις. Ἐπιπροσθέτως αἱ γενικαὶ λύσεις τῶν μὴ γραμμικῶν διαφορικῶν ἐξισώσεων δὲν εἶναι δυνατὸν πάντοτε νὰ προσδιορισθοῦν, μόνον δηλαδὴ εἰς ὡρισμένας κλάσεις τῶν ἐξισώσεων τούτων δύναταί τις νὰ ἀνεύρῃ μίαν ἢ περισσοτέρας γενικὰς λύσεις.

Εἰς τὴν παροῦσαν ἀνακοίνωσιν ὁ κ. Δημ. Μάγειρος μελετᾷ τὴν ὕπαρξιν καὶ τὸν προσδιορισμὸν πολλῶν γενικῶν λύσεων συνήθων μορφῶν μὴ γραμμικῶν διαφορικῶν ἐξισώσεων.

Πρὸς τοῦτο χρησιμοποιεῖ ὡς μεθόδους ἐρεύνης ἀφ' ἑνὸς μὲν τὸν κατάλληλον περιορισμὸν τῆς λύσεως τῆς ἐξισώσεως, ἀφ' ἑτέρου δὲ τὴν ἀναγωγὴν τῆς ἐξισώσεως εἰς γινόμενον παραγόντων. Πρέπει νὰ σημειώσῃ τις ὅτι αἱ γενικαὶ λύσεις τῶν μὴ γραμμικῶν διαφορικῶν ἐξισώσεων δὲν δύνανται ἐνίοτε νὰ ἐκφράσουν τὴν φυσικὴν πραγματικότητα καὶ ὁ προσδιορισμός των, ἀπὸ ἀπόψεως ἐφαρμογῶν, εἶναι ἐνίοτε ἄσκοπος. Τὸ τελευταῖον τοῦτο ἀποδεικνύεται ἀπὸ τὴν διαφορικὴν ἐξίσωσιν τοῦ Emden. Πράγματι ὁ Emden μελετώντας κατὰ τὰς ἀρχὰς τοῦ παρόντος αἰῶνος τὴν συμπεριφορὰν μιᾶς σφαιρικῆς ἀεριώδους μάζης ὑπὸ τὴν ἐπίδρασιν τῶν ἀμοιβαίων ἕλξεων τῶν μορίων της καὶ τὸν κλασσικὸν νόμον τῆς θερμοδυναμικῆς κατέληξε εἰς τὴν μὴ γραμμικὴν διαφορικὴν ἐξίσωσιν:

$$y'' + \frac{2}{x} y' + y^n = 0$$

Ἡ ἐξίσωσις αὕτη διὰ τὰς τιμὰς τοῦ ἐκθέτου $n = 0, 1 \ldots\ldots$ καὶ 5, δέχεται τρεῖς διαφόρους λύσεις ἐκ τῶν ὁποίων αἱ δύο πρῶται διὰ $n = 0$ καὶ $n = 1$ εἶναι γενικαὶ λύσεις, ἐνῷ ἡ τρίτη $n = 5$ εἶναι ἓν μέρος μιᾶς ἀγνώστου γενικῆς λύσεως.

Συγκρίνοντας τὰ θεωρητικὰ ἐξαγόμενα μὲ τὰ φυσικὰ δεδομένα τοῦ προβλήματος ὁ Emden εὗρε ὅτι τὰ θεωρητικὰ ταῦτα ἐξαγόμενα δὲν ἀντιπροσωπεύουν τὴν φυσικὴν πραγματικότητα.

Ἀπὸ φυσικῆς ἀπόψεως, ὁ ἐκθέτης $n$ πρέπει νὰ λαμβάνῃ τιμὰς μεταξὺ 0 καὶ 5.

Οἱ ἀστροφυσικοὶ τῆς σήμερον κάμνουν εὐρεῖαν χρῆσιν τῶν συναρτήσεων Emden καταλλήλως τροποποιημένων ὑπὸ μορφὴν ἀριθμητικῶν πινάκων διὰ τὴν ἐκτίμησιν τῆς πυκνότητος καὶ τῆς θερμοκρασίας εἰς τὸ ἐσωτερικὸν τῶν ἀστέρων.

ΜΑΘΗΜΑΤΙΚΑ.—**The general solutions of nonlinear differential equations as functions of their arbitrary constants,** *by Demetrios G. Magiros**. Ἀνεκοινώθη ὑπὸ τοῦ Ἀκαδημαϊκοῦ κ. Ἰω. Ξανθάκη.

## INTRODUCTION

One of the major difficulties encountered in the general solutions of nonlinear differential equations, in contrast to the solutions of linear ones, is the manner in which the arbitrary constants enter these solutions. In this paper information is given and results are found in connection with this subject.

In nonlinear DE, where the nonlinearity character is an essential factor, we have not, according to the present status, much information concerning the general solutions of these DE, as well as the manner in which the arbitrary constants enter to these solutions.

Given such a DE, the first subject for its solution is to attempt to find a general solution expressed in terms of the classical functions. To do this one must discover transformations, which may reduce the DE to some types that are known to have solutions of the desired kind, but this is, in general, either difficult or impossible.

It is customary to regard a linear DE as solved, if its solution can be reduced to the quadrature of a known function even though the quadrature can not be expressed in terms of classical functions.

In the same sense, one may regard a nonlinear DE as solved, if it can be reduced to the solution of a linear DE, even though the solution is not explicitly reducible to the classical functions.

In many cases of the nonlinear DE the classical functions are inadequate to express the solutions. Numerous nonclassical functions have been defined and partially explored in various ways recently, so there exists today an impressible collection of them from which one attempts to construct the general solutions of nonlinear DE.

The arbitrary constants enter the general solutions of nonlinear DE either «linearly» in a few cases, or «rationally linearly» in some other

---

* Δ. ΜΑΓΕΙΡΟΥ, **Αἱ γενικαὶ λύσεις τῶν μὴ γραμμικῶν διαφορικῶν ἐξισώσεων συναρτήσει τῶν αὐθαιρέτων σταθερῶν.**

Reprinted from the *Proceedings of the Athens Academy of Sciences* **52** (1977), 524–532.

cases. Nevertheless, in the majority of the cases, the arbitrary constants enter the general solutions in a «complicated nonlinear way», which characterizes the nature of the general solutions and their corresponding DE.

The singularities, especially the essential singularities, appearing in the general solutions, make these solutions, as well as the manner in which the arbitrary constants enter them, very complicated.

## I. GENERAL SOLUTIONS AS LINEAR FUNCTIONS OF THEIR ARBITRARY CONSTANTS

In a linear DE, the general solution is a linear combination of the fundamental set of solutions, and this general solution contains the arbitrary constants linearly. Such a general solution interprets the «principle of superposition», which holds only in linear DE.

As an example, we take the «Bessel DE»:

$$x^2y'' + xy' + (x^2 - n^2)y = 0$$

of which the general solution is of the form:

$$y = c_1 J_n(x) + c_2 Y_n(x)$$

where $c_1$, $c_2$ are the two arbitrary constants, and $J_n(x)$, $Y_n(x)$ the «Bessel functions» of the first and second kind, respectively, which are «special functions» particular solutions of the DE.

We remark that the principle of superposition is sometimes difficult to apply, as, e.g. the general solution of the Mathieu equation:

$$y'' + (a + b\cos 2x)y = 0$$

is not known containing two arbitrary constants in the above way **[2]**.

In the following we find classes of nonlinear DE of which the general solutions are linear functions of their arbitrary constants.

Some special methods are applied, illustrated by proper examples.

### I. 1. **The Factorization Method.**

By applying the factorization method to a nonlinear DE, if this method is applicable, one can have several general solutions of the DE, which contain the arbitrary constants linearly.

A general example is the first order DE: $F(x, y, y') = 0$ in case it is a polynomial for $y'$ of order m In this case, one can write:

$$F(x, y, y') \equiv y'^m + P_1(x, y) \cdot y'^{m-1} + \ldots + P_{m-1}(x, y) \cdot y' + P_m(x, y) = 0. \quad \text{(a. 1)}$$

This DE can be solved under the restriction $F_{y'}(x, y, y') \neq 0$, and if $y_1', \ldots, y_m'$ are the m simple roots, one can get F in the product form:

$$F \equiv [y_1' - p_1(x, y)] \cdot [y_2' - p_2(x, y)] \ldots [y_m' - p_m(x, y)] = 0, \quad \text{(a. 2)}$$

which is equivalent to the m DE:

$$y_1' = p_1(x, y), \quad y_2' = p_2(x, y), \quad \ldots, \quad y_m' = p_m(x, y). \quad \text{(a. 3)}$$

Integrating, the m general solutions of (a. 1) are:

$$\varphi_1(x, y, c_1) = 0, \quad \varphi_2(x, y, c_2) = 0, \quad \ldots, \quad \varphi_m(x, y, c_m) = 0. \quad \text{(a. 4)}$$

There are forms of the functions $p_i$ in (a. 3) which lead to functions $\varphi_i$ as linear functions of the arbitrary constants; for instance, in the cases where DE (a. 3) are with separable variables, or they are exact DE, or $p_i$ are homogeneous functions of, say, degree n, etc.

*Example 1:* $$y'^2 + y'(x^2y - xy) - x^3y^2 = 0. \quad (1.\ 1)$$

The factorization gives: $(y' - xy) \cdot (y' + x^2y) = 0$, when the two distinct general solutions of (1. 1), containing the arbitrary constants linearly, are:

$$y = c_1 e^{x^2/2}, \quad y = c_2 e^{-x^3/3}. \quad (1.\ 2)$$

*Example 2:* $$x^3y''y''' + x^2y''^2 - 2xy'y'' + 2yy'' = 0. \quad (2.\ 1)$$

The factorization gives: $y''(x^3y''' + x^2y'' - 2xy' + 2y) = 0$, when the two general solutions of (2. 1) are:

$$\text{(a)} \quad y = a_1 x + a_2, \qquad \text{(b)} \quad y = c_1 x + c_2 x^2 + c_3 x^{-1}. \quad (2.\ 2)$$

These two general solutions contain the family of straight lines through the origin ($a_2 = c_2 = c_3 = 0$, $a_1 = c_1$) as a «common part».

*Example 3:* $$y'^2 y''' + y^2 y''' + y''' = 0, \quad (3.\ 1)$$

$y'''$ is a common factor of the terms, then $y'''(y'^2 + y^2 + 1) = 0$, and since $y'^2+y^2+1\neq 0$, the DE (3.1) is equivalent to $y'''=0$, of which the general solution is:

$$y=c_1x^2+c_2x+c_3, \tag{3.2}$$

which consists of two different families of curves, the parabolas ($c_1\neq 0$) and the straight lines ($c_1=0$).

*Example 4:*
$$y'^3-\frac{y^2-x^2}{2xy}y'^2-y'+\frac{y^2-x^2}{2xy}=0. \tag{4.1}$$

Factorizing, one can get:

$$(y'^2-1)\cdot\left(y'-\frac{y^2-x^2}{2xy}\right)=0 \tag{4.2}$$

which is equivalent to:

$$\text{(a)}\quad y'=\pm 1, \qquad \text{(b)}\quad y'=\frac{y^2-x^2}{2xy} \tag{4.3}$$

The first of (4.3) gives two general solutions of (4.1):

$$\text{(a)}\quad y=x+c_1, \qquad \text{(b)}\quad y=-x+c_2. \tag{4.4}$$

The second DE of (4.3) has its right hand member as a homogeneous function of second degree, and can be written as:

$$y'=\frac{(y/x)^2-1}{2(y/x)}. \tag{4.5}$$

By using the transformation $y=xu$, the DE (4.5) can be solved and its general solution can be found to be:

$$x^2+y^2-cx=0. \tag{4.6}$$

The three functions (4.4, a, b) and (4.6) are the three general solutions of the DE (4.1), containing the arbitrary constants linearly.

## I. 2. The Method of Restricting Quantities of the DE.

By restricting quantities of the DE one may have several general solutions of the DE containing the arbitrary constants linearly. We give two examples.

*Example 5:*
$$y'(1+y'^2)-2y'y''=0. \tag{5.1}$$

We restrict the unknown function y of (5. 1) as follows:

(a): If y is such that $y''=0$, which implies $y'''=0$, then the general solution of (5. 1) is:

$$y = a_1 x + a_2. \tag{5. 2}$$

(b): If y is such that $y'' \neq 0$, then one can integrate (5. 1) exactly, and its new general solution is:

$$x^2 + y^2 + c_1 x + c_2 y + c_3 = 0. \tag{5. 3}$$

Both functions (5 2) and (5. 3) are distinct general solutions of (5. 1) and contain the arbitrary constants linearly.

*Example 6:*

$$y'' + \frac{2}{x} y' + y^n = 0. \tag{6. 1}$$

This is the famous Emden equation, coming from his investigation on basic problems of astrophysics. The solutions of this DE in a closed form are [**1**]:

$$\left.\begin{array}{lll} (a) & n = 0: & y = a_1 + \dfrac{a_2}{x} - \dfrac{x^2}{6} \\ (b) & n = 1: & y = c_1 \dfrac{\sin x}{x} + c_2 \dfrac{\cos x}{x} \\ (c) & n = 5: & y = \left(\dfrac{3a}{x^2 + 3a^2}\right)^{1/2}. \end{array}\right\} \tag{6. 2}$$

$a$, $a_1$, $a_2$, $c_1$, $c_2$ are arbitrary constants. The first two functions are the only known general solutions of the corresponding linear DE of (6. 1) containing the arbitrary constants lineatly. The third function of (6. 2), containing one arbitrary constant nonlinearly, is a «part» of the unknown general solution of (6. 1) in case $n = 5$.

On the occasion of this DE, we remark that any attempt to find the general solutions of (6. 1) for other values of n will be governed rather by a theoretical curiosity than by its usefulness, since, even if we know it, it has, according to Emden, no physical meaning in the Emden problems of astrophysics. Emden and his followers in astrophysics found a solution of (6. 1) in the form of a Taylor series, which interprets the reality very adequately.

### I. 3. **A class of nonlinear DE with general solutions containing the arbitrary constants linearly.**

We can find a class of nonlinear DE of which the general solution contains the arbitrary constants linearly.

Starting from the simple primitive:

$$y^2 = 4x\,; \quad y \neq 0, \quad x > 0$$

by differentiation one gets the DE: $yy' - 2 = 0$ of which the general solution is $y^2 = 4x + c_1$. Another differentiation gives: $yy'' + y'^2 = 0$ with general solution: $y^2 = 4x + c_1 x + c_2$. Continuing the differentiation up, say, to the order n, a nonlinear DE of order n results of which the general solution is:

$$y^2 = 4x + c_1 x^{n-1} + \ldots + c_{n-1} x + c_n\,,$$

where the arbitrary constants $c_1, \ldots, c_n$ enter linearly.

Generalizing the above, one can see that a nonlinear DE of order n:

$$F(y, y', \ldots, y^{(n)}) = \Phi(x), \tag{7.1}$$

where:

$$F = \frac{d^n}{dx^n} f(y), \qquad \Phi(x) = \frac{d^n}{dx^n} \varphi(x) \tag{7.2}$$

has as general solution the function:

$$f(y) = \varphi(x) + c_1 x^{n-1} + \ldots + c_{n-1} x + c_n\,. \tag{7.3}$$

We remark that the «principle of superposition» may be applicable to some nonlinear DE of which the general solutions are linear functions of the arbitrary constants, and this is an important problem.

## II. GENERAL SOLUTIONS AS RATIONALLY LINEAR FUNCTIONS OF THE ARBITRARY CONSTANTS

We distinguish a class of nonlinear DE of which the general solutions are rationally linear functions of the arbitrary constants, that is the general solutions are ratios of functions, where the arbitrary constants enter linearly in the numerators and denominators.

This class of nonlinear DE is the class of «Riccati equations of any order».

II. 1. Let us start from the simple case:

$$y = \frac{a_1 + cb_1}{a_2 + cb_2} \tag{8.1}$$

where c is the arbitrary constant, and $a_1$, $a_2$, $b_1$, $b_2$ functions of x.

Eliminating c between (8.1) and its derivative, the resulting DE has (8.1) as a general solution. The differentiation gives:

$$y' = \frac{a_1' + cb_1' - y(a_2' + cb_2')}{a_2 + cb_2}. \tag{8.2}$$

From (8.1) and (8.2) one can get, respectively:

$$c = -\frac{a_1 - ya_2}{b_1 - yb_2}, \qquad c = -\frac{a_1' - ya_2' - y'a_2}{b_1 - yb_2' - y'b_2} \tag{8.2}$$

and equating these values of c, one has:

$$(b_1 a_2 - a_1 b_2) y' + (a_1' b_2 - a_1 b_2' + b_1 a_2' - b_1' a_2) y + $$
$$+ (a_2 b_2' - a_2' b_2) y^2 = a_1' b_1 - a_1 b_1',$$

which is of the form:

$$y' = A_0(x) + A_1(x) y + A_2(x) y^2, \tag{8.3}$$

where:

$$\left.\begin{array}{ll} A_0 = (a_1' b_1 - a_1 b_1') / D, & A_1 = (a_1 b_2' - a_1' b_2 + b_1' a_2 - b_1 a_2') / D \\ A_2 = (a_2' b_2 - a_2 b_2') / D, & D = b_1 a_2 - a_1 b_2 \neq 0. \end{array}\right\} \tag{8.4}$$

The DE (8.3) is the «Riccati DE of order first», and the primitive (8.1) is its general solution, which contains the arbitrary constant c rationally linearly. We remark that the transformation:

$$y = \frac{u'}{A_3 u}, \tag{8.5}$$

applied to (8.3), leads to a linear DE of order two in u, when the DE (8.3) can be regarded as a solved DE.

As a simple example of the above is the DE: $y' = -2xy + xy^2$, which is of the form (8. 3), and which has as a general solution the function: $y = 2/(1 + ce^{x^2})$ of the form (8. 1).

II. 2. A natural generalization of (8. 1) is the function:

$$y = \frac{c_1 v_1 + \ldots + c_n v_n}{c_1 w_1 + \ldots + c_n w_2} \qquad (8.6)$$

where $c_i$ are the arbitrary constants, and $v_i$ and $w_i$ arbitrary functions of x. This generalization was introduced by E. Vessiot (1895) and G. Wallenberg (1899) [**1**]. The elimination of $c_i$ gives a nonlinear DE of order n, called a «Riccati DE of order n», of which (8. 6) is its general solution.

By a proper transformation, the solution (8. 6) of the Riccati DE of order n can be expressed in terms of the solutions of a linear DE of order (n + 1), which corresponds to this Riccati DE.

The function (8. 6) in case all of $w_i$ are zero, except one of them, say $w_n \neq 0$, becomes:

$$y = \bar{c}_1 \frac{v_1}{w_n} + \ldots + \bar{c}_{n-1} \frac{v_{n-1}}{w_n} + \frac{v_n}{w_n}, \qquad (8.7)$$

which contains the arbitrary constants $\bar{c}_i = (c_i / c_n)$, $i = 1, \ldots, (n-1)$ linearly, and it is the general solution of a linear DE of order $(n-1)$.

The polynomial DE:

$$y' = A_0(x) + A_1(x)\, y + \ldots + A_n(x)\, y^n, \qquad (8.8)$$

which is a natural generalization of the Riccati DE (8. 3), in case all of A's are constants, can be integrated exactly and its general solution contains the arbitrary constant linearly.

## III. THE GENERAL CASE OF THE GENERAL SOLUTIONS

In the general case of nonlinear DE the arbitrary constants enter into their general solutions in a «complicated nonlinear way», which characterizes the nature of the general solutions and their corresponding DE.

The singularities appearing in the general solutions make these solutions, as well as the manner in which the arbitrary constants enter them, very complicated.

An investigation on this line of problems is of theoretical and practical interest.

The writer has arrived at some results on these problems, but he considers these results as not yet sufficiently decisive to be communicated.

ΠΕΡΙΛΗΨΙΣ

Εἰς τὴν παροῦσαν ἐργασίαν ἐρευνᾶται ὁ τρόπος μὲ τὸν ὁποῖον αἱ αὐθαίρετοι σταθεραὶ εἰσέρχονται εἰς τὰς γενικὰς λύσεις τῶν μὴ γραμμικῶν διαφορικῶν ἐξισώσεων.

Αἱ μὴ γραμμικαὶ ΔΕ δύνανται νὰ ὑπαχθοῦν εἰς τρεῖς κατηγορίας.

Ἡ πρώτη ἀποτελεῖται ἀπὸ τὰς κλάσεις τῶν ΔΕ τῶν ὁποίων αἱ γενικαὶ λύσεις περιέχουν τὰς αὐθαιρέτους σταθερὰς «γραμμικῶς».

Ἡ δευτέρα περιέχει τὰς ΔΕ εἰς τῶν ὁποίων τὰς γενικὰς λύσεις αἱ αὐθαίρετοι σταθεραὶ ὑπεισέρχονται «ρητῶς - γραμμικῶς».

Ἡ τρίτη περιέχει ὅλας τὰς ἄλλας ΔΕ εἰς τῶν ὁποίων τὰς γενικὰς λύσεις αἱ αὐθαίρετοι σταθεραὶ ὑπεισέρχονται κατὰ «πολύπλοκον μὴ γραμμικὸν τρόπον».

Ἡ διερεύνησις τῶν συνθηκῶν, ὑπὸ τὰς ὁποίας ἡ «ἀρχὴ τῆς ἐπιπροσθέσεως» (principle of superposition) τῶν λύσεων τῶν μὴ γραμμικῶν ΔΕ δύναται νὰ ἐφαρμοσθῇ, ἕνα σημαντικὸν πρόβλημα, ὑπάγεται εἰς τὴν πρώτην κατηγορίαν.

Ἡ δευτέρα κατηγορία τῶν ΔΕ ἀναφέρεται εἰς τὰς γενικὰς λύσεις τῶν ἐξισώσεων τοῦ Riccati οἱασδήποτε τάξεως.

REFERENCES

1. H. Davis, «Introduction to the nonlinear differential and integral equations», Dover Publications, Inc. New York (1962), pp. 76.
2. E. Halle, «Lectures on ordinary differential equations», Addison - Wesley Publications Co, Reading, Mass, U.S.A., (1969), pp. 358 - 370.

ΠΡΑΚΤΙΚΑ ΤΗΣ ΑΚΑΔΗΜΙΑΣ ΑΘΗΝΩΝ

ΣΥΝΕΔΡΙΑ ΤΗΣ 26ΗΣ ΝΟΕΜΒΡΙΟΥ 1970

ΠΡΟΕΔΡΙΑ ΛΕΩΝ. Θ. ΖΕΡΒΑ

ΑΝΑΚΟΙΝΩΣΙΣ ΜΗ ΜΕΛΟΥΣ

ΜΑΘΗΜΑΤΙΚΑ.— **Actual Mathematical Solutions of Problems Posed by Reality, I. (A classical Procedure)***, *by D. G. Magiros***.

Ἀνεκοινώθη ὑπὸ τοῦ Ἀκαδημαϊκοῦ κ. Ἰω. Ξανθάκη.

INTRODUCTION

We discuss here phenomena or situations posed by reality which are changes of variable quantities that influence and interact to each other in an organized behavior. We call such a situation a «real system». A real system, either physical or social, corresponds to a «physical or social problem», mathematically expressed.

To describe quantitatively and explain the entire spectrum of the functional behavior of a real system, a «theory» of the system is needed, that is a set of statements concerning the behavior of the objects of the system. This theory, in general, implies a mathematical expression of the relationships between certain quantities of the system, that is a «mathematical model» of the system, associated with the «data» of the system.

The main requirement that the solution of a model is expected to satisfy is: «to interpret the real system in an adequate way». Such a

* Δ. Γ. ΜΑΓΕΙΡΟΥ, **Δεκταὶ μαθηματικαὶ λύσεις φυσικῶν προβλημάτων, I.**

** Consulting scientist, General Electric Company (RESD), Philadelphia, Pa., U.S.A.

Reprinted from the *Proceedings of the Athens Academy of Sciences* **45** (1970), 179–187.

solution of the model we call an «actual solution» of the system. There are non-actual solutions, called «formal solutions», which only satisfy the equations of the model and the data.

The above requirement would be fulfilled by the actual solution if this solution satisfies certain mathematical restrictions, called «Hadamard's restrictions». If this is so, the correspondent model must have an appropriate mathematical structure. This structure of the model shows the existence of an actual solution, although this solution is unknown.

Models with actual solutions are called: «well-posed», and with formal solutions: «non-well-posed».

In this paper we discuss a procedure by which «well-posed-models» may be constructed, and «actual solutions» be determined. Three important phases in the procedure can be distinguished: (a) the formulation of a theory concerning the system, which will lead to the construction of a model of the system; (b) the selection of a «well-posed-model», by applying the «Hadamard's restrictions»; and (c) the application of mathematical methods to find the solution of the «well-posed-model». The discussion, interpretation and evaluation of the results will complete the cycle of the research.

## THE CLASSICAL PROCEDURE

**1. Theory and models of real systems.** (3), (4), (9).

The construction of a mathematical model is, in many systems, the most important step of the study of the systems. It presupposes a «theory» of the system, that is, a set of statements some of which are verified by the experimental observations, while some others may be postulated. The objects and the formulation of the statements are the two constituents of the theory.

For a real system, the investigator needs a detailed knowledge of the observational and experimental facts, the pertinent laws, a penetrating insight, a mature judgment.

The data of the system, that is the results of observation, experiments and measurements related to the system, must be completely stated and known approximately within an accepted error, thus called «admissible data».

The domain on which the system operates must be known. This is related to the selection of the «major» variables, among the variables of the system, which define the process of the motion of the system, that is the main subject of the theory. The other variables of the system, the «minor» variables, are either «parameters» of the system, thus defining the «environment» of the system, or a «noise». A good theory of the system depends on the appropriate selection of the major and minor variables of the system.

All the above considerations, and many others of special interest, constitute the theory of a real system, and a satisfactory theory of the system makes the system «correctly stated».

Correctly stated systems can be represented in a mathematical form, called «model of the system», which gives the functional behavior of the system in a quantitative way.

The «strong interactions» between the variables of the system can be expressed explicitly in a mathematical form and give the basis of a completely deterministic (non-statistical) model. The «weak interactions» are not represented explicitly in the equations, but they are an essential part of the system related to statistical mechanics.

Arbitrary assumptions, decisions and choices in developing a theory and the model of the system, have as a result the construction of various models to the same real system.

The model summarizes the data of the system, and if one repeats an experiment or gets numbers, by using the model, and these results agree with the data assumed in constructing the model, the model is acceptable.

The applicability of a model depends on the possibility of estimating their parameters from the data.

Models, parts of which cannot explain given experimental facts of the system, are not acceptable. But if these parts of the model are cancelled, then every feature of it is related in some way to the experimental data and the model becomes acceptable.

Usually, the models are «boundary and/or initial value problems».

**2. Hadamard's restrictions, well-posed-models.**

Among the possible models of a real system correctly stated, one can distinguish the «well-posed» ones of which the solutions are «actual solution» of the system (**7**).

As we know, if the equations of the system and its data are such that:

a. the model has a solution corresponding to the data,

b. the solution is unique, and

c. the solution depends continuously on the data,

we say that the solution satisfies these three «Hadamard's restrictions», and that the equations of the model and the data give a «well-posed» or «reasonable» problem, and we have a definition of «correctness» of the problem in the sense of Hadamard (**2a**).

If the solution fails to satisfy even one of the «Hadamard's restrictions», it is a formal solution, and the problem a «non-well-posed» or «non-reasonable», or «improperly posed» problem.

Much attention has been given to such problems in recent years (**6**), (**8**).

Appropriate continuity and differentiability properties of the mathematical expression of the «well-posed-problem», by means of known existence, uniqueness and continuity theorems, assure the existence of a solution which satisfies the Hadamard's restrictions.

It remains now to see that the solutions which satisfy the Hadamard's restrictions are actual solutions of the real problem.

We remark that (**3**), (**5**):

*First:* The well-posed-problem must necessarily possess a solution.

The existence of a solution and its determination are different concepts, and we try to determine a solution only if we know it exists.

*Second:* The solution must be just one.

The existence and uniqueness properties of the solution express our belief «in causality» or «in determinism», a principle according to which one can repeat experiments with the expectation to get consistent results.

*Third:* The continuous dependence of the solution on the data has as a result that small changes in the data imply small changes in the solution.

The data, as results of observation, experiments and measurements, are given with small errors, when the solution has an uncertainty and its estimation an error (**2b**). If the above continuity property of the solution holds, then «the smaller the error in the data, the smaller the error in the estimation of the solution», and «admissible solutions» correspond to «admissible data» (**1**). The above continuity property holds for finite time. If it holds for any time, this property becomes a «stability property» of the solution in the sense of Liapunov in a «parameter-space», which is here the «data-space» (**5**).

The above remarks make clear that solutions which satisfy the Hadamard's restrictions are actual solutions of real systems.

### 3. The determination of the actual solution.

Having now found the «well-posed-problem» corresponding to the real system, we try, next, to determine its solution. Heuristic scientific reasoning toward the ultimate solution can be frequently used. The «principle of approximation» can be introduced to achieve the solution needed. The continuity property of the Hadamard's restrictions helps to apply this principle.

To the «well-posed-problem» M we try to find an appropriate «approximate problem» $M_n$, of which the solution $A_n$, containing the index n, will be determined. The limit A of the solution $A_n$ as $n \to \infty$, $A = \lim_{n\to\infty} A_n$, is the solution of the well-posed-problem, that is the actual solution of the real system (**2a**).

### 4. Summary and Conclusion.

Summarizing the preceding we see that the research for finding

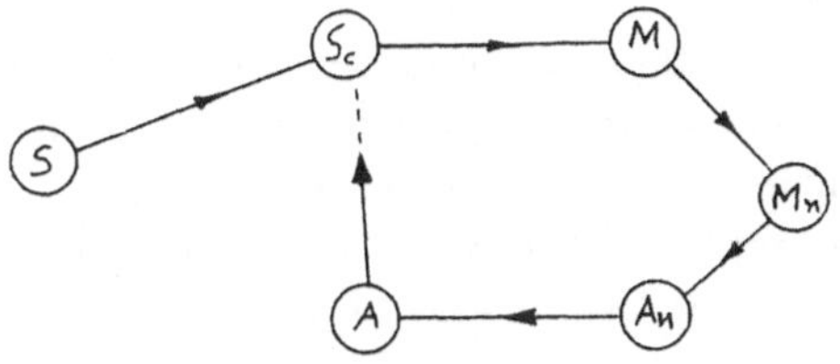


S : real system
$S_c$ : system correctly stated
M : well-posed-model
$M_n$ : approximate problem to M
$A_n$ : solution of $M_n$
$A = \lim_{n\to\infty} A_n$ : solution of M

Figure 1.

actual solutions of real systems is constituted by a set of steps, which makes a complete cycle, shown schematically in Figure 1.

| | | | |
|---|---|---|---|
| (i) | Step : | $S \rightarrow S_c$ ; | We need the theory of the system, which will make the system correctly stated and the modeling possible. |
| (ii) | Step : | $S_c \rightarrow M$ ; | We use the Hadamard's restrictions, which will help to get the well-posed-model of the system. |
| (iii) | Step : | $M \rightarrow M_n$ ; | We select the appropriate approximation to «w - p - m». |
| (iv) | Step : | $M_n \rightarrow A_n$ ; | We find the solution $A_n$ the problem $M_n$. |
| (v) | Step : | $A_n \rightarrow A$ ; | We find the solution required of the system, by using the limiting process, as $n \rightarrow \infty$. |

The step (i) is a decisive step. It requires ingenuity, experience and knowledge of the field to which the real system belongs;

The step (ii) helps to get the reasonable model on which the whole investigation is based;

The steps (iii) and (iv) need a constructive imagination and a complete knowledge of mathematics to be applied;

The step (v) is an application of the principle of approximation, which can be succeeded by applying the continuity property of the Hadamard's restrictions.

**5. Remarks on the classical procedure.**

The classical procedure for finding actual solutions of real systems shows a general orientation of thought on the problem. It is based on the Hadamard's definition of «correctness», of «well-poseness».

It was found that the Hadamard's definition of «correctness» rules out as «non-well-posed-problems» important real problems, as, e. g., problems of geophysics, and some authors at the present time present different notions of «correctness» and various approaches to the formulation and investigation of the «non-well-posed-problems» (**6**).

There are difficulties in the application of the classical procedure in some scientific fields. Problems of modern physics especially present these difficulties (**2b**).

The exact laws governing the behavior of the system under consideration may not be known, and, in this case, a complete theoretical description of the system is impossible.

Problems of fundamental particle structure and reaction present the difficulties in an acute form.

More often the situation is analogous to that of problems of atomic and molecular structure, where the interactions are known, but where the structure in a particular problem may be complicated.

In problems of classical dynamics of compressible fluids, the differential equations supplemented by boundary conditions are not always a sufficiently complete framework for an adequate description of physical reality.

The above remarks must be considered, when one tries to present a mathematical description of real systems.

Selected applications from different scientific fields should show the importance of the above procedure and remarks.

## REFERENCES

1. Brillouin, L.: «Scientific Uncertainty and Information», Academic Press, New York, 1964.
2. Courant, R.: (a) «Methods of Applied Mathematics», in the book: «Recent Advances in Science»; Editors: M. H. Shammos and G. M. Murphy, New York University Press, New York, 1956.
   (b) «Boundary Value Problems in Modern Fluid Dynamics», Proc., Intern. Congress of Mathematicians, Harvard University, Volume II, 1956.
3. «Education in Applied Mathematics»: Proc. of Conference, May 24-27, 1966. Siam Review, Volume 9, No. 2, April 1967.
4. Goodwin, B. C.: «Temporal Organization in Cells», Academic Press. New York, 1963.
5. Hale, J. K.: «Ordinary Differential Equations», Wiley-Interscience, New York, 1963.
6. Lavrentiev, M.: «Some Improperly Posed Problems of Mathematical Physics», Springer-Verlag, New York, 1967.
7. Magiros, D. G.: «Physical problems discussed mathematically», Bulletin, Greek Mathematical Society, New Series, Volume 6, II, No. 1, 143-156, 1965.
8. Miranker, W.: «A Well-Posed-Problem for a Backward Heat Equation», Proc., American Mathem. Society, April, 1961.
9. Mesarovič, M. D., Editor: «System Theory and Biology», Proc., III Systems Symposium, Springer-Verlag, New York, 1968.

## ΠΕΡΙΛΗΨΙΣ

Εἰς τὴν παροῦσαν ἐργασίαν περιγράφεται μέθοδος ἐρεύνης ἡ ὁποία ἀφορᾷ εἰς τὴν εὕρεσιν «φυσικῶς δεκτῶν» μαθηματικῶν λύσεων εἰς φυσικὰ καὶ κοινωνικὰ προβλήματα. Αἱ κύριαι φάσεις τῆς μεθόδου εἶναι: (α) Ἡ ἔρευνα ὅπως καθιερωθῇ μία «θεωρία τοῦ συστήματος», ἡ ὁποία νὰ καθιστᾷ τὸ σύστημα ἀφ᾽ ἑνὸς μὲν καλῶς ἐκπεφρασμένον, ἀφ᾽ ἑτέρου δὲ νὰ ὑποβοηθῇ εἰς τὴν εὕρεσιν μαθηματικῶν προτύπων τοῦ συστήματος· (β) Ἡ ἐκλογὴ ἑνὸς «καλῶς τεθειμένου προτύπου» τοῦ συστήματος, τὸ ὁποῖον νὰ ὁδηγῇ εἰς ἓν «καλῶς τεθειμένον μαθηματικὸν πρόβλημα», τοῦ ὁποίου νὰ προσδιορισθῇ ἡ λύσις· καὶ (γ) Ἡ εὕρεσις τῆς λύσεως τοῦ ἀρχικοῦ συστήματος, βάσει τῆς εὑρεθείσης λύσεως τοῦ καλῶς τεθέντος μαθηματικοῦ προβλήματος.

Ὑπάρχουν πεδία ἐρεύνης, ὅπου ἡ μέθοδος εἶναι ἢ δύσκολον ἢ ἀδύνατον νὰ ἐφαρμοσθῇ, ἀλλ᾽ ὅμως ὅπου αὕτη δύναται νὰ ἐφαρμοσθῇ τὰ ἀποτελέσματα δύνανται νὰ ἑρμηνεύσουν τὴν φυσικὴν πραγματικότητα κατὰ πολὺ ἱκανοποιητικὸν τρόπον.

Εἰς ἑπομένην ἐργασίαν θὰ δοθοῦν ἐφαρμογαὶ — ἀπὸ διάφορα πεδία ἐρεύνης — διὰ τῶν ὁποίων θὰ διασαφηνίζεται ἡ πορεία τῆς μεθόδου καὶ θὰ καταδεικνύεται ἡ σημασία της.


*

Ὁ Ἀκαδημαϊκὸς κ. **Ἰω. Ξανθάκης** κατὰ τὴν ἀνακοίνωσιν τῆς ἐργασίας τοῦ κ. Δ. Μαγείρου εἶπε τὰ ἑξῆς:

Ἔχω τὴν τιμὴν νὰ παρουσιάσω εἰς τὴν Ἀκαδημίαν ἐργασίαν τοῦ κ. Δημ. Μαγείρου ὑπὸ τὸν τίτλον: «Δεκταὶ Μαθηματικαὶ λύσεις φυσικῶν προβλημάτων». Εἰς τὴν ἐργασίαν ταύτην ὁ συγγραφεὺς ἐκθέτει μίαν μέθοδον ἐρεύνης ἀφορῶσαν εἰς τὴν εὕρεσιν μαθηματικῶν λύσεων φυσικῶν προβλημάτων ἀποδεκτῶν ἀπὸ φυσικῆς ἀπόψεως. Αἱ κύριαι φάσεις τῆς μεθόδου εἶναι αἱ ἑξῆς:

α) Ἡ διατύπωσις μιᾶς «θεωρίας τοῦ συστήματος» ἡ ὁποία νὰ καθιστᾷ ἀφ᾽ ἑνὸς μὲν καλῶς ἐκπεφρασμένον τὸ σύστημα, ἀφ᾽ ἑτέρου δὲ νὰ ὑποβοηθῇ εἰς τὴν εὕρεσιν μαθηματικῶν προτύπων τοῦ συστήματος.

β) Ἡ ἐκλογὴ ἑνὸς καταλλήλου προτύπου τοῦ συστήματος, τὸ ὁποῖον νὰ ὁδηγῇ εἰς ἓν καλῶς τεθειμένον μαθηματικὸν πρόβλημα καὶ

γ) Ἡ εὕρεσις τῆς λύσεως τοῦ ἀρχικοῦ συστήματος βάσει τῆς εὑρεθείσης λύσεως τοῦ καλῶς τεθέντος μαθηματικοῦ προβλήματος.

Ὁ συγγραφεὺς ἀναφέρει ὅτι ὑπάρχουν πεδία ἐρεύνης, ὅπου ἡ μέθοδος αὕτη εἶναι εἴτε δύσκολον εἴτε ἀδύνατον νὰ ἐφαρμοσθῇ. Ὅπου ὅμως αὕτη δύναται νὰ ἐφαρμοσθῇ τὰ ἀποτελέσματα, κατὰ τὸν συγγραφέα, δύνανται νὰ ἑρμηνεύσουν τὴν φυσικὴν πραγματικότητα κατὰ τρόπον λίαν ἱκανοποιητικόν.

Τέλος ὁ κ. Δ. Μάγειρος προτίθεται εἰς προσεχῆ ἀνακοίνωσίν του νὰ παρουσιάσῃ ἐφαρμογὰς ἐπὶ διαφόρων πεδίων ἐρεύνης διὰ τῶν ὁποίων θὰ διαγράφεται ἡ πορεία τῆς μεθόδου καὶ θὰ καταδεικνύεται ἡ σημασία της.

ΣΥΝΕΔΡΙΑ ΤΗΣ 28 ΙΑΝΟΥΑΡΙΟΥ 1971 21

ΜΑΘΗΜΑΤΙΚΑ.— **Actual Mathematical Solutions of Problems Posed by Reality, II. (Applications)***, *by D. G. Magiros***. Ἀνεκοινώθη ὑπὸ τοῦ Ἀκαδημαϊκοῦ κ. Ἰω. Ξανθάκη.

## INTRODUCTION

In the previous paper **3(a)**, we discussed a classical procedure for finding actual mathematical solutions of real systems in many physical or social fields. The main phases of the procedure were:

A. The creation of a theory of the system, which helps its modeling;
B. The selection of a «well-posed-model» of the system, which gives a well-posed mathematical problem, and
C. The construction of the solution of this problem, which is the «actual solution» of the system.

In the present paper we give some applications of the above classical method, by which we can see the difficulties of its application and its advantages in case this method can be applied. We select the applications from thermodynamics, astrodynamics, non-linear mechanics, biology, etc.

## 1st Application Problem of Thermodynamics: (4)

Forward and Backward Heat Flow Problem.

The Step: «$S_c \to M$» of the classical method characterizes the whole study of the problem. The problem is: *«To study the heat flow in a given medium»*. To make this physical problem correctly stated, one accepts for the medium to be homogenious and isotropic with respect to the heat flow, and that the heat flow is towards the decreasing temperature. Based on these hypotheses, the mathematical idealization, the model, is the partial differential equation:

---

* Δ. Γ. ΜΑΓΕΙΡΟΥ, **Δεκταὶ μαθηματικαὶ λύσεις φυσικῶν προβλημάτων, II. (Ἐφαρμογαί)**.

** Consulting scientist, General Electric Company (RESD), Philadelphia, Pa., U.S.A.

Reprinted from the *Proceedings of the Athens Academy of Sciences* 46 (1971), 21–31.

$$U_{xx} + U_{yy} + U_{zz} = U_t \tag{1}$$

where: $u = u(x, y, z, t)$ is the temperature in the x, y, z -space and t-time. In the equation (1) there is a coefficient depending on density, specific heat, and thermal conductivity and this coefficient is here taken equal to unity.

In case of a «one-dimensional medium», if the «data-initial condition» is:

$$u(x, 0) = n \, . \sin nx, \qquad n = \text{integer} \tag{2}$$

one can check that the solution of equation (1), satisfied by (2), is:

$$u(x, t) = n \, . \, e^{-n^2 t} \, . \sin nx \tag{3}$$

and it is unique, when the first two Hadamard's restrictions are satisfied. We distinguish here two cases:

a. It $t>0$, when one has the «forward heat problem», the solution (3) $\to 0$ and the condition (2) $\to \infty$, as $n \to \infty$, then the solution (3) satisfies also the third Hadamard's restriction, when the function (3) is accepted as an «actual solution» of the «forward heat problem», which is a «well-posed-problem».

b. If $t<0$, when one has the «backward heat problem», the solution (3) and the condition (2) $\to \infty$, as $n \to \infty$, then the solution (3) violates the third Hadamard's restriction, when the function (3) is a «formal solution» of the «backward heat problem», which is a «non-well-posed-problem».

**2nd Application** **Problem of Orbital Mechanics: 3 (b)**

An artificial celestial body is moving under the influence of a central force obeying the inverse square Newton's law toward the attractive center. A general force is applied, acts for an interval of time, then it is removed. Find the motion of the body during the action of the general force.

A model of this problem is:

$$\left.\begin{aligned} &\ddot{\underline{r}} = -\frac{\mu}{r^3(\tau)}\underline{r}(\tau) + \underline{T}(\tau) \\ &\underline{r}(0) = \underline{r}_0, \quad \dot{\underline{r}}(0) = \dot{\underline{r}}_0 + \underline{I}_0 \\ &D_1 : |\underline{r}(\tau)| < M_1, \quad |\dot{\underline{r}}| < M_2 \\ &D: \ 0 \leqslant \tau \leqslant \tau' \end{aligned}\right\} \tag{4}$$

where $\underline{T}$ the general force, $\underline{r}$ the radial vector from the attractive center to the center of mass of the body, $\underline{I}_0$ the impulse, which is given by:

$$\underline{I}_0 = \int_0^{t_0} T(t)\,dt, \qquad \tau = t - t_0. \tag{4. 1}$$

If we take the function:

$$\underline{r}(\tau) = a_1(\tau)\,\underline{r}_0^* + a_2(\tau)\,\underline{s}_0^* + a_3(\tau)\,\underline{T}_0^* \tag{5}$$

as a «trial solution», where $\underline{r}_0^*$, $\underline{s}_0^*$, $\underline{T}_0^*$ are special unit vectors, the coefficients $a_1$, $a_2$, $a_3$ must satisfy the following conditions in order that the function (5) is a «formal solution» of (4):

$$\left.\begin{aligned} &\ddot{a}_1 + \frac{\mu}{r^3}a_1 = T_1; \quad a_1(0) = r_0, \quad \dot{a}_1(0) = 0 \\ &\ddot{a}_2 + \frac{\mu}{r^3}a_2 = T_2; \quad a_2(0) = 0, \quad \dot{a}_2(0) = s_0 \\ &\ddot{a}_3 + \frac{\mu}{r^3}a_3 = T_3; \quad a_3(0) = 0, \quad \dot{a}_3(0) = 0 \end{aligned}\right\} \tag{6}$$

If $T_1$, $T_2$, $T_3$ are differentiable, $\dot{T}_1$, $\dot{T}_2$, $\dot{T}_3$ continuous, $r \neq 0$; $a_1$, $a_2$, $a_3$ twice differentiable, and $\ddot{a}_1$, $\ddot{a}_2$, $\ddot{a}_3$ continuous, we see that equations (6) satisfy the Hadamard's restrictions, when the functions: $a_1(\tau)$, $a_2(\tau)$, $a_3(\tau)$ can be uniquely determined from equations (6), and are continuous functions of the initial conditions of (6). Therefore, the solution (5) of equation (4), after the above restrictions of the force T and its derivative, is unique and depends continuously on the initial conditions of (4), then it can be accepted as an actual solution of the equation (4).

**3rd Application** **Problem of Non-Linear Mechanics: 3 (c)**

The Problem of Principal Modes of Non-Linear Systems.

The concept of «principal modes» of linear systems plays a predominant role in the analysis of the oscillatory systems of many fields.

The principal modes in linear systems are, by definition, the fundamental set of solutions of which a linear combination gives the general solution of the linear system; or, physically speaking, they are the special modes of oscillations of the linear system in terms of which we can discuss any kind of oscillations of the system.

Since the «principle of superposition» does not hold in non-linear systems, the concept of principal modes, as given above, is meaningless in non-linear systems, and the following problem may arise: *«Has the problem of principal modes of non-linear systems a physical meaning?»*; or *«How one can make the problem of principal modes of non-linear systems a well-posed problem?»*

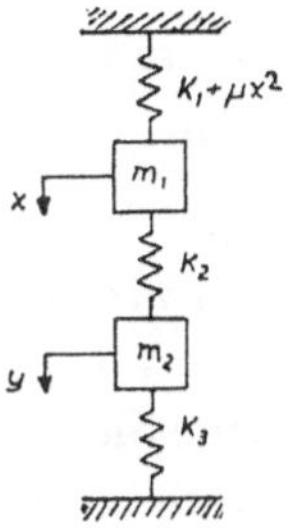

Figure 1.

The writer has publisched some papers in connection with this important problem, and transfers here some appropriate thoughts, techniques and results in order to give this problem as an example of the classical approach of the preceding paper.

We can find a new definition of the concept of principal modes for both the linear and non-linear systems, and such that the known definition in linear systems comes as a result from the new definition. The writer gave two new definitions which, under some conditions, are equivalent.

After that we try to make the physical problem correctly stated and the mathematical idealization well-posed.

We take a trial solution and make it formal, first, and then actual.

If we restrict ourselves to a «two-degrees-of-freedom» mechanical non-linear system, as shown in Figure 1, the equations of motion of the «two-masses-three springs» non-linear system are:

$$\left.\begin{aligned} \ddot{x} + \omega_1^2 x - \lambda_2 y + \lambda_1 x^3 = 0 \\ \ddot{y} + \omega_2^2 y - \lambda_3 x = 0 \end{aligned}\right\} \qquad (7)$$

where :

$$\omega_1^2 = \frac{K_1 + K_2}{m_1}, \quad \omega_2^2 = \frac{K_2 + K_3}{m_2}, \quad \lambda_1 = \frac{\mu}{m_1}, \quad \lambda_2 = \frac{K_2}{m_1}, \quad \lambda_3 = \frac{K_2}{m_2} \tag{7a}$$

and $\mu$ characterizes the non-linearity of one anchor spring.

By using the transformation :

$$x = x_1, \quad \dot{x} = x_2, \quad y = x_3, \quad y = x_4 \tag{8}$$

the system (7) can be reduced to its normal form :

$$\left.\begin{aligned} & x_i = f_i(x_1, x_2, x_3, x_4), \quad i = 1, 2, 3, 4 \\ & f_1 = x_2, \quad f_2 = -\omega_1^2 x_1 + \lambda_2 x_3 - \lambda_1 x_1^3, \quad f_3 = x_4, \quad f_4 = \lambda_3 x_1 - \omega_2^2 x_3 \end{aligned}\right\} \tag{9}$$

valid in a region R :

$$R: \ |x_i| < h, \quad i = 1, 2, 3, 4 \tag{9a}$$

The appropriate initial conditions for «principal modes» are in R :

$$x_1(0) = x_{10}, \quad x_2(0) = 0, \quad x_3(0) = x_{30}, \quad x_4(0) = 0 \tag{9b}$$

where $x_{10}$ and $x_{30}$ are appropriately related to each other.

Now we remark that the nature of the functions $f_i$ of (9) are such that all Hadamard's restrictions are satisfied. These functions $f_i$ are continuous in R, then bounded; they have continuous partial derivatives $\partial f_i / \partial x_k$ in R, when they satisfy Lipschitz conditions with respect to $x_i$ in R for a Lipschitz constant $s = \text{l.u.b} /\partial f_i / \partial x_k/$. The above properties assure the unique existence of the solution of (9) and (9b) in a region $R' \subset R$. As the initial point $x_{i0}$, $i = 1, 2, 3, 4$ varies in R′, the solution satisfies the three Hadamard's restrictions, and the problem is «well-posed».

## 4th Application **A Problem of Underwater Warfare** :

The Problem of Domes. 1

The problem of domes arose in the winter of 1942 - 1943 in connection with «underwater warfare». As is known, underwater sound ranging depends on sending out a sound beam in water and, attached to a fast-moving ship, the water steaming around the plate causes serious disturbances. For elimination of these disturbances, the projector is closed in a so-called «dome», Figure 2, which is a convex shell of metal or other material filled with water. Such domes interfere only slightly with the formation of a concentrated sound beam. During 1942 - 1943, a large number of small submarines chases were built and equipped with sound gear similar to, but smaller than, the gear used before. While the manufacture of domes to fit this smaller gear was underway, it was discovered that these smaller domes led to an intolerable diffusion on the sound beam. At that time, a quick remedy was imperative, and a mathematical analysis of the problem was needed to support and speed-up experimented work.

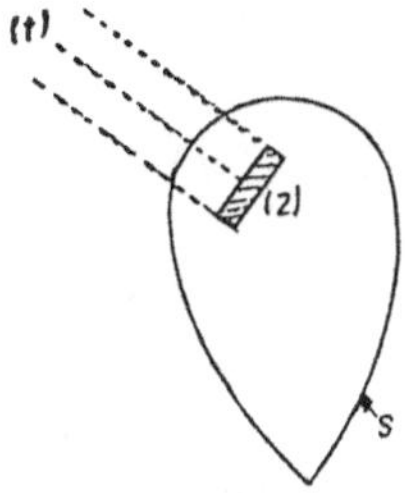


Figure 2.

(1) Axis of beam sound
(2) Projector
S Surface of Dome

The mathematical problem, related to the above physical problem, was to solve the differential equation:

$$\left.\begin{aligned} &\nabla^2 P + K^2 P = 0 \\ &\nabla^2 = \frac{\partial^2}{\partial x^2} + \frac{\partial^2}{\partial y^2} + \frac{\partial^2}{\partial z^2} \end{aligned}\right\} \quad (10)$$

in which $K = \omega/c$, $\omega$ the frequency, c the sound velocity and K has for our problem, unfortunately, different values within the shell of the dome and outside.

This mathematical idealization was not a suitable one for the problem.

They found the suitable mathematical idealization by the following process. The actual dome of small finite thickness was replaced by an extremely thin surface, then the influence of the dome was simply replaced by conditions for jump discontinuities of the disturbance q of the beam across the surface.

These conditions are:

$$\left.\begin{aligned} &[q] = \frac{p_1}{p_0^{-1}} \cdot \frac{\partial p}{\partial n} \\ &\left[\frac{\partial q}{\partial n}\right] = \frac{p_0}{p}(K_0^2 - K_1^2)\,p - \left(1 - \frac{p_0}{p}\right)\left(\frac{\partial^2 p}{\partial n^2} + 2H\frac{\partial p}{\partial n}\right) \end{aligned}\right\} \quad (11)$$

where the symbol $[\cdot]$ means jump of the quantity of the symbol across the surface, q is the disturbance of the acoustic pressure p caused by the dome and the normal derivatives $\frac{\partial}{\partial n}$ are to be evaluated on the surface S of the dome. The quantity H is the mean curvature of S, i. e. the average of the curvature of any two normal plane sections at right angles to each other. In addition to conditions (11) to be satisfied by q on S, q should be a solution of the equation :

$$\nabla^2 q + K_0^2 q = 0 \tag{12}$$

same behavior as P at $\infty$. This problem possess the unique solution :

$$q = -\frac{1}{4\pi}\iint_S \left[\frac{\partial q}{\partial n}\right]\frac{e^{ik_0 r'}}{r'}\,ds + \frac{1}{4\pi}\iint_S [q]\,\frac{\partial}{\partial n}\left(\frac{e^{ik_0 r'}}{r'}\right) dS \tag{13}$$

The quantities in the brackets are given by conditions (11), r′ is the distance from a fixed point (x, y, z) at which q (x, y, z) is to be determined to the point of integration on S. This formula yields the disturbance as the effect due to a layer of point sources and a layer of dipoles disturbed on S with intensities which are known as soon as the original pressure p is known, since the quantitites in brackets are fixed in value due to conditions (11). The relative directional disturbance :

$$\left|\frac{p_1}{p}\right| R.c.h\left(\frac{q_1}{p_1} - \frac{q_0}{p_0}\right)$$

would, finally, be obtained from (13). The solution (13) is valid for a shell of constant thickness, but it could be extended without essential error to cases in which the dome shell is made up of a not too large number of pieces, each of which is of constant thickness. All that would be necessary would be to insert a numerical factor d in the integrands on the right-hand side of expression (13), which would be precise constant on S. By this formula, one can analyze the contribution to the distortion of various factors, such as the curvature of the dome and the density and sould velocity within it.

The above kind of mathematical idealization, even without detailed numerical computation, proved helpful to the designing engineer.

**5th Application** **Biology, Ecology, Economics: 2, 5**

The Problem of Mixed Populations: Two Species Competing for a Common Food Supply.

For the study of the growth of two mixed populations of species in mutual interdependence of any kind, e. g. in competing for a common food supply, several models have been proposed. One of these models of which the formulation is based on determinizing the time-rate of change of quantities as a function of the quantities and some parameters, is :

$$\left.\begin{aligned} \dot{x} &= a\,[b - x - f_1(y)]\,x \\ \dot{y} &= c\,[d - y - f_2(x)]\,y \end{aligned}\right\} \qquad (14)$$

x and y are the numbers (or masses) of individuals of the species present at any time, and a, b, c, d parameters of which the domain of possible change define the environment of the model.

The model (14) is either «well-posed» or «non-well-posed», depending on properties of the functions $f_1(y)$ and $f_2(x)$.

Physically, the quantities x and y are non-negative, when the region D of the validity of the model (14) is the first quadrant of the x, y - plane. The initial conditions $x_0$, $y_0$ of (14) is the starting point of the solution, if this solution exists, and this point lies in the region D.

If the functions $f_1(y)$ and $f_2(x)$ are defined, single-valued and continuous in the region D, then the right-hand numbers of (14) are continuous functions of all their arguments, when a solution of (14) necessarily exists through the point $(x_0, y_0)$, and the first Hadamard's restriction is satisfied.

If, in addition, the functions $f_1(y)$ and $f_2(x)$ have continuous derivatives in y and x, respectively, then the right-hand numbers of (14) have continuous partial derivatives with respect to all their arguments, and the solution through $(x_0, y_0)$ is unique and depends continuously on the $x_0$, $y_0$, when the second and third Hadamard's restrictions are satisfied, and the model (14) is a «well-posed» one.

We remark that: the solution of the model (14), which starts from the inital point $(x_0, y_0)$, tends, as t increases, to a point $\bar{x}$, $\bar{y}$, and we may have three cases. First, the point $(\bar{x}, \bar{y})$ may be a point inside the region D, both $\bar{x}$ and $\bar{y}$ positive, when one can speak about the «co-existence of the species». Second, the point $(\bar{x}, \bar{y})$ may be identical with the origin, when one can speak about the «extinction of the species». Third, one of $\bar{x}$ and $\bar{y}$ may be zero and the other positive, and this case corresponds to the «principle of competitive exclusion», a principle much used in ecology, but which has been much criticized.

We remark that if the variables x and y of the model (14) are numbers of individuals of the populations they are restricted to be (positive) integers, when they are «step functions» of time, and the functions $f_1$ and $f_2$ of (14) are restricted to assume values according to permitted values of x and y. The functions $f_1$ and $f_2$ in this case have no properties, as mentioned above, which make the model (14) a «well-posed» one. In this case, the model (14) is not a «continuous system», but a «discrete system». If we assume that x and y in the model (14) are the masses of the populations, we can remove the above restriction of x and y and the function $f_1$ and $f_2$ regain the properties needed in order for the model to be a «well-posed» one.

All the above remarks and results can be applied to different social problems, if the competitive species and the limiting resources are appropriately specified.

To apply the above in the field of economics, the variables x and y must denote the size or extent of two commercial enterprises competing for common sources and for a common market.

**6th Application** **Modern Physics, Dynamic Meterology : 1**

The classical procedure, discussed in the preceding, and especially the step to find the «well-posed-model», combined with numerical analysis and the use of high-speed computers, gave and may give much success in the investigation of problems of great contemporary interest.

The «Synchrotron» and the «weather prediction» can be used as examples.

a. *Synchrotron.* The recently discovered «strong-focusing-principle» is the basis for the study of the multibillion-volt proton accelerators. This principle is related to the stability of solutions of ordinary linear differential equations of second order with periodic coefficients. The actual orbits, because of unavoidable imperfections of magnets and other causes, follow, approximately, linear periodic differential equations, and a modified nonlinear model is not possible. Experimental studies, under various assumptions, the use of computing and mathematical analysis, give encouragement to the designers for success.

b. *Weather Prediction.* According especially to Bjerkness, one may formulate the laws of atmospheric phenomena by models which are partial differential equations. Based on the today's data and using the Bjerkness model as a «well-posed» one, the prediction of tomorrow's weather would require qualified computer men with desk computing machines for much time.

## REFERENCES


1. Courant, R.: «Methods of Applied Mathematics» in the book: «Recent Advances of Science»; Editors: M. H. Shammos and G. M. Murphy, New York University Press, New York, 1956.
2. Cunningham, W.: «Simultaneous Non-Linear Equations of Growth», Bulletin, Mathematical Biophysics, Volume 17, 101 - 110, 1955.
3. Magiros, D. G.: (a) «Actual Mathematical Solutions of Problems Posed by Reality, I», Proc., Athens Academy of Sciences, Volume 45, 179-187, 1970.
   (b) «The Motion of an Artificial Celestial Body under the Influence of a Newtonian Center and a General Force», Proc., XV Intern. Astronautical Congress, Warsaw, Poland, 1964.
   (c) «Methods for Finding Principal Modes of Non-Linear Systems Utilizing Infinite Determinants», Journal of Mathematical Physics, 2, No. 6, 869 - 875, 1961.
4. Miranker, W.: «A Well-Posed Problem for a Backward Heat Equation», Proc., American Mathematical Society, April, 1961.
5. Volterra, V.: «Leçons sur la Théorie Mathématique de la lutte pour la vie», Gauthier-Villars et Cie, Paris, 1931.


## ΠΕΡΙΛΗΨΙΣ

Εἰς προηγουμένην ἐργασίαν **3(α)** ἀνεπτύχθη μέθοδος ἐρεύνης «φυσικῶς δεκτῶν» μαθηματικῶν λύσεων εἰς φυσικὰ καὶ κοινωνικὰ συστήματα καὶ ὑπεδείχθησαν δυσκολίαι ἐφαρμογῆς τῆς μεθόδου, ὅπως καὶ τὰ πλεονεκτήματά της. Εἰς τὴν παροῦσαν ἐργασίαν δίδονται ἐφαρμογαὶ τῆς μεθόδου εἰς διάφορα πεδία ἐρεύνης, ὡς, λ.χ., εἰς τὴν θερμοδυναμικήν, Ἀστροδυναμικήν, μή-γραμμικὴν μηχανικήν, βιολογίαν, κ.λ.π.

★

Ὁ Ἀκαδημαϊκὸς κ. **Ἰω. Ξανθάκης** κατὰ τὴν ἀνακοίνωσιν τῆς ἀνωτέρω ἐργασίας εἶπε τὰ κάτωθι :

Ἔχω τὴν τιμὴν νὰ παρουσιάσω εἰς τὴν Ἀκαδημίαν Ἀθηνῶν τὸ δεύτερον μέρος τῆς ἐργασίας τοῦ κ. Δημητρίου Μαγείρου, ὑπὸ τὸν τίτλον : «Δεκταὶ Μαθηματικαὶ Λύσεις Φυσικῶν Προβλημάτων».

Εἰς τὴν προηγηθεῖσαν ἀνακοίνωσίν του ὁ κ. Μάγειρος παρουσίασε μίαν μέθοδον ἐρεύνης μαθηματικῶν λύσεων φυσικῶν καὶ κοινωνικῶν συστημάτων «φυσικῶς ἀποδεκτῶν». Ὑπέδειξε δὲ τὰς δυσκολίας ἐφαρμογῆς τῆς μεθόδου ταύτης, ὅπως καὶ τὰ πλεονεκτήματά της, εἰς τὰς περιπτώσεις καθ' ἅς δύναται νὰ ἐφαρμοσθῇ.

Εἰς τὴν παροῦσαν ἀνακοίνωσιν παρέχονται αἱ ἐφαρμογαὶ τῆς ἐν λόγῳ μεθόδου εἰς διάφορα πεδία ἐρεύνης, ὅπως λ. χ. εἰς τὴν θερμοδυναμικήν, εἰς τὴν ἀστροδυναμικήν, ἐπὶ τοῦ προβλήματος τῆς κινήσεως τεχνητοῦ δορυφόρου ὑπὸ τὴν ἐπίδρασιν μιᾶς κεντρικῆς δυνάμεως πληρούσης τὸν νόμον τοῦ Νεύτωνος καὶ μιᾶς ὠστικῆς τοιαύτης ἐπενεργούσης ἐπί τι χρονικὸν διάστημα, καθὼς καὶ ἐπὶ προβλημάτων μή-γραμμικῶν Μηχανικῆς καὶ Βιολογίας.

# PHYSICAL PROBLEMS DISCUSSED MATHEMATICALLY

By

DEMETRIOS G. MAGIROS (in Philadelphia U.S.A.)

## INTRODUCTION

Applied mathematics within the last few years has been developed into a distinguished branch of science with completely established purpose, principles, and approach. The human need created the effort of man for a penetration of mathematical methods for study of physical and human phenomena, that is to use mathematics in real life, and this gave rise to this branch of science.

The purpose of this paper is to provide a brief discussion of what the writer conciders of special interest as an introduction to Applied Mathematics.

In the first part of the report a general approach is indicated for reasonable mathematical solutions of physical problems, while in the second part examples illustrated show the approach in application.

## PART A

## CLASSICAL APPROACH

### I. Pure and Applied Mathematics.

It is a scientific fact today that «*applied mathematics*» is distinct from «*pure mathematics*», the distinction being a matter of attitude and motivation and not a subject matter.

Pure mathematics is directed towards logical crystallization, abstraction, generalization. Applied mathematics is an inter-connection of mathematical methods for interpretation of phycical phenomena.

Only great scientists were able in the past, as e.g. Newton, Maxwell, Gauss, Riemann, Poincaré, etc, to work in pure and at the same time in applied mathematics, although at their time *«the principles of applied Mathematics»* were not completely established.

The mathematical needs during World Way II helped to distinguish applied mathematics from the other scientific pursuits of man and to get their principles clearly stated.

Applied mathematics is based on two principles; *«the principle of idealization»* and *«the principle of approximation»*. The principle of idealization is fundamental even for the formulation of the basic concepts and laws of nature. The definition of *«the density of a fluid»* at a point p can be considered as an example of this principle. The total fluid mass in a sphere of small radius $\varepsilon$ about **P**, divided by the volume of the sphere, by forcing $\varepsilon$ to tend to zero, is, by definition, the density of the fluid at the point P. The limiting process associated to this definition is an idealization of unrealistic type, since in a very small volume the fluid molecules are irregularly distributed. Idealized concepts of the above type are necessary in physical sciences. The physical laws deal with idealized concepts and they by themselves are idealizations valid «in a limited field of application» and correct «within certain possible errors.»

## II. Statement of A Physical Problem. Correctly stated Problems.

Any physical problem posed by reality presupposes a physical situation with given data, and questions to be answered by using the data.

For the physical problem one must know the pertinent physical laws, the appropriate coordinate system, especially for astronomical problems, and the data, that is the initial and / or boundary conditions.

The initial conditions prescribe the state of a system at a particular time, while the boundary conditions prescribe the physical behavior of a system at the frontier of a region independently of time. The data must be admissible, i.e. completely stated and known approximately within an accepted error.

Any incorrect assumption in the problem will be preserved as incorrect in the procedure for the solution.

If all the above requirements for the problem are satisfied, one has a «*physical problem correctly stated*».

Only for such a physical problem one tries to get a mathematical solution.

Problems of this type are the subject matter of applied mathematics

## III. Mathematical Idealization of Physical Problems Correstly Stated. Formal Solution.

A physical problem correctly stated may be capable of being idealized in different mathematical ways, and it is important to distinguish the reasonable idealization. Usually one has as a result of the idealization differential equations with initial and/or boundary conditions.

For physical problems related to motion of a system, one has «initial value problems», while for equilibrium problems one has «bounbary value problems.» In the «mixed problems» the data are initial and boundary conditions.

The differential equations of a problem are local restrictions of a function - physical quantity, local in space and local in time. It is the task of mathematicians to deduce from these local restictions a picture of the phenomenon in the large. Infinitely many functions satisfy a differential equation. The particular solution desired, which represents a particular physical phenomenon, will be selected from the manifold of all solutions, the selection depending on the data.

Incorrect mathematical formulation of a correctly stated physical problem will lead to a non - accepted mathematical answer, so that one must get a «*well formulated mathematical problem*» corresponding to the physical problem.

By purely mathematical operations in the well formulated mathematical problem certain mathematical results and a solution will be obtained, that is the «*Formal solution.*»

By having the proper mathemartical idealization $A$, one first finds an appropriate approximation $A_n$ of A, and if $S_n$ is the solution of $A_n$ and by using the limiting process one has $S_n \rightarrow S_f$ as $n \rightarrow \infty$, then $S_f$ is the solution of $A$, the formal solution.

The discussion of convergense of the formal solution in the form of series aud the region of the convergence of the solution give a deep insight to the problem.

Not only convergent solutions but some non-convergent ones are sometimes accepted, as e.g., «asymptotic expansions».

In the formal process we do not ask whether there exists a counterpart in nature, or an interpretation into the physical order of things. Such an isolation of the proccess of mathematics is called *«pure mathematics.»*

It is the endeavor which keeps in closer touch with the inherent order of the physical nature which distinguishes the field of *«applied mathematics»*.

## IV. **Physical Interpretation of the Mathematical Results.**

The mathematical idealization selected for the physical problem must be *«reasonable idealization»*, and the requirements for the reasonability are: (a) *«the existence of the solution»*, (b) *«the uniqueness of the solution»*, and (c) *«the continuous dependence on the data»*.

We discuss these criteria in detail.

(a) *«the existence»*, The well formulated mathematical problem must possess a solution. We remark that the existence of a solution does not necessarily imply the possibility of the solution. Example: *«the three-body problem»*. The existence of the solution of this problem has been proved by Sundman, but the problem has not been solved yet in a form which permits one to deal with questions of stability, in spite of all the tremendous efforts of mathematicians during past centuries. The question of stability in Celestial mechanics is closely related to the form of the solution and to the divergence of the series employed as its solution.

(b) *«The uniqueness»*. The solution of the mathematical problem must be unique.

The existence of a unique solution expresses the belief of causality or determinism without which experiments could not be repeated with the expectation of consistent results.

(c) *«The continuous dependence on the data»*: The solution must be such that small changes in the data must produce small changes in the solution.

The data, as results of experiments or measurements, contain certain small errors and then one always has an uncertainty in the solution, and error in its estimation, and it is required that «the smaller the error in the data the smaller the error in the estimation of the solution».

Also, the solution $S_n$ of the approximating problem $A_n$ for large $n$ must be a good approximation of the «true» solution. These requirements are satisfied if «the solution is continuous function of the data».

This postulate corresponds to the Liapunov stability definition in parameter space, where the parameter space is the *«initial conditions space»*.

The above requirements give a perfect mathematical description of the reality and are called the *«Hadamard's postulates»*.

The solution which satisfies these postulates is an *«actual solution»* of the problem, which is called *«well posed* or *properly posed* or *reasonable problem»*.

Violation of any postulate of the above makes the solution *«nonactual»*, when the problem is *«unreasonable»* or *«improperly posed»*.

In recent years much attention has been given to these problems, especially when the violated postulate is the last one.

The above *classical approach* for finding actual solutions of physical problems shows a general orientation of thought. We remark that there are difficulties in the application of this approach for a mathematical description of a phenomenon in some scientific fields. Problem of modern physics show these difficulties.

The exact physical laws governing the behavior of a system may not be known making it impossible to arrive at a complete theoretical description of the system.

Problems of fundamental particle structure and reaction present the difficulties in an acute form. More often the situation is analogous to that of problems of atomic and molecular structure, where the interactions are known, but where the structure in a particular problem may be so complicated.

In problems of classical dynamics of compressible fluids the differential equation supplemented by boundary conditions are not always a sufficiently complete framework for an adequate description of physical reality.

The above remarks must be always remembered when one tries to present a mathematical description of physical systems of this kind or to solve physical problems.

## V. **Conclusion**.

The above classical approach for actual solutions of physical problems is shown schematically in Fig. 1.

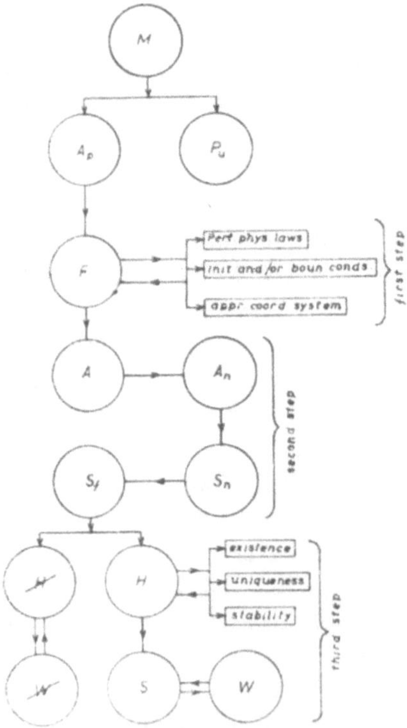


Fig. 1. Schematical Procedure For An Actual Solution Of A Physical Problem.

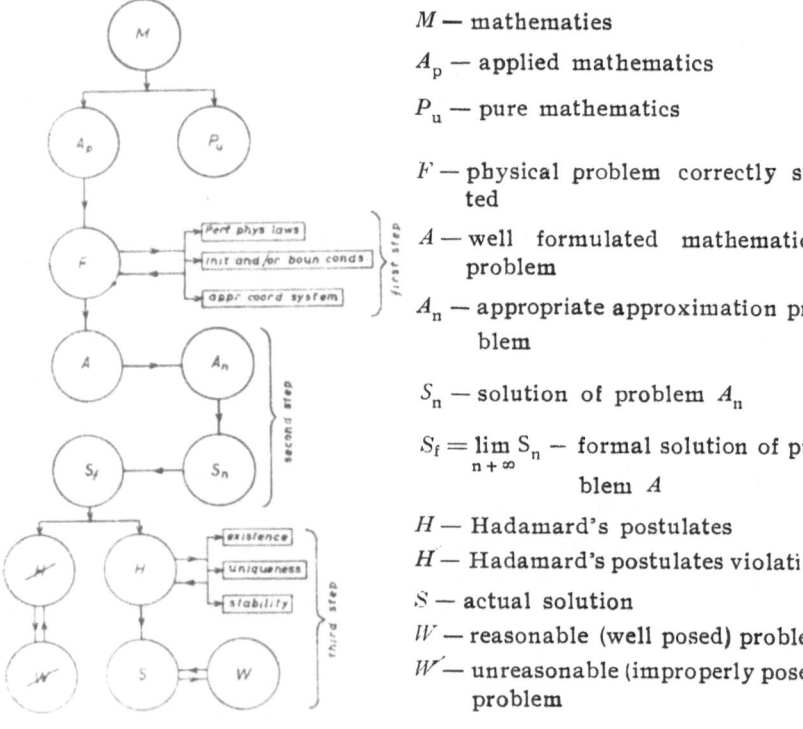

$M$ — mathematies

$A_p$ — applied mathematics

$P_u$ — pure mathematics

$F$ — physical problem correctly stated

$A$ — well formulated mathematical problem

$A_n$ — appropriate approximation problem

$S_n$ — solution of problem $A_n$

$S_f = \lim_{n+\infty} S_n$ — formal solution of problem $A$

$H$ — Hadamard's postulates

$\not{H}$ — Hadamard's postulates violation

$S$ — actual solution

$W$ — reasonable (well posed) problem

$\not{W}$ — unreasonable (improperly posed) problem

One may have three steps in the approach.

The «*first step*» is to get the physical problem correctly stated.

The «*second step*» is to get the well formulated mathematical problem and to determine the formal solution.

The «*third step*» to check whether the formal solution is an actual solution.

The first step is a decisive step and requires experience, intuitive understanding of the realities of physics, engineering and other fields.

The second step leaves open many possibilities for mathematical success.

The third step characterizes the procedure of applied mathematics.

## PART B

### ILLUSTRATED EXAMPLES

The classical approach of the preceding section for finding reasonable mathematical solutions of physical problems is successfully applied to many fields of science and technology. We give here some examples.

**Example 1.** *Forward and backward heat flow problem* Ref. (5).

This classical problem of Thermodynamics clarifies especially the third step of the approach.

The problem is «*to study the heat flow in a given medium*».

To make this physical problem correctly stated, one accepts for the medium to be «homogenious» and «isotropic» with respect to the heat flow, and that the heat flow is towards the decreasing temperature.

Based on these hypotheses, the mathematical idealization is the partial differential equation:

$$u_{xx}+u_{yy}+u_{zz}=u_t \tag{1}$$

where $u=u(x, y, z, t)$ is the temperature in the $x, y, z$—space and $t$—time.

In the equation (1) there is a coefficient, taken equal to unity, depending on density, specific heat thermal conductivity of the medium.

If, for one-dimensional medium, one takes an initial condition

$$u(x, 0) = x \sin nx, \quad n = \text{integer} \tag{2}$$

one can check that the system (1) and (2) accepts as a unique solution the function:

$$u(x, t) = n \cdot e^{-n^2 t} \cdot \sin nx \tag{3}$$

then the first two Hadamard's postulates are satisfied by this function.

a) For $t > 0$, when one has the *«forward heat problem»*, the solution (3) satisfies also the third Hadamard's postulates, and the function (3) in this case is accepted as an *«actual solution»* of the problem, which is a *«reasonable problem»*.

b) For $t < 0$, when one has the *«backward heat problem»*, the solution (3) violates the third Hadamard's postulate, then the function (3) in this case is a *«formal solution»*, and not an *«actual solution»*, of the Backward heat problem, which is *«unreasonable problem»*.

**Example 2.** *Principal modes of nonlinear systems* Ref. (4a).

The concept of *«principal modes»* of linear systems plays a predominant role in the analysis of the oscillatory systems of many fields.

The principal modes in linear systems are, by definition the fundamental set of solutions of which a linear combination gives the general solution of the linear system; or, physically speaking, they are that special modes of oscillation of the linear system in terms of which we can discuss any kind of oscillation of the system.

Since the *«principle of superposition»* does not hold in nonlinear systems, the concept of principal modes, as given above is meaningless in nonlinear systems, and the following problem may arise: *«Has the problem of principal modes of nonlinear systems a physical meaning?»*; or *«How one can make the problem of principal modes of nonlinear systems a reasonable problem?»*

The writer has published some papers in connection with this important problem, and transfers here some appropriate thoughts, techniques and results example of the classical approach of the previous section.

We can find a new definition of the concept of principal modes for both the linear and nonlinear systems, and such that the known definition in linear systems is a property resulting from the new definition. The writer gave two new definitions which, under some conditions, are equivalent.

Then we can make the physical problem correctly stated and the mathematical idealization reasonable.

We take a trial solution and make it formal first and then actual.

If we restrict ourselves to a «*two-degrees-of-freedom*» mechanical nonlinear system, as shown in Fig. 2, the equation of motion of the «*two-masses-three springs*» nonlinear system are:

$$\left.\begin{aligned} \ddot{x} + \omega_1^2 x - \lambda_2 y + \lambda_1 x^3 = 0 \\ \ddot{y} + \omega_2^2 y - \lambda_3 x = 0 \end{aligned}\right\} \quad (1)$$

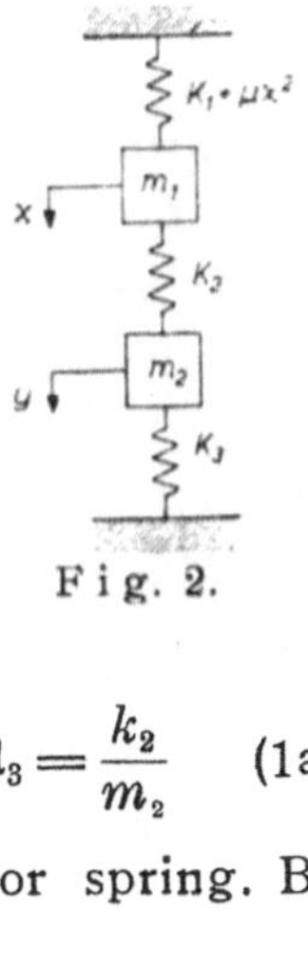


Fig. 2.

where:

$$\omega_1^2 = \frac{k_1 + k_2}{m_1}, \ \omega_2^2 = \frac{k_2 + k_3}{m_2}, \ \lambda_1 = \frac{\mu}{m_1}, \ \lambda_2 = \frac{k_2}{m_1}, \ \lambda_3 = \frac{k_2}{m_2} \quad (1a)$$

and $\mu$ characterizes the nonlinearity of one anchor spring. By using the transformation:

$$x = x_1, \quad x = x_2, \quad y = x_3, \quad \dot{y} = x_4 \quad (2)$$

the system (1) can be reduced to its normal form:

$$\left.\begin{aligned} x_i = f_i(x_1, x_2, x_3, x_4), \ i = 1, 2, 3, 4 \\ f_1 = x_2, \ f_2 = - \omega_1^2 x_1 + \lambda_2 x_3 - \lambda_1 x_1^3, \ f_3 = x_4, \ f_4 = \lambda_3 x_1 - \omega_2^2 x_3 \end{aligned}\right\} \quad (3)$$

valid in a region $R$:

$$R: |x_i| < h, \ i = 1, 2, 3, 4. \quad (3a)$$

The appropriate initial conditions for «principal modes» are in $R$:

$$x_1(0)=x_{10},\quad x_2(0)=0,\quad x_3(0)=x_{30},\quad x_4(0)=0 \tag{3b}$$

where $x_{10}$ and $x_{30}$ are appropriately related to each other.

Now we remark that the nature of the functions $f_i$ of (3) are such that all Hadamard's postulates are satisfied. These functions $f_i$ are continuous in $R$, then bounded; they have continuous partial derivatives $\partial f_i / \partial x_k$ in $R$, when they satisfy Lipschitz conditions with respect to $x_i$ in $R$ for a Lipschitz constant $s = l.u.b \,|\, \partial f_i / \partial x_k \,|$. The above properties quarantee the uniquely existing solution of the system (3) and (3b) in a region $R' \subset R$. As the initial point $x_{i0}$, $i = 1, 2, 3, 4$ varies in $R'$, the solution satisfies the three Hadamard's postulates, and the problem is *«well posed»*.

**Example 3.** *The motion of an artificial celestial body under the Influence of a Newtonian Center and a general force* (Ref. 4).

We take this problem from orbital mechanics. It is a special two-body problem treated by the writer in his own way and presented to the «*15*th *International Astronautical Congress*» Warsaw, Poland, Sept. 1964.

The problem is: «*An artificial celestial body is moving under the influence of a central force obeying the inverse square Newton's law toward an attractive center. A general force is applied, acts for an interval of time, then it is removed. Find the motion of the body during the action of the general force*».

The mathematical formulation of the problem is:

$$\left.\begin{aligned} &\ddot{\underline{r}}(\tau) = -\frac{\mu}{r^3(\tau)}\underline{r}(\tau) + \underline{T}(\tau) \\ &\underline{r}(0) = \underline{r}_0,\ \dot{\underline{r}}(0) = \dot{\underline{r}}_0 + \underline{I}_0 \\ &D:\ 0 \leqslant \tau \leqslant \tau' \\ &D_1:\ |\underline{r}(\tau)| < M_1,\quad |\dot{\underline{r}}(\tau)| < M_2 \end{aligned}\right\} \tag{1}$$

$T$ is the general force, $\underline{I}_0$ the impulse, is given by

$$\left.\begin{aligned} &\underline{I}_0 = \int_0^{t_0} T(t)dt \\ &\tau = t - t_0 \end{aligned}\right\} \tag{1a}$$

If we take the function:

$$\underline{r}(\tau)=a_1(\tau)\underline{r}_0^*+a_2(\tau)\underline{s}_0^*+a_3(\tau)\underline{T}_0^* \quad (2)$$

as a trial solution, where $\underline{r}_0^*$, $\underline{s}_0^*$, $\underline{T}_0^*$ are special unit vectors, the coefficients $a_1$, $a_2$, $a_3$ must satisfy the following conditions in order the function (2) is a formal solution of (1):

$$\left.\begin{array}{lll} \ddot{a}_1+\dfrac{\mu}{r^3}a_1=T_1; & a_1(0)=r_0, & \dot{a}_1(0)=0 \\ \ddot{a}_2+\dfrac{\mu}{r^3}a_2=T_2; & a_2(0)=0, & \dot{a}_2(0)=s_0 \\ \ddot{a}_3+\dfrac{\mu}{r^3}a_3=T_3; & a_3(0)=0, & \dot{a}_3(0)=0 \end{array}\right\} \quad (3)$$

The system of equations (3) satisfy the Hadamard's postulates, as easily can be proved, when the functions $a_1(\tau)$, $a_2(\tau)$, $a_3(\tau)$ can be uniquely determined and are continuous functions of the initial conditions of (3).

Therefore the solution (2) of (1) is a unique solution and continuous function of the initial conditions of (1).

**Example 4.** *The problem of Domes of Underwater Warfare in 1942–1943* Ref. (2a).

By this problem one can illustrate the diffilculties during the transition from the physical problem to a reasonable mathematical idealization. The problem arose in the winter *1942–1943* in connection with «underwater warfare». As is known, underwater sound ranging depends on sending out a sound beam in water and attached to a fast-moving ship, the water steaming around the plate causes serious disturbances. For elimination of these disturbances the projector is closed in a so called «*dome*», Fig. 3, which is a convex shell of metal or other material filled with water. Such domes interfere only slightly with the formation of a concentrated sound beam.

During *1942–1943* a large number of small submarines chases were built and equipped with sound gear similar to but smaller than the gear used before. While the manufacture of domes to fit this smaller gear was under way, it was discovered that these smaller domes led to an intolerable diffusion of the sound beam. At that time a quick remedy was imperative, and a mathematical analysis of the problem was needed to support and speed - up experimented work.

The mathematical problem, related to the above physical problem, was to solve the differential equation:

$$\left.\begin{aligned} &\nabla^2 P + k^2 P = 0 \\ &\nabla^2 = \frac{\partial^2}{\partial x^2} + \frac{\partial^2}{\partial y^2} + \frac{\partial^2}{\partial z^2} \end{aligned}\right\} \quad (1)$$

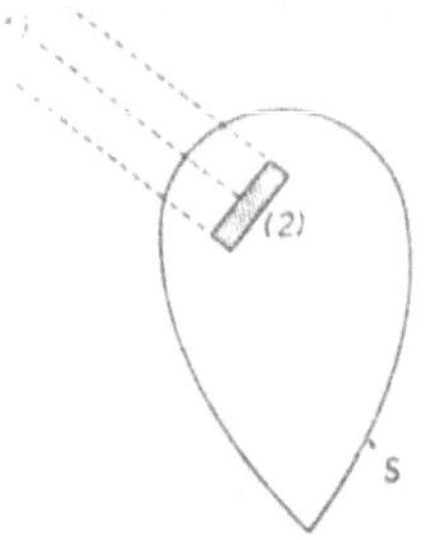


(1) axis of beam sound
(2) projector
$S$ surface of dome
Fig. 3. Dome.

in which $k = \omega / c$, $\omega$ the frequency, $c$ the sound velocity, and $k$ has for our problem unfortunately different values within the shell of the dome and outside.

This mathematical idealization was not a suitable one for the problem.

They found the suitable mathematical idealization by the following process.

The actual dome of small finite thickness was replaced by an extremally thin surface, then the influence of the dome was simply replaced by conditions for jump discontinuities of the disturbance $q$ of the beam across the surface.

These conditions are:

$$\left.\begin{aligned} &[q] = \frac{\varrho_1}{\varrho_0 - 1} \cdot \frac{\partial p}{\partial n} \\ &\left[\frac{\partial q}{\partial n}\right] = \frac{\varrho_0}{\varrho_1}(k_0^2 - k_1^2)p - \left(1 - \frac{\varrho_0}{\varrho_1}\right)\left(\frac{\partial^2 p}{\partial n^2} + 2H\frac{\partial p}{\partial n}\right) \end{aligned}\right\} \quad (2)$$

where the symbol $[\ ]$ means jump of the quantity of the symbol across the surface, $q$ is the disturbance of the acoustic pressure $p$ caused by the dome and the normal derivatives $\frac{\partial}{\partial n}$ are to be evaluated on the surface $S$ of the dome. The quantity $H$ is the mean curvature of $S$, i.e. the average of the curvature of any two normal plane sections at right angles to each other. In addition to conditions (2) to be satisfied by $q$ on $S$, $q$ should be a solution of the equation:

$$\nabla^2 q + k_0^2 q = 0. \quad (3)$$

which is regular everywhere except on $S$ and which has the same behavior as $P$ at $\infty$. This problem possesses the unique solution:

$$q = -\frac{1}{4\pi}\iint_S \left[\frac{\partial q}{\partial n}\right]\frac{e^{ik_o r'}}{r'}\, ds + \frac{1}{4\pi}\iint_S [q]\frac{\partial}{\partial n}\left(\frac{e^{ik_o r'}}{r'}\right) dS. \quad (4)$$

The quantities in the brackets are given by condition (2), $r'$ is the distance from a fixed point $(x, y, z)$ at which $q\ (x, y, z)$ is to be determined to the point of integration on $S$. This formula yields the disturbance as the effect due to a layer of point sources and a layer of dipoles disturbed on $S$ with intensities which are known as soon as the original pressure $p$ is known, since the quantities in brackets are fixed in value due to conditions (2). The relative directional disturbance:

$$\left|\frac{p_1}{p}\right| Rch\left(\frac{q_1}{p_1} - \frac{q_0}{p_0}\right)$$

would, finally, be obtained from (4). The solution (4) is valid for a shell of constant thickness, but it could be extended without essential error to cases in which the dome shell is made up of a not too large number of pieces, each of which is of constant thickness. All that would be necessary would be to insert a numerical factor $d$ in the integrands on the right-hand side of expression (4), which would be precise constant on $S$. By this formula one can analyse the contribution to the distortion of various factors, such as the curvature of the dome and the density and sound velocity within it.

The above kind of mathematical idealization, even without detailed numerical computation, proved helpful to the designing engineer.

## REFERENCES

[1] *L. Brillouin*: «Scientific Uncertainty and Information», (1964).

[2] *R. Courant*: (a) «Methods in Applied Mathematics», Rec. Adv. in Science, 1—14 (1956).

(b) «Boundary value probleme in Modern Fluid Dynamics», Rroc. Intern Congress of Math., Vol. II (1956).

156 DEMETRIOS G. MAGIROS

[3] *R. Courant* and *C. Friedrichs*: «Supersonic Flow and Shock Waves», 367 ft, (1948).

[4] *D. Magiros*: (a) «Method for Defining Principal Modes of Nonlinear Systems Utilizing Infinite Determinants», J. Math. Phys. 2, No 6, 869—875 (1961).

(b) «The Motion of an Artificial Celestial Body under the Influence of a Newtonian Center and a General Force». Proc. XVth International Astr. Congress, Warsaw, Poland (1964).

[5] *W. Miranker*: «A Well Posed Problem for a Backward Heat Equation» Proc. Am. Math. Soc. (Apr. 1961).

[6] *F. Weyl* «Applied Mathematic in U. S.», Monographs of the Soc. for Ind. and Appl. Math., No 1 (1956).

Mathematical Consultant of General Electric Comp.
M.S.D. Re-entry Systems Department, Philadelphia, PA.

*K. O. Friedrichs anniversary issue*

# Diffraction by a Semi-Infinite Screen With a Rounded End*

JOSEPH B. KELLER and DEMETRIOS G. MAGIROS

## 1. Introduction

Suppose a cylindrical wave is incident upon a semi-infinite thin screen with a rounded end, such as that shown in Figure 1. The rounded end is obtained by placing a cylindrical tip of radius $a$ on the end of the screen.

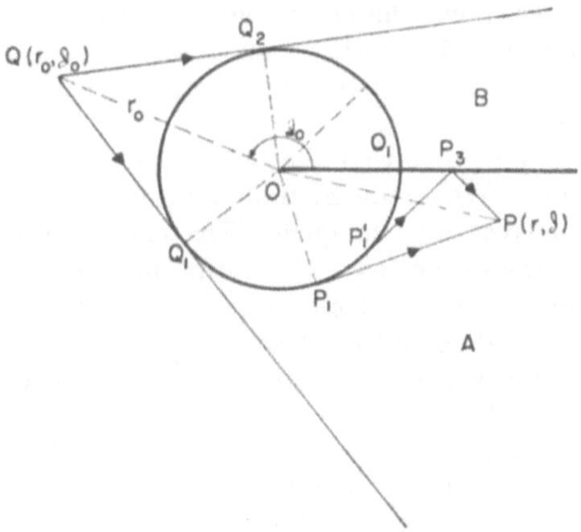


Figure 1

A semi-infinite thin screen, represented by the heavy line with its end at 0, is attached to a circular cylinder of radius $a$, represented by the circle. A line source parallel to the cylinder is at $Q$, which has coordinates $r_0$ and $\theta_0$. Two rays from $Q$ are tangent to the cylinder at $Q_1$ and $Q_2$. Their extensions bound the two parts of the shadow, $A$ below the screen and $B$ above it. The tangent ray at $Q_1$ produces a diffracted surface ray which proceeds along the cylinder in the counterclockwise direction. The two diffracted rays which it sheds at $P_1$ and $P_1'$ are shown passing through $P$, the second ray having been reflected from the screen at $P_3$.

If $a$ is large compared to the wavelength $\lambda$ of the incident radiation, then the resulting field can be determined by means of the geometrical theory of

*The research reported in this paper has been sponsered by the Electronics Research Directorate of the Air Force Cambridge Research Laboratories, Air Research and Development command, under Contract No. AF 19(604)5238. Reproduction in whole or in part permitted for any purpose of the United States Government.



Reprinted from the *Communications on Pure and Applied Mathematics* **XIV** (1961), 457–471.

diffraction [2, 3]. This theory provides a simple geometrical procedure for constructing the field. The construction is carried out in Section 2. In making this construction, it is assumed that the field $u$ is a scalar which satisfies the reduced wave equation and either it or its normal derivative vanishes on the screen. Thus $u$ may represent either the component of electric or magnetic field parallel to the generators of the cylinder, and the screen may represent a perfect conductor. In the second case, $u$ may also represent acoustic pressure and the screen may represent a rigid body.

Another method for constructing the field is that of expansion in radial eigenfunctions. This method is employed in Section 3. It is then shown that for large values of $ka$, where $k = 2\pi/\lambda$, the asymptotic form of the field so constructed coincides exactly with the diffracted field given by the geometrical theory of diffraction. This agreement confirms the geometrical theory, which is based on several hypotheses about wave propagation. However, it is not a conclusive proof of the correctness of that theory because the eigenfunctions are not complete.

In Section 4, we prove that the radial eigenfunction expansion does represent the field $u$. We do this by first expanding $u$ in angular eigenfunctions, which are known to be complete. This expansion was first given by S. N. Karp [4]. Then we convert this series into the radial eigenfunction series by applying the Watson transformation. In this way we prove that the latter series does represent $u$. Therefore, the result of the geometrical theory of diffraction, which is asymptotic to it, is indeed the asymptotic form of the exact solution.

The diffracted field in the shadow has been computed [1] from the geometrical solution and also from the exact solution as a function of $ka$. The two results agree well for $ka \geqq 2$ for both boundary conditions. In the case in which $u = 0$ on the boundary, the field in the shadow is found to decrease as $ka$ increases, being greatest when $ka = 0$, i.e. when the cylindrical tip is absent. In the other case, in which $\partial u/\partial n = 0$ on the boundary, the field in the shadow is found to oscillate as $ka$ increases, but ultimately, for sufficiently large values of $ka$, it also decreases monotonically. These results are interpreted in terms of the rays of the geometrical theory of diffraction.

It is to be noted that the problem under consideration is two dimensional, so that we may confine our attention to a single plane.

## 2. Geometrical Solution

According to the geometrical theory of diffraction, the field $u$ can be represented as the sum of a geometrical optics field $u_g$ and a diffracted field $u_d$. The field $u_g(P)$ at a point $P$ is the sum of the fields on the incident and reflected rays through $P$. The field $u_i(P)$ on an incident ray through $P$ is given by

$$u_i(P) = (8\pi kR)^{-1/2} \exp\left\{ikR + \frac{i\pi}{4}\right\}. \tag{1}$$

Here $R$ denotes the distance from the source $Q$ to the point $P$. The constants in (1) have been chosen to correspond to a source of unit strength for which the exact field would be given by $(i/4)H_0^{(1)}(kR)$. The field on a reflected ray can be constructed by a well known procedure, so we need not consider it further. Instead, we turn to the diffracted field $u_d(P)$ which is the sum of the fields on all the diffracted rays through $P$.

The diffracted rays are produced by the two rays from $Q$ which are tangent to the tip of the screen. The two points of tangency are denoted by $Q_1$ and $Q_2$ (see Figure 1). The surface ray produced at $Q_1$ will shed a diffracted ray at $P_1$ which will travel along the tangent from $P_1$ to $P$. It will also shed a diffracted ray at $P_1'$ which will be reflected at $P_3$ and then reach $P$. The surface ray itself will be reflected from the screen at $O_1$, encircle the tip and be reflected at $O_1$ again, and repeat this process infinitely many times. Each time it passes through $P_1$ and $P_1'$ in the clockwise direction, the rays it sheds will pass through $P$. Thus infinitely many diffracted rays will pass through $P$, all originally produced by the ray incident at $Q_1$. In a similar manner, the surface ray produced at $Q_2$ will be reflected at $O_1$ and then shed rays at $P_1$ and $P_1'$ which will pass through $P$. It will also give rise to infinitely many rays through $P$ as it continues to travel back and forth along the surface of the tip.

As a consequence of the preceding considerations, we see that there are four types of diffracted rays which pass through a point $P$ in the shadow. We may describe them as follows:

I. Rays produced at $Q_1$ and diffracted at $P_1$. The ray $QQ_1P_1P$ is the simplest ray of this type. For each non-negative value of the integer $n$, there is one of these rays which makes $n$ trips around the circle. This ray is reflected $2n$ times at $O_1$. Now the reflection coefficient $R$ has the value $-1$ for the case in which $u = 0$ on the screen and $+1$ for the case in which $\partial u/\partial n = 0$ on the screen. Thus in either case $R^{2n} = 1$. The path length of such a ray along the circle is $4\pi an+t$, where $t = Q_1P_1$.

II. Rays produced at $Q_1$ and diffracted at $P_1'$. The ray $QQ_1P_1P_1'P_3P$ is the simplest ray of this type. For each $n \geq 0$, the ray which makes $n$ round trips is reflected $2n+1$ times, including the reflection at $P_3$. Thus $R^{2n+1} = -1$ if $u = 0$ on the screen and $R^{2n+1} = 1$ if $\partial u/\partial n = 0$. The path length of such a ray along the circle is $4\pi an+t+t_1$, where $t_1 = P_1P_1'$.

III. Rays produced at $Q_2$ and diffracted at $P_1$. The ray $QQ_2O_1Q_2Q_1P_1P$

is the simplest ray of this type. For each $n \geqq 0$, the ray which makes $n$ round trips is reflected $2n+1$ times. Its path length on the circle is $4\pi an+t_2$, where $t_2 = Q_2O_1+O_1Q_2Q_1P_1$.

IV. Rays produced at $Q_2$ and diffracted at $P_1'$. The ray $QQ_2O_1Q_2Q_1P_1P_1'P_3P$ is the simplest ray of this type. The ray which makes $n \geqq 0$ round trips is reflected $2n+2$ times, including the reflection at $P_3$. Its path length along the circle is $4\pi an+t_2+t_1$.

According to equation (10a) of [3], the field on a diffracted ray of type I is given by

$$u_i(Q_1)\frac{e^{iks}}{s^{1/2}}\sum_m D_m^2 \exp\{(ik-\alpha_m)T_n\}. \tag{2}$$

In (2), $s$ denotes distance along the diffracted ray from the point where it leaves the surface, $T_n = 4\pi an+t$ denotes its arclength along the surface, and $\alpha_m$ and $D_m$ are certain constants. The field on the rays of other types is also given by (2) with the appropriate values of $s$ and $T_n$ in each case, and with an appropriate power of the reflection coefficient $R$.

Let the coordinates of $Q$ be $(r_0, \theta_0)$ and those of $P$ be $(r, \theta)$ in a polar coordinate system with its origin at the center of the tip and the polar axis along the screen. Then $R = (r_0^2-a^2)^{1/2}$ and $s = (r^2-a^2)^{1/2}$. We insert these values into (2) using (1) for $u_i(Q_1)$, and we sum over $n$. This summation merely involves a geometric series, which can be summed explicitly. Then we add together the corresponding sums for the four types of rays and obtain the diffracted field at a point $P$ in the shadow

$$\begin{aligned} u_d(P) &= \frac{i(8\pi k)^{1/2}\exp\left\{ik(r_0^2-a^2)^{1/2}+ik(r^2-a^2)^{1/2}-\dfrac{i\pi}{4}\right\}}{(r_0^2-a^2)^{1/4}(r^2-a^2)^{1/4}} \\ &\times \sum_m \frac{D_m^2}{1-\exp\{4\pi a(ik-\alpha_m)\}}\left[\exp\{(ik-\alpha_m)t\}\mp\exp\{(ik-\alpha_m)(t+t_1)\}\right. \\ &\left.\mp\exp\{(ik-\alpha_m)t_2\}+\exp\{(ik-\alpha_m)(t_2+t_1)\}\right]. \end{aligned} \tag{3}$$

In (3) the upper signs apply in case $u = 0$ on the screen while the lower signs apply if $\partial u/\partial n = 0$ on the screen.

In terms of the coordinates of $Q$ and $P$, the distances $t$, $t_1$ and $t_2$ are given by

$$\begin{aligned} t &= a\left[\theta-\theta_0-\cos^{-1}\frac{a}{r_0}-\cos^{-1}\frac{a}{r}\right], \\ t_1 &= 2a(2\pi-\theta), \\ t_2 &= a\left[\theta+\theta_0-\cos^{-1}\frac{a}{r_0}-\cos^{-1}\frac{a}{r}\right]. \end{aligned} \tag{4}$$

The diffraction coefficients $D_m$ and the decay constants $\alpha_m$ are given by [3, 5]

$$\alpha_m = e^{-\pi i/6}\left(\frac{k}{6a^2}\right)^{1/3} q_m, \tag{5}$$

$$D_m^2 = \left[\frac{\pi^9}{2^5 3^8}\frac{a^2}{k}\right]^{1/6} e^{\pi i/12}[A'(q_m)]^{-2}, \tag{6a}$$

$$D_m^2 = \left[\frac{\pi^9}{2^5 3^2}\frac{a^2}{k}\right]^{1/6} e^{\pi i/12} q_m^{-1}[A(q_m)]^{-2}. \tag{6b}$$

In these equations $A(q)$ denotes the Airy function. For the boundary condition $u = 0$, $q_m$ is the $m$-th positive zero of $A(q)$ while for the boundary condition $\partial u/\partial n = 0$, $q_m$ is the $m$-th positive zero of $A'(q)$.

We now insert the preceding results into (3) and obtain in the cases $u = 0$ and $\partial u/\partial n = 0$ the respective results for a point $P$ in the shadow

$$\begin{aligned} u_d(P) &= \frac{\pi}{6}\left(\frac{a}{6k^2}\right)^{1/3} \frac{\exp\left\{ik(r_0^2-a^2)^{1/2}+ik(r^2-a^2)^{1/2}+\frac{5\pi i}{6}\right\}}{(r_0^2-a^2)^{1/4}(r^2-a^2)^{1/4}} \\ &\times \sum_m \frac{\sin[(ka+i\alpha_m a)(2\pi-\theta)]\sin[(ka+i\alpha_m a)\theta_0]}{[A'(q_m)]^2 \sin[(ka+i\alpha_m a)2\pi]} \\ &\times \exp\left\{-i(ka+i\alpha_m a)\left[\cos^{-1}\frac{a}{r_0}+\cos^{-1}\frac{a}{r}\right]\right\}, \end{aligned} \tag{7a}$$

$$\begin{aligned} u_d(P) &= \frac{\pi}{18}\left(\frac{a}{6k^2}\right)^{1/3} \frac{\exp\left\{ik(r_0^2-a^2)^{1/2}+ik(r^2-a^2)^{1/2}+\frac{5\pi i}{6}\right\}}{(r_0^2-a^2)^{1/4}(r^2-a^2)^{1/4}} \\ &\times \sum_m \frac{\cos[(ka+i\alpha_m a)(2\pi-\theta)]\cos[(ka+i\alpha_m a)\theta_0]}{q_m A^2(q_m)\sin[(ka+i\alpha_m a)2\pi]} \\ &\times \exp\left\{-i(ka+i\alpha_m a)\left[\cos^{-1}\frac{a}{r_0}+\cos^{-1}\frac{a}{r}\right]\right\}. \end{aligned} \tag{7b}$$

These results hold when $\theta > \theta_0$. When $\theta < \theta_0$ we obtain the same results with $\theta$ and $\theta_0$ interchanged. Slightly different results can be obtained in the same way for a point $P$ in the illuminated region.

## 3. Radial Eigenfunction Expansion

We shall now formulate a boundary value problem for the total field

$u$ and solve it by the method of expansion in radial eigenfunctions. The conditions to be satisfied by $u(r, \theta)$ are

$$(\Delta+k^2)u = r^{-1}\delta(r-r_0)\delta(\theta-\theta_0), \tag{1}$$

$$\lim_{r\to\infty} r^{1/2}(u_r-iku) = 0, \tag{2}$$

$$u(a, \theta) = u(r, 0) = u(r, 2\pi) = 0, \tag{3a}$$

$$u_r(a, \theta) = u_\theta(r, 0) = u_\theta(r, 2\pi) = 0. \tag{3b}$$

In order to construct $u$ we introduce the radial eigenfunctions $H^{(1)}_{\nu_m}(kr)$. These Hankel functions are the radial factors of product solutions of the homogeneous equation corresponding to (1). They also satisfy the radiation condition (2). The eigenvalues $\nu_m$ are determined by requiring these functions to satisfy the first of the boundary conditions (3a) or (3b) which become, respectively,

$$H^{(1)}_{\nu_m}(ka) = 0, \tag{4a}$$

$$H^{(1)\prime}_{\nu_m}(ka) = 0. \tag{4b}$$

For large values of $ka$, and fixed $m$, the roots of (4a) and of (4b) are given by

$$\nu_m = ka+q_m(\tfrac{1}{6}ka)^{1/3}e^{i\pi/3}+O((ka)^{-1/3}). \tag{5}$$

In the case of (4a), $q_m$ is the $m$-th positive zero of the Airy function $A(q)$ while, in the case of (4b), $q_m$ is the $m$-th positive zero of $A'(q)$. We now assume that $u$ can be represented by the series

$$u = \sum_{m=1}^{\infty} f_m(\theta)H^{(1)}_{\nu_m}(kr). \tag{6}$$

To determine the coefficient functions $f_m(\theta)$ we insert (6) into (1) and make use of Bessel's equation to obtain

$$\sum_{m=1}^{\infty} r^{-2}H^{(1)}_{\nu_m}(kr)[f_m''(\theta)+\nu_m^2 f_m(\theta)] = r^{-1}\delta(r-r_0)\delta(\theta-\theta_0). \tag{7}$$

It is easy to prove that the radial eigenfunctions $H^{(1)}_{\nu_m}(kr)$ are mutually orthogonal, which may be expressed in the form

$$\int_a^\infty r^{-1}H^{(1)}_{\nu_m}(kr)H^{(1)}_{\nu_n}(kr)dr = N_n\delta_{mn}. \tag{8}$$

An analysis similar to that of Sommerfeld [6] yields for the constant $N_n$ the following results in the respective cases $u = 0$ and $\partial u/\partial n = 0$ on the screen:

(9a) $$N_n = \left[\frac{2i}{\pi\nu}\frac{\frac{\partial}{\partial\nu}H_\nu^{(1)}(ka)}{H_\nu^{(2)}(ka)}\right]_{\nu=\nu_n},$$

(9b) $$N_n = \left[\frac{2i}{\pi\nu}\frac{\frac{\partial}{\partial\nu}H_\nu^{(1)\prime}(ka)}{H_\nu^{(2)\prime}(ka)}\right]_{\nu=\nu_n}.$$

We now multiply (7) by $r^{-1}H_{\nu_n}^{(1)}(kr)$ and integrate the resulting equation from $a$ to $\infty$ with respect to $r$. Then by using (8) we obtain

(10) $$f_n''(\theta)+\nu_n^2 f_n(\theta) = N_n^{-1}H_{\nu_n}^{(1)}(kr_0)\delta(\theta-\theta_0).$$

In order to determine $f_n(\theta)$ from (10), we must impose the second and third of the boundary conditions (3a) or (3b) on $u$, which is given by (6). By again utilizing the orthogonality condition (8) these boundary conditions yield, in the respective cases,

(11a) $$f_n(0) = f_n(2\pi) = 0,$$

(11b) $$f_n'(0) = f_n'(2\pi) = 0.$$

The solutions of (10) which satisfy (11a) or (11b), respectively, are

(12a) $$f_n(\theta) = -\nu_n^{-1} \csc 2\pi\nu_n N_n^{-1} H_{\nu_n}^{(1)}(kr_0) \sin \nu_n\theta_< \sin \nu_n(2\pi-\theta_>),$$

(12b) $$f_n(\theta) = +\nu_n^{-1} \csc 2\pi\nu_n N_n^{-1} H_{\nu_n}^{(1)}(kr_0) \cos \nu_n\theta_< \cos \nu_n(2\pi-\theta_>).$$

In (12a) and (12b), $\theta_<$ and $\theta_>$ denote, respectively, the smaller and larger of the two angles $\theta$ and $\theta_0$, both of which lie between 0 and $2\pi$.

We now insert (12a) into (6) and use (9a) for $N_n$. Thus we obtain for the solution $u$ in the case of the boundary condition $u = 0$, the result

(13a) $$u = +\frac{i\pi}{2}\sum_{n=1}^{\infty} H_{\nu_n}^{(1)}(kr_0)H_{\nu_n}^{(1)}(kr)\frac{\sin \nu_n\theta_<}{\sin 2\pi\nu_n}\sin \nu_n(2\pi-\theta_>)\frac{H_{\nu_n}^{(2)}(ka)}{\frac{\partial}{\partial\nu}H_{\nu_n}^{(1)}(ka)}.$$

In the case of the boundary condition $\partial u/\partial n = 0$ we obtain instead, from (12b), (9b) and (6), the result

(13b) $$u = -\frac{i\pi}{2}\sum_{n=1}^{\infty} H_{\nu_n}^{(1)}(kr_0)H_{\nu_n}^{(1)}(kr)\frac{\cos \nu_n(2\pi-\theta_>)}{\sin 2\pi\nu_n}\cos \nu_n\theta_<\frac{H_{\nu_n}^{(2)\prime}(ka)}{\frac{\partial}{\partial\nu}H_{\nu_n}^{(1)\prime}(ka)}.$$

Let us now expand these expressions asymptotically for large values of $ka$. If $r > a$ and $r_0 > a$, we may use the Debye asymptotic expansions of the Hankel functions with arguments $kr$ and $kr_0$. However, for the Hankel functions with argument $ka$, which is nearly equal to the order $\nu_n$, we must use the expansion in terms of Airy functions. When we insert these expansions into (13a) and (13b), we obtain the results

$$
\begin{aligned}
u(r,\theta) \sim \frac{\pi}{6^{4/3}}(ka)^{1/3}\sum_{n=1}^{\infty}\frac{\exp\left\{i(k^2r^2-\nu_n^2)^{1/2}+i(k^2r_0^2-\nu_n^2)^{1/2}+\frac{5\pi i}{6}\right\}}{(k^2r^2-\nu_n^2)^{1/4}(k^2r_0^2-\nu_n^2)^{1/4}} \\
\times\frac{\sin\nu_n\theta_<\sin\nu_n(2\pi-\theta_>)}{[A'(q_n)]^2\sin 2\pi\nu_n}\exp\left\{-i\nu_n\left(\cos^{-1}\frac{\nu_n}{kr}+\cos^{-1}\frac{\nu_n}{kr_0}\right)\right\},
\end{aligned}
\tag{14a}
$$

$$
\begin{aligned}
u(r,\theta) \sim \frac{\pi}{3\cdot 6^{4/3}}(ka)^{1/3}\sum_{n=1}^{\infty}\frac{\exp\left\{i(k^2r^2-\nu_n^2)^{1/2}+i(k^2r_0^2-\nu_n^2)^{1/2}+\frac{5\pi i}{6}\right\}}{(k^2r^2-\nu_n^2)^{1/4}(k^2r_0^2-\nu_n^2)^{1/4}} \\
\times\frac{\cos\nu_n\theta_<\cos\nu_n(2\pi-\theta_>)}{q_nA^2(q_n)\sin 2\pi\nu_n}\exp\left\{-i\nu_n\left(\cos^{-1}\frac{\nu_n}{kr}+\cos^{-1}\frac{\nu_n}{kr_0}\right)\right\}.
\end{aligned}
\tag{14b}
$$

By using (5) for $\nu_n$ in these equations, we can simplify them to the form

$$
\begin{aligned}
u(r,\theta) \sim \frac{\pi}{6^{4/3}}\frac{(k^{-2}a)^{1/3}e^{5\pi i/6}}{(r^2-a^2)^{1/2}(r_0^2-a^2)^{1/2}}\exp\{ik[(r^2-a^2)^{1/2}+(r_0^2-a^2)^{1/2}]\} \\
\times\sum_{n=1}^{\infty}\frac{\sin\nu_n\theta_<\sin\nu_n(2\pi-\theta_>)}{[A'(q_n)]^2\sin 2\pi\nu_n}\exp\left\{-i\nu_n\left(\cos^{-1}\frac{a}{r}+\cos^{-1}\frac{a}{r_0}\right)\right\},
\end{aligned}
\tag{15a}
$$

$$
\begin{aligned}
u(r,\theta) \sim \frac{\pi}{3\cdot 6^{4/3}}\frac{(k^{-2}a)^{1/3}e^{5\pi i/6}}{(r^2-a^2)^{1/2}(r_0^2-a^2)^{1/2}}\exp\{ik[(r^2-a^2)^{1/2}+(r_0^2-a^2)^{1/2}]\} \\
\times\sum_{n=1}^{\infty}\frac{\cos\nu_n\theta_<\cos\nu_n(2\pi-\theta_>)}{q_nA^2(q_n)\sin 2\pi\nu_n}\exp\left\{-i\nu_n\left(\cos^{-1}\frac{a}{r}+\cos^{-1}\frac{a}{r_0}\right)\right\}.
\end{aligned}
\tag{15b}
$$

The results (15) coincide exactly with the results (7) of Section 2 which were obtained by the geometrical theory of diffraction for a point in the shadow. This verifies that the geometrical theory is correct in this case since (13), from which (15) was obtained, does indeed represent the solution as we shall show in the next section. However, the asymptotic evaluation of (13), which led to (15), is correct only for a finite number of terms in the series. This is so because in the later terms $\nu_n$ becomes large compared to $ka$, $kr$ and $kr_0$ and this was not taken into account in the asymptotic evaluation. Furthermore, for $n$ large $\nu_n$ is not given by (5). But as we shall see, (15) converges very rapidly for points in the shadow so that the error it contains in the later terms is asymptotically negligible. For points in the illuminated region, however, (15) does not converge although (13) does.

Therefore a different asymptotic evaluation must be made in this region. This will be done in the next section.

### 4. Angular Eigenfunction Expansion

The eigenvalue problem for the radial eigenfunctions involves the radiation condition and, therefore, it is not of Sturm-Liouville type. Consequently, we do not know whether these eigenfunctions are complete. Therefore, we cannot be sure that the solution $u$ can be represented by the series (6) of Section 3. We shall now prove that it can be so represented and that it is given by (13a) or (13b).

We begin with a representation of $u$ in terms of the angular eigenfunctions which are complete, since they are solutions of a Sturm-Liouville problem. This representation was given by S. N. Karp [4]. By modifying his result to correspond to our problem (1)—(3) of Section 3, or by direct expansion of $u$ in angular eigenfunctions, we find, in the two cases,

$$\text{(1a)}\quad u=\frac{i}{2}\sum_{n=1}^{\infty}H^{(1)}_{n/2}(kr_>)\left[J_{n/2}(kr_<)-\frac{J_{n/2}(ka)}{H^{(1)}_{n/2}(ka)}H^{(1)}_{n/2}(kr_<)\right]\sin\frac{n}{2}\theta_<\sin\frac{n}{2}\theta_>,$$

$$\text{(1b)}\quad u=-\frac{i}{2}\sum_{n=1}^{\infty}H^{(1)}_{n/2}(kr_>)\left[J_{n/2}(kr_<)-\frac{J'_{n/2}(ka)}{H^{(1)\prime}_{n/2}(ka)}H^{(1)}_{n/2}(kr_<)\right]\sin\frac{n}{2}\theta_<\sin\frac{n}{2}\theta_>.$$

These series are useful for small values of $ka$ but not for large values.

To obtain representations of $u$ useful for large $ka$ we apply the Watson transformation to the series. First we use in (1a) and (1b) the identities

$$\sin\frac{n}{2}\theta_> = \frac{-\sin\frac{n}{2}(2\pi-\theta_>)}{\cos n\pi},$$

$$\cos\frac{n}{2}\theta_> = \frac{\cos\frac{n}{2}(2\pi-\theta_>)}{\cos n\pi}.$$

Then we represent each series as a contour integral. This is accomplished by first choosing a contour $L$ in the $\mu$-plane which encircles the non-negative integers $\mu=n$, $n=0,1,2,\cdots$, and then an integrand which has as its only singularities within $L$ poles at these integers. In addition for each $n$, the residue at the $n$-th pole must be the $n$-th term in the series. In this way we arrive at the following integral representations:

$$
\text{(2a)}\quad u = \tfrac{1}{4}\int_L \frac{1}{\sin \mu\pi}\left[J_{\mu/2}(kr_<)H^{(1)}_{\mu/2}(ka) - J_{\mu/2}(ka)H^{(1)}_{\mu/2}(kr_<)\right] \times \frac{H^{(1)}_{\mu/2}(kr_>)}{H^{(1)}_{\mu/2}(ka)} \sin\frac{\mu\theta_<}{2}\sin\frac{\mu}{2}(2\pi-\theta_>)\,d\mu,
$$

$$
\text{(2b)}\quad u = \tfrac{1}{4}\int_L \frac{1}{\sin \mu\pi}\left[J_{\mu/2}(kr_<)H^{(1)\prime}_{\mu/2}(ka) - J'_{\mu/2}(ka)H^{(1)}_{\mu/2}(kr_<)\right] \times \frac{H^{(1)}_{\mu/2}(kr_>)}{H^{(1)\prime}_{\mu/2}(ka)} \cos\frac{\mu\theta_<}{2}\cos\frac{\mu}{2}(2\pi-\theta_>)\,d\mu.
$$

The contour $L$ encircles the positive real axis of the $\mu$-plane in the clockwise direction as is shown in Figure 2.

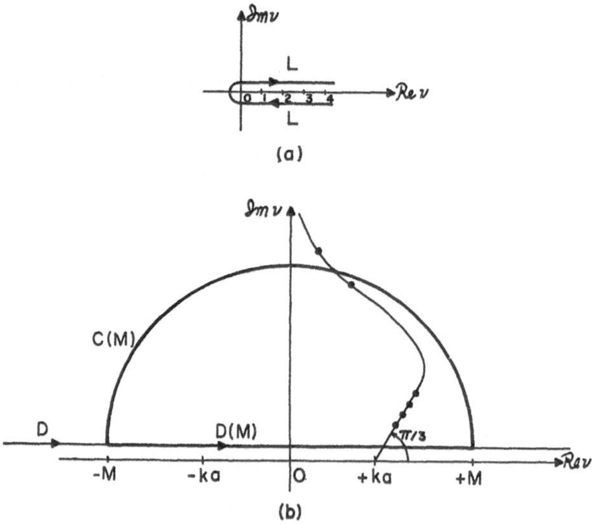


Figure 2

a. The contour $L$ in the $\nu$-plane encircles the positive real axis in the clockwise direction.
b. The contour $D$ is shown just above the real axis of the $\nu$-plane. The segment $D(M)$ of it extends from $\mathscr{R}e\,\nu = -M$ to $\mathscr{R}e\,\nu = M$. The semicircle $C(M)$ into which $D(M)$ is deformed, is also shown. The poles of the integrand in the half plane $\mathscr{I}m\,\nu > 0$ lie on the curved line shown in the first quadrant.

Let us replace $\mu$ by $\nu = \mu/2$ in (2). Let us also change $\nu$ to $-\nu$ in that part of the integral for which $\mathscr{I}m\,\nu < 0$. Then that part of the contour is transformed into a path from $-\infty$ to $0$ lying just above the negative real axis. The integrand is unchanged as we see from the relations

$$H^{(1)}_{-\nu} = e^{i\pi\nu} H^{(1)}_{\nu}, \qquad H^{(2)}_{-\nu} = e^{i\pi\nu} H^{(2)}_{\nu}. \tag{3}$$

Thus the contour $L$ can be replaced by the contour $D$ which is a line from $-\infty$ to $\infty$ lying just above the real axis of the $\nu$-plane and (2) becomes

$$u = \tfrac{1}{2}\int_D \frac{1}{\sin 2\pi\nu}[J_\nu(kr_<)H^{(1)}_\nu(ka) - J_\nu(ka)H^{(1)}_\nu(kr_<)] \times \frac{H^{(1)}_\nu(kr_>)}{H^{(1)}_\nu(ka)} \sin\nu\theta_< \sin\nu(2\pi-\theta_>)d\nu, \tag{4a}$$

$$u = \tfrac{1}{2}\int_D \frac{1}{\sin 2\pi\nu}[J_\nu(kr_<)H^{(1)\prime}_\nu(ka) - J'_\nu(ka)H^{(1)}_\nu(kr_<)] \times \frac{H^{(1)}_\nu(kr_>)}{H^{(1)\prime}_\nu(ka)} \cos\nu\theta_< \cos\nu(2\pi-\theta_>)d\nu. \tag{4b}$$

Let $D(M)$ denote that part of the contour from $-M$ to $+M$ just above the real axis (see Figure 2). Let $C(M)$ denote a semi-circle in the upper half plane with radius $M$ and center at the origin. We wish to deform the partial contour $D(M)$ into $C(M)$. This is permissible since the integrands in (4) are analytic functions of $\nu$. They are regular in the upper half of the $\nu$-plane except for poles at the zeroes of their denominators. These zeroes are the values of $\nu$ for which either

$$H^{(1)}_\nu(ka) = 0, \tag{5a}$$

or

$$H^{(1)\prime}_\nu(ka) = 0. \tag{5b}$$

In each case we shall denote these values of $\nu$ by $\nu_n$, $n = 1, 2, \cdots$. Let $N(M)$ denote the number of these values whose distance from the origin is less than $M$ and let $R_n$ denote the residue of the integrand at $\nu_n$. Then we have in each case, omitting the integrands, and choosing $M$ so that no zero lies on $C(M)$,

$$\int_{D(M)} = \sum_{n=1}^{N(M)} R_n + \int_{C(M)} . \tag{6}$$

From (4) and (6) we have in each case

$$u = \lim_{M\to\infty}\int_{D(M)} = \sum_{n=1}^{\infty} R_n + \lim_{M\to\infty}\int_{C(M)} . \tag{7}$$

We shall now assume that the limit of the last integral in (7) is zero. Then (7) yields for $u$ the following series, in which the residues $R_n$ have been written explicitly:

(8a) $$u(r,\theta) = \frac{i\pi}{2}\sum_{n=1}^{\infty} H_{\nu_n}^{(1)}(kr_<)H_{\nu_n}^{(1)}(kr_>)\frac{\sin\nu_n\theta_<}{\sin 2\pi\nu_m}\sin\nu_n(2\pi-\theta_>)\frac{H_{\nu_n}^{(2)}(ka)}{\frac{\partial}{\partial\nu}H_{\nu_n}^{(1)}(ka)},$$

(8b) $$u(r,\theta) = -\frac{i\pi}{2}\sum_{n=1}^{\infty} H_{\nu_n}^{(1)}(kr_<)H_{\nu_n}^{(1)}(kr_>)\frac{\cos\nu_n\theta_<}{\sin 2\pi\nu_n}\cos\nu_n(2\pi-\theta_>)\frac{H_{\nu_n}^{(2)\prime}(ka)}{\frac{\partial}{\partial\nu}H_{\nu_n}^{(2)\prime}(ka)}.$$

These series are identical with the radial eigenfunction expansions of $u$ given by (13) of Section 3. Thus the validity of those expansions is proved.

Another representation of $u$ can be obtained by using in (4) the identities

(9a) $$\sin\nu(2\pi-\theta_>) = -e^{i2\pi\nu}\sin\nu\theta_> + e^{i\nu\theta_>}\sin 2\pi\nu,$$

(9b) $$\cos\nu(2\pi-\theta_>) = e^{i2\pi\nu}\cos\nu\theta_> - ie^{i\nu\theta_>}\sin 2\pi\nu.$$

The resulting integrands contain one term proportional to $e^{i\nu\theta_>}$ and another proportional to $\sin\nu\theta_>$ or $\cos\nu\theta_>$. The integral of the former term is kept as an integral over $D$, while the integral of the latter term is converted into a residue series as before. In this way we obtain the alternative representations

(10a) $$u = \frac{i\pi}{2}\sum_{n=1}^{\infty} H_{\nu_n}^{(1)}(kr_<)H_{\nu_n}^{(1)}(kr_>)\frac{\sin\nu_n\theta_<\sin\nu_n\theta_>}{\sin 2\pi\nu_n}e^{i2\pi\nu_n}\frac{H_{\nu_n}^{(2)}(ka)}{\frac{\partial}{\partial\nu}H_{\nu_n}^{(1)}(ka)} - \frac{1}{4}\int_D H_\nu^{(1)}(kr_<)H_\nu^{(1)}(kr_>)\sin\nu\theta_< e^{i\nu\theta_>}\frac{H_\nu^{(2)}(ka)}{H_\nu^{(1)}(ka)}d\nu,$$

(10b) $$u = \frac{-i\pi}{2}\sum_{n=1}^{\infty} H_{\nu_n}^{(1)}(kr_<)H_{\nu_n}^{(1)}(kr_>)\frac{\cos\nu_n\theta_<\cos\nu_n\theta_>}{\sin 2\pi\nu_n}e^{i2\pi\nu_n}\frac{H_{\nu_n}^{(2)\prime}(ka)}{\frac{\partial}{\partial\nu}H_{\nu_n}^{(1)\prime}(ka)} + \frac{i}{4}\int_D H_\nu^{(1)}(kr_<)H_\nu^{(1)}(kr_>)\cos\nu\theta_< e^{i\nu\theta_>}\frac{H_\nu^{(2)\prime}(ka)}{H_\nu^{(1)\prime}(ka)}d\nu.$$

We shall see in the next section that this representation is more useful in the illuminated region, while (8) is more useful in the shadow region.

## 5. Convergence of the Series

To investigate the convergence of the various series, we follow the procedure of W. Franz [7]. The Hankel function $H_\nu^{(1)}(x)$ of large complex order $\nu$ and fixed real argument $x$ has the asymptotic form [7]

$$H_\nu^{(1)}(x) \sim -2\sqrt{\frac{2}{\pi\nu}}\sinh\left[\nu\log\frac{2\nu}{x}-\nu+i\frac{\pi}{4}\right]. \tag{1}$$

Consequently the values of $\nu$ at which $H_\nu^{(1)}(x)$ vanishes are asymptotically determined by

$$\nu\left(\log\frac{2\nu}{x}-1\right) = (n-\tfrac{1}{4})i\pi. \tag{2}$$

In (2), $n$ denotes a large integer. To solve (2) we set $\nu = \rho e^{i\phi}$. Then since the real part of the left side of (2) must be zero, we find that for large $\rho$ $\phi$ and $\rho$ must be related by

$$\phi \sim \frac{\pi}{2}\left(1-\frac{1}{\log 2\rho x^{-1}}\right). \tag{3}$$

From the imaginary parts in (2) we then find, denoting $\rho$ by $\rho_n$,

$$\rho_n \sim \frac{(n-\frac{1}{4})\pi}{\log 2(n-\frac{1}{4})\pi x^{-1}}. \tag{4}$$

Thus for large positive $n$ the zeros $\nu_n$ of $H_\nu^{(1)}(x)$ are

$$\nu_n \sim \frac{(n-\frac{1}{4})\pi}{\log 2(n-\frac{1}{4})\pi x^{-1}}\exp\left\{i\frac{\pi}{2}\left(1-\frac{1}{\log[2(n-\frac{1}{4})\pi/x\log(2(n-\frac{1}{4})\pi x^{-1})]}\right)\right\}. \tag{5}$$

From (1) we have

$$H_\nu^{(1)\prime}(x) \sim \frac{2}{x}\sqrt{\frac{2\nu}{\pi}}\cosh\left[\nu\log\frac{2\nu}{x}-\nu+\frac{i\pi}{4}\right]. \tag{6}$$

Then (2)—(5) hold for the zeroes of $H_\nu^{(1)\prime}(x)$ if $n$ is replaced by $n+\frac{1}{2}$.

By using (1) and (5) with $x = kr_>$, we find

$$|H_{\nu_n}^{(1)}(kr_>)| \sim \sqrt{\frac{2}{\pi\nu_n}}\exp\left\{\rho_n\frac{\pi}{2}\frac{\log\left(\frac{a}{r_>}\right)}{\log\left(\frac{2\rho_n}{ka}\right)}\right\}. \tag{7}$$

To evaluate $H_{\nu_n}^{(2)}(ka)$ we use the relation $H_\nu^{(2)}(x) = \overline{H_{\bar\nu}^{(1)}(x)}$ valid for $x$ real. Then we see that (1) holds for $H_\nu^{(2)}(x)$ with $i\pi/4$ replaced by $-i\pi/4$. Therefore we have

$$H_{\nu_n}^{(2)}(ka) \sim 2\sqrt{\frac{2}{\pi\nu_n}}(-1)^n. \tag{8}$$

From (1) we also obtain

$$(9) \qquad \frac{\partial}{\partial \nu} H^{(1)}_{\nu_n}(ka) \sim -2\sqrt{\frac{2}{\pi\nu_n}} \log\left(\frac{2\nu_n}{ka}\right) \cdot (-1)^n.$$

Upon combining (7)—(9) we obtain for the asymptotic form of the absolute value of the $n$-th term in (8a) of Section 4 the expression

$$(10) \qquad \frac{1}{2\rho_n \log \dfrac{\rho_n}{ka}} \exp\left\{-\rho_n\left[\theta_> - \theta_< - \frac{\pi}{2}\,\frac{\log \dfrac{a^2}{r_< r_>}}{\log \dfrac{2\rho_n}{ka}}\right]\right\}.$$

This expression diminishes sufficiently rapidly as $n$ increases to guarantee that the series (8a) of Section 4 converges absolutely for $\theta_> > \theta_<$, i.e. for $\theta \neq \theta_0$. The same result applies to the series (8b) of Section 4.

We shall now examine the leading terms in these series in order to show how rapidly they diminish. For any fixed value of $n$ and $ka$ large, $\nu_n$ is asymptotically given by (5) of Section 3. Therefore, for fixed $n$ the order and argument of the Hankel functions in the series (8) of Section 4 are both large. Thus for large $ka$ we may evaluate the Hankel functions with argument $kr_<$ or $kr_>$ by using the Debye asymptotic form which is

$$(11) \qquad H^{(1)}_\nu(x) \sim \sqrt{\frac{2}{\pi}}\,(x^2-\nu^2)^{-1/4} \exp\left\{i(x^2-\nu^2)^{1/2} - i\nu\cos^{-1}\frac{\nu}{x} - \frac{i\pi}{4}\right\}.$$

For the Hankel functions of argument $ka$ we must use instead the Airy function representation

$$(12) \qquad H^{(1)}_\nu(x) \sim \frac{2}{\pi}\, e^{-i\pi/3}\left(\frac{6}{x}\right)^{1/3} A\left[\left(\frac{6}{x}\right)^{1/3} e^{-i\pi/3}(\nu - x)\right].$$

By using these expansions in (8) of Section 4 we obtain the series (15) of Section 3. If we further retain only the leading terms in the trigonometric factors in (15a) the summand becomes

$$(13) \qquad \frac{1}{2i[A'(q_n)]^2} \exp\left\{i\nu_n\left[\theta_> - \theta_< - \cos^{-1}\frac{a}{r} - \cos^{-1}\frac{a}{r_0}\right]\right\}.$$

The absolute value of (13) is asymptotically

$$(14) \qquad \frac{1}{2[A'(q_n)]^2} \exp\left\{-q_n\left(\frac{ka}{6}\right)^{1/3} \sin\frac{\pi}{3}\left[\theta_> - \theta_< - \cos^{-1}\frac{a}{r} - \cos^{-1}\frac{a}{r_0}\right]\right\}.$$

The $q_n$ are positive and increase with $n$ like $n^{2/3}$. Therefore, the absolute value (14) decreases with increasing $n$ provided that

$$(15) \qquad \theta_> - \theta_< - \cos^{-1}\frac{a}{r} - \cos^{-1}\frac{a}{r_0} > 0.$$

For a point in the shadow region $B$ shown in Figure 2, $\theta_> = \theta_0, \theta_< = \theta$ and

$$\theta_0 > \theta + \cos^{-1}\frac{a}{r} + \cos^{-1}\frac{a}{r_0}. \tag{16}$$

Therefore, (15) is satisfied for points in $B$. Similarly for points in the shadow region $A$ of Figure 2, $\theta_> = \theta, \theta_< = \theta$ and

$$\theta > \theta_0 + \cos^{-1}\frac{a}{r} + \cos^{-1}\frac{a}{r_0}. \tag{17}$$

Thus (15) is also satisfied for points in $A$. We conclude that (15) is satisfied for points in the shadow region but not for points in the illuminated region. The same results apply to the series (15b). Consequently, for points in the shadow the terms in the radial eigenfunction series (13) of Section 3 or (8) of Section 4 decrease rapidly like $\exp\{-n^{2/3}(ka)^{1/3}c\}$, where $c$ is a positive constant. Hence the radial eigenfunction series converges rapidly only in the shadow although it also converges in the illuminated region. The series (7) of Section 2, which results from the geometrical theory of diffraction for points in the shadow, also converges rapidly. In the shadow it coincides with the asymptotic expansion (15) of the radial eigenfunction series. Outside the shadow the series (15) diverges. The representation (10) of Section 5 is useful in the illuminated region since the series in it converge rapidly there, as can be shown by an analysis like that we have just given. In fact the asymptotic expansion of the series in (10) yields exactly the same result as that given by the geometrical theory of diffraction for points in the illuminated region. The asymptotic evaluation of the integrals in (10) yields the incident and reflected fields. However, we shall not carry out these calculations.

## Bibliography

[1] Keller, J. B., *How dark is the shadow of a round-ended screen?* J. Appl. Phys., Vol. 30, 1959, pp. 1452–1454.

[2] Keller, J. B., *A geometrical theory of diffraction, Calculus of Variations and its Applications*, Proc. Symposia Appl. Math., VIII, McGraw-Hill, New York, 1958, pp. 27–52.

[3] Keller, J. B., *Diffraction by a convex cylinder*, I.R.E. Trans. Antennas and Prop., AP-4, 1956, pp. 312–321.

[4] Karp, S. N., *Diffraction by a tipped wedge*, New York Univ., Math. Res. Group, Wash. Square Coll., Res. Rep. No. EM-52, 1953.

[5] Levy, B. R., and Keller, J. B., *Diffraction by a smooth object*, Comm. Pure Appl. Math., Vol. 12, 1959, pp. 159–209.

[6] Sommerfeld, A., *Partial Differential Equations*, Academic Press, New York, 1949.

[7] Franz, W., *Über die Greenschen Funktionen des Zylinders und der Kugel* Z. Naturforschung, Vol. 9a, 1954, pp. 705–716.

Received February, 1961.

ΣΥΝΕΔΡΙΑ ΤΗΣ 3 ΙΟΥΝΙΟΥ 1976 659

ΕΦΗΡΜΟΣΜΕΝΑ ΜΑΘΗΜΑΤΙΚΑ.— **On the linearization of nonlinear models of the phenomena. First part: Linearization by exact methods,** *by Demetrios G. Magiros* *. Ἀνεκοινώθη ὑπὸ τοῦ Ἀκαδημαϊκοῦ κ. Ἰω. Ξανθάκη.


## ABSTRACT

This paper deals with remarks on the linearization methods of nonlinear mathematical models of the phenomena.

We can classify the linear methods into two distinct types, the «exact methods» and the «approximate methods».

The advantages and disadvantages of these methods are clarified by appropriately selected examples.


## 1. INTRODUCTION

The phenomena, either physical or natural or social, are interrelated variations of certain variable quantities and their rate of change, and their nature is usually discussed by mathematical models, where dominant concepts are the linearities and nonlinearities of the models, which usually are differential equations nonlinear in their variables (NLDE).

By «linearization» of a NLDE we mean a reduction to a linear differential equation (LDE), which is either «equivalent» or «almost equivalent» to the NLDE, that is the solution of the LDE may give the solution of the NLDE either «exactly» or «approximately» by an error of small order.

The linearization is a tool for a simplified and easy discussion of the nonlinear phenomena. This tool is helpful in a few cases, but in many cases, especially of practical importance, the linearized models leave out essential features of the NLDE, or they contain properties which are not properties of the NLDE, when the linearization is not acceptable for an adequate description of the real phenomena.

---

* ΔΗΜΗΤΡΙΟΥ Γ. ΜΑΓΕΙΡΟΥ, **Γραμμικοποίησις μὴ γραμμικῶν μαθηματικῶν μοντέλων τῶν φαινομένων. Μέρος I: Γραμμικοποίησις μὲ ἀκριβεῖς μεθόδους.**

Reprinted from the *Proceedings of the Athens Academy of Sciences* 51 (1976), 659–668.

The aim of this paper is to provide a sketch of ideas and techniques associated with the notion of linearization of NLDE and to exhibit advandages and disadvantages of the linearization methods by using examples of practical significance.

We distinguish two classes of linearization methods : the «exact methods», and the «approximate methods».

The exact methods give exact solutions, essentially general solutions in closed form, and the approximate methods give approximate solutions within an accepted error.

In this part of the paper we deal with «exact methods».

## 2. LINEARIZATION BY EXACT METHODS

NLDE may be reduced to LDE by «exact transformations of the variables». By this linearization we may obtain general solutions of the NLDE in a «closed form», by using the solutions of the LDE and the transformation formulae of the variables.

Also by exact linearizations, corresponding to some restrictions of the variables or of the parameters, it is possible to find general solutions of special cases of the original NLDE.

The following examples may be sufficient to show the essence and the applicability of the exact linearization methods and their important results.

In some of the examples use is made of appropriate transformations of the variables, and in some other examples appropriate restrictions of the variables or of the parameters.

E x a m p l e 1. The Bernoulli equation :

$$y' + ay = by^n \tag{1}$$

where a and b are functions of x, and $b \neq 0$, $n \neq 0$, $\neq 1$, is a monumental example of exact linearization.

We change the variable y into a new variable z, keeping the same independent variable x, by the transformation formula :

$$z = y^{1-n} \tag{1.1}$$

when the linear equation in z results :

$$z' + (1-n)\alpha z = (1-n)b \tag{1.2}$$

By solving (1.2), then using (1.1), we can get the solution of (1) in an exact and closed form.

In the specific example:

$$y' - \frac{2}{x}y = 4x^2y^{1/2}, \quad x \neq 0$$

which is of Bernoulli type, the transformation $z = y^{1/2}$ leads to the LDE:

$$z' - \frac{1}{x}z = zx^2$$

of which, by using $\mu = \frac{1}{x}$ as an integrating factor, we find the general solution: $z = (cx + x^3)$, when the general solution of the original NL equation is: $y = (cx + x^3)^2$.

As we see, the single constant of integration enters the general solution of (1) in a nonlinear way.

Example 2. The Ricatti equation:

$$y' = \alpha y^2 + by + c \tag{2}$$

where $\alpha$, b, c are functions of x, and $\alpha \neq 0$.

Liouville proved (1841) that this simple NLDE in its general form can not be solved by elementary exact methods.

(a): This equation can be linearized to a LDE of second order.

Transforming (2) by

$$y(x) = \frac{v(x)}{\alpha(x)} \tag{2.1}$$

we get:

$$v' = v^2 + \left(\frac{\alpha'}{\alpha} + b\right)v + \alpha c \tag{2.2}$$

which is NL in the new vrriable v. Now, transforming (2.2) by:

$$v(x) = -\frac{u'(x)}{u(x)} \tag{2.3}$$

we get the LDE of second order in u(x) with variable coefficients:

$$u'' - \left(\frac{\alpha'}{\alpha} + b\right)u' + \alpha c u = 0 \tag{2.4}$$

of which the solution can not, in general, be expressed in exact form in term of a finite number of elementary functions.

.(b): We can linearize the Ricatti equation (2) by specific «restrictions», e.g. when we know a particular solution $y_1(x)$.

By the transformation:

$$y(x) = y_1(x) + z(x) \tag{2.5}$$

the equation (2) leads to the Bernoulli equation in z:

$$z' - (b + 2\alpha y_1) z = \alpha z^2 \tag{2.6}$$

which can be solved by linearization, and from its solution and the transformation (2.5) we can get the general solution of (2).

Example 3. The NLDE:

$$y = \frac{X + yZ}{Y + xZ} \tag{3}$$

where X, Y, Z are homogeneous functions of x, y, the X, Y of degree $\mu$ of homogeneity, and Z of degree $\nu$, can be solved by an exact linearization. Transforming by:

$$y = xz \tag{3.1}$$

and using the homogeneity property of X, Y, Z, equation (3) becomes:

$$\frac{dx}{dz} + A(z)x = B(z)x^{\nu-\mu+2} \tag{3.2}$$

which is of Bernoulli type. The functions A(z) and B(z) are known functions, calculated in the process to find (3.2).

Example 4. The Lagrange equation:

$$y = x\varphi(y') + \psi(y') \tag{4}$$

which is linear in x and y, but nonlinear in y', can be solved by an exact linearization.

Differentiating (4) with respect to x, putting $y' = p$, then considering x as dependent variable and p as independent variable, we get:

$$(\varphi(p) - p)\frac{dx}{dp} + x\varphi'(p) + \psi'(p) = 0 \tag{4.1}$$

which is linear in x. If $x = x(p)$ is the solution of (4. I), inserting $x = x(p)$ into (4), we have the function:

$$y = x(p)\varphi(p) + \psi(p) \tag{4. 2}$$

which is the general solution of (4) in an exact closed form with p as a parameter.

E x a m p l e s 5. The NLDE: (a): $f(y') = 0$, (b): $f(y'') = 0, \ldots$ can be solved by an exact linearization.

.(a): If in (a) we put $y' = k$, when $y = kx + c$, that is $k = \frac{y-c}{x}$ then the general solution of (a) is: $f\left(\frac{y-c}{x}\right) = 0$, $c =$ parameter. E. g. the equation: $(y')^4 - 1 = 0$ has $\left(\frac{y-c}{x}\right)^4 - 1 = 0$ as general solution. We have:

$$\left(\frac{y-c}{x}\right)^4 - 1 = \left\{\left(\frac{y-c}{x}\right)^2 + 1\right\} \cdot \left(\frac{y-c}{x} - 1\right)\left(\frac{y-c}{x} + 1\right).$$

The first factor is not zero, the other factors may be zero, then the general solution of this specific example is the couple of the two families of lines: $y = x + c$, and $y = -x + c$ that is all lines in the x, y — plane parallel to the first and second bisector.

.(b): If in (b) we put $y'' = k$, when $y = \frac{k}{2}x^2 + c_1 x + c_2$, or $k = \frac{2}{x^2}(y - c_1 x - c_2)$ then the general solution of (b) is:

$$f\left(\frac{2}{x^2}(y - c_1 x - c_2)\right) = 0, \quad c_1 \text{ and } c_2 \text{ parametrs.}$$

E x a m p l e 6. We may have a linearization of a complicated NLDE by factorization, e. g., if the NLDE is factorized as:

$$(y^2 + y'^2 + 1) \cdot y'' \cdot (x^3 y''' + x^2 y'' - 2xy' + 2y) = 0 \tag{6}$$

This equation is equivalent to the system:

$$\text{(a)}: \quad y'' = 0, \qquad \text{(b)}: \quad x^3 y''' + x^2 y'' - 2xy' + 2y = 0 \tag{6. 1}$$

The first factor can not be zero. The second equation of (6.1) is of Euler type. The solutions of (6.1) are :

$$\text{(a)}: \quad y = c_1 x + c_2, \qquad \text{(b)}: \quad y = c_3 x + c_4 x^{-1} + c_5 x^2 \tag{6.2}$$

where $c_1, \ldots, c_5$ are parameters. Therefore the general solution of (6) is given by both equations (6.2), and from each point of the x, y -plane two solution curves of (6.2) pass, one from the family (6.2.a) and one from the family (6.2.b).

We may have other cases of exact linearization of NLDE [1], [2].

In the following two examples the exact linearization is of different nature:

Example 7. The problem of mechanics of the free rotation of a rigid body with any «mass distribution» is governed by the Euler system of NLDE :

$$\left.\begin{aligned} I_1 \dot{\omega}_1 + (I_3 - I_2)\,\omega_3 \omega_2 &= 0 \\ I_2 \dot{\omega}_2 + (I_1 - I_3)\,\omega_1 \omega_3 &= 0 \\ I_3 \dot{\omega}_3 + (I_2 - I_1)\,\omega_2 \omega_1 &= 0 \end{aligned}\right\} \tag{7}$$

where $\omega_1, \omega_2, \omega_3$ are the unknown angular velocity components, and the parameters $I_1, I_2, I_3$ are the moments of inertia of the given rigid body about the body coordinate system with origin the mass center of the body. We can linearize this nonlinear system, if we accept a «special mass distribution», which corresponds to a «special restriction of the parameters» $I_1, I_2, I_3$. If the mass distribution of the rotating body is such that the body has an axis of symmetry, say the $\omega_1$ -axis, then $I_2 = I_3 = I$, and (7) leads to the linear coupled system :

$$\dot{\omega}_2 = c_1 \omega_3, \qquad \dot{\omega}_3 = - c_1 \omega_2 \tag{7.1}$$

where : $\quad c_1 = \bar{\omega}_1 (I - I_1)/I = \text{constant}.$

The above linearization, occurring by a restriction of the parameters, corresponds to a physical problem, which is a special case of the initial problem expressed by (7), and the solution of the linearized system (7.1) is a special case of (7).

Example 8. The «pure Keplerian motion» of a body of mass m around a central body of mass M is due to the attractive Newtonian force of the bodies, in the absence of perturbing forces. This motion is governed by the NLDE in vector form:

$$\ddot{x} + \frac{k^2}{r^3} x = 0 \tag{8}$$

where $k = K(M+m) =$ constant, $x = (x_1, x_2, x_3)$, $r^2 = x_1^2 + {}_2^2 + x_3^2$.

In the following we see two types of exact linearizations of (8), one by «regularization», and the other by «restriction of the variables» (3).

(a): Regularization of (8). We use an auxiliary equation, and change the independent variable.

The equation (8) is «singular» with the origin as the singularity.

When the motion of m is close to M, it is a «near collision» motion, when large gravitational forces appear and sharp bends of the orbit.

Such a phenomenon occurs when, e.g., an artificial space vehicle is at its start or at its destination.

By appropriate transformation of (8), it is possible to get a regular equation free ot singularities. This is called: «regularization». In the one-dimension case, (8) becomes:

$$\ddot{x} + \frac{k^2}{x^2} = 0 \tag{8.1}$$

and the energy function of the pure Keplerian motion is:

$$h_k = \frac{k}{x} - \frac{1}{2}\dot{x}^2 \tag{8.2}$$

where the energy $h_k$ is a negative constant.

If, in these equations, instead of the «natural time» t, the «artificial time» τ is taken according to:

$$dt = x\, d\tau \tag{8.3}$$

the equations (8.1) and (8.2) become:

$$xx'' - x'^2 + k^2 x = 0 \tag{8.4}$$

$$x'^2 = 2(k^2 x + h_k x^2) \tag{8.5}$$

where τ is the independent variable.

Inserting now (8.5) into (8.4) we have the LDE

$$x'' + 2h_k x = k^2 \tag{8.6}$$

which is the «regularization» of (8).

(b): Restriction of the Variables in (8). By restricting the variables $x_1$, $x_2$, $x_3$ according to:

$$r^2 = x_1^2 + x_2^2 + x_3^2 = \text{constant} \tag{8.7}$$

we linearize (8), when it reduces to the LDE

$$\ddot{x} + vx = 0 \tag{8.8}$$

with $v = kr^{-3/2}$. The solution of (8.8) is:

$$x = a\cos vt + b\sin vt \tag{8.9}$$

with a and b are constant vectors.

The restriction (8.7) specializes the motion of m to be a motion on the surface of the sphere (8.7), and, since the motion of m is only under the influence of a central force, the motion of m is circular on a plane through the origin with period of revolution: $T = \frac{2\pi}{v} = \frac{2\pi}{k} r^{3/2}$ and velocity

$$\dot{x} = v(-c_1 \sin vt + c_2 \cos vt) \tag{8.10}$$

of magnitude:

$$U = vr = \frac{k}{\sqrt{r}} = \text{constant}. \tag{8.11}$$

By considering r of (8.7) as a parameter, the above linearization gives various circular motions of m around M inside a sphere of radius the maximum of the parameter r.

## SUMMARY

The mathematical models of the phenomena are usually NLDE, which is very difficult to be solved. Some classes of these equations can be treated by special methods, among which are the linearization methods, either exact or approximate.

In the exact linearization methods one may use appropriate transformation of the variables, when one may have general solutions of the NLDE in closed form, which is an ideal case.

Also, one may use restrictions of the variables or of the parameters of the NLDE, when one may have some special subclasses of the general solutions of the NLDE.

For a single NLDE one may have more than one formula for its general solution, and a NLDE may have, in addition, singular solutions.

## ΠΕΡΙΛΗΨΙΣ

Τὰ μαθηματικὰ μοντέλα τῶν διαφόρων φαινομένων εἶναι, συνήθως, μὴ γραμμικαὶ διαφορικαὶ ἐξισώσεις, τῶν ὁποίων ἡ λύσις εἶναι, ἐν γένει, εἴτε ἀδύνατον εἴτε πολὺ δύσκολον νὰ εὑρεθῇ. Διὰ μερικὰς κατηγορίας τῶν μὴ γραμμικῶν ἐξισώσεων δύνανται νὰ ἐφαρμοσθοῦν εἰδικαὶ μέθοδοι, μεταξὺ τῶν ὁποίων εἶναι καὶ ἡ «γραμμικοποίησις», εἴτε ἀκριβὴς (exact) εἴτε κατὰ προσέγγισιν (approximate). Παλαιότερα, ἡ σπουδὴ τῶν μὴ γραμμικῶν ἐξισώσεων διὰ γραμμικοποιήσεως (διαγραφῆς τῶν μὴ γραμμικῶν ὅρων τῶν ἐξισώσεων) ἦτο ἡ δεσπόζουσα μέθοδος, ἰδίως εἰς προβλήματα ἐφαρμογῶν. Πρὸ μερικῶν δεκαετηρίδων ὅμως ἀπεδείχθη ὅτι ἡ γραμμικοποίησις ἐξαλείφει βασικὰς ἰδιότητας τῶν λύσεων τῶν μὴ γραμμικῶν ἐξισώσεων καὶ ὅτι οἱ μὴ γραμμικοὶ ὅροι τῶν ἐξισώσεων παίζουν δεσπόζοντα ρόλον εἰς τὴν ἔρευναν τῶν φαινομένων.

Διὰ τῆς παρούσης ἐργασίας, ὅπως καὶ διὰ μιᾶς ἄλλης ποὺ θὰ ἐπακολουθήσῃ, δίδονται παρατηρήσεις ἐπὶ τῶν μεθόδων γραμμικοποιήσεως, κυρίως ἐπὶ τῆς καταλληλότητος ἢ μὴ τῆς γραμμικοποιήσεως ὡς μεθόδου ἐρεύνης τῶν φαινομένων. Εἰδικῶς, εἰς τὴν παροῦσαν ἐργασίαν ἐξετάζονται ἀκριβεῖς μέθοδοι γραμμικοποιήσεως. Εἰς αὐτὰς τὰς μεθόδους γίνεται χρῆσις καταλλήλων μετασχηματισμῶν τῶν μεταβλητῶν, ὁπότε εἶναι δυνατὸν νὰ ἐπιτευχθοῦν «γενικαὶ λύσεις» τῶν μὴ γραμμικῶν ἐξισώσεων ὑπὸ «κλειστὴν μορφὴν» (closed form), τὸ ὁποῖον εἶναι ἰδεῶδες ἐπίτευγμα.

Ἐπίσης, μὲ κατάλληλον περιορισμὸν τῶν μεταβλητῶν ἢ τῶν συντελεστῶν τῶν μὴ γραμμικῶν ἐξισώσεων, εἶναι δυνατὸν νὰ ἐπιτευχθοῦν γενικαὶ λύσεις, αἱ ὁποῖαι εἶναι εἰδικαὶ περιπτώσεις τῆς γενικῆς λύσεως τῶν μὴ γραμμικῶν ἐξισώσεων.

Μία μὴ γραμμικὴ ἐξίσωσις ἐνδέχεται νὰ ἔχῃ περισσοτέρας τῆς μιᾶς γενικὰς λύσεις, ὅπως ἐπίσης ἐνδέχεται νὰ ἔχῃ, ἐπὶ πλέον, καὶ ἀνωμάλους (singular) λύσεις.

## REFERENCES

1. E. Kamke, «Differentialgleichungen reeler Functionen», 2nd edition, Leipzig (1952).
2. D. Magiros, «Exact solutions of nonlinear differential equations», General Electric Co., 73A5240, 11/26/73.
3. E. Stiefel and G. Scheifele, «Linear and regular celestial mechanics», Springer - Verlag, New York (1971).

*

Ὁ Ἀκαδημαϊκὸς κ. **Ἰω Ξανθάκης**, παρουσιάζων τὴν ἀνωτέρω ἀνακοίνωσιν, εἶπε τὰ ἑξῆς:

Ἔχω τὴν τιμὴν νὰ ἀνακοινώσω εἰς τὴν Ἀκαδημίαν ἐργασίαν τοῦ κ. Δημητρίου Μαγείρου διὰ τὴν γραμμικοποίησιν μὴ γραμμικῶν Μαθηματικῶν Προτύπων τῶν Φαινομένων.

Ὡς γνωστὸν τὰ μαθηματικὰ πρότυπα (μοντέλα) διαφόρων φυσικῶν φαινομένων μᾶς ὁδηγοῦν συνήθως εἰς μὴ γραμμικὰς διαφορικὰς ἐξισώσεις, τῶν ὁποίων αἱ λύσεις εἶναι, κατὰ γενικὸν κανόνα, εἴτε ἀδύνατοι εἴτε πολὺ δύσκολοι. Λόγῳ τῆς δυσχερείας ταύτης, μέχρι πρό τινων ἐτῶν, διεγράφοντο οἱ μὴ γραμμικοὶ ὅροι τῶν ἀντιστοίχων διαφορικῶν ἐξισώσεων ἰδίως εἰς προβλήματα ποὺ ἀφεώρων ἐφαρμογάς· ἀλλὰ ἡ γραμμικοποίησις τῶν διαφορικῶν ἐξισώσεων διὰ τῆς μεθόδου ταύτης, δηλαδὴ τῆς διαγραφῆς τῶν μὴ γραμμικῶν ὅρων, ἐξαλείφει βασικὰς ἰδιότητας τῶν λύσεων τῶν μὴ γραμμικῶν διαφορικῶν ἐξισώσεων, διότι διαγραφόμενοι μὴ γραμμικοὶ ὅροι διαδραματίζουν ὡς ἐπὶ τὸ πλεῖστον δεσπόζοντα ρόλον εἰς τὴν ἔρευναν τῶν φαινομένων. Εἰς ὡρισμένας κατηγορίας μὴ γραμμικῶν διαφορικῶν ἐξισώσεων δύνανται νὰ ἐφαρμοσθοῦν εἰδικαὶ μέθοδοι, ποὺ στηρίζονται εἰς γραμμικοποποίησιν κατὰ τρόπον ἀκριβῆ ἢ κατὰ προσέγγισιν.

Εἰς τὴν παροῦσαν ἀνακοίνωσιν ὁ κ. Μάγειρος ἀσχολεῖται μὲ ἀκριβεῖς μεθόδους γραμμικοποιήσεως, τὰς ὁποίας ἐφαρμόζει εἰς τὰς ἐξισώσεις Bernouilli, Ricatti, καθὼς καὶ εἰς διαφόρους ἄλλας μορφὰς διαφορικῶν ἐξισώσεων.

ΠΡΑΚΤΙΚΑ ΤΗΣ ΑΚΑΔΗΜΙΑΣ ΑΘΗΝΩΝ

ΣΥΝΕΔΡΙΑ ΤΗΣ 10ΗΣ ΙΟΥΝΙΟΥ 1976

ΠΡΟΕΔΡΙΑ ΝΙΚ. Κ. ΛΟΥΡΟΥ

ΕΦΗΡΜΟΣΜΕΝΑ ΜΑΘΗΜΑΤΙΚΑ.— **On the linearization of nonlinear models of phenomena. Second part: Linearization by approximate methods,** *by Demetrios G. Magiros* *. Ἀνεκοινώθη ὑπὸ τοῦ Ἀκαδημαϊκοῦ κ. Ἰ. Ξανθάκη.

## INTRODUCTION

In a previous paper, contained in this volume, we discussed exact linearization techniques for solving nonlinear differential equations, when we get exact general solutions of NLDE in a closed form.

These ideal methods can be applied to only a few cases, when approximate linearization methods are suggested for approximate particular solutions of NLDE.

By a variety of examples, we point out different approximate linearization methods, cases where the linearization is not permitted, and cases where the linearization gives useful results.

In many fields of research, we see advantages and disadvantages of the linearization methods, as well as the importance and the influence of the nonlinearities.

---

* Δ. Γ. ΜΑΓΕΙΡΟΥ, **Γραμμικοποίησις μὴ γραμμικῶν μαθηματικῶν μοντέλων τῶν φαινομένων. Μέρος II: Γραμμικοποίησις μὲ κατὰ προσέγγισιν μεθόδους.**— Scientific Consultant, General Electric Company, (RESD) Philadelphia, PA., U. S. A.

Reprinted from the *Proceedings of the Athens Academy of Sciences* **51** (1976), 669–683.

## LINEARIZATION BY APPROXIMATE METHODS

Whenever exact methods are not applicable to NLDE, and this case is the usual one in applications, approximate methods are suggested, by which one may get approximations of particular solutions.

The approximate methods are either geometrical, or analytical, or numerical, and the concept of linearization may enter to any of these methods.

The linearization by approximate methods leads to results, which are acceptable in some cases, but not acceptable in some other cases, this depending mainly on the nature of the problem which is associated with the linearization. In some linearization approximate methods the concepts of the «error» and the «norm», which measures the error, play essential role and characterize the methods, but in some other methods the concepts of the «error» and «norm» are not necessary. By the following examples and appropriate remarks we try to clarify the above statements.

Example 1. Let us pose the following problem:

«Find a curve such that the product of the distances from two different points E and E′ to anyone of its tangents is a non-zero constant, $b^2$». We take the line EE′ as x-axis, the middle point of EE′ as the origin, and $EE' = 2c$. The property of the curve of the problem leads to DE:

$$(y - cy')^2 - c^2y'^2 = b^2(1 + y'^2)$$

then we have:

$$y = xy' \pm (b^2 + \alpha^2y'^2)^{1/2}, \qquad \alpha^2 = b^2 + c^2 \tag{1}$$

The equation (1) is of Clairaut type and by exact methods (not exact linearization) its solution can be found to be:

General Solution: $y = cx \pm (b^2 + \alpha^2c^2)^{1/2}$ (1.1)

Singular Solution: $\dfrac{x^2}{\alpha^2} + \dfrac{y^2}{b^2} = 1$ (1.2)

The true solution of the problem is the ellipse (1.2), and not any particular solution coming from the general solution (1.1).

If we try to solve the NLDE (1) by linearization, the only way is to omit the nonlinear term in (1), when the corresponding linear equation is $y = xy' + b$ of which the solution is $y = cx/b$, that is the family of straight lines in the x, y - plane, but not the solution (1.2) (ellipse).

By this example we see that the linearization by omitting the nonlinearities leads in general to unaccepted results.

Example 2. The NLDE.

$$y' = (x^2 + \varepsilon^2 y^2)^{1/2}, \quad \varepsilon - \text{parameter} \tag{2}$$

can not be solved by exact methods, but by a linearization coming from an «appropriate restriction of the variables», which leads to an approximate geometrical method, called «method of isoclines».

The linearization of (2) by omitting the nonlinearity leads to the linearized equation: $y' = \pm x$, of which the general solution is the family of parabolas $y = c \pm \frac{1}{2} x^2$, c = parameter. Such a linearization is not accepted.

The linearization of (2) by restricting the variables x and y according to the restriction:

$$(x^2 + \varepsilon^2 y^2)^{1/2} = k, \quad k = \text{constant} \tag{2.1}$$

gives the linearized equation:

$$y' = k \tag{2.2}$$

with general solution:

$$y = kx + c \tag{2.3}$$

The restriction (2.1) of the variables of (2) is written in the form:

$$\frac{x^2}{k^2} + \frac{y^2}{(k/\varepsilon)^2} = 1 \tag{2.4}$$

The above linearization of (2) is equivalent to taking curves in the x, y - plane through any point of which the unknown solution of (2) has the same slope, that is the same inclination angle with x-axis.

This idea gives a «graphical method» for construction of the solution of a NLDE, called «method of isoclines».

The «isocline curves» of (2) are the ellipses (2.4), by which one can get approximately the solution of (2) in the form of a «directed field».

Example 3. The study of the multivibrator, a basic electronic circuit, with nonlinear resistance and in the absence of external forces, leads to the famus Van der Pol equation :

$$\ddot{x} - \varepsilon(1 - x^2)\dot{x} + x = 0 \tag{3}$$

which is equivalent to the normal system :

$$\dot{x}_1 = x_2, \qquad \dot{x}_2 = \varepsilon(1 - x_1^2)x_2 - x_1 \tag{3.1}$$

The time t is the independent variable.
This equation can not be solved by exact methods.
Van der Pol found graphically that this equation has one isolated periodic solution, its «limit cycle».

We can linearize the system (3. 1) by considering $x^2$ as an infinitesimal stronger than x, that is by taking the «nonlinearity condition» :

$$x^2 = o(x) \tag{3.2}$$

when the linearized system is :

$$\dot{x}_1 = x_2, \qquad \dot{x}_2 = -x_1 + \varepsilon x_2 \tag{3.3}$$

The origin of (3. 3) is an «unstable equilibrium» for $\varepsilon>0$, and it is a «globally asymptotically stable» for $\varepsilon<0$.

According to a theorem of Liapunov, the stability situation of the origin of (3. 1) is «topologically» the same with the stability situation of the origin of the linearized system (3. 3), and, then, if the point of interest in our investigation is the stability situation of the origin of the NLS (3. 1), the above linearization by the condition (3. 2), is permitted.

But, the NLS (3. 1) has one «limit cycle», while the linearized system (3. 3) has no such solution. In addition, if the origin of (3. 3) is «asymptotically stable», it will be «globally asymptotically stable», while the origin of (3. 1) will be «asymptotically stable» in a finite region, the «region of attraction», of which the determination of the boundary needs the knowledge of the nonlinearity of (3. 1). Therefore, the linearization, from the above point of interest, is not permitted, and the nonlinearity is necessary to be taken into account.

**Linearization and the Stability of Equilibrum Points of Dynamical Systems.**

The results of the previous example (the Van der Pol equation) can be generalized and supplemented by considering the general system of NLDE in its normal form :

$$\dot{x}_i = f_i(t, x_1, \ldots, x_n); \quad i = 1, \ldots, n \tag{a}$$

where the functions $f_i$ have the properties that guarantee the existence and uniqueness of the solution of (a) through any point $x_o$ in the region of the validity of (a) :

$$D: \quad 0 \leqslant t, \quad -\infty < x_1, \ldots, x_n < \infty \tag{a. 1}$$

In a large number of cases, the system (a) can be written in the form :

$$\dot{x} = Ax + X \tag{a. 2}$$

where $x$, $\dot{x}$, $X$ are n-column matrices, A is a $(n \times n)$ - matrix either constant or time-dependent; X does not contain linear terms in the variables $x_1, \ldots, x_n$ then X represents the set of the nonlinearities of (a), and $X(t, 0) \equiv 0$. The system :

$$\dot{x} = Ax \tag{a. 3}$$

is the linear part of (a. 2), that is its «first approximation».

For the applications, especially of engineering kind, the deduction of the NLDE (a. 2) to the LDE (a. 3), that is the linearization of (a. 2), has been universally prevailed in the past. This lasted until some decades ago, when stringent requirements and serious demands of new phenomena, as, e. g., phenomena of vacuum tubes, made clear the inadequacy of the linearization and the importance of the nonlinearities [**1**].

To study properties of the solutions of the NLDE (a. 2), (e. g., their continuity, periodicity, boundedness, stability, oscillation) the linearization of (a. 2) helps only in a few cases, and its acceptance is permitted only «under restrictions of the nature of the matrix A and of the nature and smallness of the nonlinearities X».

The study of the stability of the trivial solution of (a. 2) gives a typical area of research for a verifi cation of the above statements and remarks. Let us first state some results and apply them to appropriate examples in order to show possibilities for a linearization and the influence of the nonlinearities.

The systems (a. 2) and (a. 3) may be either «critical» or «noncritical». In the «noncritical systems» the real parts of the eigenvalues of the matrix A are all nonzero; and in the «critical systems» there are eigenvalues of A with zero real part.

## a. Criteria of the First Approximation.

The following classical «Liapunov criteria of the first approximation» are based on the nature of the eigenvalues of the system.

Criterion a. «If in a noncritical linear system (a. 3) all eigenvalues have negative real part, then this system is «asymptotically stable» at the origin. If, in addition, the nonlinearity X of the system (a. 2) satisfies the condition:

$$\lim_{|x| \to 0} \frac{|X|}{|x|} = 0 \qquad \text{(a. 4)}$$

then, (a. 2) is also «asymptotically stable» at the origin».

Criterion b: «If in a noncritical linear system (a. 3) there are eigenvalues, at least one, with positive real part, then (a. 3) and (a. 2) are «unstable» at the origin, even if the nonlinearity condition (a. 4) holds».

Criterion c: «In a critical system one can not decide about the stability or instability on the system at the origin without taking into account the nonlinearities of the system».

As we can see, the linearization of the system in the first two cases gives results accepted for the nonlinear system, but in the third case, the «undecided case», the linearization is not permitted.

The above criteria are referred to «autonomous systems». In «nonautonomous systems», the linearization in the case of the first criterion necessitates nonlinearity conditions different than (a. 4).

E.g., if the matrix A is either constant or periodic function of time t, and $X = X(t, x)$, the «nonlinearity condition» will be:

$$\lim_{|x| \to 0} \frac{|X(t, x)|}{|x|} \leqslant m\, e^{at} |x|^{b} \qquad (a.5)$$

where $a$, $b$, $m$ are positive constants and $t \geqslant 0$.

In implicit equations, e.g., if $X = X(x, \dot{x})$, the «nonlinearity condition» will be:

$$\lim_{\substack{|x| \to 0 \\ |\dot{x}| \to 0}} \frac{|X(x_1 \dot{x})|}{|x| + |\dot{x}|} = 0 \qquad (a.6)$$

There are «nonlinearity conditions» of integral type.

## b. Criteria by Using a Liapunov Function.

The use of Liapunov functions, which gives the «second Liapunov method» in stability theory, answers the question of the stability situation of the origin in critical systems.

A function $V = V(t, x_1, \ldots, x_n)$, of which the time derivative is calculated by using the NLDE (a):

$$\dot{V} = \frac{\partial v}{\partial t} + \sum_{i=1}^{n} \frac{\partial v}{\partial x_i} \frac{\partial x_i}{\partial t} = \frac{\partial v}{\partial t} + \sum_{i=1}^{n} \frac{\partial v}{\partial x_1} f_i \qquad (a.7)$$

is called a «Liapunov function».

To find a «Liapunov fuction» for a nonlinear system is a very important problem. The following criteria by using Liapunov functions consist of the classical «second group of Liapunov criteria».

Criterion d: «If in a region D, where the origin of the system is included, there is a Liapunov function V such that V is positive and its time derivative $\dot{V}$ is negative, except at the origin where both V and $\dot{V}$ are zero, then the origin (as an equilibrium point of the system) is «asymptotically stable».

Criterion e: «If in a region D, which includes the origin, there exists a Liapunov V such that V is positive in D, except at the origin where it is zero, and $\dot{V}$ is either zero everywhere in D or negative in D with the exception of some points of D (the origin included) where it is zero, then the origin is «stable» in D».

Criterion f: «If V and $\dot{V}$ are both positive in D, but both zero at the origin, then the origin is «unstable» in D».

Of all the above criteria there are modifications and extensions, but these criteria are sufficient to show possibilities for a successful linearization of nonlinear systems, the influence of the nonlinearities for the stability of the origin of the systems, and indicate an approximate calculation of the «stability regions» of the origin.

By the following examples we analyze and accomplish all these [**3**].

Example 4. The Duffing equation:

$$\ddot{x} + \alpha_1 \dot{x} + \alpha_2 x + b x^3 = 0 \tag{4}$$

which in a normal form, can be written as:

$$x = x_1, \quad \dot{x}_1 = x_2, \quad \dot{x}_2 = -\alpha_2 x_1 - \alpha_1 x_2 - b x_1^3 \tag{4.1}$$

or in a matrix form:

$$\begin{pmatrix} \dot{x}_1 \\ \dot{x}_2 \end{pmatrix} = \begin{pmatrix} 0 & 1 \\ -\alpha_2 & -\alpha_1 \end{pmatrix} \begin{pmatrix} x_1 \\ x_2 \end{pmatrix} = \begin{pmatrix} 0 \\ -b x_1^3 \end{pmatrix} \tag{4.2}$$

has three equilibrium points, among which is the origin of which we ask for the stability situation.

The nonlinearity condition (a. 4) is satisfied, for

$$\lim_{|x| \to 0} \frac{|X|}{|x|} = \lim_{|x_1| \to 0} \frac{|b|\,|x_1^3|}{|x_1|} = \lim_{|x_1| \to 0} |b|\, x_1^2 = 0 \tag{4.3}$$

The eigenvalues are given by:

$$\lambda = \frac{1}{2}\left(-\alpha_1 \pm \sqrt{\alpha_1^2 - 4\alpha_2}\right) \tag{4.4}$$

then the origin of (4), that is its state $(x, \dot{x})$ at the origin, is

(a): «asymptotically stable», if $\alpha_1 > 0$, $\alpha_2 > 0$

(b): in the «undecided case», if $\alpha_1 = \alpha_2 = 0$, or $\alpha_1 = 0$, $\alpha_2 > 0$, or $\alpha_1 \neq 0$, $\alpha_2 = 0$

(c): «unstable» in the other cases of $\alpha_1$ and $\alpha_2$.

Example 5. The equation:

$$\ddot{x} + \dot{x} + x - \sqrt{x} = 0, \qquad x > 0 \tag{5}$$

can be written in the form:

$$x = x_1, \qquad \dot{x}_1 = x_2, \qquad \dot{x}_2 = -x_1 - x_2 + \sqrt{x_1} \tag{5.1}$$

or:

$$\begin{pmatrix} \dot{x}_1 \\ \dot{x}_2 \end{pmatrix} = \begin{pmatrix} 0 & 1 \\ -1 & -1 \end{pmatrix} \begin{pmatrix} x_1 \\ x_2 \end{pmatrix} = \begin{pmatrix} 0 \\ \sqrt{x_1} \end{pmatrix} \tag{5.2}$$

The equilibrium points of this system are the origin and ($x = 1$, $x_2 = 0$), and we are interested in the stability of the origin.

The eigenvalues are:

$$\lambda = \frac{1}{2}(-1 \pm \sqrt{3}\, i), \qquad i = \sqrt{-1} \tag{5.3}$$

The nonlinearity condition (a. 4) is not satisfied, for:

$$\lim_{|x| \to 0} \frac{|X|}{|x|} = \lim_{|x_1| \to 0} \frac{\sqrt{x_1}}{|x_1|} = \lim_{|x_1| \to 0} \frac{1}{\sqrt{x_1}} = \infty \tag{5.4}$$

Then, the state ($x = 0$, $\dot{x} = 0$) is «asymptotically stable» for the linear system corresponding to (5), but this state is «unstable» for the system (5) because of its nonlinearity.

Example 6. The linear equation:

$$\ddot{x} + k\dot{x} + (\omega^2 + \varepsilon \cos t)\, x = 0 \tag{6}$$

with k positive constant, and $\varepsilon$ small parameter, is equivalent to:

$$\dot{x} = y, \qquad \dot{y} = -(\omega^2 + \varepsilon \cos t)\, x - ky \tag{6.1}$$

Its eigenvalues are:

$$\lambda = \frac{1}{2}\left(-k \pm \sqrt{k^2 - 4(\omega^2 + \varepsilon \cos t)}\right) \tag{6.2}$$

and the origin of (6) is «asymptotically stable».

If we add to (6) a nonlinearity $X = X(x)$ such that:

$$\lim_{|x| \to 0} (|X| / |x|) = 0$$

then the state $x = 0$, $\dot{x} = 0$ will be «asymptotically stable».

Also, if we add a nonlinearity $X = X(t, x)$ such that:

$$\lim_{|x|\to 0} (|X(t,x)| / |x|) = 0$$

holds uniformly in t for $t \geqslant 0$, then the state $(x, \dot{x})$ is «asymptotically stable» at the origin.

Example 7.

$$\ddot{x} + \alpha x - e^t x^2 = 0 \tag{7}$$

The nonlinearity $X = -e^t x^2$ satisfies the condition (a. 5) for $m = \alpha = b = 1$, for:

$$\lim_{|x|\to 0} \frac{-e^t x^2}{|x|} = \lim_{|x|\to 0} e^t |x| = 0 \tag{7.1}$$

The eigenvalues are $\lambda = \pm\sqrt{-\alpha}$, then we have two cases:

(a): for $\alpha > 0$, $\lambda = \pm\sqrt{\alpha}\, i$, (b): for $\alpha < 0$, $\lambda = \pm\sqrt{\alpha}$

Therefore, although the nonlinearity is favorable for stability of the origin, if $\alpha > 0$ we have the «undecided case», and if $\alpha < 0$ the origin is «unstable».

Example 8. We consider the nonlinear singular equation:

$$\ddot{x} + \left(\frac{1}{1+x}\right)\dot{x} + \left(\frac{1}{1+x}\right)x = 0 \tag{8}$$

For the discussion of this equation by linearization, we need to separate in it the linear and nonlinear parts. We can write (8) as:

$$\ddot{x} + \left(1 - \frac{x}{1+x}\right)\dot{x} + \left(1 - \frac{x}{1+x}\right)x = 0$$

when:

$$\ddot{x} + \dot{x} + x - \left[\frac{x}{1+x}(x+\dot{x})\right] = 0 \tag{8.1}$$

The nonlinearity: $X = -\left[\frac{x}{1+x}(x+\dot{x})\right]$ satisfies the condition needed for stability of the origin, for:

$$\lim_{\substack{|x|\to 0 \\ |\dot{x}|\to 0}} \frac{\frac{x}{1+x}\,(|x|+|\dot{x}|)}{|x|+|\dot{x}|} = \lim_{|x|\to 0} \frac{|x|}{|1+x|} = 0 \tag{8.2}$$

The eigenvalues of the equation (8.1) are :

$$\lambda = \frac{1}{2}\left(-1 \pm \sqrt{3}\, i\right) \tag{8.3}$$

then the state $(x, \dot{x})$ is «asymptotically stable» at the origin.

The previous examples are treated by using the criteria of the «first approximation», that is, by using the nature of the eigenvalues. In the following two examples we use the criteria of the «Liapunov second method», which are related to Liapunov functions.

Example 9.

$$\left.\begin{aligned} \dot{x} &= -y + \alpha x^3 \\ \dot{y} &= \phantom{-}x + \alpha y^3 \end{aligned}\right\} \tag{9}$$

The constant $\alpha$ characterizes the nonlinearity of the system. The origin is an equilibrium point, and the eigenvalues are $\lambda = \pm i$, then the stability of the origin can not be decided by linearization. We use a Liapunov function and apply the «criteria of the second group».

The function :

$$V = x^2 + y^2 \tag{9.1}$$

can be taken as a Liapunov function.

(a) : For the linear system $(\alpha = 0)$,

$$\dot{x} = -\dot{y}, \qquad y = x \tag{9.2}$$

the derivative of V becomes :

$$\dot{V} = 2(x\dot{x} + y\dot{y}) = 2(-xy + xy) = 0 \tag{9.3}$$

Both V and $\dot{V}$ are zero at the origin, V is positive in the x, y - plane, and $\dot{V}$ is everywhere zero, then the origin is «stable» (but not «asymptotically stable») for the linear system (9.2).

(b): The function $\dot{V}$ for the nonlinear system (9) becomes:

$$\dot{V} = 2\{x(-y+\alpha x^3) + y(x+\alpha y^3)\} = 2\alpha(x^4+y^4) \qquad (9.4)$$

Both V and $\dot{V}$ are zero at the origin, $V > 0$; and $\dot{V} > 0$ for $\alpha > 0$, $\dot{V} < 0$ for $\alpha < 0$.

Then, for $a > 0$ the origin is «unstable» for the system (9), and for $\alpha < 0$ it is «asymptotically stable».

This example shows the great influence of the nonlinearity, even if it is very small in magnitude. The stability of the origin of a linear system is completely changed if we add to the linear system a small nonlinearity, and in the nonlinear system which results, the nature of the stability depends not only on the magnitude of the nonlinearity but also on the sign of the nonlinearity.

Example 10. The linearization is unable to help the calculation of the «region of asymptotic stability» of the origin of a nonlinear system. The nonlinearities play a decisive role in this problem.

The Liapunov functions help to find an approximation to the region. The Van der Pol equation can be used as an example.

The Van der Pol equation, written in a normal form, is:

$$\dot{x}_1 = x_2, \qquad \dot{x}_2 = -x_1 + \varepsilon(1-x_1^2)x_2 \qquad (10)$$

and the origin of this system is, for $\varepsilon < 0$, «asymptotically stable». An appropriate Liapunov fuction for this system is:

$$V = \frac{1}{3}(x_1^2 + x_2^2) \qquad (10.1)$$

when, by using (10), its derivative is:

$$\dot{V} = \varepsilon x_2^2(1-x_1^2) \qquad (10.2)$$

The function V is zero at the origin and positive everywhere in the x, y - plane. The function $\dot{V}$ is zero at the origin; and since $\varepsilon < 0$, it is $\dot{V} < 0$ for $1 - x_1^2 > 0$, that is inside the unit circle:

$$x_1^2 + v_2^2 = 1 \qquad (10.3)$$

This circle is a part of the unknown «region of attraction» of the origin of the Van der Pol equation, that is this circle is an approximation of this region.

Example 11. Linearization of Almost Linear Systems.

In the previous examples, we saw different types of approximate linearization methods, which are not explicitely related to the concept of the «error of the approximation». In this example of «linearization of almost linear systems» we have a case of linearization which is characterized by the «error of approximation».

The nonlinear systems of the special form:

$$m\ddot{x} + kx = \varepsilon f(t, x, \dot{x}) \tag{11}$$

where m and k are positive constants, ε a small parameter, and f a nonlinear function, is called «almost linear» or «quasi - linear» system.

It appears in many fields and its treatment is very difficult.

By the «method of Krylor-Bogaliubov» one can replace this system by an «equivalent linear system» [**2**].

The linearization by this method is succeeded by the use of a combination of restrictions and transformations of variables and coefficients, and introducing some integrator operators.

This method can be successfully applied to modern control systems, where certain linear loops are interlinked with nonlinear ones.

The starting point of the method are formulae known from the «first approximation» of the system.

Let us see the essentials of the method.

By a series of mathematical operations during the process of linearization two parameters $\lambda_1$ and $\lambda_2$ are introduced, and, by using them, the equations (11) can be reduced to the linear system:

$$m\ddot{x} + \lambda_1 \dot{x} + \lambda_2 x = 0(\varepsilon^2) \tag{11.1}$$

The solution of (11.1) gives the solution of (11) with accuracy of $\varepsilon^2$, and (11.1) and (11) are «almost equivalent» with error $\varepsilon^2$.

$\lambda_1$ and $\lambda_2$ are given in integral form by:

$$\left.\begin{aligned} \lambda_1 &= \frac{\varepsilon}{\pi\alpha\omega}\int_0^{2\pi} f(\alpha\cos\psi, \ -\alpha\omega\sin\psi)\sin\psi\, d\psi \\ \lambda_2 &= k - \frac{\varepsilon}{\pi\alpha}\int_0^{2\pi} f(\alpha\cos\psi, \ -\alpha\omega\sin\psi)\cos\psi\, d\psi \end{aligned}\right\} \tag{11.2}$$

where $\alpha = \alpha(t)$ and $\psi = \psi(t)$ are the $\alpha$ and $\psi$ of $x = \alpha \cos \psi$, which is the solution of the first approximation of (11), that is when $\varepsilon = 0$ and $\omega = \sqrt{k/m}$ is the linear frequency.

$\lambda_1$ is the equivalent coefficient of damping, and $\lambda_2$ the equivalent coefficient of restoring force. $\lambda_1$ and $\lambda_2$ depend on the amplitude $\alpha$.

The integral representation (11.2) of $\lambda_1$ and $\lambda_2$ is found formally, but one can give a justification of these formulae by using the «principle of energy balance», and the «principle of harmonic balance».

The calculation of $\lambda_1$ and $\lambda_2$ is the main subject of the above linearization «method of Krylov - Bogaliubov».

## ΠΕΡΙΛΗΨΙΣ

Εἰς προηγουμένην ἐργασίαν, ἡ ὁποία περιέχεται εἰς τὸν παρόντα τόμον, ἔχομεν ἐξετάσει «ἀκριβεῖς μεθόδους γραμμικοποιήσεως» μὴ γραμμικῶν μοντέλων τῶν φαινομένων, αἱ ὁποῖαι δίδουν «γενικὰς λύσεις» τῶν μοντέλων.

Εἰς τὴν παροῦσαν ἐργασίαν ἐξετάζομεν «κατὰ προσέγγισιν μεθόδους γραμμικοποιήσεως», ὁπότε λαμβάνομεν προσεγγίσεις «μερικῶν λύσεων» τῶν μοντέλων. Ἡ σπουδὴ τῶν ἰδιοτήτων τῶν λύσεων μὴ γραμμικῶν διαφορικῶν ἐξισώσεων μόνον εἰς ὀλίγας περιπτώσεις ὑποβοηθεῖται μὲ τὴν γραμμικοποίησιν. Εἰς τὰς περισσοτέρας περιπτώσεις ἡ γραμμικοποίησις ὁδηγεῖ εἰς συμπεράσματα ἀσύμφωνα μὲ τὴν πραγματικότητα. Διὰ μίαν γραμμικοποίησιν, ἡ ὁποία ἐνδέχεται νὰ εἶναι ἐπωφελής, εἶναι ἀναγκαῖον νὰ λαμβάνεται ὑπ' ὄψιν ἡ φύσις τοῦ γραμμικοῦ μέρους τῶν μοντέλων, καθὼς καὶ κατάλληλοι περιορισμοὶ τῶν μὴ γραμμικοτήτων τῶν μοντέλων.

Τὸ πρόβλημα τῆς εὐσταθείας καταστάσεων ἰσορροπίας δυναμικῶν συστημάτων, αἱ ὁποῖαι δύνανται νὰ θεωροῦνται ὡς ἡ ἀρχὴ τῶν συστημάτων ἀναφορᾶς τῶν μοντέλων, δίδει ἓν τυπικὸν παράδειγμα, ποὺ ὑποδεικνύει περιπτώσεις δεκτῆς γραμμικοποιήσεως, καθὼς καὶ τὴν ἐπίδρασιν τῶν μὴ γραμμικοτήτων εἰς τὴν ἐξέτασιν τῶν φαινομένων.

Ἀκόμη καὶ εἰς μὴ γραμμικὰ συστήματα, τὰ ὁποῖα εἶναι «σχεδὸν γραμμικά», ἡ γραμμικοποίησις ἐμφανίζει ἀφαντάστους δυσκολίας.

Ἡ ἀλήθεια τῶν ἀνωτέρω παρατηρήσεων δεικνύεται μὲ κατάλληλα παραδείγματα.

REFERENCES

1. S. Lefschetz, «Geometric Differential Equations: Recent Past and Proximate Future», Differential Equations and Dynamical Systems, Academic Press, New York, pp. 1-14, 1967.
2. N. Minorsky, «Nonlinear Oscillations», D. Van Nostrum and Co., New York, pp. 349-355, 1962.
3. R. Struble, «Nonlinear Differential Equations», McGraw-Hill Book Co., New York, Chapter 5, 1962.

★

Ὁ Ἀκαδημαϊκὸς κ. **Ἰω. Ξανθάκης**, παρουσιάζων τὴν ἀνωτέρω ἐργασίαν, εἶπε τὰ ἑξῆς:

Ἔχω τὴν τιμὴν νὰ παρουσιάσω ἐργασίαν τοῦ κ. Δημητρίου Μαγείρου ὑπὸ τὸν τίτλον: «Κατὰ προσέγγισιν Μέθοδοι γραμμικοποιήσεως μὴ γραμμικῶν διαφορικῶν ἐξισώσεων».

Εἰς προγενεστέραν ἀνακοίνωσιν ὁ κ. Μάγειρος ἀναφέρεται εἰς ἀκριβεῖς μεθόδους γραμμικοποιήσεως μὴ γραμμικῶν διαφορικῶν ἐξισώσεων προτύπων τῶν φαινομένων, αἳ ὁποῖαι μᾶς ὁδηγοῦν εἰς γενικὰς λύσεις τῶν ἐν λόγῳ προτύπων.

Εἰς τὴν παροῦσαν ἐργασίαν ἐξετάζονται «Κατὰ προσέγγισιν μέθοδοι αἳ ὁποῖαι μᾶς ὁδηγοῦν εἰς μερικὰς λύσεις τῶν ἐν λόγῳ διαφορικῶν ἐξισώσεων». Ἡ σπουδὴ τῶν ἰδιοτήτων τῶν λύσεων μὴ γραμμικῶν διαφορικῶν ἐξισώσεων εἰς ὀλίγας μόνον περιπτώσεις ὑποβοηθεῖται μὲ τὴν γραμμικοποίησιν. Πράγματι εἰς τὰς περισσοτέρας τῶν περιπτώσεων ἡ γραμμικοποίησις μᾶς ὁδηγεῖ εἰς συμπεράσματα ποὺ δὲν συμφωνοῦν μὲ τὴν πραγματικότητα. Εἴς τινας ὅμως περιπτώσεις αἳ κατὰ προσέγγισιν μέθοδοι γραμμικοποιήσεως μᾶς παρέχουν συμπεράσματα ἀποδεκτά.

Ὁ συγγραφεὺς παρέχει μίαν ποικιλίαν ἐφαρμογῶν, ὅπου αἳ κατὰ προσέγγισιν μέθοδοι γραμμικοποιήσεως δὲν εἶναι ἐπιτρεπταί, καθὼς καὶ περιπτώσεις ὅπου αἳ ἐν λόγῳ μέθοδοι μᾶς ὁδηγοῦν εἰς χρήσιμα συμπεράσματα.

ΜΑΘΗΜΑΤΙΚΑ.— **Characteristic Properties of Linear and Nonlinear Systems,** *by Demetrios G. Magiros* *. Ἀνεκοινώθη ὑπὸ τοῦ Ἀκαδημαϊκοῦ κ. Ἰω. Ξανθάκη.

## INTRODUCTION

In two previous papers (published in: Practica of Athens Academy, June 1976; and as GE Reports: (a) 76SDRO26, 6/28/76; and b) 76SDRO27, 7/1/76, we examined exact and approximate methods of linearization of nonlinear systems.

In the present paper the main characteristic properties of linear and nonlinear systems will be discussed. Some of these properties characterize only the linear systems, and some others only the nonlinear systems. Some properties of NLS disappear by its reduction to a LS, and, therefore, the nature of the problem associated with the NLS, accompanied by the knowledge of the properties of the systems, decide whether the linearization of the NLS is permitted or not.

Nonlinear systems, that is phenomena whose behavior can be described by models which are nonlinear differential equations (NLDE), become increasingly important in many fields, as in astronomy, space flight, automatic control, biology, economics.

Since by linearization of nonlinear systems important features of the phenomena are neglected, it is necessary to know general features, basic striking properties and characteristic peculiarities of linear and nonlinear systems.

We discuss this subject here, and the discussion is illustrated by simple examples.

## I. THE PRINCIPLE OF SUPERPOSITION

This principle, first stated by D. Bernoulli (1775) and used by Fourier (1822) in his theorem, holds in linear systems (LS) and characterizes them, but in general, does not hold in nonlinear systems (NLS).

* Δ. Γ. ΜΑΓΕΙΡΟΥ, **Χαρακτηριστικαὶ Ἰδιότητες Γραμμικῶν καὶ μὴ Γραμμικῶν Συστημάτων.**

Reprinted from the *Proceedings of the Athens Academy of Sciences* **51** (1976), 907–935.

This principle consists of two properties :

(a): The sum of any number of linearly independent particular solutions of a DE is also a solution of the DE ; and

(b): Any constant multiple of a solution is also a solution.

In homogeneous LS the above principle holds and it characterizes completely these systems. By using this principle, one can obtain the general solution of these systems as a linear combination of some easy to get special solutions, the fundamental set of solutions.

In nonhomogeneous LS:

$$\dot{x}_i = Ax_i + h_i(t) \tag{1}$$

where A is a $(n \times n)$ matrix and $h_i(t)$ a «forcing function» or «input», the above principle means that: if $x_1$ is a solution of system (1) with input $h_1$, and $x_2$ another solution with input $h_2$, then $c_1 x_1 + c_2 x_2$ is solution of (1) with input: $c_1 h_1 + c_2 h_2$. The $c_1$ and $c_2$ are arbitrary constants.

• In NLS the principle of superposition does not hold, and, then, its general solution, if it exists, can not be formulated in the simple way as in L. S.

There are NLDE where only the property (b) of the above holds, but not the property (a). For example, for the equation :

$$y''^2 + yy' + y'^2 = 0 \tag{1.1}$$

ot which the terms have the same degree (two) in $y, y', y''$ we see that:

(i) : If $y_1$ and $y_2$ are «linearly independent» solutions, then $y=y_1+y_2$ is not a solution, but $y = c_1 y_1$ and $y = c_2 y_2$ are solutions ;

(ii): If $y_1$ and $y_2$ are «Linearly dependent» solutions, that is $y_2 = cy_1$, then this equation has $y = y_1 + y_2$ as a solution.

## II. THE GLOBAL PROPERTY

The global (or predictability, or provincial) property characterizes the LS, but not the NLS.

• In LS the local behavior of the solutions implies their global behavior. That is, the global behavior can be predicted from the local behavior, then the LS, which may be defined for all values of time, are by nature «provincial».

• In NLS this property does not hold, the global behavior of NLS can not be implied from their local behavior, that is, the «unpredictability» characterizes the NLS. In NLS it may not be possible to extend the solutions beyond a certain time, or these solutions need not be defined for all values of time.

The linearization of a NLS may help to get local properties of NLS. By the following simple example the above are clarified. [4]

Let us take a LS and a NLS:

$$(a):\ \dot{x} = -x,\quad x(0) = x_0;\qquad (b):\ \dot{x} = -x + \varepsilon x^2,\quad x(0) = x_0 \tag{2}$$

where $\varepsilon$ is a parameter. Their solutions are, respectively:

$$(a):\ x(t) = x_0 e^{-t};\qquad (b):\ x(t) = \frac{x_0}{\varepsilon x_0 - (\varepsilon x_0 - 1)e^{t}} \tag{2.1}$$

graphically shown in Figure 1.

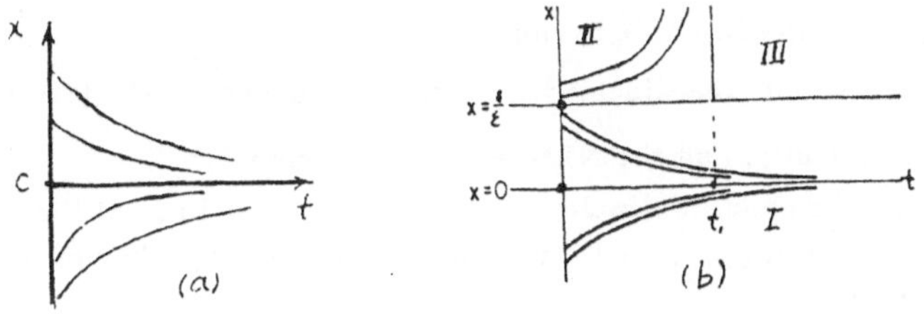


Fig. 1.

The equilibrium point of (2.a) is $x = 0$, and of (2.b) $x = 0$, $x = \frac{1}{\varepsilon}$.

For both systems (2), $x = 0$ is «asymptotically stable», but $x = \frac{1}{\varepsilon}$ is «unstable».

The solution (2.1b) for $x > \frac{1}{\varepsilon}$ becomes infinite for the finite time:

$$t_1 = \log \frac{\varepsilon x_0}{\varepsilon x_0 - 1}. \tag{2.2}$$

In Figure (1.b) we distinguish three regions: I, II, III.

In region I $\left(0 \leqslant t,\ x < \frac{1}{\varepsilon}\right)$ all solutions have the t-axis as an asymptote;

In region II $\left(0 \leqslant t \leqslant t_1,\ x > \frac{1}{\varepsilon}\right)$ the solutions become unbounded for the finite time $t_1$; and

In region III $\left(t_1 \leqslant t,\ x > \frac{1}{\varepsilon}\right)$ there are no solutions.

We see that:

(i): In region I and close to origin, the solutions (2.1.b) have the same topological behavior as in the linear case (2.1.a), which means that the linearization of (2.b) in the region I gives useful information, and the local behavior of the solutions of both system (2) in I implies their global behavior.

(ii): In region II, the nonlinear term $\varepsilon x^2$ causes the existence of a new phenomenon, and the linearization makes this phenomenon disappear.

From the local behavior of the solutions close to $x = \frac{1}{\varepsilon}$ one can not predict the future behavior.

As $\varepsilon \rightarrow 0$, the line $x = \frac{1}{\varepsilon}$ and the regions II, III tend to disappear at infinity, and the NLS tends to become a LS.

By the above example it becomes clear that in NLS the linearization has limitations, and, in general, the future behavior cannot be predicted.

## III. LIMIT CYCLES

Limit cycles may be a phenomenon of NLS, but never a phenomenon of LS. Periodic phenomena of LS or NLS correspond to closed trajectories, called «cycles» with a period or frequency a finite number.

The cycles may constitute a «continuum spectrum», but may be isolated cycles, called «limit cycles», when in a neighborhood of them no other cycles exist.

The limit cycles must not be confused with some nonperiodic closed trajectories, which are members of a special class of solution-curves of systems, the «separatrices».

If, in a LS, a periodic solution y exists, then, due to the principle of superposition, which holds in LS, cy must be also a periodic solution, and since c is an arbitrary constant, no limit cycles exist in LS.

In some NLS, due to the special nature of their nonlinearities, limit cycles may exist, and for each NLDE the problems of existence of one or several limit cycles, their uniqueness and stability, as well as the construction of their boundaries and the calculation of their periods are important and, in general, difficult problems.

Stable limit cycles correspond to important physical phenomena, as, for example, to self-excited oscillations.

E x a m p l e 1. The system: [8]

$$\left.\begin{aligned} \dot{x} &= \quad y + \frac{x}{\sqrt{x^2+y^2}}\{1-(x^2+y^2)\} \\ \dot{y} &= -x + \frac{y}{\sqrt{x^2+y^2}}\{1-(x^2+y^2)\} \end{aligned}\right\} \tag{3}$$

has the unit circumference $x^2+y^2=1$ as a «stable limit cycle». Indeed transforming (3) into polar coordinates and using: $x\dot{x}+y\dot{y}=r\dot{r}$, we have:

$$\dot{r} = 1 - r^2, \quad \dot{\theta} = 1 \tag{3.1}$$

of which the solution is:

$$r = \frac{ce^{2t}-1}{ce^{2t}+1}, \quad \theta = t + c_1 \tag{3.2}$$

where $c = \dfrac{1+r_0}{1-r_0}$ and $c_1$ are the integration constants, $r_0$ the initial condition. If $r_0 < 1$, then, as $t \to \infty$, $r \to 1$ from inside. If $r_0 > 1$, then, as $t \to \infty$, also $r \to 1$ but from outside, when $x^2+y^2=1$ is a «stable limit cycle» of (3).

E x a m p l e 2. The system:

$$\left.\begin{aligned} \dot{x} &= -y + x(x^2+y^2-1) \\ \dot{y} &= \quad x + y(x^2+y^2-1) \end{aligned}\right\} \tag{3.3}$$

has the unit circumference $x^2+y^2=1$ as an «unstable limit cycle».

This system, in polar coordinates, is written in the form:

$$\dot{r} = r(r^2-1), \quad \dot{\theta} = -1 \tag{3.4}$$

with a solution :

$$r = \frac{1}{\sqrt{1 - ce^{2t}}}, \quad \theta = t + c_1 \tag{3.5}$$

where $c = \frac{t_0^2 - 1}{r_0^2}$ and $c_1$ are the integration constants, and $t_0$, $r_0$ the initial conditions. We can check that $x^2 + y^2 = 1$ is an «unstable limit cycle» of (3.3).

Example 3. The NLDE ;

$$\ddot{x} + 3x - 4x^3 + x^5 = 0 \tag{3.6}$$

has infinitely many cycles, and some closed trajectories of special type (separatrices), but no limit cycles. We explain this statement.

The points : $x = 0$, $x = \pm 1$, $x = \pm \sqrt{3}$ on the x-axis are singular points of (3.6); the points : $x = 0$, $x = \pm \sqrt{3}$ are centers, and $x = \pm 1$ saddle points. [7.b]

The phase portrait of (3.6) is shown in Figure 2. There are special solution curves from a saddle point to another one, the «separatrices»

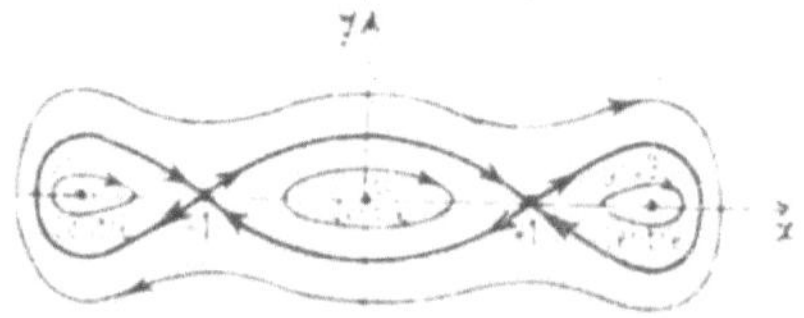

Fig. 2.

of (3.6), which separate the whole x, y-plane into four distinct regions in each of which there is a continuum spectrum of cycles. To go from one saddle point to the other, following a separatrice, theoretically infinite time is needed, and (3.6) has no limit cycles.

## IV. SELF-EXCITED OSCILLATIONS

Self-excited (or self-sustained) oscillations are special periodic phenomena corresponding to stable limit cycles. They do not exist in LS, but may exist in NLS. They can be produced in NLS, where the nonlinearities appear in the damping forces, without the influence of external forces, that is in NLS of the form: $\ddot{x} + \varepsilon\varphi(x, \dot{x}) + kx = 0$.

In particular, the form:

$$\ddot{x} + \varepsilon\varphi(\dot{x}) + kx = 0 \tag{4}$$

is very useful. $\varepsilon$ and $k$ are constants.

To the nonlinearity of this equation we can give another useful form. We differentiate (4) with respect to the independent variable $t$ then put $\dot{x} = y$, when (4) reduces to:

$$\ddot{y} + \varepsilon\psi(y)\dot{y} + \varkappa y = 0 \tag{4.1}$$

where $\psi(y)$ is the derivative of $\varphi(\dot{x})$ with respect to $\dot{x} = y$.

The above NLE can be transformed into another form where $k = 1$, by changing the independent variable $t$ into a new one $\tau$, according to $\tau = kt$, when, e.g., (4) can be written in the form:

$$x'' + \varepsilon\varphi_1(x') + x = 0 \tag{4.2}$$

where $\varphi_1(x') = \frac{1}{k^2}\varphi(kx')$, and the derivatives are taken with respect to $\tau$.

Electrical systems involving vacuum tubes, mechanical systems of action of solid friction, the Froude's pendulum, and other systems, which can be formulated as special cases of the above NLDE, can execute selfexcited oscillations.

Rayleigh (1883) first studied this kind of oscillations in connection with acoustical phenomena, then Van der Pol (1927) in connection with electrical phenomena.

The Rayleigh equation is:

$$\ddot{x} + (-\alpha + b\dot{x}^2)\dot{x} + kx = 0 \tag{4.3}$$

which is of the form (4). The Van der Pol equation is:

$$\ddot{y} - \varepsilon(1 - y^2)\dot{y} + y = 0 \tag{4.4}$$

wich is of the form (4.1).

The Rayleigh equation (4.3) can be reduced to the Van der Pol equation (4.4), by changing the variables $t$ and $x$ in (4.3) into new variables $\tau$ and $y$, respectively, according to formulae:

$$\tau = \sqrt{k}t, \quad y = \sqrt{\frac{3bk}{\alpha}}\dot{x} \tag{4.5}$$

and getting $\varepsilon = \frac{\alpha}{\sqrt{k}}$.

- We give some special examples.

Example 1. We take a special case of the Rayleigh equation (4.3) with $\alpha = 1$, $b = \frac{1}{3}$, $k = 1$, that is the NLDE:

$$\ddot{x} - \dot{x} + x + \frac{1}{3}\dot{x}^3 = 0. \tag{4.6}$$

The linear part of (4.6):

$$\ddot{x} - \dot{x} + x = 0 \tag{4.7}$$

has $\lambda = \frac{1}{2} \pm i\frac{\sqrt{3}}{2}$ as eigenvalues, then its origin is unstable and the solutions around the origin are spirals that wind away from the origin.

If we add to (4.7) the nonlinearity $\frac{1}{3}\dot{x}^3$, we have (4.6) which in the phase plane can be written in the form:

$$\frac{dy}{dx} = \frac{-x + y - \frac{1}{3}y^3}{y}. \tag{4.8}$$

Applying the Liénard's graphical method, Figure (3a), we find a single stable limit cycle as a closed solution, Figure (3b), corresponding to self-excited oscillations.



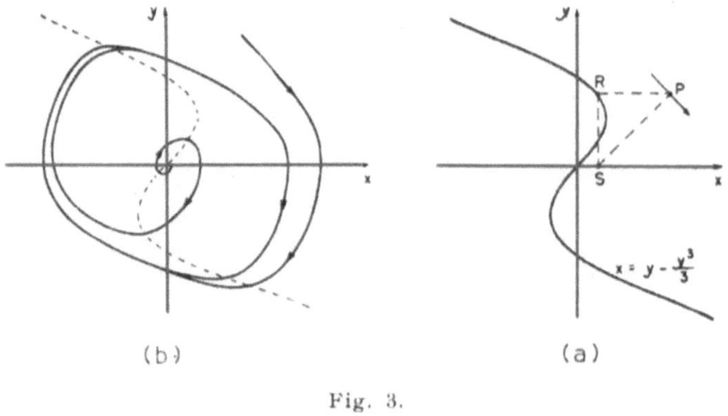


Fig. 3.

Example 2. The Van der Pol equation:

$$\ddot{x} - \varepsilon(1 - x^2)\dot{x} + x = 0 \tag{4.9}$$

is equivalent to the equation:

$$\frac{dy}{dx} = \frac{-x + \varepsilon(1 - x^2)y}{y}. \tag{4.10}$$

For $\varepsilon = 0$, the general solution of (4.10) is the family of concentric circles with center the origin. For $\varepsilon = 1$, application of isocline method shows the limit cycle as in Figure 4, corresponding to self-excited oscillations.

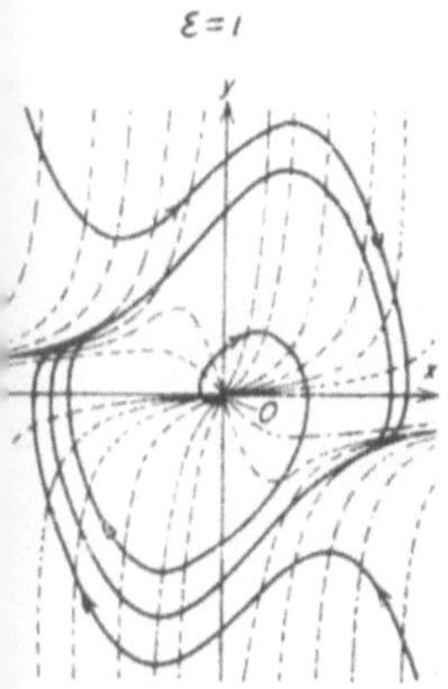


Fig. 4.

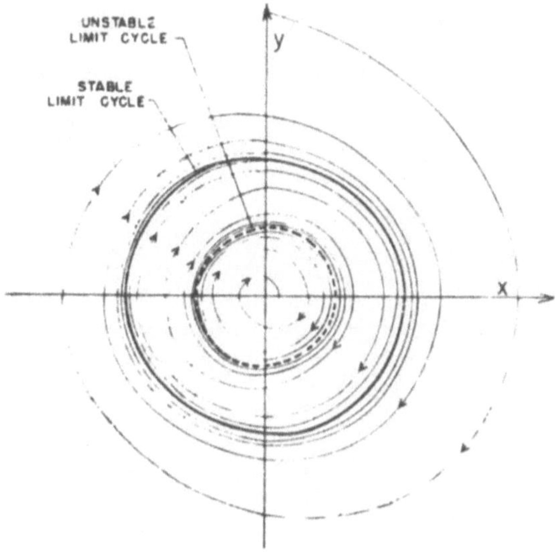


Fig. 5.

Example 3. The equation: [2]

$$\ddot{x} + \varepsilon(1 - \alpha\dot{x}^2 + b\dot{x}^4)\dot{x} + x = 0 \tag{4.11}$$

for $\varepsilon = 1$ is equivalent to:

$$\frac{dy}{dx} = \frac{-x - (1 - \alpha y^2 + by^4)y}{y} \tag{4.12}$$

of which the graphical solution, as shown in Figure 5, has two limit cycles, one unstable and the other stable, corresponding to self-excited oscillations.

Example 4. The equation:

$$\ddot{x} + x = \alpha\dot{x} + bx\dot{x} + c\dot{x}^2 + dx^2 \tag{4.13}$$

can be produced in the theory of a common cathode generator, taking into account the anode reaction, if the value characteristic is represented by a quadratic polynomial. [11]

This equation in case: $\alpha = 0.2$, $b = 1$, $c = -1$, $d = c$ becomes:

$$\ddot{x} = -x + (0.2 + x - \dot{x})\dot{x} \tag{4.14}$$

which, in the phase plane, is equivalent to

$$\frac{dy}{dx} = \frac{-x + (0.2 + x - y)y}{y}. \tag{4.15}$$

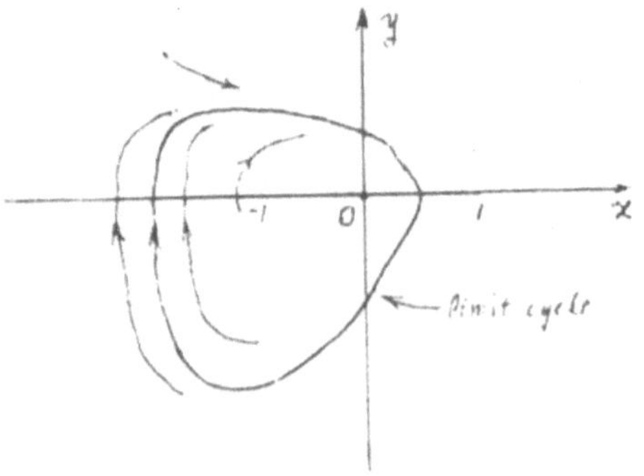


Fig. 6.

Equation (4.15) has one equilibrium point, the origin, which is unstable, and the method of isoclines, applied to (4.15), leads to Figure 6, where we see that a stable limit cycle exists, corresponding to self-excited oscillations. By using a graphical method, we can check that the shape of the limit cycle is distorted with the increase of the parameter $\alpha$, becoming more and more non-sinusoidal.

## V. THE PHENOMENA OF SUBHARMONIC RESPONSE

Subharmonic phenomena occur in some NLS but, in general, not in LS. They appear when the systems are subjected to external periodic forces, say sinusoidal.

If the frequency of the external force is $\omega$, the system, under the influence of such a force, may exhibit periodic motions with frequency $\frac{\omega}{n}$, where $n = 2, 3, \ldots$, and such motions are, by definition, «subharmonic oscillations» or simply «subharmonics» of order $\frac{1}{n}$, $n = 2, 3, \ldots$

The existence of subharmonics can be found theoretically and checked experimentally.

Mechanical, electrical, acoustical, aerodynamical phenomena, and so on, exhibit subharmonic response. We refer some physical examples.

- The «loudspeaker» can be considered as a physical example of subharmonics of order $\frac{1}{2}$. A sinusoidal current in the coil causes the loudspeaker diaphragm to vibrate axially about a central position. These vibrations may, under certain circumstances, be with frequency half of that of the driving current.
- An aerodynamical model of subharmonics could be based on the fact that cartain parts of an airplane can be excited to violent oscillations by an engine running with a frequency much larger than the natural frequency of the oscillating parts.
- An electrical model of subharmonics might be an electrical oscillatory circuit in which the nonlinear oscillation take place because of a saturablecore inductance under the impression of an alternating electromotive force of sinusoidal type.

In connection with subharmonics of a nonlinear forced system, important problems of current interest are: to find conditions for the existence of one or several subharmonics and their appropriate order, to calculate their amplitudes, to discuss their stability and determine their region of stability. These problems lead to restrictions of the coefficients of the system and of its nonlinearities, and to restrictions of the amplitude and frequency of the external forces.

By the following examples we clarify some of the above statements.

Example 1. The Linear Forced System of One Degree of Freedom.

We consider the linear system of one degree of freedom under the influence of an external sinusoidal force in two cases, one with constant coefficients and the other with variable coefficients.

Case A: Linear System With Constant Coefficients. In this case we have:

$$\ddot{x} + 2r\dot{x} + p^2 x = B\cos(\omega t + \varphi) \tag{5}$$

where $r, p, B, \omega, \varphi$ are constants; $B$, $\omega$ and $\varphi$ are the amplitude, the frequency and the phase of the external force.

We investigate the possibility of the existence of subharmonics of this system.

The general solution of (5) is of the form:

$$x = A_1 e^{\lambda_1 t} + A_2 e^{\lambda_2 t} + B_1 \cos(\omega t + \varphi - \delta) \tag{5.1}$$

where $\delta$ is the phase shift, $A_1$ and $A_2$ are arbitrary constants, and the amplitude $B_1$ of the last term is due to the external force.

The eigenvalues $\lambda_1$ and $\lambda_2$ are given by:

$$\lambda_{1,2} = -r \pm \sqrt{r^2 - p^2} = -r \pm iq, \qquad q = \sqrt{p^2 - r^2}. \tag{5.2}$$

The subharmonics should come from the first two terms of (5.1) which must be periodic, then we exclude $q = 0$, and $q =$ imaginary and we accept the case of a real $q$, that is $p^2 > r^2$, then, by calculating $B_1$, the solution (5.1) will have the form:

$$x = e^{-rt}(c_1 \cos qt + c_2 \sin qt) + \frac{B\cos(\omega t + \varphi - \delta)}{\sqrt{(p^2 - \omega^2)^2 + 4r^2\omega^2}} \tag{5.3}$$

with $c_1$, $c_2$ arbitrary constants.

The first part of (5.3) is the «free oscillations», and the second part the «forced oscillations» of the system (5).

The amplitudes of these two oscillations must be bounded, then we exclude for the first term of (5.3) $r < 0$, and for the second term $p^2 = \omega^2$, $r = 0$. Also, we exclude the case $r > 0$, because, in this case, the free oscillations of (5.3) are damped out and only the forced oscillations can be observed. Therefore, it remains the only case, the undamped case, $r = 0$, when (5.3) assumes the form :

$$x = c_1 \cos pt + c_2 \sin pt + \frac{B \cos(\omega t + \varphi - \delta)}{|p^2 - \omega^2|} \tag{5.4}$$

with the restriction $p \neq \omega$. $\delta$ in this formula is either zero (if $\omega < p$), or $2\pi$ (if $\omega > p$), when the «free oscillations» are either «in phase», or «$180^0$ out of phase» with the «forced oscillations». We can select $\omega$ and $p$ such that $\omega = np$, $n =$ integer, when the free oscillations of (5.4) should be «subharmonics of order $\frac{1}{n}$» of the undamped system (5).

But in the actual cases no system is undamped, and we can tell that we have «unstable subharmonics», which are not acceptable in practice. As a result, the system (5) with constant coefficients has no subharmonic oscillations.

Case B. Linear Systems With Variable Coefficients. In this case, subharmonics may exist even in the presence of viscous damping, but they rest on hypotheses which are not always met exactly by reality.

The systems with coefficients varying periodically in time are of great interest, and the external forces may depend not only on time, but on displacement and velocity as well.

We give two physical examples.

B.1: The Problem of Transverse Vibrations of a Rod Under the Action of a Longitudinal Periodic Force. [1]

This problem, after some hypotheses and transformations, leads to the «Mathieu equation»

$$\ddot{x} + \omega^2 (1 - h \cos vt) x = 0 \tag{5.5}$$

where

$$\omega^2 = \frac{g\pi^4 EI}{\gamma A l^4}, \quad h = \frac{Pl^2}{\Pi EI} \ll 1, \tag{5.6}$$

l is the length of the rod, A its cross-section, $\gamma$ its density, EI its rigidity $\omega$ the free frequency of (5. 5), and the external force is $F(t) = P\cos\nu t$. We can calculate a solution of the equation (5. 5) which is a subharmonic of order $\frac{1}{2}$, of the form :

$$x = \alpha\cos\left(\frac{\nu}{2}t + \theta\right) \tag{5. 7}$$

by calculating $\alpha$ and $\theta$ as appropriate functions of time.

The calculation shows that oscillations of the form (5. 7) will be automatically excited, if the parameters $\omega$, h of (5. 5) and the frequency $\nu$ of the external force are restricted according to :

$$2\omega\left(1 - \frac{h}{4}\right) < \nu < 2\omega\left(1 + \frac{h}{4}\right) \tag{5. 8}$$

which represents a zone for the existence of such a subharmonic.

The solution (5. 7) is in its «first approximation».

The solution of (5. 5) in its «second approximation» is :

$$x = \alpha\cos\left(\frac{\nu}{2}t + \theta\right) - \frac{\alpha h\omega}{8\left(\omega + \frac{\nu}{2}\right)}\cos\left(\frac{3}{2}\nu t + \theta\right) \tag{5. 9}$$

under the restriction :

$$2\omega\left(1 - \frac{h}{4} - \frac{h^2}{64}\right) < \nu < 2\omega\left(1 + \frac{h}{4} + \frac{h^2}{64}\right) \tag{5. 10}$$

We remark that $\omega$ and h are taken as parameters of (5. 5), and the subharmonic of (5. 5) may be called «parametric subharmonic».

B. 2. A Conical Loudspeaker Diaphragm. [7a]

The mechanism, shown in Figure 7, of which a simplified version may be a device for a conical loudspeaker diaphragm, can execute a parametric subharmonic oscillation of order $\frac{1}{2}$ of the driving force.

The mass m slides over a frictionless horizontal plane. The links and the spring are massless. The pin-joints at O, A, B are frictionless

OA and AB are long enough for motion parallel to BD to be negligible in comparison with that along the axis of the spring. The driving force $F_{0}= lf_0 \cos 2vt$ is applied to the cross-head B. It may be resolved into

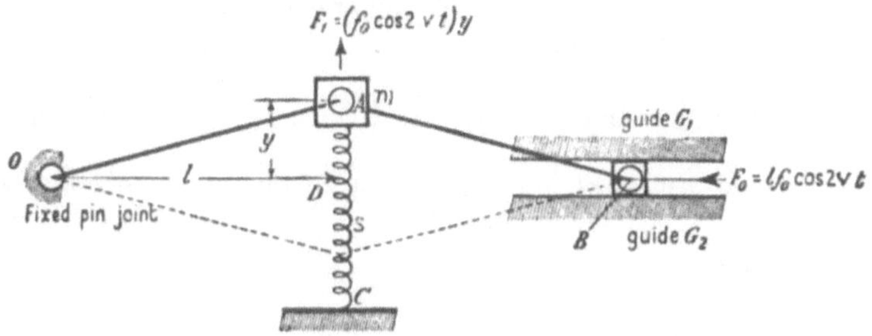


Fig. 7.

two components, one along AB, the other along DA. The latter is nearly $(f_0 \cos 2vt)y$ and it causes m to slide along the line CDA. There are three forces associated with m: one the inertia force $m\ddot{y}$, one the constraint sy due to the spring, and the driving force $(f_0 \cos 2nt)y$. The equation of the motion of m is:

$$m\ddot{y} + sy = (f_0 \cos 2v)\, y \tag{5.11}$$

which, if $\omega^2 = \frac{s}{m}$, $h = \frac{f_0}{s}$, becomes

$$\ddot{y} + \omega(1 - h\cos 2vt)\, y = 0 \tag{5.12}$$

a «Mathieu equation». As we can check, this equation enhibits a parametric subharmonic oscillation of order $\frac{1}{2}$. Figure 7 can be considered as a schematic plan of mechanism illustrating Mathieu equations.

We now refer to some examples of subharmonics of NLS.

Example 2. Existence of Subharmonics of NL Nondissipative Systems.

The motion of a material point along a line is under the influence of a nonlinear restoring force, g(x), and an external time-dependent force, f (t), in the absence of resistance, when the equation of its motion is:

$$\ddot{x} + g(x) = f(t) \tag{5.13}$$

f (t) is periodic, $f(t) = f(t+T)$, and g (x) satisfies a Lipschitz condition. From the theorems of existence of subharmonic of (5. 13) we refer only to the following two: [9]

I: «If f (t) is an even function, $f(t) = f(-t)$, n a natural number, and $\dot{x}(0) = \dot{x}\left(\frac{n}{2}T\right) = 0$, then (5. 13) has a subharmonic of order $\frac{1}{n}$».

II: «If f (t) and g (x) are odd functions, $f(-t) = -f(t)$, $g(-x) = -g(x)$, and $x(0) = x\left(\frac{n}{2}T\right) = 0$, then (5. 13) has a subharmonic of order $\frac{1}{n}$».

Example 3. The NLDE:

$$\ddot{y} + \omega_0^2 y - 2\varepsilon(1 - by^2)\dot{y} = -\frac{4\varepsilon\omega}{\sqrt{b}}\cos 3\omega t \tag{5.14}$$

with NL damping has a subharmonic of order $\frac{1}{3}$.

Indeed, by using the trigonometric identity:

$$4\cos^3\omega t = 3\cos\omega t + \cos 3\omega t$$

we can check that the function

$$y = \frac{2}{\sqrt{b}}\sin\omega t$$

satisfies (5. 14).

Example 4. A generalization of the equation (5. 14) is:

$$\ddot{y} + \omega_0^2 y + \varepsilon f(y)\dot{y} = A\cos(n\omega t + \varphi) \tag{5.15}$$

with $\varepsilon \ll 1$. By applying the «Poincaré method» and restricting A, f(y) and the integer n appropriately, we can find stable subharmonics of this equation. [3]

Example 5. We consider the NLS [6a, b]

$$\ddot{Q} + \bar{k}\dot{Q} + \bar{c}_1 Q + \bar{c}_2 Q^2 + \bar{c}_3 Q^3 = A \sin 2t \tag{5.16}$$

coming from electrical problems. By changing the coefficients according to:

$$\bar{k} = \varepsilon k, \quad 1 - \bar{c}_1 = \varepsilon c_1, \quad \bar{c}_2 = \varepsilon c_2, \quad \bar{c}_3 = \varepsilon c_3 \tag{5.17}$$

the system can be written in the useful form:

$$\ddot{Q} + Q = \varepsilon f(Q, \dot{Q}) + A \sin 2t \tag{5.18}$$

where:

$$f(Q, \dot{Q}) = -k\dot{Q} + c_1 Q - c_2 Q^2 - c_3 Q^3 \tag{5.19}$$

Case A: For $\varepsilon = 0$, the solution of (5.18) is:

$$Q = \bar{x}_1 \sin t + \bar{x}_2 \cos t - \frac{A}{3} \sin 2t \tag{5.20}$$

where $\bar{x}_1$ and $\bar{x}_2$ are arbitrary constants. The first two terms of (5.20) give the «subharmonic component» of order $\frac{1}{2}$ of the solution (5.20) with amplitude: $\bar{r} = \sqrt{\bar{x}_1^2 + \bar{x}_2^2}$.

This subharmonic is, as we know, without practical importance.

Case B: For $\varepsilon \neq 0$, we try to establish a solution of (5.18) of the form (5.20), where, instead of the constants $\bar{x}_1$ and $\bar{x}_2$, we calculate appropriate functions $x_1 = x_1(\varepsilon, t)$ and $x_2 = x_2(\varepsilon, t)$, such that their limits, as $\varepsilon \to 0$, are the constants $\bar{x}_1$ and $\bar{x}_2$, respectively.

The calculation of the amplitude $r = \sqrt{x_1^2 + x_2^2}$ of the subharmonic of order $\frac{1}{2}$ in the nonlinear case leads to appropriate restrictions for the existence of a real and stable amplitude r, and for the existence of two subharmonics of order $\frac{1}{2}$ with two amplitudes $r_1$ and $r_2$, which can be calculated.

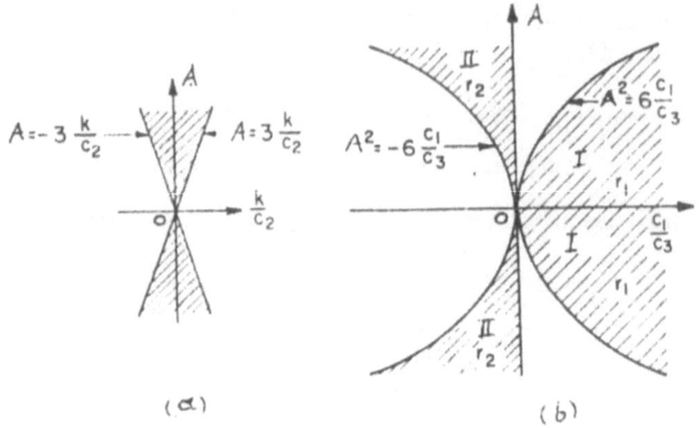


Fig. 8.

In Figure 8(a), the shaded regions in the $\left(\frac{k}{c_2}, A\right)$ -plane corresponds to a real amplitude r of subharmonics.

In Figure 8(b), the shaded regions I and II in the $\left(\frac{c_1}{c_3}, A\right)$ -plane correspond to real amplitudes $r_1$ and $r_2$, respectively, of the subharmonics

E x a m p l e 6. The forced system which is linearly damping and has a nonlinear restoring force expressed by : [5]

$$\ddot{y} + y = \varepsilon\{-k\dot{y} - f(y) + A\sin(\omega t + \varphi)\} \qquad (5.21)$$

where $\varepsilon \ll 1$, $f(y)/y \geqslant 0$, and $\varepsilon$, A, k positive, can be studied by applying the «Cartwright-Littlewood method», when important results in connection with its subharmonics can be found. Some of these results are

(a) : Subharmonics of a given order exist in certain frequency bands

(b) : If, at a given frequency, several states of subharmonics are possible, the amplitude is largest and the frequency band smallest for the lower order of subharmonics.

(c) : For a given form of the non-linearity f(y), the highest order obtainable is a function of the ratio A / k, which also determines the width of the frequency bands.

We remark that the subharmonic oscillations of a NLS belong to a general class of periodic motions characterized by the property that the ratio of the frequencies $\omega_0$ (free frequency) and $\omega$ (frequency of the external force) is a rational number, that is : $\frac{\omega_0}{\omega}=\frac{n_0}{n}$, where $n_0$ and $n$ are mutually prime integers.

We may have the following cases :

(a) If $n_0 = n = 1$, then: $\omega_0 = \omega$: harmonic oscillation of the system

(b) If $n = 1$, then: $\omega_0 = n_0\omega$: harmonic oscillation of order $n_0$

(c) If $n_0 = 1$, then: $\omega_0 = \frac{\omega}{n}$: subharmonic oscillation of order $\frac{1}{n}$

(d) If $1 < n_0 < n$, then: $\omega_0 = n_0\left(\frac{\omega}{n}\right)$: $n_0$ multiple of subharmonic of order $\frac{1}{n}$.

## VI. AMPLITUDE AND FREQUENCY OF PERIODIC SOLUTIONS OF FREE LINEAR AND NON-LINEAR SYSTEMS

The periodic solutions of free (unforced) LS have amplitude independent of the frequency, and the frequency is the same for all trajectories. On the contrary, in the periodic solutions of free NLS the amplitude depends on the frequency, and the frequency changes from trajectory to trajectory. We see examples.

Example 1. The free LS :

$$\ddot{x} + 2r\dot{x} + p^2x = 0 \tag{6}$$

has as general oscillatory solution the function :

$$x = e^{-rt}(c_1 \cos \beta t + c_2 \sin \beta t), \quad \beta = \sqrt{p^2 - r^2} \tag{6.1}$$

where $c_1$ and $c_2$ arbitrary constants.

(a): In the damped case, $r \neq 0$, the factor $e^{-rt}$ of (6.1), which characterizes the amplitude of the oscillatory motion (6.1), either tends to infinity (for $r < 0$), or tends to zero (for $r > 0$) as $r \to \infty$, and the motion is not periodic.

(b): In the undamped case, $r = 0$, (6.1) becomes:

$$x = c_1 \cos pt + c_2 \sin pt \tag{6.2}$$

which is a family of periodic motions with the same frequency p, but amplitude $c = \sqrt{c_1^2 + c_2^2}$ calculated from the initial conditions. These amplitudes are independent of the frequency.

Example 2. The free NLS: [10 a]

$$\ddot{x} + p^2 x + b x^3 = 0 \tag{6.3}$$

can be solved exactly, and the period of the closed trajectories can be calculated. (6.3) is equivalent to:

$$\dot{x} = y, \qquad \dot{y} = -(p^2 x + b x^3)$$

then to:

$$\frac{dy}{dx} = -\frac{p^2 x + b x^3}{y} \tag{6.4}$$

of which the solution is the family of the closed trajectories:

$$y^2 + p^2 x^2 + \frac{b}{2} x^4 = c \tag{6.5}$$

c is the integration constant.

For the period T of the trajectories (6.5), we insert $y = \dot{x}$ and we take into account the symmetry of these trajectories, when:

$$T = 4 \int_0^{a} \frac{dx}{\sqrt{c - \left(p^2 x^2 + \frac{b}{2} x^4\right)}} . \tag{6.6}$$

Transforming according to:

$$x = A \sin \theta \tag{6.7}$$

we can find:

$$T = 4\sqrt{2} \int_0^{\pi/2} \frac{dx}{\sqrt{2p^2 + bA^2 + bA^2 \sin^2 \theta}} . \tag{6.8}$$

As a result, the period T, or the frequency ($T^{-1}$), corresponding to the periodic motion on a trajectory is dependent on the amplitude of this motion, and the period, or the frequency, changes from trajectory to trajectory. In the special case $b=0$ the system (6.3) becomes linear, and its period, which comes from (6.8) for $b=0$, is a constant for all trajectories.

## VII. THE RESONANCE PHENOMENA

The resonance phenomena occur in forced LS and NLS, in case the free frequency of the system is equal or very close to the frequency of the external force.

The damping in a LS and the nonlinearities in a NLS play a very important role in the resonance phenomena.

By the following examples is shown the influence of the damping and of the nonlinearities to the resonance phenomena.

Example 1. The general oscillatory solution of a damped forced LS:

$$\ddot{x} + 2r\dot{x} + p^2 x = \alpha \cos \omega t \tag{7}$$

is the function:

$$x = e^{-rt}(c_1 \cos qt + c_2 \sin qt) + A \cos(\omega t + \varphi) \tag{7.1}$$

where:

$$q = \sqrt{p^2 - r^2}, \qquad A = \frac{\alpha}{\sqrt{(p^2-\omega^2)^2 + 4r^2\omega^2}} \tag{7.2}$$

$c_1$ and $c_2$ in (7.1) are arbitrary constants.

The system (7) is in «resonance» with the external force, if $\omega = p$, and «near resonance» is $\omega - p \ll 1$.

We distinguish two cases:

(a): If (7) is «undamped», $r=0$, then the solution (7.1) becomes:

$$x = c_1 \cos pt + c_2 \sin pt + \frac{\alpha}{|p^2-\omega^2|} \cos(\omega t + \varphi). \tag{7.3}$$

If, in (7.3), $\omega$ is equal or close to $p$, the amplitude of the term of forced oscillations of (7.3) becomes infinite or very large, when the undamped system (7) is in «resonance» or «near resonance» with the external force.

(b): If (7) is positively damped, $r>0$, the free oscillations of (7.1) are oscillatory motions but not periodic, since for $t \to \infty$, are damped out, when in the case of «resonance», $p = \omega$, the solution (7.1) becomes:

$$x = \frac{\alpha}{2\omega r} \cos \omega t. \tag{7.4}$$

This is a periodic motion with finite amplitude, which becomes very large if the damping coefficient r becomes very small.

By the above example is shown that the damping in LS can prevent resonance, and that a weak damping force can be capable of sustaining oscillations of large amplitude.

Example 2. Consider the LS and NLS:

$$\text{(a)}: \ \ddot{x} + x = 0, \qquad \text{(b)}: \ \ddot{x} + x + \frac{1}{6} x^3 = 0 \tag{7.5}$$

of which the general solutions are:

$$\text{(a)}: \ x^2 + y^2 = c^2, \qquad \text{(b)}: \ x^2 + y^2 + \frac{1}{12} x^4 = c^2 \tag{7.6}$$

All solutions of the LS are periodic with the same period.

But the period of the solutions of the NLS changes from trajectory to trajectory, and the period varies with amplitude. A periodic disturbance in the NLS will become out of phase with the free motion, and the forcing function should be an obstacle to increasing amplitude. The period varies with amplitude and non periodic solutions are possible.

As a result, the nonlinearity can prevent resonance, even in the absence of damping.

Due to the nonlinearity, the frequency will be changed, then resonance will be stopped.

The nonlinear terms exert, in general, a stabilizing influence until the motion has passed.

«Resonance phenomena» are in many cases unavoidable. They are dangerous, but sometimes controllable, and, although uncomfortable, they are not in all cases undesirable.

We refer to some physical examples related to resonance.

• If an elastic machine part vibrates in resonance with a sinusoidal force, it may become the source of vibrations with large amplitudes, which, in turn, may produce excessive stresses and lead to possible failure. It is, then, vital to design machine parts, or other engineering structures, in such a way as to avoid resonance with periodic forces.

• Tacoma Narrows bridge offers an example of a big failure in engineering history, due to resonance. This suspension bridge, just after its opening, started to exhibit a marked flexibility and a series of torsicnal oscillations, the amplitude of which steadily increased until the convolutions tore several suspenders loose, and the span of the bridge broke up (November 7, 1940) four months after its building. The wind created aerodynamical forces, which, at the time, were insufficiently understood.

• When a group of soldiers marches in step over a suspension bridge, the feet of the group exert a periodic force on the road bed. If the period of marching is equal to the natural period of the bridge resonance occurs and the sustained bridge oscillations may become dangerous.

• Resonance is sometimes not undesirable. One can in fact utilize it to produce large vibrations by means of small forces. E.g., the vibrations of a string can be sustained by means of an electro-magnet which is activated from a weak alternating current.

## VIII. JUMP, OR HYSTERESIS, PHENOMENA

Jump discontinuities are phenomena in damped forced NLS and not of LS. They are found and explained mathematically and checked experimentally, especially in electrical and mechanical systems.

There are frequency regions where the amplitude of the oscillations jumps discontinuously and, in these regions, the oscillations have a kind of instability.

Let us take, as an example, the system : **[10b]**

$$\ddot{x} + c\dot{x} + p^2x + bx^3 = F\cos(\omega t + \varphi) \tag{8}$$

The investigation of a periodic solution of this equation of the form : $x = A\cos\omega t$ loads to the formula :

$$\left\{(p^2 - \omega^2)A + \frac{3}{4}bA^3\right\}^2 + cA^2\omega^2 = F^2 \tag{8.1}$$

where we consider A versus ω, p and c constants, and F as a parameter.

In the undamped case, $c = 0$, Figure 9 shows the amplitude curves which are curves without closed branches.

In the damped case, $c > 0$, Figure 10 shows these curves having a single branch for each value of F.

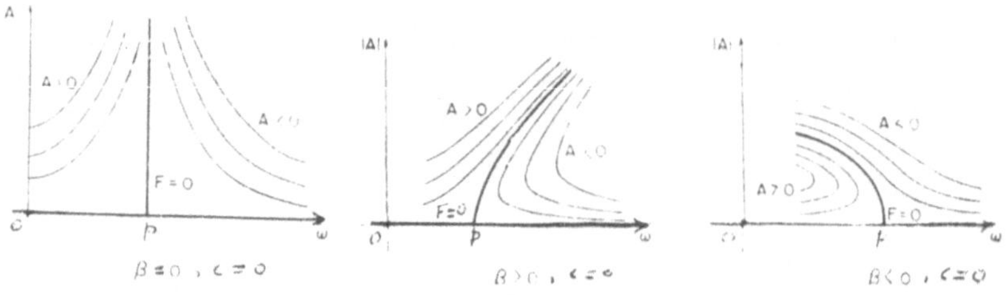


Fig. 9.

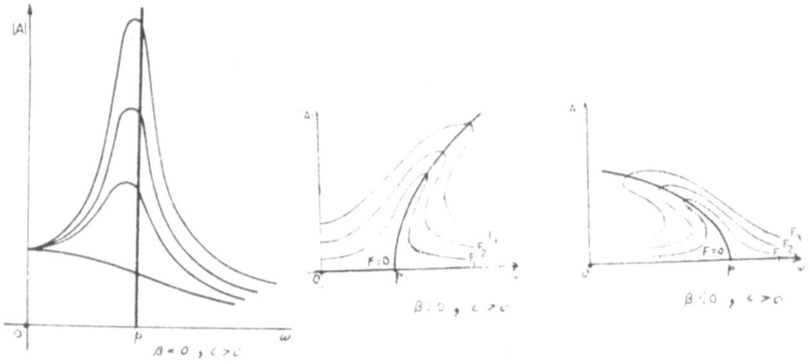


Fig. 10.

Suppose we keep $F = \text{constant}$ and vary $\omega$ from large values to smaller values, starting, say, at point 1, Figure 11 (a). As $\omega$ decreases, A increases slowly until point 3 (tangent point to the curve), when a further decrease of $\omega$ causes a jump up to the amplitude from point 3 to point 5 of the curve, and after that the amplitude decreases with $\omega$.

If we start increasing ω from a value corresponding to point 6, the amplitude follows the portion $6 \to 5 \to 4$ of the curve, when, if 4 is the tangent point to the curve, the amplitude jumps down to point 2, and after that decreases slowly with increasing ω.

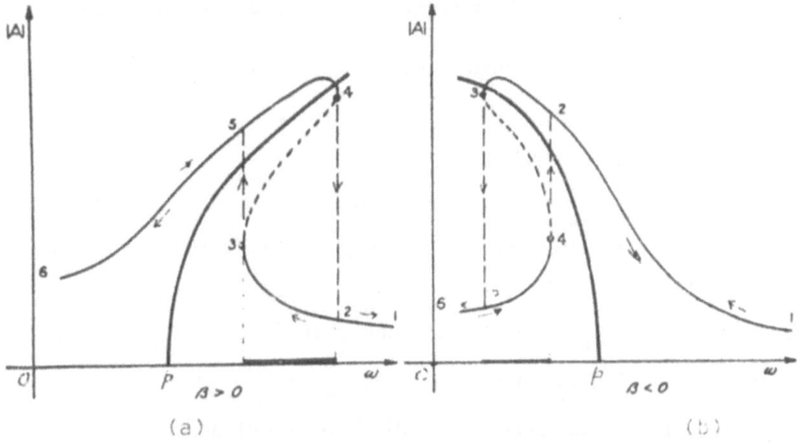


Fig. 11.

The same phenomenon occurs in Figure 11 (b), but in the reverse direction. The above is the jump, a hysteresis, phenomenon, corresponding to an interval of ω, in which the oscillations are unstable.

The portion $3 \to 4$ of the curve is a «dead portion».

## IX. COMBINATION FREQUENCIES

Helmholtz in his acoustical studies, and Poincaré in his studies of NLDE established that, in addition to certain fundamental frequencies $\omega_1$ and $\omega_2$ in a NLS, there exist solutions of the same DE with the frequencies: $\omega = m\omega_1 + n\omega_2$, where m and n are integers.

These are called «combination frequencies», or «combination tones» of the system, and the phenomena of combination frequencies are phenomena of NLS.

Example 1. The oscillations of the LS:

$$\ddot{x} + c\dot{x} + x = F_1 \cos \omega_1 t + F_2 \cos \omega_2 t \equiv H(t) \tag{9}$$

where $\omega_1 \neq \omega_2 \neq 1$, are superposition of a damped free oscillation and a forced oscillation, which, in turn, is a superposition of the fundamental oscillations due to each of the two separate tones of the excitation H (t) individually. That is essentially the oscillations of the LS (9) are simply a superposition of the two harmonies with frequencies $\omega_1$ and $\omega_2$.

Example 2. Consider now the NLS: [10c]

$$\ddot{x} + c\dot{x} + x - bx^3 = H(t). \tag{9.1}$$

We have two cases:

Case a: $\frac{\omega_1}{\omega_2}$ = rational number. In this case the excitation H (t) is a periodic function of time, and (9. 1) has periodic solutions of various kinds.

Case b: $\frac{\omega_1}{\omega_2}$ = irrational number. In this case H (t) is an almost periodic function of time, and since $\frac{\omega_1}{\omega_2} \neq \frac{p}{q}$, where p and q are mutually prime integers, we will have solutions with frequency: $\omega = q\omega_1 - p\omega_2 \neq 0$.

In this case, by using approximate methods for calculation of approximations of the solutions of (9. 1), we find that these approximations of the solutions contain terms with denominators powers of $(\pm m\omega_1 \pm n\omega_2)$, where m and n are integers.

But, by virtue of the «Kronecker theorem» it is known that the expressions $(\pm m\omega_1 \pm n\omega_2)$ are arbitrarily close to zero for infinitely many different integers m and n, when the approximations of the solutions of (9. 1) are divergent, because of the «difficulty of small divisors», which was first pointed out by Poincaré in discussing perturbations methods with problems in Celestial Mechanics.

The «difficulty of small divisors» can be circumvented in some cases if we use viscous damping in the system.

In case the excitation of the NLS has a single frequency, which is an irrational multiple of the free frequency of the system, we have the same situation as above.

## RESULTS AND REMARKS

• The linear and nonlinear systems have striking properties which characterize them. Some properties are properties of LS and some others only of NLS.

• The principle of superposition of the solutions, the global property, the independence of amplitude and frequency of an oscillation, characterize LS.

The existence of limit cycles, the change of frequency from trajectory to trajectory, the jump discontinuities of the amplitude of oscillations, the development of stable self-excited oscillations, the phenomenon of combination tones, characterize NLS.

The resonance phenomena are related to LS as well as to NLS. The damping in LS can prevent resonance, and the nonlinearities in NLS can stop resonance, and by a proper selection of nonlinearities the system can become stabilized.

• The reduction of a NLS to a LS implies that some properties of NLS will be lost, and, therefore, if a problem in connection with the NLS is related to a property which will be lost by linearization, then the linearization as a method to solve the problem is not applicable.

## ΠΕΡΙΛΗΨΙΣ

Τὰ γραμμικὰ (ΓΣ) καὶ μὴ - γραμμικὰ (ΜΓΣ) συστήματα ἔχουν ἰδιότητας αἱ ὁποῖαι ἄλλαι μὲν χαρακτηρίζουν τὰ ΓΣ ἄλλαι τὰ ΜΓΣ.

• Εἰς τὰ ΓΣ ἡ γενικὴ λύσις δύναται νὰ δοθῇ ὡς γραμμικὸς συνδυασμὸς μερικῶν ἁπλῶν λύσεων, καὶ αἱ σταθεραὶ ὁλοκληρώσεως εἰσέρχονται εἰς τὰς γενικὰς λύσεις γραμμικῶς. Εἰς τὰ ΜΓΣ ἢ δὲν ὑπάρχουν γενικαὶ λύσεις, ἤ, ἐὰν ὑπάρχουν, ἔχουν πολύπλοκον μορφήν, ὅπου αἱ σταθεραὶ εἰσέρχονται ἐν γένει μὴ - γραμμικῶς.

• Εἰς τὰ ΓΣ αἱ λύσεις ἔχουν ὁλικὸν χαρακτῆρα, ἐνῷ εἰς τὰ ΜΓΣ ἔχουν τοπικὸν χαρακτῆρα.

• Εἰς τὸ ΓΣ τὸ πλάτος περιοδικῶν κινήσεων εἶναι ἀνεξάρτητον τῆς συχνότητος, καὶ εἰς ἓν συνεχὲς πεδίον περιοδικῶν κινήσεων αἱ τροχιαὶ ἀντιστοιχοῦν εἰς τὴν αὐτὴν συχνότητα. Τοὐναντίον, εἰς τὰ ΜΓΣ τὸ πλάτος ἐξαρτᾶται ἀπὸ τὴν συχνότητα, καὶ ἡ συχνότης ἀλλάζει ἀπὸ τροχιᾶς εἰς τροχιάν.

• Ἡ ὕπαρξις ἀνωμαλιῶν εἰς τὰ πλάτη περιοδικῶν κινήσεων συναρτήσει τῆς συχνότητος, εἶναι φαινόμενον τῶν ΜΓΣ, καὶ τοιοῦτον φαινόμενον δὲν ὑπάρχει εἰς τὰ ΓΣ.

• Ἡ ὕπαρξις μεμονωμένων περιοδικῶν κινήσεων (limit cycles), ὅπως καὶ «εὐσταθῶν ταλαντώσεων αὐτοαναπτυσσομένων» χαρακτηρίζει τὰ ΜΓΣ, καὶ τοιαῦτα φαινόμενα δὲν ὑφίστανται εἰς τὰ ΓΣ.

• Τὸ φαινόμενον τοῦ «συνδυασμοῦ συχνοτήτων» (combination frequencies) εἶναι φαινόμενον τῶν ΜΓΣ καὶ ὄχι τῶν ΓΣ.

• Αἱ δυνάμεις «damping» εἰς τὰ ΓΣ δύνανται νὰ ἐμποδίσουν φαινόμενα «resonance», ἐνῷ εἰς τὰ ΜΓΣ αἱ μὴ-γραμμικότητες δύνανται νὰ χρησιμοποιηθοῦν δι' εὐστάθειαν ἔναντι «resonance».

## REFERENCES


1. N. Bogoliubov and Y. Mitropolsky, «Asymptotic method in the theory of nonlinear oscillations», Hindustan Publishing Corp., India, Delhi-6 (1961), pp. 267-280.
2. F. Clauser, «The behavior of nonlinear systems», J. of the Aeronautical Sciences (May 1956), pp. 411-434.
3. H. Cohen, «On subharmonic synchronization of nearly linear systems», Quart. Appl. Math., Vol. XIII (April 1955), pp. 102-105.
4. J. Hale and J. La Salle, «Differential equations: linearity vs. nonlinearity», SIAM Review, Vol. 5 (July 1963), pp. 249-273.
5. S. Lundqwist, «Subharmonic oscillations in a nonlinear system with positive damping», Quart. Appl. Math., Vol. XIII (Oct. 1955), pp. 305-310.
6. D. Magiros, (a) «Subharmonics of any order of nonlinear systems of one degree of freedom: Application to subharmonics of order 1/3», J. Information and Control, Vol. 1, No. 3 (September 1958), pp. 198-227.
   (b) «On a problem of nonlinear mechanics», J. Information and Control, Vol. 2, No. 3 (September 1959), pp. 297-309.
7. N. McLachlan, «Ordinary nonlinear differential equations», 2nd edition, Oxford (1955), (a) pp. 130-131, (b) p. 196.
8. N. Minorsky, «Nonlinear oscillations», D. Van Nostrand Co., New York (1962), pp. 72-73.
9. V. Pliss, «Nonlocal problems of the theory of oscillations», Acad. Press, New York (1966), pp. 219-227.

10. J. Stoker, «Nonlinear vibrations», Interscience Publishers, Inc., New York (1950), (a) pp. 20-23, (b) pp. 94-96, (c) pp. 112-114.
11. N. Bautin, «On a Certain Differential Equation Having a Limit Cycle», Gorkij Physico-Technical Institute of the University (1937), pp. 229-243.

★

Ὁ Ἀκαδημαϊκὸς κ. **Ἰωάννης Ξανθάκης,** παρουσιάζων τὴν ἀνωτέρω ἀνακοίνωσιν, εἶπε τὰ ἑξῆς :

Εἰς τὴν παροῦσαν ἀνακοίνωσιν ὁ κ. Μάγειρος μελετᾷ τὰς χαρακτηριστικὰς ἰδιότητας τῶν γραμμικῶν καὶ μὴ γραμμικῶν συστημάτων. Ἡ διερεύνησις αὕτη τὸν ὡδήγησεν εἰς τὰ ἑξῆς γενικὰ συμπεράσματα :

1. Εἰς τὰ γραμμικὰ συστήματα ἡ γενικὴ λύσις δύναται νὰ εἶναι ἕνας γραμμικὸς συνδυασμὸς μερικῶν ἁπλῶν λύσεων, ὅπου αἱ σταθεραὶ ὁλοκληρώσεως εἰσέρχονται εἰς τὰς γενικὰς λύσεις γραμμικῶς. Ἀντιθέτως εἰς τὰ μὴ γραμμικὰ συστήματα ἢ δὲν ὑπάρχουν γενικαὶ λύσεις ἢ ἐὰν ὑπάρχουν ἔχουν πολύπλοκον μορφήν, αἱ δὲ σταθεραὶ εἰσέρχονται μὴ γραμμικῶς.

2. Εἰς τὰ γραμμικὰ συστήματα τὸ πλάτος τῶν περιοδικῶν κινήσεων εἶναι ἀνεξάρτητον τῆς συχνότητος, αἱ τροχιαὶ δέ, εἰς ἓν συνεχὲς πεδίον περιοδικῶν κινήσεων, ἀντιστοιχοῦν εἰς τὴν αὐτὴν συχνότητα. Ἀντιθέτως εἰς τὰ μὴ γραμμικὰ συστήματα τὸ πλάτος ἐξαρτᾶται ἀπὸ τὴν συχνότητα, ἡ ὁποία ἀλλάζει ἀπὸ τροχιᾶς εἰς τροχιάν.

3. Ἡ ὕπαρξις ἀνωμαλιῶν εἰς τὰ πλάτη περιοδικῶν κινήσεων συναρτήσει τῆς συχνότητος παρατηρεῖται μόνον εἰς τὰ μὴ γραμμικὰ συστήματα, οὐδόλως δὲ εἰς τὰ γραμμικὰ τοιαῦτα.

4. Ἡ ὕπαρξις μεμονωμένων περιοδικῶν κινήσεων καθὼς καὶ εὐσταθῶν ταλαντώσεων αὐτοαναπτυσσομένων εἶναι χαρακτηριστικὸν τῶν μὴ γραμμικῶν συστημάτων, δὲν παρατηρεῖται δὲ τὸ φαινόμενον τοῦτο εἰς τὰ γραμμικὰ συστήματα.

Τέλος τὸ φαινόμενον τοῦ «Συνδυασμοῦ συχνοτήτων» παρατηρεῖται μόνον εἰς τὰ μὴ γραμμικὰ συστήματα.

# MATHEMATICAL MODELS OF

# PHYSICAL AND SOCIAL SYSTEMS

By: Demetrios G. Magiros*

ABSTRACT

In this paper a "classical procedure" is discussed for building mathematical models of physical, technological and real-life phenomena. The method is accompanied by appropriate remarks and applications.

In Chapter 1 the procedure is discussed for the model building, which is characterized by some steps of the investigation.

In Chapter 2 remarks are given related to concepts involved, to applicability of the method, to reasonability of the problem, etc. By these remarks the investigator faces views for examination and problems for solution.

In Chapter 3 applications are illustrated, by which one can see how the method for model building can be applied.

The applications are distinguished into applications of "complete cycle" and of "incomplete cycle".

The paper is dedicated to Leon Brillouin, the outstanding scientist and human, whose memory will always be a source for inspiration in my work, and for whom my admiration, respect and gratitude are unlimited.

---

*Scientific Consultant
General Electric Co., RSD, Philadelphia, PA.
September, 1980

Reprinted from a *Technical Report, Genl. Electric Co.,* RSD, Philadelphia (Sept. 1980), 1–56.

## TABLE OF CONTENTS

TABLE OF CONTENTS

TABLE OF CONTENTS

-1-

## INTRODUCTION

"Model building" is a significant part of applied mathematics and of great importance in science and technology.

Mathematical methods and reasoning, properly applied to the investigation of a phenomenon, may present theoretical conclusions, which, by an empirical verification, may lead to new insights of the phenomenon and to important contributions to human knowledge.

Research in an empirical field may reach a state of "axiomatization."

This state of the field is of such a perfection that it is possible to formalize in the field a set of propositions, which satisfy certain conditions free of contradictions. In such a situation of the field mathematics is indispensable for the investigation.

For a successful study of a phenomenon, one may follow a set of steps of research. The description of the phenomenon, its understanding and explanation, its prediction and control, are "steps of the modeling process."

Facts and data, hypotheses and axioms, basic laws and theories, tables and diagrams, all these related to the phenomenon, serve as "tools for modeling" the phenomenon.

Many phenomena of different nature, phenomena with a variety of style, purpose, effect, have factors in common, and lead to the same mathematical structure, so the modeling may be used as a "unification method" for the treatment of groups of phenomena.

In modeling, heuristic and non-rigorous reasoning are often employed, and, therefore, the results might be incomplete and of doubt. But, realistic problems often require techniques that cannot, at the moment, be rigorously justi-

-2-

fied. A rigorous reasoning, which is able to penetrate the essentials of the subject matter in question, is always needed, but, for the sake of progress, the mathematical and experimental analysis must proceed, even if, at the moment, the logical structure is incomplete.

In this paper the building of mathematical models of phenomena or systems will be discussed. An exposition of the modeling process shall be given first, and, then, through remarks and applications, attempts will be made to clarify the steps of the process and the concepts entering into the process.

## CHAPTER ONE

## A CLASSICAL PROCEDURE FOR THE CREATION OF MATHEMATICAL MODELS OF SYSTEMS

[Ref. 7(a), 18(a, d), 20]

The building of mathematical models is an important method of studying the behavior of complicated systems and forecasting their future. In developing the procedure to find mathematical models one may have the following steps.

### STEP I: MAKING THE SYSTEM WELL EXPRESSED PHYSICALLY.

The starting point for the investigation of models of systems is the knowledge of the original data of the system, that is the knowledge of correct observations, measurements and experimental facts related to the system, which must be completely stated and known approximately within an accepted error, thus being: "admissible data". It is also necessary to know the "pertinent laws" to the system and all the information needed in order to make the system physically clear, when we call it "well expressed".

- 3 -

Abstractions and generalizations, which violate the essence of reality, must be avoided in this step.

## STEP II: MAKING THE SYSTEM CORRECTLY STATED MATHEMATICALLY

This step is the most decisive for the investigation because it helps to go from the reality to mathematics. It often requires ingenuity, experience, intuitive understanding of the scientific field to which the system belongs, and ability of the investigator to apply his mathematical knowledge.

The selection of the main variables, among all the variables of the system, and the creation of the theory of the system characterize this step.

### 2.1 Major and Minor Variables of the System.

For a system which is clearly stated physically, one must select an appropriate number of appropriate variables of the system suitable for the description of the process of the system. The selected variables are the "major variables", or "generalized coordinates" of the system, also called "state variables", because they are used as components for the description of the "state" of the system.

The other variables of the system, the "minor variables", called "parameters", define the "environment" of the system.

Restrictions on the magnitude of the state variables and of the parameters give the "regions" of the state variables and the parameters, and the system is considered operating in these regions.

### 2.2 Theory and Model of the System.

After the selection of the state variables of the system, one must try to seek out a set of statements in connection with the behavior of these

- 4 -

variables of the system, based on appropriate hypotheses. If the investigator is successful to get relationships between the variables of the system, based on the statements and the hypotheses, he has by all these created a "theory" of the system, and a theory leads, in general, to mathematical expressions called "mathematical model" of the system, which gives a functional behavoir of the system in a quantitative way.

Well-expressed systems, for which one was successful to create a theory and a correspondent model, we call "correctly stated" systems.

Correctly stated models are, in general, differential equations, because, in general, rates of change of the variables enter the models.

These differential equations are associated with appropriate admissible data, and these data are the initial and/or boundary conditions.

STEP III: GETTING A REASONABLE MODEL.

The hypotheses, decisions and choices in developing a theory and the corresponding model of a real system referred in the preceding step, were arbitrary, then, one may build various models for the same real system, and these models do not necessarily give actual solutions, that is, solutions which interpret the reality in an adequate way.

One needs to find a way to distinguish, among the correctly stated models of the system, a model which ought to be the most preferable from the point of view of interpreting the reality, that is a model with actual solution, called "reasonable model", or "properly posed", or "well posed".

Model reasonable in the Sense of Hadamard. A way to check whether a correctly stated model is a "reasonable" one, is to examine whether the solution of the constructed model satisfies the conditions:

- 5 -

(i) the solution of the model exists,

(ii) the solution is unique, and

(iii) the solution depends continuously on the data.

These conditions for the solution of the model are called "Hadamard restrictions", and the model with solution satisfying the Hadamard restrictions is called: "reasonable in the sense of Hadamard".

If the solution of a model fails to satisfy even one of the Hadamard restrictions, it is a formal solution of a model called "non-reasonable", or "non-properly posed", or "non-well posed" in the sense of Hadamard.

In a "reasonable model", the "existence" and "uniqueness" of its solution express the belief in "determinism", a principle according to which one can repeat experiments with the expectation to get consistent results.

For the "reasonability" of the model the third Hadamard restriction is necessary. The data, as results of observations, experiments,and measurements, are given with small errors, when the solution has an uncertainty and its estimation an error. The acceptance of the "continuity property" of the solution with the data implies that "the smaller the error in the data, the smaller the error in the estimation of the solution", then "admissible data" will correspond to "admissible solutions", that is solutions "physically accepted".

We remark that: in the case where a model is expressed by a system of ordinary differential equations in its normal form, it is possible to check the reasonability of the model, without knowing its solution, by considering the form of the functions of the right-hand members of the equations. If these functions and their first partial derivatives with respect to the unknown variables are continuous, the solution satisfies the three Hadamard restrictions, and the model is "reasonable".

- 6 -

## STEP IV: DETERMINING THE ACTUAL SOLUTION.

The determination of the solution of a "reasonable model" is a mathematical step leading in many cases to possibilities for constructive imagination. Sometimes it is possible to solve the model equations explicitely and to get the solution in a closed form in terms of simple functions, when the behavior of the system is given in a general form. But more often it happens not to be able to do it, and in such circumstances one uses approximate methods for solutions. In many cases the wide availability of higher speed computers eliminates tne need for a solution to be in closed form.

Heuristic reasoning toward the ultimate solution in many cases may be used.

The principle of approximation to achieve the solution is in some cases needed, and the continuity property of the solution, that is the third Hadamard restriction, helps to apply this principle.

In some cases, the reasonable model $M$ is reduced to an auxiliary "approximate model" $M_n$ of which the solution $A_n$ can be found more easily, when the required solution is the $\lim_{n \to \infty} A_n = A$.

## STEP V: COMPARING THE RESULTS COMING FROM THE MODEL WITH THE EMPIRICAL RESULTS.

The theoretical results coming from the reasonable model must agree with the empirical results, and this agreement guarantees the establishment of the model.

-7-

SUMMARY

Summarizing the preceding, one can see that the procedure to get a physically accepted mathematical solution of a real system must follow some steps, shown in Figure 1. These steps are:

I. The real system ($S_r$) must become physically complete, that is "well-expressed" ($S_w$), by taking into account all the physical information needed.

II. The well-expressed system ($S_w$) must be done "correctly stated" mathematically ($S_c$), by selection of the appropriate state variables, and creation of the theory and the corresponding model of the system.

III. The most "reasonable model" (M), among the correctly stated models ($S_c$) of the system, must be selected, by imposing the Hadamard restrictions to the solution of the model.

IV. The (actual) solution (A) of the reasonable model (M) must be determined, and

V. The theoretical results coming from the reasonable model must be compared with empirical results, when their agreement establishes reasonable model ($M_e$).

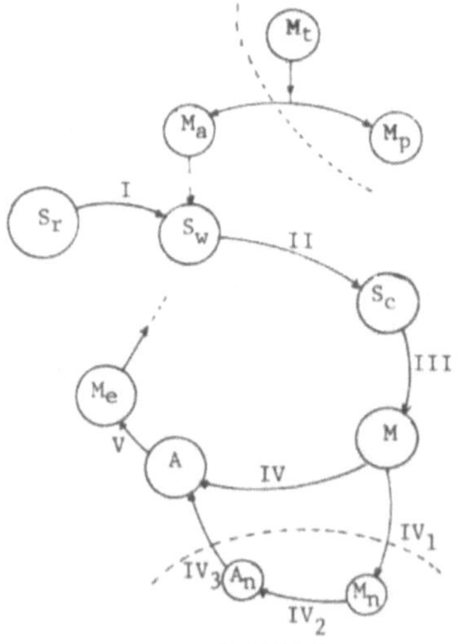


FIGURE 1.

$M_t$ : mathematics

$M_p$ : pure mathematics

$M_a$ : applied mathematics

$S_r$ : real system

$S_w$ : well-expressed system (physically)

$S_c$ : correctly stated model (mathematically)

$M$ : reasonable model

$M_n$ : auxiliary model

$A_n$ : solution of $M_n$

$A$ : solution of $M$

$M_e$ : model empirically checked

- 8 -

# CHAPTER TWO

## REMARKS ON THE CREATION OF MATHEMATICAL MODELS

[Ref. 1(a), 2, 4, 7(b), 8, 12, 13, 15, 16, 18(d), 19, 20, 23, 24, 25]

In the preceding chapter we discussed an organized procedure of a method for building mathematical models of real systems.
But, many points of the method need a clarification and the concepts involved in the method, as well as the results given by it, require a thorough analysis.
By the remarks of this chapter we try to accomplish these necessities.
In this chapter remarks are given on the variables of the system and its data, on controllable systems, on the theory and the corresponding model of the systems and on the hypotheses on which theory and model are based, on deterministic and stochastic systems and models, on the applicability of the method, on the reasonability of problems, etc.
These remarks may put to the investigator views for examination and problems for solution.

### 1. Remarks on the nature of a system and its isolation

The definitions of systems and models, given in the preceding, help the investigator's attempts for modeling. The nature of a system is characterized by the nature of its elements and parts. So, a watch is a mechanical system, a radio an electrical system, a university an education system, a family a social system.

The system under investigation was, in the preceding, silently accepted as isolated, that is, under no external influence.
But this is rarely an actual situation. The coupling of the system under investigation with other systems, which are not of an immediate interest, may be unsuspected and uncontrollable.
Such a situation exists, for example, in biological and economic systems. Biological components may have different performance characteristics in isolation than they do when they are connected within the system of interest. In economic systems, where the data are poor, unforeseen changes: Political, social, technological, psychological, climatic, make the prediction of the economic situation not possible.

## 2. On the steps of modeling.

By the step I one makes the system "well-expressed," that is "physically complete." Data, inputs, assumptions and conjectures of qualitative nature, are constituents of this step. By this step, one obtains, in general, to get the system as a specified order set of interconnected operations performed by its elements, which participate in the process of transformation. Historical realization and the concept of regularization are included in the above situation. The well-expressed system is a formalization of the physical knowledge, and it expresses the so-called "structure" of the system, which will lead to the quantitative analysis of the next steps.

By the steps II and III, one has a mathematical idealization of completely given physical systems, namely, one has "correctly stated" and "reasonable" systems and models. The model describes the special manner of transformation of inputs to outputs, that is, shows the "mode of functioning" of the system, that is its "activity", which involves a manner of transformation of inputs to outputs and the limits of this process.

It is the nature of this activity which defines distinct systems.

The invariance in the structure of a system implies invariant laws of modes of the transformation.

By the step IV, the solution of a "reasonable" model is given and this solution is the time path of the "response" of the system. From this response many conclusions can be obtained, especially a prediction is available.

3. On controlled systems

If a system A, in its native form, does not operate in a desirable form, that is, its outputs x deviate from the demands, it is possible sometimes to modify or "control" the system A by introducing appropriate inputs u, that is, special "control functions", which are either "forcing functions" (time dependent) or "feedback controls" (functions of state variables). We may use a "controller" B, Figure 2, of which the "operating law" produces the appropriate inputs u of the system A. The "driving action" x*, input on the controller B, realizes the purpose of the control, that is, represents the instruction of what must be the product x of the system A. $z$ is the noise.

- 11 -

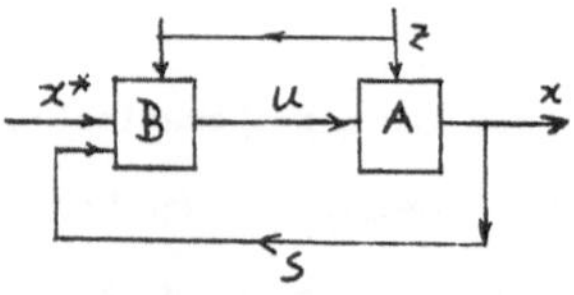


FIGURE 2

If the controller B does not get information about the actual state x of the controlled system A, the system A is an "open-loop" system; and if there exists a "feedback path" S by which the controller B receives the information x of A, the system is a "closed-loop" or "feedback" system.

The study of control systems, although of engineering discipline, is a mathematically oriented science. Probability theory, calculus of variations, dynamic programming, maximum principle and other mathematical branches, give mathematical methods applicable to control problems and especially to problems of "optimization of controls", the most interesting problems.

By "optimization of controls" we mean a mathematical description of control systems in such a way that the application of rigorous procedures of this representation results in the best performance of the systems within the framework of the relevant limitations and variables.

The solution of optimal problems permits a maximum effectiveness in manufacturing, it increases the productivity, it provides the quantity of production, the economy of electric energy, and so on.

The "completely controllable" and "completely observable" controlled systems are the most important systems.

In a "completely controllable" system it is possible to move any state of the system to any other state during any finite time interval by a suitably chosen input;and in a "completely observable" system, the knowledge of the input and output on a finite time interval is sufficient to determine uniquely the state of the system.

4. On the hypotheses

The hypotheses play a basic role in the investigation and they are an important weapon in the investigator's hands.

But, the useful hypotheses must be always under empirical control, that is, an agreement is needed between the results coming from the hypotheses and the observed facts.

It is an important method of research to form hypotheses and then to compare the results with the reality.

Hypotheses on a subject having empirical support are of active interest in the subject. In many cases they lead to discoveries, to theories, to laws.

Hypotheses which are not under experimental control cannot be accepted for scientific purposes.

It is by experiments that one hopes to get confirmation of hypotheses and new ideas for better theories,and such interplay constitutes a most desirable development.

- 13 -

Hypotheses are a mental technique, a product of imagination and they are productive, if it is possible to be a basis for discovery of the truth and lead to observations and experiments.

Fruitful hypotheses lead, in general, to a generalization, and a generalization leads to a mathematical model able for prediction.

5. On the variables of a system

The major variables of a system are significant varieties of the system and omission of an important major variable from their selection will result a wrong notation of the relationships, then a wrong model of the system.

In systems of physical sciences, the selection of the major variables is not a difficult subject, for the quantities in these sciences are expressible in terms of well-defined and standard units which are time-invariant.

But, in systems of, say, biology, economics, sociology, one meets a difficult situation for the selection of the major variables, since the variables in these fields are, in general, time-dependent, when the selection of major variables is an elusive subject. For the above case, one may have an indication for the selection of the major variables if one has information about the dependence of the increments of the variables with time.

If a system, after a small disturbance, can get its original situation, this will be in a finite time called a "relaxation time" of the system. A system with small (considerable) relaxation time is called "fast" (slow) system.

The relaxation time of the variables of a system is not the same for each variable.

Variables with fast (slow) increment can be taken as major variables (parameters).

Two selections of the major variables for the same phenomenon will lead to two different models with different relaxation time,when the selection of the major variables is more successful in the faster system. In other words, a successful selection of the major variables makes the system as faster as possible.

The specification of the number of major variables is also very important. The smaller the number of the major variables, the simpler the model of the system. If the number of the major variables is inappropriately small, the corresponding model will be oversimplified and, as such, the model loses the essentials of the system and it is unrealistic. To know where and when to simplify a complicated system, this is a very important subject for an investigator.

6. On the data, and deterministic and stochastic systems

The original empirical facts of a system, that is the data, were considered in the preceding as completely given and accurately observed

and measured in order to state the problem at the beginning and to give a preliminary study of the system.
By such data and same hypotheses on the parameters of the system, the procedure leads to models with the possibility of exact prediction and the models are termed "deterministic".

But in the actual scientific work, the above situation is unrealistic. In scientific fields, as e.g., in biology, economics, social sciences, unusual uncertainties occur and variabilities involve, when the deterministic concepts and techniques are, in many cases, inappropriate.
There appear inevitable errors and difficulties. Errors in the experiments and measurements; errors and difficulties in getting good quantifiable data, or non-quantifiable ones, whenever needed; errors in the determination of the parameters and difficulties in fitting them to the data. Also, in some cases, the actual values of the variables emerge from the chaos of predestination, when the variables are considered as "random variables".
Due to the above, the "determinism" must be replaced by the "probabilistic causality", and the investigation will, in general, be concerned with probabilistic theories, stochastic processes, statistical methods, methods of statistical mechanics, decision analysis.

7. On the theories of a system

The theory of a system is a set of statements and a collection of propositions about the behavior of the system, based on the data and some hypotheses. Some of the statements are verified by experimental observations, but some others are postulated or based on hypotheses.

The theory is a human creation much of which is a work of imagination, and nothing human is permanent. As a product of human brain, the theory is open to discussion and possible revision or modification, and sometimes rejection, if new empirical facts are contradictory to the theory. Experiments and theory must be intimately related and none can make significant progress without the other.

A theory on a system may be considered acceptable, if

(i) the theory is able to produce rules for modeling;

(ii) the region of the applicability of the theory, which is useful for predictions, is known;

(iii) the limits of the errors in this region are known.

The bigger the region of applicability of a theory, the more opportunities for practical uses of the theory.

8. <u>On the models of the systems</u>

The construction of a model deals with finding, in mathematical terms, the interrelationships between the variables or parts of the system and its inputs and outputs, and this construction presupposes a theory of the system to which the model corresponds.

Without a conjunction of systematic quantitative data and a general framework of the theoretical orientation and theoretical ideas about description and processes, the development of models should be impossible. The model usually explains the anatomy of the system.

The procedure of modeling is sometimes to start building a simple and descriptive model, then penetrating more and more into the causal relationship, and, then, modifying the starting model according to new hypotheses and new empirical facts. One selects such a construction of the model that will facilitate deductive investigation of the properties of interest. One builds models for specific purposes, and one may have for the same system different models for different purposes.
Due to the complexity of the real systems, the building of their models must be necessarily restricted only to important features of the system and some specific details will have to be ignored for a manageable description.

Among the different types of models of systems, one can distinguish the "time-delay" systems, and the "deterministic" and "stochastic" systems.

Systems of which the process at a time depends not only on the state at that time but also on the past history of the systems, that is the process is characterized by "after-effects", are called "time-delay" or "time-lag" systems, when we will have "time-delay" or "time-lag" models. For example, in problems of long-range planning in economics one has "after-effects" situations.
This kind of systems is expressed by "differential equations with deviating arguments", of which important classes are the "difference-differential equations" and the "retarded differential equations" with "delay periods" or "retardations".

In deterministic systems the effect of any change in the system can be predicted with certainty.

In practice, especially in social sciences, this is not the case, and, either because the system is not fully specified, or because of the unpredictable character of much human behavior, there are usually elements of uncertainty in any prediction. These uncertainties can be accommodated if we introduce probability distributions into the model in place of the variables, or, more precisely, if the equations of the model include "random variables". Such a model is called "stochastic".
"Stochastic processes" occur widely in nature, say in economics, in psychology, in social sciences, and they are developed in time according to probabilistic laws.
By the stochastic processes one cannot predict the future behavior of the system with constancy, but only attach probabilities to various possible future states.
By probabilistic methods one can specify the state of an individual only in terms of its probabilities of choices, and go no further, thus probabilistic methods form no basis for prediction in cases where two responses contradict.
Any model describing human behavior must be formulated in stochastic terms.
In case one needs the solution of the models, it may be advisable to use deterministic approximations.
The stochastic models give a way of testing the fit of the model to data.
The deterministic models are very simple compared to stochastic ones, and the stochastic equations are usually cumbersome.

- 19 -

There are relationships between deterministic and probabilistic points of view.

The deterministic form is the same as the mean probability of distribution. We can pass on to more precise stochastic formulation by giving special attention to special features suggested by deterministic models.

The models are mixed, deterministic and stochastic, that is, sometimes we have stochastic behavior out of deterministic models, and sometimes the models are partly stochastic.

If in a deterministic model, one admits random variables, the model becomes a stochastic one, when it raises problems of identifiability, that is we may write down a model when certain constants cannot be estimated, however many observations we have, unless we introduce extraneous information.

Deterministic treatments may be regarded as approximately valid in certain circumstances and worth examining in any given situation. The deterministic models are preferred in case they give the expected values of the distribution.

In chemical kinetics there exists a set of mathematical theories or models to account for reaction rates. Problems in physical chemistry deal with models, and, since physical chemists, dealing with aggregates of billions of molecules, do not need to make their model a stochastic one, they will treat the phenomena as deterministic.

Finally, we remark that the model is not the reality but it is an instrument, a method, to investigate the reality.

A model is for the reality what a geographical map is for the part of the surface of the earth that it maps. Both, models and maps, give methods for the investigation of the reality, in order to satisfy the eagerness and attempts of the investigator.

9. On the applicability of the classical method

The classical method for finding models of systems, wherever this method is applicable, gives good results.

But, there are scientific fields where the method either cannot be applicable, or it is difficult to be applied.

If the exact laws, which govern the system under investigation, are not completely known, then a theoretical description of the system is not possible and the classical method is inapplicable. This occurs e.g., in problems of atomic and molecular structure, where the interactions are known, but where the structure may be complicated.

In problems of classical dynamics of compressible fluids, the differential equations, even if supplemented by boundary conditions, are not always a complete framework for an adequate description of the physical reality.

10. On the reasonability of models

The classical method for finding "reasonable" models and "actual" solutions of real systems, shows a general orientation of thought on the problem and it is not always literally binding. It is based on the "Hadamard restrictions" which are necessary and sufficient for deterministic models.

The concept of reasonability of a problem plays an important role in the solution of a system. This concept was introduced by Hadamard at the beginning of this century and a variety of very interesting problems, reasonable in the sense of Hadamard, can be successfully treated.
But, it was found that the Hadamard concept of reasonability rules out, as unreasonable problems, important real problems.
Various developments in physics point up the need for restrictions of the validity of Hadamard definition of reasonability, and this indicates an important consequence for natural philosophy as well as for practical applicability.
There are available today different notions of reasonability, and various approaches to the formulation of an investigation of problems which are unreasonable in the sense of Hadamard.
The Russians Tychonov and Lavrentiev have a success in treating such problems. Probability methods are used in their study.

## 11. On the Concept of Probability

In case probabilistic models are used, the role of probability definition is basic for the investigation.

To a real stochastic event $\alpha$ , one can assign a probability $p(\alpha)$, for the concept of which one may have four essentially definitions

(a) The "classical probability definition", that is the ratio of a favorable to total number of alternations,

(b) The "probability definition as a measure of belief", which is of the form of inductive reasoning,

(c) The "axiomatic probability definition", for which measure theory is needed, and which, according to Kolmogoroff, is a superior definition, and

(d) The "relative frequency probability definition", which is very popular and most used among physicists and engineers.

If an experiment is repeated $n$ times, and the event $\alpha$ is occurring $n_\alpha$ times, the relative frequency of occurrence is $\frac{n_\alpha}{n}$ and it is directly observable.

To make statements about relative frequency, one uses probability and defines the probability of the event $\alpha$ as the limit of the relative frequency as $n \to \infty$, $p(\alpha) = \lim_{n \to \infty} \frac{n_\alpha}{n}$.

Such a limit is not an experimental quantity.

The existence of this limit presupposes the possibility of repeating experiments infinitely many times, which is unrealistic.

Therefore, the relative frequency definition of probability presupposes that the "principle of approximation", a basic principle of applied mathematics, and the "law of large numbers", a law of probability theory, are workable.

# CHAPTER THREE

## APPLICATIONS

The creation of models is, in general, very difficult work, and the establishment of a model needs in many cases investigation of many years. The applications of the classical method, which was developed in Chapter I, can be classified into two classes: Applications of "complete cycle" and of "incomplete cycle".

For applications of "complete cycle" it is necessary:

(i) to confront the physical problem, and construct its mathematical model,

(ii) to compare the results coming from the model with experimental results, and these results must agree.

We can consider as belonging to this class models of great historical interest, developed after years of effort by brilliant men.

In applications of "incomplete cycle" either the model is not yet completely established, or the interests of the investigator are only on some steps of the procedure.

### A. APPLICATIONS OF COMPLETE CYCLE. (Ref. 3, 27)

We give some applications of models of complete cycle of deterministic type.

#### 1. The Bernoulli Model of a Vibrating Beam.

Daniel Bernoulli (1700-1782), working with the problem of a vibrating beam:

(i) derived the model of the vibrating beam in the form of differential equations,

(ii) solved the resulting equation in the case of vibrating cantilever beam,

(iii) calculated the frequencies,

(iv) sketched out the shapes of the model,

(v) made experiments with vibrating splines,

(vi) traced their shapes, and finally.

(vii) compared the experimental results with the results predicted by the model.

Bernoulli did all these approximately two centuries ago, and, until today, the theoretical results, coming from his model, are in a complete agreement with the experimental results, and the Bernoulli model is a perfect example of complete cycle model. We notice that the work of Bernoulli plays an important role in the "theory of music" and the "theory of musical instruments".

We now mention some models of complete cycle, which are of high originality, created by talented persons.

2. The Newton Mathematical Model of the Laws of Gravitation, which gives a deep explanation of the Kepler empirical laws of the motion of planets, and it is the basis for astronomical and astronautical investigations. During the seventeenth and eighteenth centuries the Newton model, according to the observations and measurements of this time, has been established as the only exact and true model. But, during the nineteenth century, new perfections and modifications in the technique of observations and measurements made possible the creation of the Einstein theory of relativity, which completes the Newton model.

3. The Einstein Mathematical Model of the Relativity Theory, which completes our knowledge on time and space, and it is the basis for so many contemporary investigations.

- 25 -

4. The Maxwell mathematical formulation of electromagnetic theory, which is the basis for the study of electromagnetic phenomena, and explains the basic principles of electric power transmission and radio technology. Mathematical reasoning led Maxwell (1801-1865) to produce the existence of electro-magnetic waves including the many familiar types since discovered radio, radar, cosmic rays, etc.

5. The Yucawa mathematical theory of nuclear forces (Nobel Prize, 1949), by which Yucawa predicted the existence of mesons, and which later was confirmed by experiments.

6. The Frish and Tinbergen mathematical model in econometrics (Nobel Prize 1969), by which these economists described and predicted economical phenomena.

7. The mathematical development of electronic computer, in which Von Neuman was a deep investigator.
The electronic computer changed the functioning of today's society and it will influence significantly future life.
Many contemporary problems of great importance are expected to be solved by the creation of special models by using high-speed electronic computers.
But, for this scientists are needed with the ability to apply mathematical thinking and methods to the computer science, as exactly this was the case in physics, where scientists like Newton, Laplace, Maxwell, Einstein, a.o, by using mathematics in the physical reality, created today's progress in physics.
The weather prediction for some years is a very important problem for the solution of which the use of high-speed computers will play, as it is believed, a decisive role.

The laws of atmospheric phenomena can be expressed, according especially to Bjerkness, by mathematical models which are partial differential equations. Based on meteorological data and using these models, if they become reasonable, the prediction of tomorrow's weather for some years will become, as it is believed, a reality, by using computer machines in the hands of qualified computer men.

8. The mathematical work for the discovery of Neptune by Adams and Leverrier. Adams in England and Leverrier in France, working, independently to each other, on the problem of the "irregularities in the motion of the planet Uranus," discovered (1846) Neptune by an appropriate mathematical model.

We remark that each of the above models, which are considered as models of complete cycle, is associated with a history, by which can be shown the power of mathematics as a research method, and let us discuss the history of one of them, say the discovery of Neptune.

The discovery in 1846 of the planet Neptune was a very impressive and spectacular achievement of mathematics applied to astronomy. The existence of this new member of the solar system with specified properties and specified location in the heavens was demonstrated with pencil and paper, by finding the appropriate mathematical model to the problem. There was left to observers only the routine task of pointing their telescopes at the spot the mathematicians marked.

As early as about 1820 the astronomers noticed irregularities in the motion of the planet Uranus, that is, deviations of its observed orbit from its calculated positions.

- 27 -

Adams in England, and Leverrier in France, working independently to each other, solved the above problem by the discovery of the planet Neptune, which was the reason for the irregularities in the motion of the planet Uranus.
The Adams memoire on the above problem was a masterpiece of remarkable mathematical maturity. But Airy, the Royal astronomer in England, didn't give too much attention to Adams' work, or the observers in England didn't have appropriate astronomical maps, or didn't find favorable conditions to observe the new star at the place and in time indicated by Adams.
Leverrier, several months later of the Adams' success, finished his calculations and asked the astronomers Galle and D'Arrest in Berlin, who had good astronomical maps, to see the new star at a place and in time indicated by his calculations. On September 23, 1846, these German astronomers saw first the new star.

- 28 -

## B. APPLICATIONS OF INCOMPLETE CYCLE

In each of the following applications our interest is to emphasize appropriately some steps of the classical method for modeling and to use remarks and hypotheses for modifications of models.

### 1. APPLICATION 1 (MECHANICS). THE SIMPLE PENDULUM.

We discuss the solution of the problem of the motion of a simple pendulum in the way to see how, by changing the hypotheses we modify the model in order that the modified models satisfy practical needs.
We follow the Newton theory of "motion and force", and not the Lagrange theory of "equilibrium and work". These theories, although give different models, have the same results.

A mass $m$ suspended from a fixed point $O$ by means of a thread of length $\ell$ is a simple pendulum. The mass is displaced from its equilibrium position with an angle $\vartheta_0$ with respect to the vertical, and then is free to move. We examine the motion of such a simple system by imposing the concrete problem: "how the motion of the pendulum depends on $m$, $\ell$, $\vartheta_0$."
To solve this problem, we accept special conditions and assumptions, and we try, based on them, to make the system "well-expressed" and "correctly stated".
We assume

(i) The mass $m$ is a material point,

(ii) The mass $m$ is free to move, without initial velocity, when it is in the place with angle $\vartheta_0$,

(iii) The thread is absolutely solid with constant length and negligible weight, and fixed in its end $O$,

(iv) The resistance of the environment of the pendulum, as well as the friction at $O$, are negligible,

- 29 -

(v) The motion of $m$ is only under the influence of gravity forces,

(vi) The motion of $m$ is on a vertical plane through $O$ .

Under the above assumptions, the motion of $m$ is on an arc of the circumference with center $O$ and radius $\ell$ , and the place of $m$ , at any time $t$ , is known, if the corresponding angle $\vartheta$ is a known function of time $t$ , that is $\vartheta = \vartheta(t)$.

The data are the initial conditions: $\vartheta(0) = \vartheta_0$ , $\dot{\vartheta}(0) = 0$.

So, for the problem, the time $t$ is selected as "independent variable", the angle $\vartheta$ as the "main variable", and as "parameters" the length $\ell$ , the mass $m$ , and the acceleration of gravity $g$ .

The problem becomes a "static problem" by including "inertial forces" and using the appropriate "Newton law".

Under all the above the problem is "well expressed" and "correctly stated", and, as we know, the corresponding mathematical model, consistent with the above hypotheses, is

$$\ddot{\vartheta} + (g/\ell)\sin\vartheta = 0, \quad \vartheta(0) = \vartheta_0, \quad \dot{\vartheta}(0) = 0. \tag{1.1}$$

The model (1.1) is based on the above hypotheses, but not all these hypotheses are accepted in practice, then the results coming from this model can not be satisfied by experiments and measurements. The model (1.1) needs appropriate modifications, in order to interpret the reality adequately. The guide for the modifications is the special interests of the investigator and his physical intuition, the appropriate change of the hypotheses, and the mathematical needs. We see below some modifications.

(a) We suppose that the resistance of the environment of the pendulum is considerable. In case this resistance is considered as a force analogous to the velocity of the moving mass, the term $\left(\frac{K}{m}\dot{\vartheta}\right)$ must be added to the equation (1.1) and the modified model is:

- 30 -

$$\ddot{\vartheta} + \frac{K}{m}\dot{\vartheta} + \frac{g}{\ell}\sin\vartheta = 0 \qquad (1.2)$$

$K$ characterizes the magnitude of the resistance of the environment.

(b) By supposing that the end $O$ of the thread is moving during the motion of $m$, we will have a new model coming from (1.1). In case $O$ moves horizontally and periodically, the modified model becomes an equation of Mathieu type.

(c) Another modification of (1.1) comes, if the length $\ell$ of the thread is changed during the motion of $m$. In case $\ell$ is taken analogous to the time and the velocity, and if we have small oscillation angles of the pendulum, then, under appropriate transformations of the corresponding equation, we lead to an equation of Bessel type as the corresponding modified model.

(d) The linearization of the nonlinear equation (1.1) gives another modified model. That is, under the hypothesis of small oscillation angles of the pendulum, $\vartheta$ and $\sin\vartheta$ are equivalent infinitesimals, and the modified model is:

$$\ddot{\vartheta} + \frac{g}{\ell}\vartheta = 0. \qquad (1.3)$$

In this case we must remember that the linearization of nonlinear models eliminates, in general, important characteristics of the solution of the nonlinear model.

## 2. APPLICATION 2 (THERMODYNAMICS)

### THE HEAT FLOW PROBLEM. (Ref. 21)

The concept of reasonability and unreasonability in the sense of Hadamard is, by this application, clarified.

To make the problem of heat flow in a given medium correctly stated in the sense of Hadamard, one accepts that the medium is homogeneous and isotropic

- 31 -

with respect to the heat flow, and that the heat flow is towards the decreasing temperature. Based on these assumptions, the model of the problem is expressed by the elliptic partial differential equation:

$$u_{xx} + u_{yy} + u_{zz} = u_t \tag{2.1}$$

where $u = u(x,y,z,t)$ is the temperature. In the equation (2.1) there is a factor-coefficient depending on the density, specific heat and thermal conductivity, and this coefficient is here taken equal to unity. If the data are:

$$u(x,o) = n \sin nx \,, \quad n = \text{integer} \tag{2.2}$$

the equation (2.1), in a one-dimension medium, has as a solution the function:

$$u(x,t) = n \cdot e^{-n^2 t} \cdot \sin nx \tag{2.3}$$

which is a unique solution, then the two Hadamard restrictions are satisfied. As for the third Hadamard restriction, one can distinguish two cases

(i) For $t > o$, when one has the "forward heat problem", the solution (2.3), for any fixed $n$ and $t \to \infty$, tends to zero, and the condition (2.2) is bounded; then the third Hadamard restriction is satisfied, the problem is "reasonable" in the sense of Hadamard, and the solution (2.3) is "physically accepted".

(ii) For $t < o$, when one has the "backward heat problem", the solution (2.3) for any fixed $n$ and $t \to \infty$, tends to infinity, and the condition (2.2) is bounded; then the third Hadamard restriction is violated, the problem is "unreasonable" in the sense of Hadamard, and the solution (2.3) is a "formal solution" of the heat problem, that is "not physically accepted".

3. APPLICATION 3 (NONLINEAR MECHANICS)

THE PROBLEM OF PRINCIPAL MODES OF NONLINEAR SYSTEMS. [Ref. 18(b)]

The concept of "principal modes" of nonlinear systems plays a predominant role in the analysis of oscillatory systems in many fields.

The principal modes of linear systems are, by definition, the fundamental set of solutions of which a linear combination gives the general solution of the linear system; or, physically speaking, they are the special modes of oscillations of the linear system in terms of which one can discuss any kind of oscillations of the system.

Since the "principle of superposition" does not hold in nonlinear systems, the concept of principal modes, as is given above, is meaningless in nonlinear systems, and the following question arises:

"Has the problem of principal modes of nonlinear systems a physical meaning"?

or: "How one can make the problem of principal modes of nonlinear systems a reasonable problem"?

For the solution of this problem, one must find a new definition of the concept of principal modes valid for both the linear and nonlinear systems, and such that the known definition in linear systems to be a special case. We can find two such new definitions of principal modes of nonlinear systems which, under some special conditions, are equivalent.

After we fix the subject of the definition of principal modes, we must make the problem "well expressed", "correctly stated" and "reasonable".

Let us restrict ourselves to a "two degrees of freedom" mechanical system, Figure 3. The motion of this system is governed by the model:

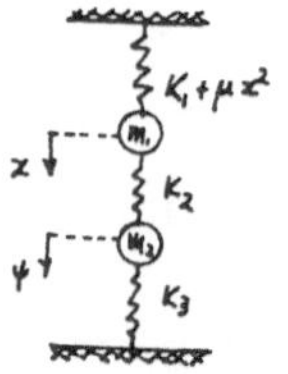


FIGURE 3.

$$\left.\begin{aligned}\ddot{x} + \omega_1^2 x - \lambda_2 \psi + \lambda_1 x^3 &= 0\\ \ddot{\psi} + \omega_2^2 \psi - \lambda_3 x &= 0\end{aligned}\right\} \qquad (3.1)$$

where:

$$\left.\begin{aligned}\omega_1^2 = \frac{K_1+K_2}{m_1}, \quad \omega_2^2 = \frac{K_2+K_3}{m_2}\\ \lambda_1 = \frac{\mu}{m_1}, \; \lambda_2 = \frac{K_2}{m_1}, \; \lambda_3 = \frac{K_2}{m_2}\end{aligned}\right\} \qquad (3.2)$$

and $m_1$ , $m_2$ are the oscillating masses, $\mu$ characterizes the nonlinearity of the first anchor spring.

By using the transformation:

$$x = x_1, \quad \dot{x} = x_2, \quad \psi = x_3, \quad \dot{\psi} = x_4 \qquad (3.3)$$

the system (3.1) takes the normal form:

$$\left.\begin{aligned}&(a): \quad \dot{x}_i = f_i(x_1, x_2, x_3, x_4), \quad i = 1,2,3,4\\ &(b): \quad f_1 = x_2, \; f_2 = -\omega_1^2 x_1 + \lambda_2 x_3 - \lambda_1 x_1^3, \; f_3 = x_4, \; f_4 = \lambda_3 x_1 - \omega_2^2 x_3\end{aligned}\right\} \qquad (3.4)$$

valid in a region $R$

$$R: \quad |x_i| < \ell, \quad i = 1, 2, 3, 4. \qquad (3.5)$$

The appropriate initial conditions for existence of principal modes in $R$ are:

$$x_1(0) = x_{10}, \quad x_2(0) = 0, \quad x_3(0) = x_{30}, \quad x_4(0) = 0 \qquad (3.6)$$

where the non-zero $x_{10}$ and $x_{30}$ are appropriately related to each other.

We now remark that the functions $f_i$ , given by (3.4, b), are such that the Hadamard restrictions are satisfied. These functions are continuous in $R$ , then bounded, they have continuous partial derivatives $\frac{\partial f_i}{\partial x_k}$ in $R$ , when they satisfy Lipschitz conditions with respect to $x_i$ in $R$ for a Lipschitz constant: $s = \text{l.u.b.} \left| \frac{\partial f_i}{\partial x_k} \right|$.

The above properties of $f_i$ assure the unique existence of the solution of the model (3.4) associated with the conditions (3.6) in a region $R' \subset R$ . As the initial point $x_{i0}$, $i=1,2,3,4$ varies in $R'$ , the solution, which is yet unknown, satisfies the three Hadamard restrictions, and the model (3.4), subject to initial conditions (3.6), is "reasonable" in Hadamard sense.

APPLICATION 4 (ASTRODYNAMICS)

We discuss the following problem

"An artificial celectial body is moving under the influence of a central force obeying the inverse square Newton law toward the attractive center. A general perturbing force is applied, acts for an interval of time, then it is removed. Find the motion of the body during the action of the perturbing force". [Ref 18(c)]

According to known reasoning, this problem can be made "correctly stated", when the model is:

$$\left.\begin{aligned} &\ddot{\underline{r}}(\tau) = -\frac{\mu}{r^3(\tau)}\,\underline{r}(\tau) + \underline{T}(\tau) \\ &\underline{r}(0) = \underline{r}_0\,, \quad \dot{\underline{r}}(0) = \dot{\underline{r}}_0 + \underline{I}_0 \\ &D_1 : \quad |\underline{r}(\tau)| < M_1 \\ &D_2 : \quad |\dot{r}| < M_1 \\ &D_3 : \quad 0 \leq \tau \leq \tau' \end{aligned}\right\} \tag{4.1}$$

$\underline{T}$ is the perturbing force, $\underline{r}$ the radial vector from the attractive center to the center of the mass of the body, and $\underline{I}_0$, the impulse, is given by:

$$\underline{I}_0 = \int_0^{t_0} \underline{T}(t)\,dt\,, \quad \tau = t - t_0 \tag{4.2}$$

Forces $\underline{T}$ can be produced, e.g., by drag, by asphericity of the central body, etc.

The force $\underline{T}$ may depend on the position and velocity of the moving body, and on the time explicitly. We accept here that the force $\underline{T}$ is "considerable" with respect to central attraction, and that it is only a "function of time".

We select appropriate coordinate axes on which the unit vectors are: $\underline{r}_o^*$ , $\underline{s}_o^*$ , $\underline{T}_o^*$ , shown in Figure 4.

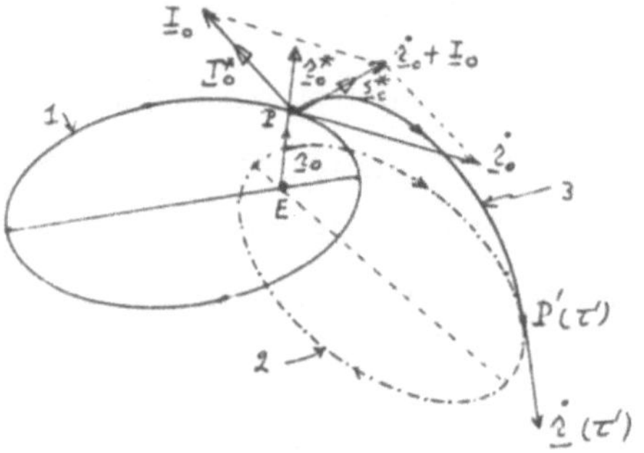


1: original orbit

2: new orbit

3: orbit during the action of $\underline{T}$ .

FIGURE 4.

Referred to this coordinate system, we take the function:

$$\underline{r}(\tau) = \alpha_1(\tau)\,\underline{r}_o^* + \alpha_2(\tau)\,\underline{s}_o^* + \alpha_3(\tau)\,\underline{T}_o^* \tag{4.3}$$

as a "trial solution" of the system (4.1), and by an appropriate calculation of the coefficients: $\alpha_1(\tau)$, $\alpha_2(\tau)$, $\alpha_3(\tau)$ , we make this function first a "formal solution", and then an "actual solution" of (4.1).

The investigation shows that the function (4.3) becomes a "formal solution" of the system (4.1), if $\alpha_1$ , $\alpha_2$ , $\alpha_3$ of (4.3) satisfy the equations

$$\left.\begin{aligned} \ddot{\alpha}_1 + \frac{\mu}{r^3}\alpha_1 &= T_1 \; ; \quad \alpha_1(0) = r_o \, , \quad \dot{\alpha}_1(0) = 0 \\ \ddot{\alpha}_2 + \frac{\mu}{r^3}\alpha_2 &= T_2 \; ; \quad \alpha_2(0) = 0 \, , \quad \dot{\alpha}_2(0) = s_o \\ \ddot{\alpha}_3 + \frac{\mu}{r^3}\alpha_3 &= T_3 \; ; \quad \alpha_3(0) = 0 \, , \quad \dot{\alpha}_3(0) = 0 \end{aligned}\right\} \tag{4.4}$$

The solutions $\alpha_1(\tau)$, $\alpha_2(\tau)$, $\alpha_3(\tau)$ of (4.4) exist and are unique if the components $T_1$, $T_2$, $T_3$ of $\underline{T}$ are continuous functions of time and differentiable, also, if $\dot{T}_1$, $\dot{T}_2$, $\dot{T}_3$ are continuous functions of time, and also $\ell \neq 0$.

Under the above restrictions of $T_1$, $T_2$, $T_3$, the functions $\alpha_1$, $\alpha_2$, $\alpha_3$ satisfy the Hadamard restrictions, and the function (4.3) is an "actual solution" of the model (4.1), then "physically accepted".

5. APPLICATION 5 (PROBLEM OF UNDERWATER WARFARE)

THE PROBLEM OF DOMES [Ref. 7(a)]

The problem of domes arose in the winter of 1942-43 in connection with "underwater warfare", and it gives a good example of the difficulties to go from reality to an acceptable idealized mathematical model.

As is known, underwater sound ranging depends on sending out a sound beam in water and, attached to a fast-moving ship, the water steaming around the plate causes serious disturbances. For elimination of these disturbances, the projector is enclosed in a "dome", Figure 5, which is a convex shell of metal, or other material, filled with water. Such domes interfere only slightly with the formation of a concentrated sound beam.

During 1942-43, a large number of small submarine chasers were built and equipped with sound gear similar to, but smaller than, the gear used before. While the manufacture of domes to fit this smaller gear was under way, it was discovered that these smaller domes led to an intolerable diffusion on the sound beam. At that time, a quick remedy was imperative, and a mathematical analysis of the problem was needed to support and speed-up experimental work. The mathematical problem, related to the above physical problem, was to solve the differential equation:

-37-

FIGURE 5.

(1) axis of beam sound

(2) projector

S surface of dome

$$\left.\begin{aligned} &\nabla^2 P + k^2 P = 0 \\ &\nabla^2 = \frac{\partial^2}{\partial x^2} + \frac{\partial^2}{\partial y^2} + \frac{\partial^2}{\partial z^2} \end{aligned}\right\} \quad (5.1)$$

in which $k = \omega / c$, $\omega$ the frequency $c$ the sound velocity.

$k$ has for our problem different values within the shell of the dome and outside.

This model was not a suitable one for the problem. A suitable model was found by the following process.

The actual dome of small finite thickness was replaced by an extremely thin surface, then the influence of the dome was simply replaced by conditions for jump discontinuities of the disturbance $q$ of the beam across the surface. These conditions are

$$\left.\begin{aligned} &[q] = \frac{\rho_1}{\rho_0 - 1} \frac{\partial p}{\partial n} \\ &\frac{\partial q}{\partial n} = \frac{\rho_0}{\rho_1}(k_0^2 - k_1^2)p - \left(1 - \frac{\rho_0}{\rho_1}\right)\left(\frac{\partial^2 p}{\partial n^2} - 2H\frac{\partial p}{\partial n}\right) \end{aligned}\right\} \quad (5.2)$$

where the symbol $[\cdot]$ means jump of the quantity of the symbol across the surface, $q$ is the disturbance of the acoustic pressure $p$ caused by the dome, and the normal derivatives $\frac{\partial}{\partial n}$ are to be evaluated on the surface $S$ of the dome. The quantity $H$ is the mean curvature of $S$, i.e. the average of the curvature of any two normal plane sections at right angles to each other. In addition to conditions (5.2) to be satisfied by $q$ on $S$, $q$ should be a

- 38 -

solution of the equation:

$$\nabla^2 q + K_0^2 q = 0 \tag{5.3}$$

which is regular everywhere except on $S$ and which has the same behavior as $P$ at $\infty$. This problem possesses the unique solution:

$$q = -\frac{1}{4\pi}\iint_S \left[\frac{\partial q}{\partial n}\right] \frac{e^{iK_0 r'}}{r'} ds + \frac{1}{4\pi}\iint_S [q] \frac{\partial}{\partial n}\left(\frac{e^{iK_0 r'}}{r'}\right) ds \tag{5.4}$$

The quantities in the brackets are given by conditions (5.2), $r'$ is the distance from a fixed point $(x, y, z)$, at which $q(x, y, z)$ is to be determined, to the point of integration on $S$. This formula yields the disturbance as the effect due to a layer of point sources and a layer of dipoles disturbed on $S$ with intensities which are known as soon as the original pressure $p$ is known, since the quantities in brackets are fixed in value due to conditions (5.2). The relative directional disturbance

$$\left|\frac{p_1}{p}\right| \quad Re. \measuredangle \left(\frac{q_1}{A} - \frac{q_0}{p_0}\right)$$

would, finally, be obtained from (5.4). The solution (5.4) is valid for a shell of constant thickness, but it could be extended without essential error to cases in which the dome shell is made up of a not too large number of pieces, each of which is of constant thickness. All that would be necessary would be to insert a numerical factor $\alpha$ in the integrands on the right-hand side of expression (5.4), which would be piecewise constant on $S$. By this formula, one can analyze the contribution to the distortion of various factors, such as the curvature of the dome and the density of sound velocity within it. The above kind of mathematical idealization, even without detailed numerical computation, proved very helpful to the designing engineer.

## 6. APPLICATION 6. (BIOLOGY, ECOLOGY, ECONOMICS, ETC.) (Ref. 1, 5, 9, 10, 11, 14, 17, 26)

There are many assemblies around us of which the elements influence each other through competition and cooperation. Some examples are: population of various biological species; components of the nervous system; coupled reacting chemical components in the atmosphere, in bodies of water, in organisms as a whole or in part; business; political parties; countries.

With the currently developing interest in the investigation of biological, social and economic mechanisms, it is of considerable importance to find mathematical models, which might be amenable to a detailed investigation.

Let us discuss

THE PROBLEM OF MODELING OF POPULATION GROWTH,

as a biological problem, but, by appropriate changes of the discussion and appropriate changes in the meaning of the variables involved, the discussion can be used for modeling in other fields.

Lofka and Volterra are pioneers in the problem and their models are the basis for modifications or generalizations of the models.

We first discuss the modeling in the case of a single population, then the case of two or more populations in coexistence.

### 6.1 GROWTH OF A SINGLE POPULATION.

Let $N(t)$ be the number of individuals in the population, or population density per unit area, at a certain time $t$. $N(t)$ is a positive function of time, a "step function" of $t$, and it is constant in an interval of time, when the population does not change, and it is discontinuous, when a birth or death changes the population.

To create a model for the growth of a single population, one may start from two assumptions

(i) $N(t)$ is a continuous function of time $t$,

(ii) The instantaneous rate of growth of $N(t)$ depends on $N(t)$ itself, that is

$$\dot{N} = f(N). \tag{6}$$

The function $f(N)$ must be specified according to new assumptions and observation facts.

We distinguish the following two cases.

(a) Single Nonisolated Population.

By nonisolated population we mean population with unlimited food supply and space.

If $n$ is the birth coefficient and $m$ the death coefficient, then, in a small interval of time $\Delta t$, by assuming that the increment of the population $\Delta N$ is analogous of $\Delta t$ and $N$ itself, one will have in the limiting case

$$\dot{N} = (n-m)\, N \tag{6.1}$$

that is we assume as function $f(N)$ in (6) the function $f(N) = (n-m)N$.
The coefficient $(n-m)$ characterizes the nature of the population.
If $N(0) = N_0$ is the initial population, then the solution of (6.1) is:

$$N = N_0\, e^{(n-m)t}. \tag{6.2}$$

In case $n < m$, the population tends to disappear. But in case $n > m$ the population increases exponentially to infinity under the hypothesis of unlimited food supply and space for the population.
The above conclusion does not agree with observations and experimental facts, then the model (6.1) must be changed.

(b) Single Isolated Population

In any population with a large number of individuals, the nourishment is bad, the food supply limited, the space also limited, the population is "isolated", and the growth of the population meets a resistance. Its mortality increases and the population can not pass an upper limit but tends to approach a maximum $N_M$ . The model (6.1), which is good for $N$ close to $N_0$ , needs a modification in order to accomplish the observations.

For a modification of the model (6.1) one needs to insert a coefficient for the resistance effect. This comes by changing the law of mortality, that is by changing the mortality coefficient $m$ into

$$m' = m + bN \tag{6.3}$$

when the new model is

$$\dot{N} = (n-m)N - bN^2 \tag{6.4}$$

when the function $f(N)$ of (6) is now $f(N) = (n-m)N - bN^2$.

The equation (6.4) is an equation of Bernoulli's type, is called a "Verhalst equation", and its general solution, called "logistic curve", is

$$N = \frac{n-m}{b + e^{-c} \cdot e^{-(n-m)t}} \tag{6.5}$$

where $c$ is a constant.

This solution can be modified in order to contain the initial $N_0$ and the maximum $N_M$ populations.

The solution (6.5) for $t=0$ gives $e^{-c} = (n-m-bN_0)/N_0$ , when (6.5) becomes

$$N = \frac{(n-m)N_0}{bN_0 + (n-m-bN_0)e^{-(n-m)t}} \tag{6.6}$$

which includes $N_0$ .

For $n>m$ , as $t \to \infty$ , $N$ becomes $N_M = (n-m)/b$ , when (6.6) becomes

$$N = \frac{N_0 N_M}{N_0 + (N_M - N_0)e^{-(n-m)t}} \tag{6.7}$$

which includes $N_0$ and $N_M$ .

- 42 -

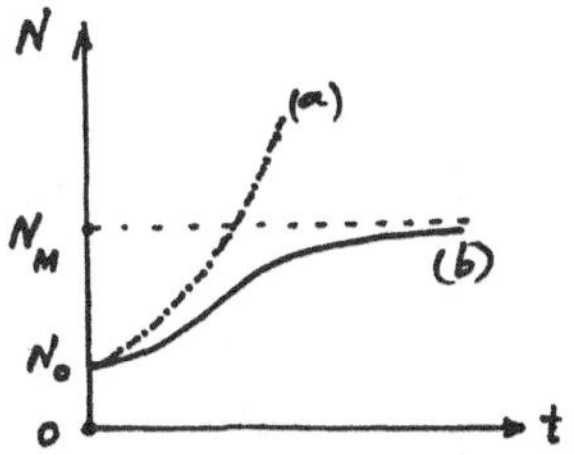


FIGURE 6. Curve (a) is the graph of (6.2), and (b) that of (6.7)

In Figure 6, the solid curve (b) is the graph of the solution (6.7), and the dotted curve (a) the graph of (6.2).

According to the formula (6.7), the population $N$ increases when $n > m$, and finally reaches its maximum $N_M$, which is the "equilibrium of the population", when $m' = m + bN_M = n$.

If $n = m$, then $N = N_0$ for any $t$; and if $n < m$, then $N \to 0$ for $t \to \infty$.

For $N_0$ small, compared to $N_M$, and $N$ close to $N_0$, then the term $bN^2$ of (6.3) is negligible compared to the other term, when the model (6.4) near $N_0$ is identical to the model (6.1).

If $N$ increases, the term $bN^2$ becomes considerable, the model (6.1) stops to be a good model, and the model (6.4) is the appropriate model.

The model (6.4) is an appropriate one for a variety of single population growth problems, and its solution (6.7), that is the curve (b) in Figure 6, fits quite well the data on population growth of many species.

For this, we mention two examples:

(i) The formula (6.7) has been applied to population growth of U.S.A. and gave accurate predictions, although the creation of the model (6.4) is based on very simplified assumptions.

(ii) The growth of yeast cells in a laboratory culture, for which the data of the following table are given, corresponds to the curve of the Figure 7, which agrees with the curve (b) of Figure 6.

- 43 -

| Time In Hours | Number of Individuals | Time In Hours | Number of Individuals |
|---|---|---|---|
| 0 | 9.6 | 10 | 513.3 |
| 2 | 29.0 | 12 | 594.4 |
| 4 | 71.1 | 14 | 640.8 |
| 6 | 174.6 | 16 | 655.9 |
| 8 | 350.7 | 18 | 661.8 |

Table of Data

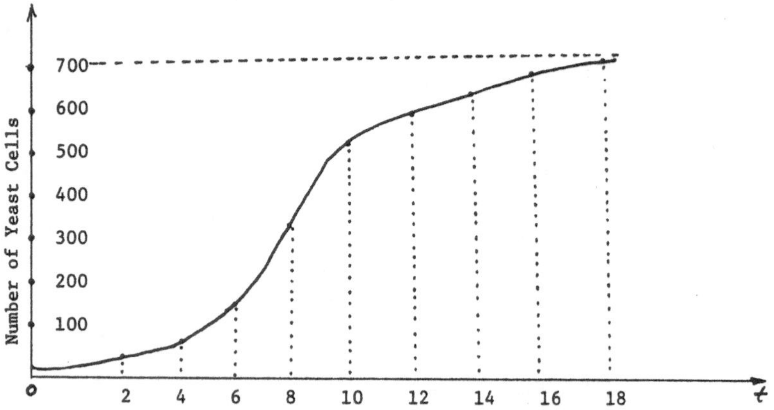


FIGURE 7. The Growth Curve of Yeast Cells in Laboratory

We remark that different assumptions for specification of the function $f(N)$ of the general original model (6) will give different models, which may be proved to be in many cases better than, the above models. E.g., if, instead of $N^2$ in the model (6.4), one takes $N \cdot \log N$, one has another useful model.

## 6.2 THE CASE OF TWO SPECIES

We first clarify some notions usually appearing in coexisting populations.

In any model of two species with population densities $N_1$ and $N_2$ in community, its solution starts from an initial point $(N_{10}, N_{20})$ of the first quadrant of the $N_1, N_2$ -plane and terminates to a point $(\bar{N}_1, \bar{N}_2)$ in finite or infinite time, and we may have the following three cases

(i) The point ( $\bar{N}_1$ , $\bar{N}_2$ ) may coincide with the origin, when one speaks about "extinction of the species";

(ii) One of $\bar{N}_1$, $\bar{N}_2$ may be zero and the other nonzero, that is the point ( $\bar{N}_1$ , $\bar{N}_2$ ) may be a point of one of the axes, and this case corresponds to the "principle of competitive exclusion", a principle much used in ecology, but which has been much critisized.

(iii) The point ( $\bar{N}_1$ , $\bar{N}_2$ ) may be a point with nonzero coordinates, then may be inside a region of the first quadrant of $N_1$ , $N_2$ -plane, that is all points ( $N_1$ , $N_2$ ) may be in the region and $N_1$ , $N_2$ may be changed periodically in the region. In this case one speaks about "coexistence of two species", "symbiosis", "parasitism".

Let us discuss now the modeling of two species of populations inhabiting the same environment and competing for limiting food supply and space.

Lotka investigated the problem in case of living species, the "host", in the presence of "parasites" capable of resisting the development of the host. Volterra discussed a similar problem, that of the evaluation in a closed sea of two kinds of fish, the first "soles" and the second "sharks", the second eating the first. Experimental and statistical facts in the Adriatic sea

- 45 -

inspired Volterra in the mathematical discussion of the problem. The Lotka and Volterra problems are equivalent from a theoretical point of view. The soles are the hosts and the sharks their parasites.

Let $n_1$, $m_1$, $b_1$ be the birth, death and competition coefficients, respectively, of the population $N_1$, the host (prey). The increase of $N_1$ in time $dt$, due to the births, is $n_1 N_1 dt$. This is an hypothesis not always true, but we consider it as true temporarily. Let $N_2$ be the population of parasites (predators). The number of meetings will be proportional to $N_1 N_2$, say, $a N_1 N_2$, when the increment of $N_1$ is

$$\dot{N}_1 = (n_1 - m_1) N_1 - b_1 N_2^2 - a N_1 N_2 \tag{6.8}$$

which is a DE corresponding to the host population. This formula differs from (6.4) by the term $-a N_1 N_2$ of deaths after the meeting of host and parasite individuals.

Now, the equation corresponding to the parasite population. The births will be proportional to $N_1$ and to $N_2$, then proportional to their product $N_1 N_2$. If $m_2$ is the coefficient of natural deaths, and $b_2$ the competition coefficient, then the evolution of parasites is given by the equation:

$$\dot{N}_2 = k a N_1 N_2 - m_2 N_2 - b_2 N_1^2. \tag{6.9}$$

The two simultaneous DE (6.8) and (6.9) is the model of the above "prey-predator system". This system has been generalized to the set of $n$-species. If the effect of the species $j$ on the growth $\dot{N}_i$ of the species $i$ is proportional to the number of meetings between them, which we assume to be the product $N_i N_j$ of their population sizes, then the model of this system is

$$\dot{N}_i = K_i N_i + \sum_{j=1}^{n} a_{ij} N_i N_j \quad , \quad i=1,\ldots,n. \tag{6.10}$$

For different sets of the values of the coefficients of (6.10) and different signs of these values, one may have different situations of the population growth in coexisting species, that is different physical problems.

The above models, called "Lotka-Volterra models", are "reasonable" in the sense of Hadamard.

## 6.3 REVISIONS, MODIFICATIONS AND GENERALIZATIONS OF THE LOTKA-VOLTERRA MODELS OF POPULATION GROWTH.

(a) Some Remarks.

The Lotka-Volterra models for the population of interacting species are created based mainly on the assumptions

(i) All the members of a species have the same age,

(ii) Any member of the species reacts instantaneously to any change in the environment, and

(iii) The prey-predator interactions affect instantaneously the population of both the prey and the predator.

Such assumptions may be realized in laboratory situations, but, in general, are not realistic. Observations and experiments contradict some results coming from the Lotka-Volterra models.

The growth of a population may be affected by temperature, humidity, age, distribution of the species, time of the year, time lags in population, and other ecological factors, and the biological interactions are competition, parasitism, symbiosis, complex interactions between the organisms and its physical environment.

The attempts to manage the population growth will profit from the use of mathematical models, even when these models were found with very simplified assumptions, as this happens with Lotka-Volterra models.

We must not expect to find a model of population growth to satisfy all the above ecological factors, and to include all the important features about the population. We must restrict ourselves to modified models, which satisfy a specific program of research and to be realistic.

There are no clear-cut guidelines for choosing one interaction over the other in population dynamics, and it is the experiment which would help the decision.

Depending on the problem and on the purpose of the investigation, one may use techniques of statistical mechanics, or one may take into account statistical aspects of the population growth of individual species and the correlation between population of different species.

Usually, one can not have information about detailed interactions between various species, then their statistical mechanics treatment is desirable. The statistical treatment provides an empirical method for calculating the interactions between two species and the stability of the population.

In many problems the interaction coefficients of the models is difficult to be evaluated, when the prediction is imprecise. One must assume in these cases that the coefficients of the models are random variables, when the populations are characterized by a probability distribution, which can be used to calculate the average population.

Each modification or generalization of existing models must have as a purpose to cover questions of specific research and to be more realistic. Models exactly solvable are preferable, because their exact solutions provide a testing ground for general principles used in the study of population dynamics.

We now give some modified and generalized models.

(b) Exactly Solvable Models.

The system of equations (6.8) and (6.9) can be solved only by using approximate methods. We see here that, by specific modifications and transformations, these equations will lead to equations exactly solvable. We have two cases.

$(b_1)$ Case Without Mutual Interactions of the Two Species.

In the equations (6.8) and (6.9), in case of no mutual interactions, $b_1 = b_2 = 0$ , when, if instead of the product $N_1 N_2$ we take $N_1 \log N_2$ in the first equation and $N_2 \log N_1$ in the second, the model reduces to an exactly solvable one.

In this case, if, for simplicity, we take $a_1$ , $a_2$ , $\lambda_1$ , $\lambda_2$ instead of $n_1 - m_1$ , $m_2$ , $\alpha$ , $k\alpha$ , respectively, the new model becomes

$$\left.\begin{aligned} \dot{N}_1 &= a_1 N_1 - \lambda_1 N_1 \log N_2 \\ \dot{N}_2 &= a_2 N_2 + \lambda_2 N_2 \log N_1 \end{aligned}\right\} \qquad (6.12)$$

with initial values $N_{10}$ and $N_{20}$ .

By the transformation

$$X = \log N_1 - \frac{a_2}{\lambda_2} \, , \quad X_2 = \log N_2 - \frac{a_1}{\lambda_1} \qquad (6.13)$$

the model (6.12) becomes

$$\dot{X}_1 = -\lambda_1 X_2 \, , \quad \dot{X}_2 = \lambda_2 X_1 \qquad (6.14)$$

of which the initial conditions are:

$$X_{10} = \log N_{10} - \frac{a_2}{\lambda_2} \, , \quad X_{20} = \log N_{20} - \frac{a_1}{\lambda_1} \qquad (6.15)$$

The solution of (6.14), subject to conditions (6.15) is

$$\left.\begin{aligned} X_1 &= X_{10} \cos \omega t - \frac{\lambda_1}{\omega} \sin \omega t \\ X_2 &= X_{20} \cos \omega t + \frac{\lambda_2}{\omega} \sin \omega t \end{aligned}\right\} \qquad (6.16)$$

where $\omega = \sqrt{\lambda_1 \lambda_2}$.

Eliminating $t$ in (6.16), one gets the family of orbits

$$\left.\begin{aligned} &\text{(a)} \quad \lambda_2 X_1^2 + \lambda_1 X_2^2 = c \\ &\text{(b)} \quad c = \lambda_2 X_{10}^2 + \lambda_1 X_{20}^2 \end{aligned}\right\} \qquad (6.17)$$

which are ellipses in the $X_1, X_2$ -plane.

- 49 -

By using (6.13) and (6.16), the solution of (6.12) can be written in the form

$$\left.\begin{aligned} N_1 &= \exp\left\{ X_{10}\cos\omega t - \frac{\lambda_1}{\omega} X_{20}\sin\omega t + \frac{a_2}{\lambda_2}\right. \\ N_2 &= \exp\left\{ X_{20}\cos\omega t + \frac{\lambda_2}{\omega} X_{10}\sin\omega t + \frac{a_1}{\lambda_1}\right. \end{aligned}\right\} \tag{6.18}$$

Then the corresponding orbits are members of the family of the complicated curves in the $N_1$, $N_2$-plane

$$\lambda_2\left(\log N_1 - \frac{a_2}{\lambda_2}\right)^2 + \lambda_1\left(\log N_2 - \frac{a_1}{\lambda_1}\right)^2 = 2c \tag{6.19}$$

where $c$ is given by (6.17.b).

Through the transformation (6.13), the equilibrium $\left(N_1 = e^{a_2/\lambda_2}, N_2 = e^{a_1/\lambda_1}\right)$ of (6.12) is transformed into the center of the ellipses (6.17), which is the origin in the $X_1$, $X_2$-plane.

By using the above exact solution of the model (6.12) and restricting suitably the values of the constants, one may find conditions for "extinction" of the species, for "competitive exclusion", or for their "stable coexistence". Some results of the investigation by using (6.12) are different from results by using the corresponding Lotka-Volterra model, and experiments decide on the choice of the model (6.12).

($b_2$) Case with Mutual Interactions of the Two Species.

The model of (6.8) and (6.9) with mutual-interactions and with self-interactions, if modified according to

$$\left.\begin{aligned} \dot N_1 &= a_1 N_1 - \lambda_{11} N_1 \log N_1 - \lambda_{12} N_1 \log N_2 \\ \dot N_2 &= a_2 N_2 - \lambda_{22} N_2 \log N_2 - \lambda_{21} N_2 \log N_1 \end{aligned}\right\} \tag{6.20}$$

Such a model can be solved exactly.

We remark that both models (6.12) and 6.20) are reasonable in the sense of Hadamard.

(c) Some Modified and Generalized Models.

An important generalization, which includes many classes of problems of population growth, is the model

$$\dot{N}_i = f_i(N_1,\dots,N_n,P_1,\dots,P_m)\,,\quad i=1,\dots,n \tag{6.21}$$

of $n$ species in community with population densities $N_i$, $i=1,\dots,n$. The parameters $P_j$, $j=1,\dots,m$ define the environment where the evolution of $N_i$ takes place.

One can incorporate the fact " $f_i \equiv 0$ for $N_i = 0$ " in the system (6.21), if one writes this system as

$$\dot{N}_i = N_i\, g_i(N_1,\dots,N_n,P_1,\dots,P_m),\ i=1,\dots,n. \tag{6.22}$$

The parameters $P_j$ in (6.22) influence the populations, their rate of reproduction, the duration of their life. They are related to topological and geographical features of the species, or to climate conditions, etc. These parameters may depend on each other, when they may exist relationships between the parameters themselves, when the number of the parameters in (6.22) may be decreased.

In the above general model, the populations $N_i$ may in general be also subject to certain relationships, which enable one to decrease the number of $N_i$. We assume that the model (6.22) is taken after using the restraint relationships, when the number of species and the number of parameters are the smallest possible to determine the phenomenon.

The model (6.22) must be a "reasonable" one, when the functions must be appropriately restricted. The $g_i$'s and their first partial derivatives with respect to $N_i$ and $P_j$ must be defined, single-valued and continuous functions of $N_i$ and $P_j$ and of the initial values $N_{0i}$ and $P_{0j}$ in the state space and the parameter space, when there exists only one solution of the model (6.22), starting from the initial state $N_{0i}$ for any initial parameter point $P_{0j}$, and this solution is physically accepted.

- 51 -

We might mention a special class of models created by relaxing the known assumptions on which the Lotka-Volteva models are based, the class of models with "time-lags" populations.

In the special case of a "single species" of this class, the following equations with "time-lag" where proposed as models

$$\dot{N}(t) = K\left\{1-(1/\theta)\,N(t-\tau)\right\}N(t) \tag{6.23}$$

and

$$\dot{N}(t) = K\left\{1-(1/\theta)\,N(t-\tau)\right\}N(t-\tau)-K'N(t) \tag{6.24}$$

of which there are some generalizations.

In these models it is assumed that the birth rate coefficient is diminished by a quantity proportional to the population of the preceding generation, the positive $\tau$ being the generation time, that is time required in going from an egg-stage to the adult-stage.

The solution of the above models shows a variety of properties depending on the parameters included. Of these properties some are more of mathematical interest than of practical usefulness.

The "birth-death process", applied to population growth, to social and physical contagion, as well as to other phenomena, illustrates well some of the advantages and disadvantages of a deterministic or probabilistic approach.

For example, let us see the deterministic and stochastic models in the case of a single population.

For a single population of size $N$ at time $t$, if $n$ and $m$ are the birth and death coefficients, respectively, the deterministic model is

$$\dot{N} = (n-m)\,N \tag{6.25}$$

which is a very simple model clearly used, as we know, in some circumstances.

For the probabilistic version of this particular process, that is for the corresponding probabilistic model, we express the rate of change $\dot{p}_i$ in the probability $p_i$ of being in state $i$ in terms of the three conditional probabilities: $n(i-1)$, $(m+n)i$, $m(i+1)$, since the probabilities of birth or death are proportional to the existing sizes: $(i-1)$, $i$, $(i+1)$, respectively, for the three states, when we have the stochastic model of the above process

$$\dot{p}_i = n(i-1)\,p_{i-1} - (m+n)\,i\,p_i + m(i+1)\,p_{i+1} \tag{6.26}$$

This stochastic model is, by neglecting some factors, an accurate one, but it is difficult to be treated analytically.

The stochastic model (6.26) gives the probability that a population will die out due to change fluctuation, an information much important in some applications, but which cannot be given by the deterministic model (6.25).

The deterministic model (6.25) is the "expected mean value" of the probability distribution, and this model is preferred in cases one only needs the expected value of the probability distribution.

In general, in spite of the obvious stochastic nature of processes, much of the existing theory is deterministic, because the mathematics of a full stochastic version of many of the models is much intractable. One must prefer the deterministic version of the model in order to make progress by taking an approximation of the model, which first is stochastically formulated.

We, finally, remark that the previous discussion, if suitably modified, that is by appropriate changes to the problem and appropriate specification of the competitive species and the limiting resources, might be useful for an investigation of a problem of nature different than the above. E.g., one can have a problem in the field of economics, if the variables denote the size or extent of commerical enterprises competing for a common source and for a common market.

- 54 -

## BIBLIOGRAPHY

1. Ayala, F.: (a) "Biological evaluation: Natural selection or random walk"? American Scientist, (Nov-Dec. 1974), pp. 672-701, U.S.A.
(b) "Competition between species", American Scientist (1972), pp. 348-357, U.S.A.

2. Bartholomew, D.: "Stochastic models for social processes", 2nd Ed., John Wiley and Sons, New York (1973), Chapter I, U.S.A.

3. Birckoff, Garrett: "Mathematics and computer science", American Scientist,(Jan-Feb. 1975) pp. 83-91, U.S.A.

4. Brillouin, L.: (a) "Scientific uncertainty and information", Academic Press, New York (1946), U.S.A.
(b) "Relativity re-examined", Academic Press, New York (1970), U.S.

5. Coleman, S.: "Introduction to mathematical sociology", The Free Press of Glencoe, Collier-Macmillan Lim., London (1964), pp. 526-528, England.

6. Cortes, F., Przeworski, A. and Sprague, J.: "Systems analysis for social scientists", John Wiley and Sons, New York (1974), U.S.A.

7. Courant, R.: (a) "Methods of applied mathematics", in the book: "Recent advances in science" of Shammos, M. and Murphy, G. Science Editions, Inc., New York (1961), U.S.A.
(b) "Boundary value problems in modern dynamics", Proc., Intern. Congress of Math, Vol. II (1950)

8. Education in Applied Mathematics: SIAM Review, Vol. 9, No. w (April 1967), U.S.A.

9. Gandolfo, G.: "Mathematical methods and models in economic dynamics," North Holland Publ. Co., Amsterdam (1971).

10. Goel, N., Maitra, S. and Montroll, E.: "On the Volterra and other nonlinear models on interacting populations", Review of Modern Physics, Vol. 43, No. 2, Part 1 (April 1971) pp. 231-276, U.S.A.

- 55 -

11. Gomatan, J.: "A new model for interacting populations",
I: "Two species system", pp. 347, 353
II: "Principle of competitive exclusion", pp. 355-364;
Bulletin of Mathematical Biology, Vol. 36 (1974), U.S.A.

12. Goodwin, B.: "Temporal organization in cells",
Academic Press, New York (1963), U.S.A.

13. Keisler, H.: "Model theory", Proc., Intern. Congress of Math, Vol. I (1970)

14. Kostitzin, V.: "Biologie mathématique," Librarie Armand Colin, Paris (1937), France.

15. Lavrentiev, M.: "Some improperly-posed problems of mathematical physics",
Springer-Verlag, New York (1967), U.S.A.

16. Lin, C. and Segel, L.: "Mathematics applied to deterministic problems in the Natural Sciences", Macmillan Publ. Co., New York (1974), U.S.A.

17. Lotka, A.: "Elements of physical biology", Baltimore, Williams (1974), U.S.A.

18. Magiros, D.: (a) "Physical problems discussed mathematically", Bulletin Greek Math. Soc., Vol. 6, II, No. 1 (1956), Athens, Greece.
(b) "Methods for defining principal modes of nonlinear systems utilizing infinite determinants", Journal of Math. Physics, Vol. 2, No. 66 (Nov-Dec., 1960), Americal Institute of Physics, U.S.A.
(c) "Motion in a Newtonian field modified by a general force", Journal of Franklin Institute (Sept. 1964), U.S.A.
(d) "The role of mathematics in the investigation of physical and social phenomena," Greek Math. Soc. "The Euclide", (May-June, 1973), Athens, Greece.

19. Marčhac, G.: "Methods and problems of computational mathematics",
Proc., Intern. Congress of Math, Vol. I (1970).

20. Mesaronič, M.: "Systems theory and biology", Proc., III Systems Symposium
Spring-Verlag, New York (1968), U.S.A.

21. Miranker, N.: "A well-posed problem for a backward heat equation",
Proc. Amer Math. Soc. (April 1961), U.S.A.

- 56 -

22. Neumann, John Von: "The mathematician", Collected Works, Vol. I, pp. 1-9, Oxford, New York (1961).

23. Rapoport, A.: "Mathematics outside the physician sciences", SIAM Review (Apr. 1973), pp. 481-502, U.S.A.

24. Papoulis, A.: "Probability random variables and stochastic processes", McGraw-Hill Book Co., New York (1965), Chapter 1, U.S.A.

25. Poincare, H.: (a) "Science and hypothesis", Dover (1952), New York, U.S.A.
(b) "The value of science", Dover, (1958), New York, U.S.A.

26. Volterra, V.: "Leçons sur la théorie mathematique de la lutte pour la vie ," Gauthier Villars et Cie, Paris (1931), France

27. Will, G.: "Gravitation theory", Scientific American (Nov. 1974), U.S.A.

# PART II

# NONLINEAR MECHANICS

This part contains papers dealing with subharmonic oscillations, principal modes, and celestial and orbital mechanics.

Papers 12 through 17 treat various problems of subharmonic oscillations of any order, especially of order 2 and 3. In these problems, the nonlinearity appears in the elastic forces. A very important fact is that the coefficients of the equations are not assumed to be necessarily small. Steady-state and transient type responses are examined. The conditions under which subharmonics exist are established, and the stability of these subharmonics in the steady-state is investigated. Various problems and applications related to subharmonics are discussed. Magiros' studies on subharmonics of order 1/3 formed the theoretical basis of the IBM research on the "Atomic clock".

Papers 18 through 21 are devoted to the study of principal modes which play the predominant role in the analysis of the oscillating systems in all fields. In these papers a method for calculating principal modes of linear and nonlinear systems is established. The author uses two definitions of principal modes in nonlinear systems which are shown to be equivalent under certain restrictions. These definitions filled-in a gap in linear and nonlinear system theory and helped in the understanding of the general behavior of solutions of nonlinear systems. The method yields the possibility of getting the principal modes in the form of a convergent series, all coefficients of which can be calculated. The problem of principal modes in Magiros' sense is a well-posed problem. The singularities of the system, i.e. the internal resonant frequencies of the components of principal modes solution, are also discussed.

Finally Papers 22 through 29 are devoted to celestial and orbital mechanics (two-body) problems which are of great interest both in theory and practice. Magiros' contribution in this field is especially significant. In papers 22, 23, 24 and 26 the motion of man-made celestial body under the influence of an attractive center, according to the inverse square of Newton's law and a "general force", is indicated. A solution of special form, referred to a specific inertial coordinate system is established. Taylor series in time are employed. It is shown that the solution found satisfies Hadamard's postulates. The region of convergence of the solution is determined and the cases of sudden or gradual application of the force are examined. In paper 25 the impulse (or impulsive force), required to effectuate a new Keplerian orbit around an attractive center at a given point in space is calculated. The solution of the problem is based on an auxiliary problem (solved geometrically), on a projection property of the "admissible impulse" and on trigonometric calculations. Finally, papers 27, 28 and 29 discuss the solution derived in the previous papers, when there are singularities. By an appropriate restriction of the new forces, the finite region of convergence of the solution is calculated. The case of collision is also examined in some instances. Furthermore, the re-entry problem is discussed as an application of the general procedure developed in paper 26.

ΠΡΑΚΤΙΚΑ
ΤΗΣ ΑΚΑΔΗΜΙΑΣ ΑΘΗΝΩΝ

ΕΤΟΣ 1957: ΤΟΜΟΣ 32ος

ΑΝΑΤΥΠΟΝ
ΣΕΛ. 77 - 85

# Subharmonics of any order in case of non linear restoring force. Part I.,

*by* ***Dem. G. Magiros******.**

Ἀνεκοινώθη ὑπό τοῦ κ. Βασ. Αἰγινήτου.

*Introduction.*

We discuss here the subharmonics of any order in the case of linear damping, sinusoidal external force, and cubic type restoring force, with coefficients not necessarily small. By using ideas of Van der Pol[1], Mandels-

* ΔΗΜ. Γ. ΜΑΓΕΙΡΟΥ, Περὶ τῶν ὑποαρμονικῶν ταλαντώσεων οἱασδήποτε τάξεως. Μέρος Α'.

[1] VAN DER POL, *Phil. Mag.*, **3**, 1927, 65.

Reprinted from the *Proceedings of the Athens Academy of Sciences* **32** (1957), 77–85.

tam - Papalexi[1], Andronow - Witt[2], we get a proper transformation of the equation, and for its «steady state» and «transient» solutions use of the Poincaré's method for periodic solutions is made.

The conditions for the existence of the subharmonics and their stability in a steady state are discussed by considering that of the singularities of the corresponding equation.

The formulae given here can be used for investigation of the subharmonics of any order.

§ 1. *The problem.*

Many problems in Physics lead to the differential equation of the form:

$$\ddot{Q} + \bar{k}\dot{Q} + \bar{c}_1 Q + \bar{c}_2 Q^2 + \bar{c}_3 Q^3 = B \sin n\tau, \qquad n = 2, 3, \dots, \tag{1}$$

where the coefficients $\bar{k}$, $\bar{c}_1$, $\bar{c}_2$, $\bar{c}_3$, B are not necessarely small. The solution of (1) is known when the coefficients are small, but it is unknown in case of not necessarily small coefficients.

We intend to find the solution in this last case.

§ 2. *Proper transformation.*

We transform (1) by taking a parameter $\varepsilon$ such that:

$$\bar{k} = \varepsilon k, \quad 1 - \bar{c}_1 = \varepsilon c_1, \quad \bar{c}_2 = \varepsilon c_2, \quad \bar{c}_3 = \varepsilon c_3 \tag{2}$$

The result of this transformation is:

$$\ddot{Q} + Q = \varepsilon f(Q, \dot{Q}) + B \sin n\tau \tag{3}$$

$$f(Q, \dot{Q}) = -k\dot{Q} + c_1 Q - c_2 Q^2 - c_3 Q^3 \tag{3α}$$

The coefficients and $\varepsilon$ in (2) are finite.

a) In case $\varepsilon = 0$, the solution of (1) is:

$$Q = x \sin\tau - y \cos\tau + \frac{B}{1 - n^2} \sin n\tau, \tag{4}$$

with period $2\pi$; $n \neq 1$. The arbitrary constants x and y can be determined by using the initial conditions of (1). x and y are the components of the subharmonic of order $\frac{1}{n}$, of which $r = (x^2 + y^2)^{1/2}$ is the amplitude. The third part in (4) is the harmonic part of the solution due to the forcing term of (1).

[1] L. MANDELSTAM - N. PAPALEXI, *Techn. Phys.*, U. S. S. R., 1935, 415.

[2] A. ANDRONOW - A. WITT, Arch. für Electrotechn., 1930.

b) In case $\varepsilon \neq 0$, we attempt to determine a periodic solution of (3) of the form (4), in which x and y are fonctions of $\varepsilon$ and $\tau$ and such that the limits:

$$\lim_{\varepsilon \to 0} x, \quad \lim_{\varepsilon \to 0} y \tag{5}$$

are the constants of the «generating solution» in case $\varepsilon = 0$.

§ 3. *Reduction to a first order system.*

We take a new variable q according to:

$$q = Q - \frac{B}{1-n^2} \sin n\tau. \tag{6}$$

Substituting (6) into (3) we get:

$$\ddot{q} + q = \varepsilon f(q, \dot{q}) \tag{7}$$

where:

$$\begin{aligned} f(q, \dot{q}) = \Big[ & -k\dot{q} + c_1 q - c_2 q^2 - c_3 q^3 - \frac{kn}{1-n^2} B \cos n\tau \\ & + c_1 \frac{B}{1-n^2} \sin n\tau - c_2 \frac{B^2}{(1-n^2)^2} \sin^2 n\tau - c_3 \frac{B^3}{(1-n^2)^3} \sin^3 n\tau \\ & - 2c_2 \frac{B}{1-n^2} q \sin n\tau - 3c_3 \frac{B}{1-n^2} q^2 \sin n\tau \\ & - 3c_3 \frac{B^2}{(1-n^2)^2} q \sin^2 n\tau \Big] . \end{aligned} \tag{7α}$$

Introduce into (7) new variables $u_1$ und $u_2$ defined by:

$$\begin{cases} u_1 = \dot{q} \cos \tau + q \sin \tau \\ u_2 = \dot{q} \sin \tau - q \cos \tau , \end{cases} \tag{8}$$

from which we get:

$$\begin{cases} q = u_1 \sin \tau - u_2 \cos \tau \\ \dot{q} = u_1 \cos \tau + u_2 \sin \tau , \end{cases} \tag{9}$$

and

$$\begin{cases} \dot{u}_1 = (\ddot{q} + q) \cos \tau \\ \dot{u}_2 = (\ddot{q} + q) \sin \tau , \end{cases} \tag{10}$$

when according to (7), we have:

$$\begin{cases} \dot{u}_1 = \varepsilon f_1(u_1, u_2, \tau) \\ \dot{u}_2 = \varepsilon f_2(u_1, u_2, \tau) , \end{cases} \tag{11}$$

where:

$$(11\alpha) \qquad \begin{cases} f_1(u_1, u_2, \tau) = f(u_1, u_2, \tau) \cdot \cos\tau \\ f_2(u_1, u_2, \tau) = f(u_1, u_2, \tau) \cdot \sin\tau ; \end{cases}$$

The function $f(u_1, u_2, \tau)$ is given from (7α) by using (9).

The equation (3) is replaced by the first order system (11), which gives advantages in the analysis.

From (6) and the first of (9) we obtain:

$$(12) \qquad Q = u_1 \sin\tau - u_2 \cos\tau + \frac{B}{1-n^2} \sin n\tau .$$

The expressions (4) and (12) are of the same form, then we ask for a determination of the limits:

$$(5\alpha) \qquad \lim_{\varepsilon\to 0} u_1 , \quad \lim_{\varepsilon\to 0} u_2$$

§ 4. *Discussion of a general system.*

We briefly discuss, as it is needed for our purpose here, the solution of the general system:

$$(N) \qquad \begin{cases} \dot{u}_1 = \varepsilon \bar{f}_1(u_1, u_2, \tau) \\ \dot{u}_2 = \varepsilon \bar{f}_2(u_1, u_2, \tau) \\ u_1(\tau_0) = u_{1\tau_0}, \; u_2(\tau_0) = u_{2\tau_0} , \end{cases}$$

where the functions $\bar{f}_1$ and $\bar{f}_2$ are analytic in $u_1$ and $u_2$, and periodic of period $2\pi$ and continuous in $\tau$, hence continuous in $u_1$, $u_2$, $\tau$, and therefore $|\bar{f}_1|$ and $|\bar{f}_2|$ have upper bounds $M_1$ and $M_2$ respectively in the domain:

$$(D) \qquad |u_1 - u_{1\tau_0}| < r_1 , \; |u_2 - u_{2\tau_0}| < r_2 , \quad \tau_0 \leq \tau \leq T ,$$

and in this domain $\bar{f}_1$ and $\bar{f}_2$ are expansible as power series in $(u_1 - u_{1\tau_0})$ and $(u_2 - u_{2\tau_0})$ and convergent.

The number $\varepsilon$ is real. The system (11) of our problem is a special case of the system (N).

α) *The formal solution.*

We want to find a solution $u_1$ and $u_2$ of the system (N) such that, if $\varepsilon \to 0$, $u_1$ and $u_2$ tend to the constants x and y. For $\varepsilon \neq 0$, $u_1$ aud $u_2$ depend on $\varepsilon$ and $\tau$, and assume that for $\tau = \tau_0$ they differ a little from x and y,

$$(13) \qquad u_{1\tau_0} = x + \xi , \quad u_{2\tau_0} = y + \eta$$

$\xi$ and $\eta$ are very small.

Take as formal solutions $u_1$ and $u_2$ of (N) the expressions:

(14)
$$\begin{cases} u_1 = X_1^{(0)} + \varepsilon X_1^{(1)} + \varepsilon^2 X_1^{(2)} + \dots \\ u_2 = X_2^{(0)} + \varepsilon X_2^{(1)} + \varepsilon^2 X_2^{(2)} + \dots \end{cases}$$

where:

(14α)
$$X_1^{(0)} = x + \xi(\varepsilon, \tau_0) \qquad X_2^{(0)} = y + \eta(\varepsilon, \tau_0).$$

The coefficients X of (14) are regular functions of $x, y, \xi, \eta, \tau$ if $|x|, |y|, |\xi|, |\eta|, |\tau-\tau_0|$ are less than certain constants.

β) *The coefficients of the formal solution.*

By presupposing the series (14) as convergent, we can find that the coefficients X of (14) are given by:

(15)
$$X_1^{(1)} = \int_{\tau_0}^{\tau} [\bar{f}_1] d\tau,$$

$$X_2^{(1)} = \int_{\tau_0}^{\tau} [\bar{f}_2] d\tau,$$

$$X_1^{(2)} = \int_{\tau_0}^{\tau} \left\{ X_1^{(1)} [\bar{f}_1]_{u_1} + X_2^{(1)} [\bar{f}_1]_{u_2} \right\} d\tau,$$

$$X_2^{(2)} = \int_{\tau_0}^{\tau} \left\{ X_1^{(1)} [\bar{f}_2]_{u_1} + X_2^{(1)} [\bar{f}_2]_{u_2} \right\} d\tau,$$

$$X_i^{(3)} = \int_{\tau_0}^{\tau} \sum_{j=1}^{2} [\bar{f}_i]_{u_j} X_j^{(2)} + \frac{1}{2} \int_{\tau_0}^{\tau} \sum_{j=1}^{2} \sum_{l=1}^{2} [\bar{f}_i]_{u_j u_l} X_j^{(1)} X_l^{(1)},$$

. . . . . . . . . . . . . . . . . . . . . . . . . .

$j, l, i = 1, 2.$

The brackets indicate that the corresponding functions and their derivatives are taken at $u_1 = u_{1\tau_0} = x + \xi$, $u_2 = u_{2\tau_0} = y + \eta$. By carrying out the integrations in the first two of (15) we find the functions $X_1^{(1)}$, $X_2^{(1)}$. Upon substituting the $X_1^{(1)} X_2^{(1)}$ into the second two of (15), the integrands become known continuous functions of $\tau$, then $X_1^{(1)}$, $X_2^{(2)}$ are defined by quadratures. With the same procedure we can find $X_1^{(3)}, X_2^{(3)}, \dots$

γ) *The convergence of the formal sotution.*

To prove the convergence and to find the domain of the validity of the series (14), we use the «method of dominants». We can prove that the condition for the convergence is:

(16)
$$|\varepsilon| < \frac{r}{4M(\tau - \tau_0)},$$

(16α)
$$M \geq M_i, \quad r \leq r_i, \quad i = 1, 2.$$

In many cases the domain of convergence of (14) is much larger than the condition (16) shows.

The inequality (16) can be satisfied by imposing restrictions upon both $\varepsilon$ and $\tau$, or by taking $\tau$ arbitrarily and restricting $\varepsilon$ or by taking $\varepsilon$ arbitrarily and then restricting $\tau$.

§ 5. *Use of the periodicity.*

Take now the solution $u_1$ and $u_2$ of (N) as periodic one in $\tau$ with period $2\pi$, that is:

(17) $$u_1(\tau_0 + 2\pi) - u_1(\tau_0) = 0, \qquad u_2(\tau_0 + 2\pi) - u_2(\tau_0) = 0.$$

Applying this periodicity condition to the series (14) we have:

(18) $$\begin{aligned} &\{X_1^{(1)}(\tau_0 + 2\pi) - X_1^{(1)}(\tau_0)\} + \varepsilon \{X_1^{(2)}(\tau_0 + 2\pi) - X_1^{(2)}(\tau_0)\} + \ldots = 0, \\ &\{X_2^{(1)}(\tau_0 + 2\pi) - X_2^{(1)}(\tau_0)\} + \varepsilon \{X_2^{(2)}(\tau_0 + 2\pi) - X_2^{(2)}(\tau_0)\} + \ldots = 0. \end{aligned}$$

If the Fourier series developments in $\tau$ of $\bar{f}_1$ and $\bar{f}_2$ are:

(19) $$\begin{aligned} \bar{f}_1 &= A_0 + A_1 \cos\tau + B_1 \sin\tau + \ldots + A_m \cos m\tau + B_m \sin m\tau + \ldots \\ \bar{f}_2 &= C_0 + C_1 \cos\tau + D_1 \sin\tau + \ldots + C_m \cos m\tau + D_m \sin m\tau + \ldots, \end{aligned}$$

where the coefficients A, B, C, D are functions of: $u_{1\tau_0} = x + \xi$, $u_{2\tau_0} = y + \eta$, and such that the developments of (19) are convergent. By taking into account the expression of X given by (15), the conditions (18) give:

(20) $$\begin{aligned} &A_0(x + \xi, y + \eta) + \varepsilon \varphi_1(x + \xi, y + \eta, \tau_0) + \ldots = 0, \\ &C_0(x + \xi, y + \eta) + \varepsilon \varphi_2(x + \xi, y + \eta, \tau_0) + \ldots = 0. \end{aligned}$$

Provided that the jakobian of (20) is not zero, i. e.

(21) $$\begin{vmatrix} \frac{\partial A_0}{\partial \xi} & \frac{\partial C_0}{\partial \xi} \\ \frac{\partial A_0}{\partial n} & \frac{\partial C_0}{\partial n} \end{vmatrix} \neq 0$$

we can solve the system (20) in $\xi$ and $\eta$ interms of $\varepsilon$ and $\tau_0$:

(22) $$\xi = \xi(\varepsilon, \tau_0), \qquad \eta = \eta(\varepsilon, \tau_0),$$

with the conditions:

$$\xi = \xi(o, \tau_0) = 0, \qquad \eta = \eta(o, \tau_0) = 0,$$

If the Jakobian is zero, we consider in (20) terms of 1, 2,... degree in $\varepsilon$, that is the functions $\varphi_1, \varphi_2, \ldots$[1]

From (22) and (13) we get:

[1] P. Fatou, *Bulletin Société Math. de France*, 1928 - 30, pp. 112 - 115.

(23) $$u_{1\tau_0} = x + \xi(\varepsilon, \tau_0), \quad u_{2\tau_0} = y + \eta(\varepsilon, \tau_0);$$

$\xi$ and $\eta$ are given in series in $\varepsilon$, $\tau_0$ in the interval $[0,2\pi]$. $\varepsilon$ is under the condition (16). Then is the series (14) the only unknown are the constants x and y.

§ 6. *The components of the amplitude of the subharmonics. Conditions of existence of the subharmonics.*

The equations (20) must be satisfied in case: $\varepsilon=\xi=\eta=0$. In this case the equations (20) give:

(24) $$A_0(x, y)=0, \quad C_0(x, y)=0.$$

The solutions (x, y) of the system (24) give the limits x and y, when the steady state solutions of the original equation are known in the form (4).

The conditions for real intersections of the curves (24) in the x, y-plane give the conditions of the existence of the subharmonics of our equation (3).

§ 7. *The stability of the steady state subharmonics.*

For the stability of the steady state subharmonics we study the stability of the singularities of the equation:

(25) $$\frac{du_2}{du_1} = \frac{\bar{f}_2}{\bar{f}_1},$$

which comes from the system (N). The difficulty is that $\bar{f}_1$ and $\bar{f}_2$ depend on time $\tau$.

But by taking into account the developements of $\bar{f}_1, \bar{f}_2$ given by (19), and that *their mean values with respect to time* $\tau$ *over the period* $2\pi$ *are* $A_0$ *and* $C_0$ *respectively,* the singularities of (25) are that of the equation:

(25α) $$\frac{dy}{dx} = \frac{C_0(x, y)}{A_0(x, y)},$$

then the singularities are given by the solutions (x, y) of the system (24).

According to the corresponding theory of Poincaré[1] and Bendixson[2] the distinction between the different kinds of the singularities depends on two numbers $\varrho_1$ and $\varrho_2$, the roots of the characteristic equation:

(26) $$\begin{vmatrix} a_1-\varrho & b_1 \\ a_2 & b_1-\varrho \end{vmatrix} = 0,$$

[1] H. Poincaré, Sur les courbes définies par une équation différentielle. *Œuvres*, Gauthier-Villars, Paris, Vol. 1892.

[2] I. Bendixson, *Acta Math.*, **24**, 1901.

where:

(26α) $$\alpha_1 = \frac{\partial A_0}{\partial x}, \quad \alpha_2 = \frac{\partial A_0}{\partial y}, \quad b_1 = \frac{\partial C_0}{\partial x}, \quad b_2 = \frac{\partial C_0}{\partial y}$$

For non zero roots $\varrho_1$ and $\varrho_2$ of (26) the «simple singularities» are classified in the following classes:

I: *«nodal points»*, when $\varrho_1$, $\varrho_2$, are real and of the same sign,

II: *«saddle points»*, when $\varrho_1$, $\varrho_2$, are real but of opposite sign,

III: *«spiral points»*, when $\varrho_1$, $\varrho_2$, are complex conjugates, aud

IV: *«spiral points* or *centers»*, when $\varrho_1$, $\varrho_2$, are pure imaginaries.

The condition of the roots being pure imaginaries, which is a necessary condition for being a center, it is not a sufficient condition. There is the Poincaré's criterion[1] for distinguishing spiral from center in this case.

We define the above singularities as *«stable»* or *«unstable»*, when any point on any integral curve moves into the said singularity or not with increasing time τ, i.e. according as the real part of the roots is negative or positive respectively.

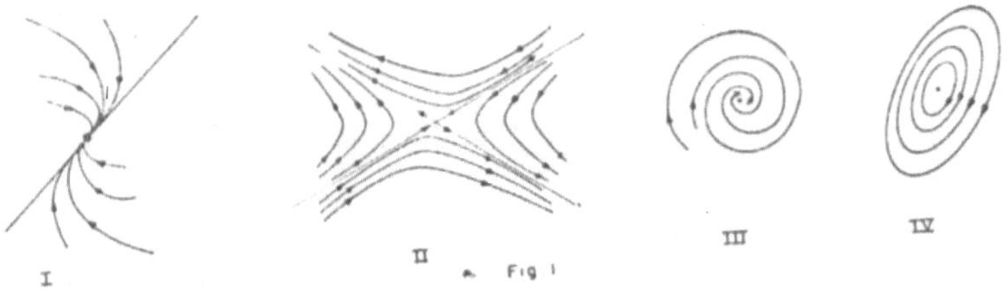


Fig 1

The singularities are shown in Fig. 1, where I is a «nodal stable», III a «stable spiral», IV a (neutral) center, and II a «saddle point intrinçically unstable».

§ 8. *Application to the system* (N).

Let us apply the previous theory in the case of the equation $\bar{f}_1$ and $\bar{f}_2$ given by (11α).

The important thing here is to give to functions $f_1$ und $f_2$ a proper form from which we can get the development in Fourier series. If we replace the

[1] J. HADAMARD, *Rice Institute Pamphlet*, **20**, 1, 1933, 9 - 28.

powers and products of sines and cosines of multiple angles in constructing the functions $f_1$ and $f_2$ from the function f, we get the following results, if we restrict ourselves to the coefficients which are useful for the construction of the functions $A_0$ and $C_0$, which depend on the number n characterizing the order of the subharmonics:

$$f_1(u_1, u_2, \tau) = \left\{ -\frac{1}{2} k u_1 - \frac{1}{2} c_1 u_2 + \frac{3}{8} c_3 u_2^3 + \frac{3}{8} c_3 u_1^2 u_2 + \frac{3}{4} c_3 \frac{B}{(1-n^2)^2} u_2 \right\} +$$

$$\tag{27} \begin{aligned} &+ \dots \\ &+ \left\{ -\frac{1}{2} c_2 \frac{B}{1-n^2} u_1 \right\} \cos(n-2)\tau + \\ &+ \left\{ \frac{3}{4} c_3 \frac{B}{1-n^2} u_1 u_2 \right\} \cos(n-3)\tau + \dots \end{aligned}$$

$$f_2(u_1, u_2, \tau) = \left\{ \frac{1}{2} c_1 u_1 - \frac{1}{2} k u_2 - \frac{3}{8} c_3 u_1^3 - \frac{3}{8} c_3 u_1 u_2^2 - \frac{3}{4} c_3 \frac{B^2}{(1-n^2)^2} u_1 \right\} +$$

$$\tag{28} \begin{aligned} &+ \dots \\ &+ \left\{ \frac{1}{2} c_2 \frac{B}{1-n^2} u_2 \right\} \cos(n-2)\tau + \dots \\ &+ \left\{ -\frac{3}{8} c_3 \frac{B}{1-n^2} (-u_1^2 + u_2^2) \right\} \cos(n-3)\tau + \dots \end{aligned}$$

All these are referred to any order $\frac{1}{n}$ of subharmonics. In a next paper we shall apply the above theory for the subharmonics of order one third.

### ΠΕΡΙΛΗΨΙΣ

Εἰς τὴν ἐργασίαν ταύτην μελετῶνται αἱ ὑποαρμονικαὶ ταλαντώσεις οἱασδήποτε τάξεως εἰς τὴν περίπτωσιν μὴ γραμμικῆς ἐλαστικῆς δυνάμεως. Ἡ ἀντίστοιχος διαφορικὴ ἐξίσωσις μετασχηματίζεται καταλλήλως διὰ χρησιμοποιήσεως ἰδεῶν τῶν Van der Pol, Mandelstam - Papalexi καὶ Andronow - Witt. Σπουδάζονται λύσεις «εὐσταθεῖς» καὶ «μεταβατικαὶ» διὰ χρησιμοποιήσεως τῆς μεθόδου περιοδικῶν λύσεων τοῦ Poincaré. Ἡ σπουδὴ τῶν συνθηκῶν διὰ τὴν ὕπαρξιν τῶν ὑποαρμονικῶν ταλαντώσεων καὶ τὴν εὐστάθειάν των ἀνάγεται εἰς τὴν σπουδὴν τῶν ἀνωμάλων σημείων τῆς ἀντιστοίχου ἐξισώσεως. Εἰς ἑπομένην ἀνακοίνωσιν θὰ ἐκτεθῇ ἡ ἔρευνα παρ' ἐμοῦ τῶν ὑποαρμονικῶν ταλαντώσεων τάξεως ἑνὸς πρὸς τρία, ὡς ἐφαρμογὴ τῶν γενικῶν σκέψεων τῆς παρούσης ἀνακοινώσεως.

ΠΡΑΚΤΙΚΑ
ΤΗΣ ΑΚΑΔΗΜΙΑΣ ΑΘΗΝΩΝ

ΕΤΟΣ 1957: ΤΟΜΟΣ 32ΟΣ

ΑΝΑΤΥΠΟΝ
ΣΕΛ. 101 - 108

# Subharmonics of order one third in the case of cubic restoring force. Part II,

*by* ***Dem. G. Magiros****.

Ἀνεκοινώθη ὑπὸ τοῦ κ. Βασιλ. Αἰγινήτου.

*Introduction*

In this paper we discuss briefly the subharmonics of order one third in the case of a cubic restoring force.

The properly transformed equations, that give the components of the amplitude of the subharmonics, contain, if the amplitude of the external

* ΔΗΜ. ΜΑΓΕΙΡΟΥ, Περὶ τῶν ὑποαρμονικῶν ταλαντώσεων τάξεως ἑνὸς πρὸς τρία.

Reprinted from the *Proceedings of the Athens Academy of Sciences* **32** (1957), 101–108.

force takes the values of an interval given in the paper, the ratios of the coefficients of the damping and the restoring force, and these ratios, under certain condition, can have any values, the coefficients themselves need not necessarily be small, a case very important in many engineering problems. We solve here the problem of subharmonics, their existence and stability, in case of given coefficients of the damping and the cubic restoring force, and the amplitude of the sinusoidal external force. An example is illustrated and the corresponding sketch of the singularities and the integral curves in the whole plane, which is separated into proper regions, is given.

§ I. *The equations for the components of the amplitude of subharmonics.*

The equation to be solved is:

$$\ddot{Q} + \bar{k}\dot{Q} + \bar{c}_1 Q + \bar{c}_2 Q^2 + \bar{c}_3 Q^3 = B \sin 3\tau. \tag{1}$$

By using:

$$\bar{k} = \varepsilon k, \quad 1 - \bar{c}_1 = \varepsilon c_1, \quad \bar{c}_2 = \varepsilon c_2, \quad \bar{c}_3 = \varepsilon c_3, \tag{2}$$

the equation is transformed into:

$$\ddot{Q} + Q = \varepsilon f(Q, \dot{Q}) + B \sin 3\tau, \tag{3}$$

$$f(Q, \dot{Q}) = -k\dot{Q} + c_1 Q - c_2 Q^2 - c_3 Q^3. \tag{3α}$$

In case $\varepsilon = 0$, the solution of (3) is given by:

$$Q = x \sin \tau - y \cos \tau - \frac{B}{8} \sin 3\tau, \tag{4}$$

where x and y are coustants, known for given initial conditions.

In case $\varepsilon \neq 0$ we try to determine the steady state solutions of the equation (3), i.e. the constant limits: $\underset{\varepsilon \to 0}{x(\varepsilon, \tau)}$, $\underset{\varepsilon \to 0}{y(\varepsilon, \tau)}$, according to the previous paper, part I.[1]

For this we have to find the functions $A_0(x, y)$ and $C_0(x, y)$, of the paper [I]. These functions come from the equations (27) aud (28) of the paper [I], if we put $n = 3$ and x and y instead of $u_1$ and $u_2$ respectively, when the result is:

---

[1] D. G. MAGIROS, Subharmonics of any order in case of nonlinear restoring forces. *Praktika of Athens Academy* 32, 1957, pp. 77.

$$(5)\quad \begin{aligned} A_o(x, y) &= \frac{1}{2}\left\{-kx - c_1 y + \frac{3}{4} c_3 y \left(x^2+y^2+\frac{B^2}{32}\right) - \frac{3}{16} c_3 B\, xy\right\}, \\ C_o(x, y) &= \frac{1}{2}\left\{c_1 x - ky - \frac{3}{4} c_3 x \left(x^2+y^2+\frac{B^2}{32}\right) + \frac{3}{32} c_3 B\,(-x^2+y^2)\right\}, \end{aligned}$$

and the equations which give the unknown x and y are:

$$(6)\quad \begin{aligned} kx + c_1 y - \frac{3}{4} c_3 y \left(x^2+y^2+\frac{B^2}{32}\right) + \frac{3}{16} Bxy = 0, \\ c_1 x - ky - \frac{3}{4} c_3 x \left(x^2+y^2+\frac{B^2}{32}\right) + \frac{3}{32} c_3 B\,(-x^2+y^2) = 0. \end{aligned}$$

In the case $c_3 \neq 0$, (6) can be written as:

$$(7)\quad \begin{aligned} \mu x + \lambda y - y\left(x^2+y^2+\frac{B^2}{32}\right) + \frac{1}{4} Bxy = 0, \\ \lambda x - \mu y - x\left(x^2+y^2+\frac{B^2}{32}\right) + \frac{1}{8} B(-x^2+y^2) = 0, \end{aligned}$$

with:

$$(7\alpha)\quad \lambda = \frac{4}{3}\frac{c_1}{c_3}, \quad \mu = \frac{4}{3}\frac{k}{c_3}.$$

We ask for «real solutions» (x, y) of the system (7).

*Remarks.* From the prescribed initial conditions of the equation (3), say $Q_o$ and $\dot{Q}_o$ at $\tau=0$, we have, according to (4),

$$(8\alpha)\quad Q_o = -y, \quad \dot{Q}_o = x - \frac{3}{8} B,$$

when the given initial conditions correspond to the point:

$$(8\beta)\quad x = \dot{Q}_o + \frac{3}{8} B, \quad y = -Q_o,$$

in the x, y—plane, which is the «starting point».

Starting from the «starting point» and following the corresponding integral curve with the lapse of time we can terminate to a «final point», which corresponds to the proper steady solution. The coordinates of the «final point» are solutions of the system (7). Given the initial conditions and the amplitude B of the external force, the «starting point» in the x, y—plane is defined; conversly, any point of x, y—plane can be taken as «starting point» by properly choosing the initial conditions and the amplitude B.

If the «starting point» is selected in coincidence with a «stable final point», no «transient phenomena» may exist.

§ II. *Restrictions to the coefficients of the equation* (1).

If:

$$(8)\quad A = r^2 + \frac{B^2}{32}, \quad r^2 = x^2 + y^2,$$

the system (7) is written as:

$$\text{(9)}\qquad \begin{aligned} \mu x+\lambda y-Ay&=-\frac{1}{8}B(2xy)\\ \lambda x-\mu y-Ax&=-\frac{1}{8}B(-x^2+y^2). \end{aligned}$$

Squaring and adding (9) we find:

$$\text{(10)}\qquad \lambda^2+\mu^2+A^2-2A\lambda=\frac{B^2}{64}r^2;$$

eliminating A between (10) and (8) we have.

$$\text{(11)}\qquad r^4+\left(\frac{3}{64}B^2-2\lambda\right)r^2+\left(\lambda^2+\mu^2+\frac{B^4}{32^2}-\lambda\frac{B^2}{16}\right)=0\,,$$

the roots of which are:

$$\text{(11}\alpha\text{)}\qquad r^2=\frac{1}{2}\left\{2\lambda-\frac{3}{64}B^2\pm\sqrt{\left(2\lambda-\frac{3}{64}B^2\right)^2-4\left(\lambda^2+\mu^2+\frac{B^4}{32^2}-\lambda\frac{B^2}{16}\right)}\right\}.$$

The reality of $r^2$ requires:

$$\text{(12)}\qquad I\equiv 7B^4-2^8\lambda B^2+2^{14}\mu^2\leq 0.$$

The roots of: $I=0$ are:

$$\text{(13)}\qquad B^2=\frac{2^7}{7}\left(\lambda\pm\sqrt{\lambda^2-7\mu^2}\right),$$

then the condition (12) requires the following conditions to be fulfilled:

$$\text{(14)}\qquad \begin{cases} \alpha) & \lambda^2-7\mu^2>0\,,\\ \beta) & \dfrac{2^7}{7}\left(\lambda-\sqrt{\lambda^2-7\mu^2}\right)\leq B^2\leq\dfrac{2^7}{7}\left(\lambda+\sqrt{\lambda^2-7\mu^2}\right). \end{cases}$$

By using (7α) and (2) we find the following restrictions for $\bar{k}$, $\bar{c}_1$, $\bar{c}_3$, B:

$$\text{(15)}\qquad \begin{cases} \alpha) & \left(\dfrac{1-\bar{c}_1}{\bar{k}}\right)^2>7\,,\\ \beta) & \dfrac{2^9}{3\cdot 7}\left(\dfrac{1-\bar{c}_1}{\bar{c}_3}-\sqrt{\left(\dfrac{1-\bar{c}_1}{\bar{c}_3}\right)^2-7\left(\dfrac{\bar{k}}{\bar{c}_3}\right)^2}\right)\leq B^2\leq\dfrac{2^9}{3\cdot 7}\left(\dfrac{1-\bar{c}_1}{\bar{c}_3}+\right.\\ & \left.+\sqrt{\left(\dfrac{1-\bar{c}_1}{\bar{c}_3}\right)^2-7\left(\dfrac{\bar{k}}{\bar{c}_3}\right)^2}\right). \end{cases}$$

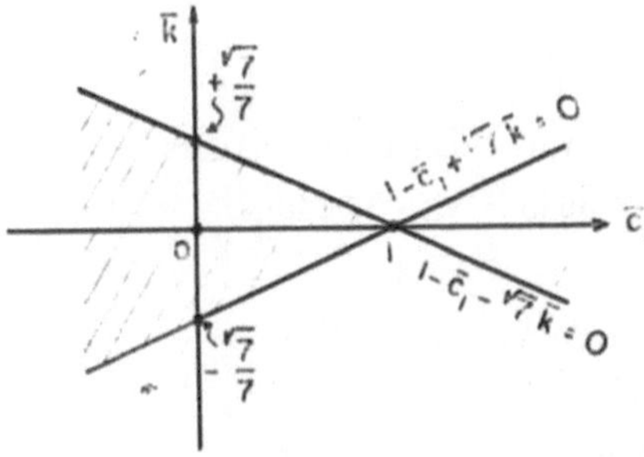


Fig. 1

The inequality (15α) can be written as:

$$\left(1-\bar{c}_1-\sqrt{7}\,\bar{k}\right)\left(1-\bar{c}_1+\sqrt{7}\,\bar{k}\right)>0,$$

then only the shaded region in Fig. 1 is valid in the $\bar{c}_1$, $\bar{k}$-plane.

From (11) or (11α) we can draw $r^2$, versus $B^2$, and by using (15β) we can have the arcs of the diagram which are valid in our problem.

§ III. *The solutions of the system (9).*

The system (9) can be written as:

$$(16)\qquad \begin{aligned} &\left(\mu+\frac{1}{4}By\right)x+(\lambda-A)y=0,\\ &\frac{1}{8}Bx^2-(\lambda-A)x+\left(\mu-\frac{1}{8}By\right)y=0\,\cdot \end{aligned}$$

The vanishing of, say, the Sylvester's eliminant, which is the condition for common roots of (16), leads, for non-zero roots, to the cubic:

$$(17)\qquad y^3-3\left(\frac{4}{B}\right)^2\left\{(\lambda-A)^2+\mu^2\right\}y-2\left(\frac{4\mu}{B}\right)^3-2\left(\frac{4}{B}\right)^3\mu(\lambda-A)^2=0$$

By knowing the coefficients of (1), we know, from (7α) and (2), λ and μ, then we know the amplitude r from (11α) in its two values. On the circumference with radius r there are one or three singularities, of which the ordinates y are the real roots of (17), when for their abscissas we apply the Pythagoras theorem. The singularities (x, y) are therefore at most seven, included the origin, which, in every case, is a singularity.

§ IV. *Example:*

Given: « $B=4,\ \bar{k}=\frac{3}{16},\ \bar{c}_1=1\frac{3}{4},\ \bar{c}_3=-\frac{1}{2}$ ».

We can find is this special case:

$$(18)\qquad \begin{cases} \dfrac{1-\bar{c}_1}{\bar{c}_3}=\dfrac{c_1}{c_3}=\dfrac{3}{2},\ \dfrac{\bar{k}}{\bar{c}_3}=\dfrac{k}{c_3}=-\dfrac{3}{8},\ \lambda=2,\ \mu=-\dfrac{1}{2},\\ r_1^2\simeq 2,\quad r_2^2\simeq 1{,}25,\quad r_1\simeq 1{,}414,\quad r_2\simeq 1{,}118,\\ A_1\simeq 2{,}5,\quad A_1^2\simeq 6{,}25,\quad A_2\simeq 1{,}75,\quad A_2^2\simeq 3{,}06, \end{cases}$$

then:

$$(19)\qquad \begin{aligned} &\alpha)\quad y_1^3-\frac{3}{2}y_1+\frac{1}{2}=0,\\ &\beta)\quad y_2^3-\frac{15}{16}y_2+\frac{5}{16}=0. \end{aligned}$$

Each of these cubic equations has three real unequal roots:

(20α) $y_{11}\simeq 0{,}9996,\quad y_{12}\simeq -1{,}3645,\quad y_{13}\simeq 1{,}3645,$

the first, and

(21α) $y_{21}\simeq 0{,}7032,\quad y_{22}\simeq -1{,}1034,\quad y_{23}\simeq 1{,}0431,$

the second, when the corresponding abscissas are:

(20β) $x_{11}\simeq 0{,}9996,\quad x_{12}\simeq 0{,}3648,\quad x_{13}\simeq -0{,}3648,$

(21β) $x_{21}\simeq 0{,}8672,\quad x_{22}\simeq 0{,}1755,\quad x_{23}\simeq -0{,}3991.$

§ V. *The stability of the solutions.*

For the study of the stability of the solutions the number $\varepsilon$ enters. The number $\varepsilon$ must be such that:

(22) $$|\varepsilon| < \frac{r}{4M(\tau - \tau_0)},$$

according to paper [1], § IV, γ.

If in (22) the initial time $\tau_0 = 0$, by taking arbitrarily $\varepsilon = 1$, this means that the max $\tau = T$ is taken according to:

(23) $$T < \frac{r}{4M}.$$

Take now the partial derivatives with respect to x and y of the functions $A_0(x,y)$ and $C_0(x,y$ given by (5). By establishing the restriction (23), which corresponds to $\varepsilon = 1$, these partial derivatives can be written as follows:

(24) $$\begin{aligned}
\frac{\partial A_0}{\partial x} &\equiv a_1 = \frac{1}{2}\bar{c}_3\left\{-\frac{\bar{k}}{\bar{c}_3} + \frac{3}{2}xy - \frac{3}{16}By\right\},\\
\frac{\partial A_0}{\partial y} &\equiv a_2 = \frac{1}{2}\bar{c}_3\left\{-\frac{1-\bar{c}_1}{\bar{c}_3} + \frac{3}{4}\left(x^2 + 3y^2 + \frac{B^2}{32}\right) - \frac{3}{16}Bx\right\},\\
\frac{\partial C_0}{\partial x} &\equiv b_1 = \frac{1}{2}\bar{c}_3\left\{\frac{1-\bar{c}_1}{\bar{c}_3} - \frac{3}{4}\left(3x^2 + y^2 + \frac{B^2}{32}\right) - \frac{3}{16}Bx\right\},\\
\frac{\partial C_0}{\partial y} &\equiv b_2 = \frac{1}{2}\bar{c}_3\left\{-\frac{\bar{k}}{\bar{c}_3} - \frac{3}{2}xy + \frac{3}{16}By\right\}.
\end{aligned}$$

The characteristic roots, which help to find the type of the singularity, according to § VII of paper [1], is:

(25) $$p_{1,2} = \frac{1}{2}\left\{a_1 + b_2 \pm \sqrt{(a_1 - b_2)^2 + 4a_2b_1}\right\}.$$

The computation for the singularities of our example, the coordinates of which are given by (20α,β) and (21α,β), gives:

(26)

| | | | | | |
|---|---|---|---|---|---|
| 0: | The origin | $x = 0$ | $y = 0$ | : | «*stable spiral*» |
| I: | The point | $x_{11} = 0{,}9996$ | $y_{11} = 0{,}9996$ | : | » » |
| II: | » » | $x_{12} = 0{,}3648$ | $y_{12} = -1{,}3645$ | : | » » |
| III: | » » | $x_{13} = -0{,}3648$ | $y_{13} = 1{,}3645$ | : | «*saddle point*» |
| IV: | » » | $x_{21} = 0{,}8675$ | $y_{21} = 0{,}7032$ | : | » » |
| V: | » » | $x_{22} = 0{,}1755$ | $y_{22} = -1{,}1034$ | : | » » |
| VI: | » » | $x_{23} = -0{,}3991$ | $y_{23} = 1{,}0431$ | : | » » |

The origin corresponds to «harmonic solution», which, as stable, is acceptable. The points I, II, III are on the circumference with radius $r_1 = 1{,}414$. The ponts I, II correspond to acceptable stable subharmonic solutions. The points IV, V, VI, which are on the circumference with radius $r_2 = 1{,}118$, are «intrinçically unstable».

§ VI. *Non-existence of limiting cycles.*

From (24) we have:

$$\frac{\partial A_0}{\partial x} + \frac{\partial C_0}{\partial y} = -\bar{k} \tag{27}$$

valid in the whole x,y-plane, and according to Bendixson's[1] criterion no limit cycles can exist in the whole x,y-plane.

For $\bar{k} = 0$, some of the singularities may be «centers», then we may have «closed integral curves».

§ VII. *Sketch corresponding to the above example.*

Applying the «method of isoclines» to the differential equation:

$$\frac{dy}{dx} = \frac{C_0(x,y)}{A_0(x,y)}, \tag{28}$$

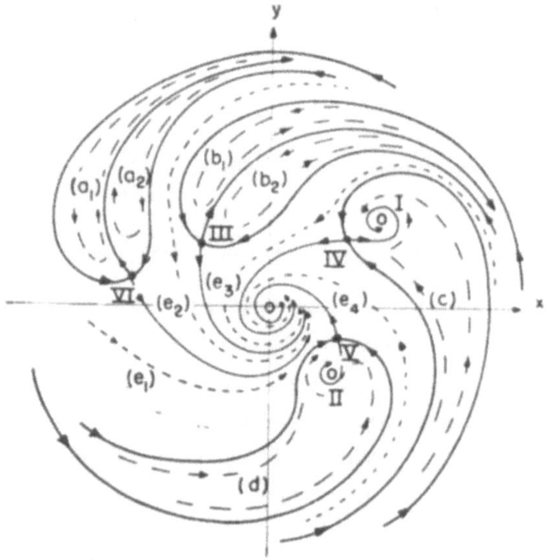


Fig. 2

[1] I. Bendixson, *Acta Math.* 24 (1901), 1-88.

where the functions $A_o(x,y)$ and $C_o(x,y)$ are given by (5), we can have a figure showing the singularities, the integral curves and the separation into regions corresponding to the above example. In Fig. 2 a sketch of these things is given.

The solid lines are the boundaries from the saddle points, the dotted lines are the integral curves. The regions: $(\alpha_1)$, $(\alpha_2)$, $(\beta_1)$ and $(\beta_2)$ correspond to no solutions of our example. The regions: $(e_1)$, $(e_2)$, $(e_3)$ and $(e_4)$ correspond to «stable harmonic solution»: $Q = -\frac{1}{2}\sin 3\tau$. The region (c) correspond to «stable solution»: $Q = 0{,}9996 \sin\tau - 0{,}9996 \cos\tau - \frac{1}{2} \sin 3\tau$; and the region (d) correspond to «stable solution»: $Q = 0{,}3648 \sin\tau + 1{,}3645 \cos\tau - \frac{1}{2}\sin 3\tau$. The amplitude of the subharmonic term in the last two stable solutions is the same: $r_1 = 1{,}414$.

## ΠΕΡΙΛΗΨΙΣ

Εἰς τὴν ἐργασίαν ταύτην συζητεῖται ἐν συντομίᾳ τὸ ζήτημα τῶν ὑποαρμονικῶν ταλαντώσεων τάξεως ἑνὸς πρὸς τρία εἰς τὴν περίπτωσιν κυβικῆς συναρτήσεως τῆς ἐλαστικῆς δυνάμεως. Οἱ συντελεσταὶ τῆς διαφορικῆς ἐξισώσεως εἶναι ὄχι κατ' ἀνάγκην μικρῶν τιμῶν.

Αἱ ἐξισώσεις αἱ δίδουσαι τὰς συνιστώσας τῶν ὑποαρμονικῶν περιέχουν τὰ πηλίκα τῶν συντελεστῶν τῆς ἐλαστικῆς δυνάμεως καὶ τῆς ἀντιστάσεως (damping) καὶ τὰ πηλίκα αὐτὰ ὑπὸ δεδομένας συνθήκας δύνανται νὰ ἔχουν οἱασδήποτε τιμάς, χωρὶς νὰ εἶναι ἀναγκαῖον νὰ δεχθῶμεν μικρὰς τιμὰς διὰ τοὺς συντελεστάς, περίπτωσις πολὺ σημαντικὴ εἰς πολλὰ προβλήματα τῶν μὴ γραμμικῶν ταλαντώσεων. Εὑρίσκονται ἐνταῦθα αἱ συνιστῶσαι τῶν ὑποαρμονικῶν, ἐρευνᾶται τὸ ζήτημα τῆς ὑπάρξεως καὶ εὐσταθείας των εἰς τὴν περίπτωσιν ἐξωτερικῆς δυνάμεως ἡμιτονοειδοῦς τύπου ὑπὸ πλάτος μεταβλητὸν εἰς δεδομένον διάστημα, διδομένων τῶν συντελεστῶν τῆς ἐλαστικῆς δυνάμεως καὶ τῆς ἀντιστάσεως. Δίδεται παράδειγμα ἀριθμητικὸν ὡς ἐφαρμογὴ τῆς θεωρίας, καθὼς καὶ σχεδιάγραμμα ἀντιστοιχοῦν εἰς τὸ παράδειγμα αὐτό.

# Remarks on a problem of subharmonics*,

*by* ***Dem. G. Magiros.***

Ἀνεκοινώθη ὑπὸ τοῦ κ. Βασιλ. Αἰγινήτου.

*Introduction.*

This paper is a supplement of the author's previous paper under the title: «*Subharmonics of order one third in the case of cubic restoring force*», contained in this volume as the author's second work on subharmonics and called in the following «paper B», and its previous «paper A». In the first chapter we discuss the conditions under which the basic equation (1) of the «paper B» accepts the harmonic solution $\left(-\frac{B}{8}\sin 3\tau\right)$ as a stable one, the subharmonic term of the solution being zero. In the second chapter the «inverse problem» of that of the «paper B» is discussed. This «inverse problem», so simple from a mathematical point of view, according to the equations found, seems to be of importance from an engineering point of view.

I. *The singularity of the origin in the general case.*

The basic equation is:

$$\ddot{Q} + \bar{k}\dot{Q} + \bar{c}_1 Q + \bar{c}_2 Q^2 + \bar{c}_3 Q^3 = B\sin 3\tau, \tag{1}$$

and it can be written in the form:

$$\ddot{Q} + Q = \varepsilon f(Q, \dot{Q}) + B\sin 3\tau, \tag{2}$$

with:

$$f(Q, \dot{Q}) = -k\dot{Q} + c_1 Q - c_2 Q^2 - c_3 Q^3, \tag{2α}$$

if:

$$\bar{k} = \varepsilon k, \quad 1 - \bar{c}_1 = \varepsilon c_1, \quad \bar{c}_2 = \varepsilon c_2, \quad \bar{c}_3 = \varepsilon c_3. \tag{3}$$

The steady state solution is:

$$Q = x\sin\tau - y\cos\tau - \frac{B}{8}\sin 3\tau, \tag{4}$$

and the components x and y of the amplitude r of the subharmonic of order one third are given, according to «paper B», by the two equations:

$$\begin{cases} \mu x + \lambda y - y\left(x^2 + y^2 + \frac{B^2}{32}\right) + \frac{1}{4}Bxy = 0, \\ \lambda x - \mu y - x\left(x^2 + y^2 + \frac{B^2}{32}\right) + \frac{1}{8}B(-x^2 + y^2) = 0, \end{cases} \tag{5}$$

---

* Παρατηρήσεις ἐπὶ προβλήματος τῶν ὑποαρμονικῶν.

Reprinted from the *Proceedings of the Athens Academy of Sciences* **32** (1957), 143–146.

where:

$$\lambda=\frac{4}{3}\frac{c_1}{c_3}=\frac{4}{3}\frac{1-\bar{c}_1}{\bar{c}_3}, \qquad \mu=\frac{4}{3}\frac{k}{c_3}=\frac{4}{3}\frac{\bar{k}}{\bar{c}_3}. \tag{5α}$$

From (5) we see that the origin is a singularity of our equation; in other wordes, the function: $Q=-\frac{B}{8}\sin 3\tau$ is always a solution of the equation, the harmonic solution.

For the stability of this harmonic solution we take the derivatives given by (24) in the «paper B», established under the conditions:

$$\varepsilon=1, \qquad T=\max\tau<\frac{r}{4M}. \tag{6}$$

From (24) of the «paper B», in the case x=0, y=0, we get:

$$\begin{aligned} a_1&=-\frac{1}{2}k, & a_2&=-\frac{1}{2}c_1+3c_3\left(\frac{B}{16}\right)^2,\\ b_1&=\frac{1}{2}c_1-3c_3\left(\frac{B}{16}\right)^2, & b_2&=-\frac{1}{2}k, \end{aligned} \tag{7}$$

when, from (26) of the «paper A», the characteristic roots are given by:

$$p_{1,2}=-\frac{1}{2}k\pm i\left\{\frac{1}{2}c_1-3c_3\left(\frac{B}{16}\right)^2\right\}. \tag{8}$$

We, therefore, have, according to the definitions on the singularities of the «paper A» the following:

A. If the imaginary part of the characteristic roots is not zero, that is if:

$$\frac{c_1}{c_3}\neq 6\left(\frac{B}{16}\right)^2, \tag{9}$$

then:

α) for $k>0$, the origin is a «stable spiral point»;

β) for $k<0$, the origin is an «unstable spiral point»;

γ) for $k=0$, the origin is either a «center» or a «spiral point».

B. If the imaginary part is zero, that is if:

$$\frac{c_1}{c_3}=6\left(\frac{B}{16}\right)^2, \tag{10}$$

then:

α) for $k>0$, the origin is a «nodal stable point»;

β) for $k<0$, the origin is an «unstable nodal point»;

γ) for $k=0$, the origin is not a simple singularity of the kind we know from «paper A» and «paper B», since the characteristic roots are zero.

The result from the above is that: the origin is a stable singularity when k>0, of the spiral type under the condition (9), and of the nodal type under the condition (10). In other words, the function $(Q=-\frac{B}{8}\sin 3\tau)$ is the harmonic stable solution of the equation (1) if $k>0$.

We plot in the: $\frac{1-\bar{c}_1}{\bar{c}_3}$, $\bar{k}$—plane the above results. The left half of this plane corresponds to the instability of the zero-subharmonic, that is to instability to the solution $Q=-\frac{B}{8}\sin 3\tau$; the right half to the stability,

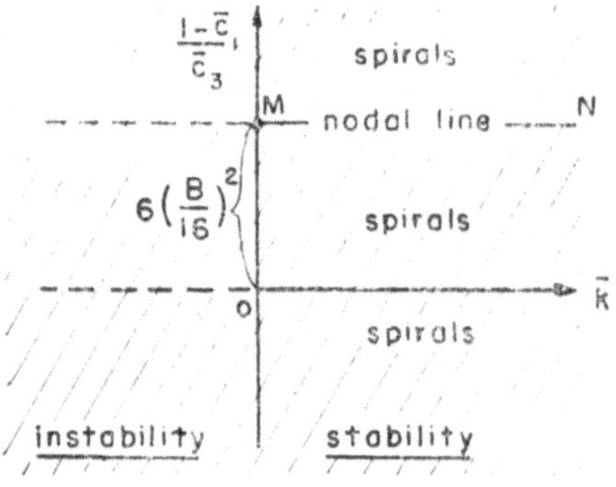


Fig 1

with all «spirals», except the points of the line $\frac{1-\bar{c}_1}{\bar{c}_3}=6\left(\frac{B}{16}\right)^2$ which are «nodals».

The points of the: $\frac{1-\bar{c}_1}{\bar{c}_p}$—axis may be either «centers» or «spirals», except the point M which is not a simple singularity but of an advanced order.

The distance of the «nodal line» MN from the $\bar{k}$—axis has a maximum and a minimum, due to the restrictions of B, given in the «paper B», and in the case of free vibrations (B=0) the «nodal line» is the $\bar{k}$—axis itself.

2. *The inverse problem.*

The inverse problem of that of «paper B» is the following: «*Given the amplitude r of the subharmonic vibrations, find the coefficients of the differential equation, and study the stability of the solutions obtained*».

The solution of this problem correspondsto determine the numbers $\lambda$ and $\mu$ in terms of r and B, by using the additional equation:

$$x^2+y^2-r^2=0. \tag{11}$$

This determination is impossible due to the form of the equations (5).

From (5) and (11) we can have $\lambda$ and $\mu$ in terms of x, y, r, B; then the determination of $\lambda$ and $\mu$ needs to know B and two of x, y, r. If we know B, x, y, then the numbers $\lambda$ and $\mu$ are known by solving the system (5) in $\lambda$ and $\mu$, (11) being a restriction between x, y, r. By knowing $\lambda$ and $\mu$, we know the ratios $\frac{1-\bar{c}_1}{\bar{c}_3}$, $\frac{\bar{k}}{\bar{c}_3}$, then any two coefficients from $\bar{c}_1$, $\bar{c}_3$, $\bar{k}$ can be determined in terms of their third one, which can have arbitrary values, and the «inverse problen» is solved, since the subject of the stability can be treated as is shown in «paper B».

## ΠΕΡΙΛΗΨΙΣ

Ἡ ἐργασία αὕτη ἀποτελεῖ συμπλήρωμα τῆς δευτέρας ἐργασίας ἡμῶν ἐπὶ τῶν ὑποαρμονικῶν ταλαντώσεων (βλ. σελ. 77 κἑξ. τοῦ παρόντος τόμου).

Εἰς ταύτην α΄) δίδονται αἱ συνθῆκαι ὑπὸ τὰς ὁποίας ἡ βασικὴ διαφορικὴ ἐξίσωσις δέχεται λύσιν συνισταμένην ἀπὸ μόνον τὸ ἁρμονικόν της μέρος (ἄνευ ὑποαρμονικοῦ) καὶ δὴ εὐσταθὲς καὶ ἑπομένως φυσικῶς δεκτὴν λύσιν· β΄) ἐξετάζεται τὸ «ἀντίστροφον πρόβλημα» τῆς δευτέρας, ἀνωτέρω μνημονευθείσης, ἐργασίας. Τὸ πρόβλημα τοῦτο ἐμφανίζεται ἐδῶ, βάσει τῆς σειρᾶς συλλογισμῶν τῶν προηγουμένων καὶ τῆς παρούσης ἐργασίας, ὡς ἁπλούστατον μαθηματικῶς, ὅμως ἀποτελεῖ πρόβλημα πολλῆς σπουδαιότητος ἀπὸ τεχνικῆς ἀπόψεως.

ΑΝΑΤΥΠΟΝ
ΣΕΛ. 448 - 451

# On the singularities of a system of differential equations, where the time figures explicitly,

*by* ***Dem. G. Magiros****.

Ἀνεκοινώθη ὑπὸ τοῦ κ. Βασ. Αἰγινήτου.

**1.** In my first paper of this volume is referred, without any explanation, that the singular points of the system:

$$\frac{du_1}{dt} = \varepsilon F_1(u_1, u_2, t), \qquad \frac{du_2}{dt} = \varepsilon F_2(u_1, u_2, t), \tag{1}$$

fulfill the equations:

$$A_0(u_1, u_2) = 0, \qquad C_0(u_1, u_2) = 0, \tag{2}$$

where $A_0$ and $C_0$ are the first terms of the Fourier series expansions of the functions $F_1$ and $F_2$ respectively.

In the following we discuss the above subject. We restrict ourselves to the system (1), although the theory can be applied to more general systems.

A *constant solution* $\{u_1, u_2\}$, of the system (1), determines a point in the $u_1, u_2$ - plane independent of the time t, and this point is, by definition, a *singular point* of the system (1).

In the following we try to find how to determine approximately the singular points of the system (1).

**2.** Suppose we are given that the functions $F_1$ and $F_2$ fulfill the *expansibility conditions* into Fourier series in t[1], according to which we have:

$$\left.\begin{aligned} \frac{du_1}{dt} &= \varepsilon\{A_0 + A_1 \cos t + B_1 \sin t + \ldots + A_m \cos mt + B_m \sin mt + \ldots\} \\ \frac{du_2}{dt} &= \varepsilon\{C_0 + C_1 \cos t + D_1 \sin t + \ldots + C_m \cos mt + D_m \sin mt + \ldots\} \end{aligned}\right\} \tag{3}$$

where the coefficients A, B, C, D are functions of $u_1$, and $u_2$.

By the above we mean that $F_1$ and $F_2$ fulfill the conditions of the *Fourier's theorem*[1], then:

a) $F_1$, $F_2$ are periodic in t of period, say, $2\pi$,

b) $F_1$, $F_2$ are integrable, say Riemann - integrable, in $[t_0, t_0 + 2\pi]$,

c) $F_1$, $F_2$ have limited total fluctuations in $[t_0, t_0 + 2\pi]$, and

d) the coefficients in (3) can be found according to the standard manner.

* ΔΗΜ. ΜΑΓΕΙΡΟΥ, Ἐπὶ τῶν ἀνωμάλων σημείων διαφορικοῦ συστήματος, ὅπου ὁ χρόνος εἰσέρχεται ἐκπεφρασμένως.

Reprinted from the *Proceedings of the Athens Academy of Sciences* **32** (1957), 448–451.

Let us take the system:

$$\frac{du_1}{dt} = \varepsilon A_o(u_1, u_2), \qquad \frac{du_2}{dt} = \varepsilon C_o(u_1, u_2), \tag{4}$$

where the time t does not figure explicitly.

Each of the above systems accepts a unique solution $\{u_1, u_2\}$, which assumes given values: $\{u_{1o}, u_{2o}\}$ at $t=t_o$, provided that the functions in their right-hand sides fulfill a Lipschitz condition, when their arguments are restricted to be in the domain:

$$D: \qquad |u_1 - u_{1o}| \leqq k_1, \qquad |u_2 - u_{2o}| \leqq k_2, \qquad |t - t_o| < T.$$

Apply Picard's method of succesive approximations for calculation of the solution $\{\bar{u}_1, \bar{u}_2\}$ of the system (4), by taking as zeroth approximation arbitrary conditions $\{\bar{u}_{1o}, \bar{u}_{2o}\}$. The successive approximations, which converge to the solution $\{\bar{u}_1, \bar{u}_2\}$ of (4), are:

$$\left.\begin{aligned}
&\overset{(1)}{\bar{u}_1} = \bar{u}_{1o} + \varepsilon\int_{t_o}^{t} A_o(\bar{u}_{1o}, \bar{u}_{2o})\,dt, \qquad \overset{(1)}{\bar{u}_2} = \bar{u}_{2o} + \varepsilon\int_{t_o}^{t} C_o(\bar{u}_{1o}, \bar{u}_{2o})\,dt, \\
&\cdots\cdots\cdots\cdots\cdots\cdots\cdots\cdots \\
&\overset{(n)}{\bar{u}_1} = \bar{u}_{1o} + \varepsilon\int_{t_o}^{t} A_o(\overset{(n-1)}{\bar{u}_1}, \overset{(n-1)}{\bar{u}_2})\,dt, \qquad \overset{(n)}{\bar{u}_2} = \bar{u}_{2o} + \varepsilon\int_{t_o}^{t} C_o(\overset{(n-1)}{\bar{u}_1}, \overset{(n-1)}{\bar{u}_2})\,dt
\end{aligned}\right\} \tag{5}$$

Apply also the above method for calculation of the solution $\{u_1, u_2\}$ of (3) by taking as zeroth approximarion arbitrary initial conditions $\{u_{1o}, u_{2o}\}$ at $t=t_o$. The successive approximations are:

$$\left.\begin{aligned}
&\overset{(1)}{u_1} = u_{1o} + \varepsilon\int_{t_o}^{t} A_o(u_{1o}, u_{2o})\,dt + \varepsilon\int_{t_o}^{t} A_1(u_{1o}, u_{2o})\cos t\,dt + \varepsilon\int_{t_o}^{t} B_1(u_{1o}, u_{2o})\sin t\,dt + \ldots \\
&\overset{(1)}{u_2} = u_{2o} + \varepsilon\int_{t_o}^{t} C_o(u_{1o}, u_{2o})\,dt + \varepsilon\int_{t_o}^{t} C_1(u_{1o}, u_{2o})\cos t\,dt + \varepsilon\int_{t_o}^{t} D_1(u_{1o}, u_{2o})\sin t\,dt + \ldots \\
&\cdots\cdots\cdots\cdots\cdots\cdots\cdots\cdots \\
&\overset{(n)}{u_1} = u_{1o} + \varepsilon\int_{t_o}^{t} A_o(\overset{(n-1)}{u_1}, \overset{(n-1)}{u_2})\,dt + \varepsilon\int_{t_o}^{t} A_1(\overset{(n-1)}{u_1}, \overset{(n-1)}{u_2})\cos t\,dt + \varepsilon\int_{t_o}^{t} B_1(\overset{(n-1)}{u_1}, \overset{(n-1)}{u_2})\sin t\,dt + \ldots \\
&\overset{(n)}{u_2} = u_{2o} + \varepsilon\int_{t_o}^{t} C_o(\overset{(n-1)}{u_1}, \overset{(n-1)}{u_2})\,dt + \varepsilon\int_{t_o}^{t} C_1(\overset{(n-1)}{u_1}, \overset{(n-1)}{u_2})\cos t\,dt + \varepsilon\int_{t_o}^{t} D_1(\overset{(n-1)}{u_1}, \overset{(n-1)}{u_2})\sin t\,dt + \ldots
\end{aligned}\right\} \tag{6}$$

which couverge to the unique solution $\{u_1, u_2\}$ of the system (3).

Let us take the same initial conditions:

$$u_{1o} = \bar{u}_{1o}, \quad u_{2o} = \bar{u}_{2o}, \tag{7}$$

in the approximations (5) and (6), and subtract properly; the result for the $n^{th}$ approximation is:

$$\left.\begin{aligned} \overset{(n)}{u}_1-\overset{(n)}{\bar{u}}_1 &= \varepsilon\int_{t_o}^{t}\{A_o(\overset{(n-1)}{u_1},\overset{(n-1)}{u_2})-A_o(\overset{(n-1)}{\bar{u}_1},\overset{(n-1)}{\bar{u}_2})\}dt+\varepsilon\int_{t_o}^{t}A_1(\overset{(n-1)}{u_1},\overset{(n-1)}{u_2})\cos t\,dt+ \\ &\quad+\varepsilon\int_{t_o}^{t}B_1(u_1,u_2)\sin t\,dt+\ldots \\ \overset{(n)}{u}_2-\overset{(n)}{\bar{u}}_2 &= \varepsilon\int_{t_o}^{t}\{C_o(\overset{(n-1)}{u_1},\overset{(n-1)}{u_2})-C_o(\overset{(n-1)}{\bar{u}_1},\overset{(n-1)}{\bar{u}_2})\}dt+\varepsilon\int_{t_o}^{t}C_1(\overset{(n-1)}{u_1},\overset{(n-1)}{u_2})\cos t\,dt+ \\ &\quad+\varepsilon\int_{t_o}^{t}D_1(\overset{(n-1)}{u_1},\overset{(n-1)}{u_2})\sin t\,dt+\ldots \end{aligned}\right\} \quad (8)$$

The integrals in (8) are bounded and the right-hand sides contain $\varepsilon$ as a common factor, then the approximations, and consequently their limits, are for small $\varepsilon$ of order $\varepsilon$, that is:

$$|u_1-\bar{u}_1|=0(\varepsilon), \qquad |u_2-\bar{u}_2|=0(\varepsilon). \quad (9)$$

In (9) the $\{u_1, u_2\}$ and $\{\bar{u}_1, \bar{u}_2\}$ are solutions of (3) and (4) respectively, then any solution of (4) can be considered as an approximation of the solution of (3) of the first order in $\varepsilon$.

**3.** The constant solutions of (3) come when, in (1), $\varepsilon F_1$ and $\varepsilon F_2$ tend to zero then, since, the time t figures explicitly in $F_1$ and $F_2$, when $\varepsilon \to 0$. But in (4) the time t does not figure explicitly, then we can get constant solutions of (4), if $\varepsilon$ is not necessarily zero, by taking proper initial conditions in the approximations (5), namely the initial conditions $\{\bar{u}_{1o}, \bar{u}_{2o}\}$ which fulfill the conditions:

$$A_o(\bar{u}_{1o}, \bar{u}_{2o})=0, \qquad C_o(\bar{u}_{1o}, \bar{u}_{2o})=0, \quad (10)$$

when the integrals in (5) are zero, and the solution $\{\bar{u}_1, \bar{u}_2\}$ of (4) is the constant $\{\bar{u}_{1o}, \bar{u}_{2o}\}$ for any $\varepsilon$, included $\varepsilon = 0$.

A constant solution $\{\bar{u}_1, \bar{u}_2\}$ of the approximate system (4), which fulfills (10), in considered as an approximate solution of the exact system (3) of first order in $\varepsilon$.

**4.** The above technique of replacing the system (3) where the time t figures explicity, by the approximate system (4), where the time t does not figure explicitly, consists essentially of substituting a function by its *«mean value»* over an interval[2], which is called *«moving average»* or *«sliding* *mean»* [4]. Since the *«moving average»* is, in general, smother than the original function, the above technique, which is known as the *«averaging principle»* [3], offers advantages in the study of the original system, and it is realized in practice with good results, say in economics, or in electrical problems, e. g. in the photoelectric reproduction of sound, in television images, etc. [4].

***Acknowledgments.***

*Most of the work of the papers Part I (above p. 77 - 85) and part II (above p. 101 - 108) was carried outwhen the author was on a project, sponsord by I. B. M. Watson Laboratory of Columbia University. The author is deeply indebted to Prof.* L. H. THOMAS *of Columbia University for his interest and many invaluable discussions concerning the problems of these papers, and to Prof.* J. B. KELLER *of New-York University for his helpful critisism.*

ΠΕΡΙΛΗΨΙΣ

Ἡ ἐργασία αὕτη ἀναφέρεται ἐπὶ τῶν ἀνωμάλων σημείων τοῦ συστήματος (1), τῶν ὁποίων ἡ σπουδὴ γίνεται διὰ τῆς σπουδῆς τῶν ἀνωμάλων σημείων τοῦ συστήματος (4), τὰ ὁποῖα πληροῦν τὰς συνθήκας (2). Διὰ τῆς χρήσεως τῆς ἀνωτέρω μεθόδου, ἡ ὁποία εἶναι γνωστὴ ὡς *«ἀρχὴ τοῦ μέσου ὅρου»*, παρακάμπτονται μεγάλαι μαθηματικαὶ δυσκολίαι, αἱ δὲ λαμβανόμεναι κατὰ προσέγγισιν λύσεις εἶναι εἰς τὴν πρᾶξιν λίαν ἱκανοποιητικαί.

REFERENCE

[1] E. WITTAKER - S. WATSON, A Treatise of Modern Analysis, 4th ed. (1952), 164.

[2] P. FATOU, *Société Math. de France,* Bulletin (1929).

[3] N. KRYLOFF & N. BONGOLIUBOFF, Introduction of Noulinear Mechanics, p. 12, *Ann. of Math. Studies,* 11.

[4] B. VANDER POL & H. BREMMER, Operational Calculus, chapt. XIV, § 2 (1955).

# Subharmonics of Any Order in Nonlinear Systems of One Degree of Freedom: Application to Subharmonics of Order 1/3*

DEMETRIOS G. MAGIROS†

*New York University, New York*

This paper consists of two parts. In the first one we discuss the subharmonics of any order in the case where the nonlinearity enters in the elastic forces. The basic differential equation is with coefficients not necessarily small. The "steady state" and "transient" solutions of the differential equation, and the conditions for the existence of the subharmonics and their stability in a steady state are examined. In the second part of the paper an application is given, namely, the investigation of the subharmonics of order 1/3 according to the theory of the first part. An illustrated example is given, the "inverse problem" is examined, and the conditions for the stability and instability of the "harmonic solution" are found.

## INTRODUCTION

The behavior of an oscillatory system with one degree of freedom is governed by an equation of the form

$$\ddot{Q} = \phi(Q, \dot{Q}, t), \tag{a}$$

$\phi$ being a function nonlinear in the variables $Q$ and $\dot{Q}$, and periodic or almost periodic in $t$. In some cases Eq. (a) leads to the equation

$$\ddot{Q} + Q = \epsilon f(Q, \dot{Q}) + kg(t), \tag{b}$$

from which we may get the first order equation

$$\frac{du_2}{du_1} = \frac{f_2(u_1, u_2, t)}{f_1(u_1, u_2, t)}, \tag{c}$$

where $u_1$ and $u_2$ are related to $Q$ and $\dot{Q}$ by proper relations.

* Most of the work of the paper was carried out when the author was employed on a project sponsored by I.B.M. Watson Laboratory at Columbia University. The author is deeply indebted to Prof. L. H. Thomas of Columbia University for his interest and the many invaluable discussions concerning the problems of the paper, and to Prof. J. B. Keller of New York University for his helpful criticism.

† Present address: Hofstra College, Hempstead, New York.



Reprinted from *Information and Control* 1 (1958), 198–227.

In this paper we treat a special case of the Eq. (a), namely, the problem of subharmonics of any order when the system is governed by an equation of the form

$$\ddot{Q} + \bar{k}\dot{Q} + \bar{c}_1 Q + \bar{c}_2 Q^2 + \bar{c}_3 Q^3 = B \sin nt, \tag{d}$$

where the coefficients are not necessarily very small. This case is very important from a mathematical point of view and very useful in nonlinear engineering problems.

A mechanical model of a system, governed by Eq. (d), might be a mass under the action of a viscous damping force linear in the velocity, of an elastic force which is a cubic function of the deflection, and of a simple harmonic forcing function of a given frequency and not necessarily small amplitude. An electrical model might be an electrical oscillatory circuit in which the nonlinear oscillations take place because of a saturable-core inductance under the impression of an alternating electromotive force of sinusoidal type (Hayachi, 1953). An aerodynamical model could be based on the fact that certain parts of an airplane can be excited to violent oscillations by an engine running with a number of revolutions much larger than the natural frequency of the oscillating parts (Von Kármán, 1940).

By using ideas of Van der Pol (1927), Mandelstam and Papalexi (1935), and Andronow and Witt (1930), we transform Eq. (d) to the forms (b) and (c). We deal with the steady state and transient solutions, by using Poincaré's method for periodic solutions (Friedrichs, 1953; Poincaré, 1890; Tsien, 1956). The condition for the existence of the subharmonics and their stability in a steady state are discussed by considering the stability of the singularities of the corresponding equation of the form (c). In the appendix (Magiros, 1957d) we discuss the fact that the singularities of an ordinary system, where the time enters explicitly, are the same as those of an "approximate system," where the time does not enter explicitly. As an application of the theory we discuss the subharmonics of order $\frac{1}{3}$. In this case, the equations that give the components of the amplitude of the subharmonics in steady state are properly transformed. The components of the amplitude are given in terms of the amplitude of the external force, and the ratios of the coefficients of the linear damping and the cubic elastic force. Under certain conditions given here, these ratios can have any values. The coefficients themselves need not necessarily be small—a case important in theoretical and practical problems. Some bounds are given for the amplitude of the external force, and some restrictions for the coefficients of the differential equation. An example is illustrated and the corresponding sketch of the integral curves in the whole plane, which is separated into appropriate regions, is given. The "converse problem" is also discussed. Finally, we discuss the conditions under which we have the stability or instability of the "harmonic part" of the solution, the subharmonic part being zero.

## PART A: THE GENERAL CASE

### 1. The Problem (Magiros, 1957a)

We discuss subharmonics of any order $1/n (n = 2, 3, \cdots)$ of the differential equation

$$\ddot{Q} + \bar{k}\dot{Q} + \bar{c}_1 Q + \bar{c}_2 Q^2 + \bar{c}_3 Q^3 = B \sin nt, \tag{1}$$

in their steady and transient states, in the case when the coefficients of (1) are not necessarily small—a case very important in engineering problems. Dots in (1) denote derivatives with respect to time $t$. By taking a parameter $\epsilon$ such that

$$\bar{k} = \epsilon k, \qquad 1 - \bar{c}_1 = \epsilon c_1, \qquad \bar{c}_2 = \epsilon c_2, \qquad \bar{c}_3 = \epsilon c_3, \tag{2}$$

Eq. (1) can be written as

$$\ddot{Q} + Q = \epsilon f(Q, \dot{Q}) + B \sin nt, \tag{3}$$

$$f(Q, \dot{Q}) = -k\dot{Q} + c_1 Q - c_2 Q^2 - c_3 Q^3. \tag{3a}$$

The parameter $\epsilon$ is a constant with respect to the variables $Q$, $\dot{Q}$. All the coefficients in (2) must be finite as well as the parameter $\epsilon$. The case $\epsilon = 0$ corresponds to that in which all the coefficients in left-hand members of (2) are zero, except $\bar{c}_1 = 1$, and the coefficients in right-hand members can take any finite value. When $\epsilon \neq 0$, even if $\epsilon$ is small, the coefficients of left-hand members of (2) are not necessarily small, if we take appropriate values for the coefficients of the right-hand members. If we know the values of $Q$ and $\dot{Q}$ at $t = t_0$, say $Q_{t=t_0} = Q_t$, $\dot{Q}_{t=t_0} = \dot{Q}_0$; and if the function $f(Q, \dot{Q})$, given by (3a), is continuous and has continuous derivatives with respect to $Q$ and $\dot{Q}$ in a domain $D$,

$$D: \ Q_0 - \xi_1 \leqq Q \leqq Q_0 + \xi_1, \quad \dot{Q}_0 - \xi_2 \leqq \dot{Q} \leqq \dot{Q}_0 + \xi_2, \quad t_0 \leqq t \leqq T;$$

then $| f(Q, \dot{Q}) |$ has an upper bound in the domain of the variables, which depends on the domain.

In case $\epsilon = 0$, Eq. (3) reduces to

$$\ddot{Q} + Q = B \sin nt, \tag{4}$$

of which the solution of period $2\pi$ is

$$Q = x \sin t - y \cos t + (B/1 - n^2) \sin nt, \tag{5}$$

where $n \neq \pm 1$ and $x$, $y$ are arbitrary constants which can be determined by using the initial conditions.

If $\epsilon \neq 0$, we attempt to determine a periodic solution of Eq. (3) of the form (5), in which $x$ and $y$ are functions of $\epsilon$ and $t$, say $x = u_1(\epsilon, t)$, $y = u_2(\epsilon, t)$, and such that when $\epsilon \to 0$ the limits

$$\lim_{\epsilon\to 0} u_1, \qquad \lim_{\epsilon\to 0} u_2 \tag{6}$$

are the constants of the "generating solution" in case $\epsilon = 0$. Our type of periodic solution, given by (5), consists of two parts. One part, the term $(B/1 - n^2) \sin nt$, with the same frequency as that of the external force, is called the harmonic part of the solution. The other part, $x \sin t - y \cos t$, with frequency a fraction $1/n$ of that of the external force, is called the "subharmonic part," and its amplitude is $\zeta = (x^2 + y^2)^{1/2}$. Given the amplitude and the frequency of the external force, the amplitude of the harmonic part of the solution is constant. But this does not happen with the components $x$ and $y$ of the amplitude of the subharmonic part of the solution.

When $\epsilon = 0$, and for given initial conditions, $x$ and $y$ are constants, and in this case we speak about "subharmonics in steady states of oscillation," either stable or unstable. If $\epsilon \neq 0$, the components $x$ and $y$ are functions of time $t$, we have a "transient phenomenon," and we speak about "subharmonics in transient states." The steady states of oscillation, i.e., the equilibrium positions of the system, are correlated with the singularities of the differential equation; the transient states with their integral curves, which do or do not terminate on the singularities, depending on the stability or instability of the singularities. The solution in the transient state yields, with the lapse of time, ultimately to steady state solution (stable) (Hayachi, 1953; Stoker, 1955).

## 2. Reduction of Eq. (3) to an Equivalent Normal System

Take a new variable $q$ according to the relation

$$q = Q - (B/1 - n^2) \sin nt. \tag{7}$$

Substituting (7) into (3) we obtain

$$\ddot{q} + q = \epsilon f(q, \dot{q}), \tag{8}$$

with

$$\begin{aligned} f(q, \dot{q}) = {} & -k\dot{q} + c_1 q - c_2 q^2 \\ & - \frac{kn}{1 - n^2} B \cos nt + c_1 \frac{B}{1 - n^2} \sin nt \\ & - c_2 \frac{B^2}{(1 - n^2)^2} \sin^2 nt - c_3 \frac{B^3}{(1 - n^2)^3} \sin^3 nt \\ & - 2c_2 \frac{B}{1 - n^2} q \sin nt - 3c_3 q^2 \sin nt - 3c_3 \frac{B^2}{(1 - n^2)^2} q \sin^2 nt. \end{aligned} \tag{8a}$$

Introduce into (8) new variables, $u_1$ and $u_2$, defined by

$$u_1 = \dot{q} \cos t + q \sin t, \qquad u_2 = \dot{q} \sin t - q \cos t. \tag{9}$$

From (9) we get

$$q = u_1 \sin t - u_2 \cos t, \qquad \dot{q} = u_1 \cos t + u_2 \sin t, \tag{10}$$

$$\dot{u}_1 = (\ddot{q} + q) \cos t, \qquad \dot{u}_2 = (\ddot{q} + q) \sin t, \tag{11}$$

when, according to (8), we have

$$\dot{u}_1 = \epsilon f_1(u_1, u_2, t), \qquad \dot{u}_2 = \epsilon f_2(u_1, u_2, t), \tag{12}$$

where

$$\begin{aligned} f_1(u_1, u_2, t) &= f(u_1, u_2, t) \cdot \cos t, \\ f_2(u_1, u_2, t) &= f(u_1, u_2, t) \cdot \sin t, \end{aligned} \tag{12a}$$

with

$$\begin{aligned} f(u_1, u_2, t) = \Big[ & -k(u_1 \cos t + u_2 \sin t) + c_1(u_1 \sin t - u_2 \cos t) \\ & - u_2(u_1^2 \sin^2 t + u_2^2 \cos^2 t - 2u_1 u_2 \sin t \cos t) \\ & - c_3(u_1^3 \sin^3 t - u_2^3 \cos^3 t + 3u_1 u_2^2 \sin t \cos^2 t \\ & - 3u_1^2 u_2 \sin^2 t \cos t) - k \frac{nk}{1 - n^2} \cos nt + c_1 \frac{B}{1 - n^2} \sin nt \end{aligned}$$

$$- c_2 \frac{B^2}{(1 - u^2)^2} \sin^2 nt - c_3 \frac{B^3}{(1 - u^2)^3} \sin^3 nt \tag{12b}$$

$$- 2c_2 \frac{B}{1 - n^2} (u_1 \sin t - u_2 \cos t) \sin nt$$

$$- 3c_3 \frac{B}{1 - n^2} (u_1^2 \sin^2 t + u_2^2 \cos^2 t - 2u_1u_2 \sin t \cos t) \sin nt$$

$$\left. - 3c_3 \frac{B^2}{(1 - n^2)^2} (u_1 \sin t - u_2 \cos t) \sin^2 nt \right].$$

The function $f(u_1, u_2, t)$ is found by inserting (10) into (8a). The system (12), which takes the place of Eq. (3), gives advantages in the analysis. From (7) and the first of (10) we obtain

$$Q = u_1 \sin t - u_2 \cos t + (B/1 - n^2) \sin nt. \tag{13}$$

The expressions (5) and (13) are of the same form, then a solution $\{u_1, u_2\}$ of the system (12) gives the components of the subharmonics in the transient states, and the limits $\lim_{\epsilon \to 0} u_1$, $\lim_{\epsilon \to 0} u_2$ give the components of the subharmonics in the steady states.

## 3. The General System

In this section we shall discuss the system

$$\begin{aligned} \dot{u}_1 &= \epsilon F_1(u_1, u_2, t), & \dot{u}_2 &= \epsilon F_2(u_1, u_2, t), \\ u_1(t_0) &= u_{10}, & u_2(t_0) &= u_{20}, \end{aligned} \tag{14}$$

in a general way, as it is needed for our purposes. The functions $F_1$ and $F_2$ are taken with the following properties: They are analytic in $u_1$ and $u_2$, continuous and periodic with period $2\pi$ in $t$, and hence continuous in $u_1, u_2, t$; therefore $|F_1|$, $|F_2|$ have least upper bounds $M_1, M_2$ respectively, in a domain

$$D_1: \quad |u_1 - u_{10}| < \zeta_1, \qquad |u_2 - u_{20}| < \zeta_2, \qquad t_0 < t < T;$$

also $F_1, F_2$ are expansible as power series in $(u_1 - u_{10})$, $(u_2 - u_{20})$ and convergent in the domain $D_1$. The number $\epsilon$ is real.

(a) Try to find a solution $\{u_1, u_2\}$ of the system (14) such that if $\epsilon \to 0$, $u_1$ and $u_2$ tend to constants $x$ and $y$. When $\epsilon \neq 0$, $u_1$ and $u_2$ depend on $\epsilon$ and $t$, and we assume that for $t = t_0$ they differ a little from $x$ and $y$; that is,

$$u_{10} = x + \xi, \qquad u_{20} = y + \eta, \tag{15}$$

where $\xi$, $\eta$ are very small. These conditions can be considered as initial conditions for the system (14).

Take as a formal solution $\{u_1, u_2\}$ of (14) the following expressions:

$$\begin{aligned} u_1 &= X_1^0 + \epsilon X_1^1 + \epsilon^2 X_1^2 + \cdots \\ u_2 &= X_2^0 + \epsilon X_2^1 + \epsilon^2 X_2^2 + \cdots \end{aligned} \tag{16}$$

where

$$X_1^0 = x + \xi, \qquad X_2^0 = y + \eta, \tag{16a}$$

and the other coefficients of (16) are regular functions of $x$, $y$, $\xi$, $\eta$, $t$ when $|x|$, $|y|$, $|\xi|$, $|\eta|$, $|t - t_0|$ are less than certain constants. We determine the coefficients $X$ of (16) by presupposing the convergence of the series (16).

(b) For the determination of the coefficients $X$, substitute (16) into the system (14) after the right members of (14) have been developed as power series in $(u_1 - u_{10})$, $(u_2 - u_{20})$, $\epsilon$. Then rearrange the terms according to powers of $\epsilon$, and equate coefficients of corresponding powers of $\epsilon$. We obtain an infinite series of systems of differential equations, from which, by integrating, we have

$$\begin{aligned} X_1^1 &= \int_{t_0}^t [F_1]\, dt, \qquad X_2^1 = \int_{t_0}^t [F_2]\, dt, \\ X_1^2 &= \int_{t_0}^t \{X_1^1[F_1]_{u_1} + X_2^1[F_2]_{u_2}\}\, dt, \\ X_2^2 &= \int_{t_0}^t \{X_1^1[F_2]_{u_1} + X_2^1[F_2]_{u_2}\}\, dt, \\ X_i^3 &= \int_{t_0}^t \sum_{j=1}^2 X_i^2[F_i]_{u_i}\, dt + \frac{1}{2}\int_{t_0}^t \sum_{j=1}^2 \sum_{l=1}^2 X_j^1 X_l^1 [F_i]_{u_j u_l}\, dt, \\ X_i^k &= \int_{t_0}^t P_i^k(X_i^1, \cdots X_i^{k-1})\, dt, \\ & j, l, i = 1, 2 \qquad k = 1, 2, \cdots \end{aligned} \tag{17}$$

The brackets indicate that the corresponding functions and their derivatives are taken at: $u_1 = u_{10} = x + \xi$, $u_2 = u_{20} = y + \eta$. The $P_i^k$ have the following properties. (i) They are polynomials in $X_1^1, X_2^1, \cdots, X_1^{k-1}, \cdots$, not in $(t - t_0)$. (ii) Their coefficients are linear functions of the coefficients of the expansion of $F_1$, $F_2$ given in power series of $(u_1 - u_{10})$, $(u_2 - u_{20})$. (iii) The numerical multipliers of $f_1$ and $f_2$ are positive numbers, namely, the numbers arising in Taylor's expansion and in forming products of various power series. All $X$, according to (17), are zero at $t = t_0$.

By carrying out the integrations in the first two of (17), we find $X_1^1$, $X_2^1$. Upon substituting the values of $X_1^1$, $X_2^1$ into the second two of (17), the integrands become known continuous functions of $t$, when $X_1^2$, $X_2^2$ are defined by quadratures. With the same procedure we can find $X_1^3, X_2^3, \cdots$.

(c) The convergence of the formal solution (16) in our domain was presupposed for the determination of the coefficients $X$, given by (17). This convergence and the domain of its validity come from using the "method of dominants," as follows. There always exists (Petrovsky, 1954) a function which dominates $F_1$, $F_2$ of (14); that is, there exists a function $F$ such that the nonnegative coefficients of its expansion in power series, in the neighborhood of any point of the domain $D_1$, are not smaller than the absolute values of the corresponding coefficients of the power series expansion of $F_1$, $F_2$ in the neighborhood of the point. Such a dominating function of $F_1$, $F_2$ in $D_1$ is the function

$$F = \frac{M}{1 - [(x_1 - u_{10}) + (x_2 - u_{20})]\zeta^{-1}}, \tag{18}$$

where

$$M \geqq M_i, \qquad \zeta \leqq \zeta_i, \qquad i = 1, 2. \tag{18a}$$

This gives a power series expansion that converges when

$$| x_i - u_{i0} | < \tfrac{1}{2}\zeta, \qquad i = 1, 2.$$

Consider the auxiliary system

$$\dot{x}_i = \epsilon \frac{M}{1 - \frac{1}{2}[(x_1 - u_{10}) + (x_2 - u_{20})]} \tag{19}$$

$$x_i(t_0) = u_{i0}, \qquad i = 1, 2.$$

The system (19) gives $\dot{x}_1(t_0) = \dot{x}_2(t_0)$. On making use of the initial conditions, we have

$$x_1 - u_{10} = x_2 - u_{20} = \bar{x}, \tag{20}$$

and $\bar{x}$ satisfies the differential equation

$$\begin{aligned} \dot{\bar{x}} &= \epsilon \frac{M}{1 - (2\bar{x}/\zeta)} \\ \bar{x}(t_0) &= 0. \end{aligned} \tag{21}$$

On separating the variables in (21) and performing the quadratures, we find that

$$\bar{x} - \bar{x}^2/\zeta = \epsilon M(t - t_0).$$

Then

$$\bar{x} = \frac{1}{2}\zeta\left\{1 - \left[1 - \frac{4\epsilon M(t - t_0)}{\zeta}\right]^{1/2}\right\}, \tag{22}$$

by selecting the negative sign before the radical, in order to satisfy the condition $\bar{x}(t_0) = 0$.

According to (20), the solution of (19) is

$$x_1 = u_{10} + \bar{x}, \qquad x_2 = u_{20} + \bar{x}, \tag{23}$$

$\bar{x}$ being given by (22).

The expansion of the right-hand member of (22) as power series in $\epsilon$ has the singularity: $\epsilon = \zeta/4M(t - t_0)$. Therefore the expansion converges for all values of $\epsilon$ such that

$$|\epsilon| < \frac{\zeta}{4M(t - t_0)}. \tag{24}$$

The expansions of $x_1$, $x_2$ as power series in $\epsilon$ converge under the same condition (24). The second members of (19) have all properties of the second members of (14), so that the system (19) can be solved directly by using power series in $\epsilon$ of the form (16). This solution is unique, and so it is identical with that given by (22) and (23), if we take the expansion of $\bar{x}$ in powers of $\epsilon$. Then the power series solution of (19) in $\epsilon$ is valid under the condition (24).

If the series solutions of (19) and (21) are

$$\bar{x} = \bar{x}^0 + \epsilon\bar{x}^1 + \epsilon^2\bar{x}^2 + \cdots$$
$$\bar{x}^0 = 0 \tag{25a}$$

$$x_i = x_i^0 + \epsilon x_i^1 + \epsilon^2 x_i^2 + \cdots$$
$$x_1^0 = u_{10}, \qquad x_2^0 = u_{20}, \qquad x_1^1 = x_2^1 = \bar{x}^1, \qquad x_1^2 = x_2^2 = \bar{x}^2, \cdots \tag{25b}$$

and we consider the formulas (17) that give the coefficients $X_1^1$, $X_2^1$, and the formulas that give the coefficients $x_1^1$, $x_2^1$, then the integrand of the former is dominated by the integrand of the latter in these definite integrals. Therefore

$$x_i^1 \geqq |X_i^1|, \qquad t_0 \leqq t \leqq T, \qquad i = 1, 2. \tag{26a}$$

The corresponding result is true successively for terms with higher indices. Thus

$$x_i^k \geqq |X_i^k|, \qquad t_0 \leqq t \leqq T, \qquad i = 1, 2, \tag{26b}$$

and the investigation of the convergence of (16) is completed.

(d) In many special cases the domain of convergence of (16) is much larger than the condition (24) shows. This is due to additional properties that the differential equations possess, which give to their solution a wider domain of convergence. The inequality (24) can be satisfied (i) by imposing restrictions upon both $\epsilon$ and $t$, or (ii) by taking $t$ arbitrarily and then, for the convergence, restricting $\epsilon$, or (iii) by taking $\epsilon$ arbitrarily and then restricting $t$.

For nonnegative $\epsilon$, (24) gives

$$t - t_0 \leqq \frac{\zeta}{4\epsilon M} = T - t_0 .$$

Then

$$T = \max t = t_0 + \frac{\zeta}{4\epsilon M}, \tag{27a}$$

and if $t_0$ varies in $0 \leqq t_0 \leqq 2\pi$, we have

$$\frac{\zeta}{4\epsilon M} \leqq T \leqq 2\pi + \frac{\zeta}{4\epsilon M}. \tag{27b}$$

4. THE USE OF THE PERIODICITY OF THE SOLUTION $\{u_1, u_2\}$

In the theory of the previous section no use of the periodicity of the solution $\{u_1, u_2\}$ of the system (14) has been made. Let us assume now that $u_1, u_2$ are periodic in $t$ with period $2\pi$; that is,

$$u_1(t_0 + 2\pi) - u_1(t_0) = 0, \qquad u_2(t_0 + 2\pi) - u_2(t_0) = 0. \tag{28}$$

Applying (28) to (16) we have

$$\begin{aligned} &[X_1^1(t_0 + 2\pi) - X_1^1(t_0)] + \epsilon[X_1^2(t_0 + 2\pi) - X_1^2(t_0)] + \cdots = 0, \\ &[X_2^1(t_0 + 2\pi) - X_2^1(t_0)] + \epsilon[X_2^2(t_0 + 2\pi) - X_2^2(t_0)] + \cdots = 0. \end{aligned} \tag{29}$$

Now take the Fourier series development in $t$ of the function $F_1$ and $F_2$ :

$$\begin{aligned} F_1 &= A_0 + A_1 \cos t + B_1 \sin t + \cdots A_m \cos mt + B_m \sin mt + \cdots \\ F_2 &= C_0 + C_1 \cos t + D_1 \sin t + \cdots C_m \cos mt + D_m \sin mt + \cdots, \end{aligned} \tag{30}$$

where the coefficients $A, B, C, D$ are functions of $u_1, u_2$ and such that the series

$$\begin{aligned} &| A_0 | + | A_1 | + \cdots + | A_m | + | B_m | + \cdots, \\ &| C_0 | + | C_1 | + \cdots + | C_m | + | D_m | + \cdots, \end{aligned} \tag{30a}$$

are uniformly convergent with least upper bound $\bar{M}$; also the series of partial derivatives with respect to $u_1, u_2$ :

$$\begin{aligned} &| \partial A_0/\partial u_1 | + | \partial A_1/\partial u_1 | + \cdots, \\ &| \partial C_0/\partial u_2 | + | \partial C_1/\partial u_2 | + \cdots, \end{aligned} \tag{30b}$$

are uniformly convergent with least upper bound $\bar{H}$.

By taking into account the expressions of $X$ given by (17) and the series (30), the conditions (29) give

$$\begin{aligned} A_0(x + \xi, y + \eta) + \epsilon\phi_1(x + \xi, y + \eta, t_0) + \cdots = 0, \\ C_0(x + \xi, y + \eta) + \epsilon\phi_2(x + \xi, y + \eta, t_0) + \cdots = 0, \end{aligned} \tag{31}$$

where we suppose

$$A_0(0, 0) = 0, \qquad C_0(0, 0) = 0. \tag{31a}$$

Provided that the Jacobian of the system (31) is not zero,

$$\begin{vmatrix} \partial A_0/\partial \xi & \partial C_0/\partial \xi \\ \partial A_0/\partial \eta & \partial C_0/\partial \eta \end{vmatrix} \neq 0, \tag{32}$$

we can solve the system (31) in $\xi$ and $\eta$ in terms of $\epsilon$ and $t$;

$$\xi = \xi(\epsilon, t_0), \qquad \eta = \eta(\epsilon, t_0), \quad \text{say,} \tag{33}$$

with the conditions

$$\xi = \xi(0, t_0) = 0, \qquad \eta = \eta(0, t_0) = 0. \tag{33a}$$

If the Jacobian is zero, it is necessary to consider in (31) terms of degree 1, 2, $\cdots$ in $\epsilon$, that is, the function $\phi_1$, $\phi_2$, $\cdots$, (Faton, 1929). $\xi$, $\eta$ in (33) are given in series in $\epsilon$, where $t_0$ can take any value of the interval $0 \leqq t_0 \leqq 2\pi$ and $\epsilon$ is in accordance with the condition (24). From (15), (16a), and (33) we get

$$X_1^0 = u_{10} = x + \xi(\epsilon, t_0), \qquad X_2^0 = u_{20} = y + \eta(\epsilon, t_0), \tag{34}$$

where $x$ and $y$ are unknown. If we find the constants $x$ and $y$, the transient solution (16) of the system (14) is completely determined.

## 5. Determination of the Constants $x$ and $y$: Conditions of Existence of Steady States Subharmonics

Equations (31) must be satisfied when $\epsilon = \xi = \eta = 0$. In that case (31) give

$$A_0(x, y) = 0, \qquad C_0(x, y) = 0. \tag{35}$$

The solution $(x, y)$ of the system (35) gives the constants $x$ and $y$, which characterize the steady state subharmonics. The conditions for real intersections of the curves (35) in the $x$, $y$-plane give the conditions of the existence of the steady state subharmonics.

## 6. The Stability of the Steady State Subharmonics

The singular points of the general system (14), where the time enters explicitly, are given, according to the appendix of the paper, by Eq. (35). The stability of these singularities gives the stability of the steady state solution of the system (14). The following definitions of the singular points shall be used in the second part of this paper.

According to Poincaré (1892) and Bendixson (1901), the distinction between the different kinds of singularities depends on two numbers $\rho_1$, $\rho_2$, the roots of the characteristic equation

$$\begin{vmatrix} a_1 - \rho & b_1 \\ a_2 & b_2 - \rho \end{vmatrix} = 0; \tag{36}$$

that is, on the numbers

$$\rho_{1,2} = \tfrac{1}{2}\{a_1 + b_2 \pm [(a_1 - b_2)^2 + 4a_2b_1]^{1/2}\}, \tag{36a}$$

where:

$$a_1 = \partial A_0/\partial x, \qquad a_2 = \partial A_0/\partial y, \qquad b_1 = \partial C_0/\partial x, \qquad b_2 = \partial C_0/\partial y. \tag{36b}$$

For nonzero roots $\rho_1$, $\rho_2$ the "simple singularities" are classified in four classes as follows:

I. "Nodal points," when $\rho_1$, $\rho_2$ are real and of same sign:

$$(a_1 - b_2)^2 + 4a_2b_1 \geqq 0, \qquad a_1b_2 - a_2b_1 > 0. \tag{37a}$$

II. "Saddle (pass) points," when $\rho_1$, $\rho_2$ are real but of opposite sign:

$$(a_1 - b_2)^2 + 4a_2b_1 \geqq 0, \qquad a_1b_2 - a_2b_1 < 0. \tag{37b}$$

III. "Spiral (focus) points," when $\rho_1$, $\rho_2$ are complex conjugates:

$$(a_1 - b_2)^2 + 4a_2b_1 < 0, \qquad a_1b_2 - a_2b_1 \neq 0. \tag{37c}$$

IV. "Spiral points" or "centers," when $\rho_1$, $\rho_2$ are pure imaginaries:

$$(a_1 - b_2)^2 + 4a_2b_1 < 0, \qquad a_1b_2 - a_2b_1 = 0. \tag{37d}$$

Under the conditions (37d) the singularity "may be a center." These conditions are necessary, but not sufficient. There is "Poincaré's criterion" in this case for distinguishing a spiral point from a center (Hadamard, 1933).

Define the above singularities as "stable" or "unstable" when, with increasing time $t$, any point on any integral curve does or does not move into the said singularity, that is, according as the real part of the roots is negative or positive, respectively. In Fig. 1, I shows a "stable node," II a saddle point which is "intrinsically unstable," III a "stable spiral point," IV a (neutral) center. By "intrinsically unstable" we mean that

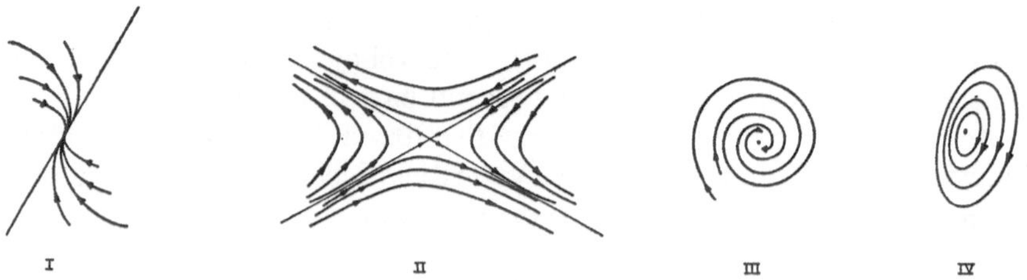


FIG. 1. The four kinds of simple singularities. I is the stable node, II the saddle, III the stable spiral, IV the center.

the saddle point is terminated by four trajectories forming two distinct integral curves. Two of these trajectories approach the saddle point with increasing time $t$, and correspond to the stable curves, while the others move away from it with increasing time and correspond to the unstable curves. There exist four regions containing continua of hyperbolically shaped integral curves which do not approach the saddle point (Hayachi, 1953).

## 7. Application to the System (14)

Now, coming back to our original problem, we see that the system (12) is a special case of the system (14). To apply the theory of the previous sections concerning the system (14) to our special case of the system (12), we give to the functions $f_1$ and $f_2$ of (12a) and (12b) an appropriate form from which we can get the development into Fourier series for any order of the subharmonics. By replacing the powers and products of sines and cosines of multiple angles in constructing $f_1$ and $f_2$, according to (12a) and (12b), we can get the following results, if we restrict ourselves to coefficients which are useful for the construction of the corresponding functions $A_0$ and $C_0$, which depend on the number $n$ characterizing the order of the subharmonics:

$$\begin{aligned} f_1(u_1, u_2, t) &= \left[-\frac{1}{2}ku_1 - \frac{1}{2}c_1u_2 + \frac{3}{8}c_3u_2^3 + \frac{3}{8}c_3u_1^2u_2 \right. \\ &\left. + \frac{3}{4}c_3\frac{B^2}{(1-n^2)^2}u_2\right] + \left[-\frac{1}{2}c_2\frac{B}{1-n^2}u_1\right]\cos(n-2)t + \cdots \qquad (38a) \\ &+ \left[\frac{3}{4}c_3\frac{B}{1-n^2}u_1u_2\right]\cos(n-3)t + \cdots \end{aligned}$$

$$\begin{aligned} f_2(u_1, u_2, t) &= \left[\frac{1}{2}c_1u_1 - \frac{1}{2}ku_2 - \frac{3}{8}c_3u_1^3 - \frac{3}{8}c_3u_1u_2^2 \right. \\ &\left. - \frac{3}{4}c_3\frac{B^2}{(1-n^2)^2}u_1\right] + \left[\frac{1}{2}c_2\frac{B}{1-n^2}u_2\right]\cos(n-2)t + \cdots \qquad (38b) \\ &+ \left[-\frac{3}{8}c_3\frac{B^2}{1-n^2}(-u_1^2 + u_2^2)\right]\cos(u-3)t + \cdots \end{aligned}$$

For any given $n$, that is, for any order of the subharmonics, the corresponding functions $A_0(x, y)$, $C_0(x, y)$ are the independent-of-time terms of the right-hand members of (38a) and (38b), respectively, if we put $x$ and $y$ instead of $u_1$ and $u_2$, respectively. If we know $f_1$ and $f_2$, the equations $A_0 = 0$ and $C_0 = 0$ are known, and their solutions give the steady states subharmonics of the initial equations. The stability of the steady state solutions comes from the study of the singularities $(x, y)$, according to what has been quoted above on the singularities. The terms of the functions $A_0(x, y)$ and $C_0(x, y)$ are, as we can see from the formulas (38a) and (38b), such that we can have $x$ and $y$, from $A_0 = 0$ and $C_0 = 0$, in terms of $B$ and the ratios $k/c_3$, $c_1/c_3$, $c_2/c_3$ for $c_3 \neq 0$.

PART B. APPLICATION: SUBHARMONICS OF ORDER ⅓ (MAGIROS, 1957b)

## 8. The Equations for the Components of the Amplitude of the Subharmonics

From Eq. (1) and its solution (5) for $n = 3$ we deduce

$$\ddot{Q} + \bar{k}\dot{Q} + \bar{c}_1 Q + \bar{c}_2 Q^2 + \bar{c}_3 Q^3 = B \sin 3t, \tag{39}$$

$$Q = x \sin t - y \cos t - \tfrac{1}{8} B \sin 3t. \tag{40}$$

For $\epsilon \neq 0$ the components $x$ and $y$ of the subharmonics depend on time $t$, and this transient state yields, with the lapse of the time, to the steady state, either stable or unstable. Since the coefficients of (39) can have arbitrary values, Eq. (39) may contain various types of periodic solutions for different initial values. For the constant values of $x$ and $y$, which characterize the subharmonics in the "steady state," we must find the functions $A_0(x, y)$ and $C_0(x, y)$. If $x$ and $y$ are substituted for $u_1$ and $u_2$, Eqs. (38a) and (38b) give for $n = 3$:

$$\begin{aligned} A_0(x, y) &= \tfrac{1}{2}[-kx + c_1 y + \tfrac{3}{4} c_3 y(x^2 + y^2 + \tfrac{1}{32}B^2) - \tfrac{3}{16} c_3 Bxy], \\ C_0(x, y) &= \tfrac{1}{2}[c_1 x - ky - \tfrac{3}{4} c_3 x(x^2 + y^2 + \tfrac{1}{32}B^2) \\ &\qquad + \tfrac{3}{32} c_3 B(-x^2 + y^2)]. \end{aligned} \tag{41}$$

Setting the right-hand members of (41) equal to zero, we have the following equations for the determination of $x$ and $y$:

$$\begin{aligned} \mu x + \lambda y - y(x^2 + y^2 + \tfrac{1}{32}B^2) + \tfrac{1}{4}Bxy &= 0, \\ \lambda x - \mu y + x(x^2 + y^2 + \tfrac{1}{32}B^2) + \tfrac{1}{8}B(-x^2 + y^2) &= 0, \end{aligned} \tag{42}$$

where

$$\lambda = \frac{4}{3}\frac{c_1}{c_3} = \frac{4}{3}\frac{1 - \bar{c}_1}{\bar{c}_3}, \qquad \mu = \frac{4}{3}\frac{k}{c_3} = \frac{4}{3}\frac{\bar{k}}{\bar{c}_3}. \tag{42a}$$

To any real solution of the system (42) corresponds a solution of (39) in the form of (40).

*Remarks*

For prescribed initial conditions of (39), say $Q_0$ and $\dot{Q}_0$ at $t = 0$, we have, according to (40),

$$Q_0 = -y, \qquad \dot{Q}_0 = x - \tfrac{3}{8}B, \tag{43}$$

and the given initial conditions correspond to the point

$$x = \dot{Q}_0 + \tfrac{3}{8}B, \qquad y = -Q_0 \tag{43a}$$

in the $x$, $y$-plane. Starting from the point (43a), the "starting point," and following the corresponding integral curve with the lapse of time, we may terminate to a "final point," which corresponds to the proper steady solution. The coordinates of this final point are solutions of the system (42) and the stability of this point can be investigated according to Section 6.

Given the initial condition and the amplitude $B$ of the external force, the "starting point" in the $x$, $y$-plane is defined; conversely, any point of the $x$, $y$-plane can be taken as a starting point by properly choosing the initial conditions and the amplitude $B$. If the "starting point" is selected in coincidence with a "stable final point," no "transient phenomena" must exist. This last remark may be of importance in engineer ing problems.

## 9. Restrictions of the Coefficients of Eq. (39)

The system (42) can be written as

$$\begin{aligned} \mu x + \lambda y - Ay &= -\tfrac{1}{8}B(2xy) \\ \lambda x - \mu y - Ax &= -\tfrac{1}{8}B(-x^2 + y^2), \end{aligned} \tag{44}$$

with

$$A = \zeta^2 + \tfrac{1}{32}B^2, \qquad \zeta^2 = x^2 + y^2. \tag{44a}$$

Squaring, and adding (44) we find

$$\lambda^2 + \mu^2 + A^2 - 2\lambda A = \tfrac{1}{64}B^2\zeta^2,$$

which, by using (44a), can give

$$\zeta^4 + \left(\frac{3}{64}B^2 - 2\lambda\right)\zeta^2 + \left(\lambda^2 + \mu^2 + \frac{B^4}{32^2} - \lambda\frac{B^2}{16}\right) = 0, \tag{45}$$

with roots

$$\zeta^2 = \frac{1}{2}\left\{2\lambda - \frac{3}{64}B^2 \pm \left[\left(2\lambda - \frac{3}{64}B^2\right)^2 - 4\left(\lambda^2 + \mu^2 + \frac{B^4}{32^2} - \lambda\frac{B^2}{16}\right)\right]^{1/2}\right\}. \tag{46}$$

The reality of $\zeta^2$ requires

$$I \equiv 7B^4 - 2^8\lambda B^2 + 2^{14}\mu^2 \leqq 0. \tag{47}$$

The roots of $I = 0$ are

$$B^2 = \frac{2^7}{7}[\lambda \pm (\lambda^2 - 7\mu^2)^{1/2}] \tag{48}$$

Then the condition (47) is fulfilled only when the following inequalities are fulfilled:

$$\lambda^2 - 7\mu^2 > 0, \tag{49a}$$

$$\frac{2^7}{7}\{\lambda - (\lambda^2 - 7\mu^2)^{1/2}\} \leqq B^2 \leqq \frac{2^7}{7}\{\lambda + (\lambda^2 - 7\mu^2)^{1/2}\}. \tag{49b}$$

The inequalities (49), by using (42a), can be written as

$$\left(\frac{1 - \bar{c}_1}{\bar{k}}\right)^2 > 7, \tag{50a}$$

$$\frac{2^9}{21}\left\{\frac{1 - \bar{c}_1}{\bar{c}_3} - \left[\left(\frac{1 - \bar{c}_1}{\bar{c}_3}\right)^2 - 7\left(\frac{\bar{k}}{\bar{c}_3}\right)^2\right]^{1/2}\right\} \leqq B^2 \leqq \frac{2^9}{21} \cdot \left\{\frac{1 - \bar{c}_1}{\bar{c}_3} + \left[\left(\frac{1 - \bar{c}_1}{\bar{c}_3}\right)^2 - 7\left(\frac{\bar{k}}{\bar{c}_3}\right)^2\right]^{1/2}\right\}. \tag{50b}$$

The inequalities (50) are restrictions of the values of the coefficients of the differential equation (39) for real amplitude of the subharmonics of order $1/3$.

From (46), if $\lambda$ and $\mu$ are given, we can draw the diagram of $\zeta^2$ versus $B^2$, taking into account the arc of this diagram which is valid for $B^2$, according to the restriction (50b).

As for the condition (50a), it can be written in the form:

$$(1 - \bar{c}_1 - \sqrt{7}\bar{k})\cdot(1 - \bar{c}_1 + \sqrt{7}\bar{k}) > 0, \tag{51}$$

which corresponds to the shaded region of the $\bar{c}_1$, $\bar{k}$-plane (Fig. 2).

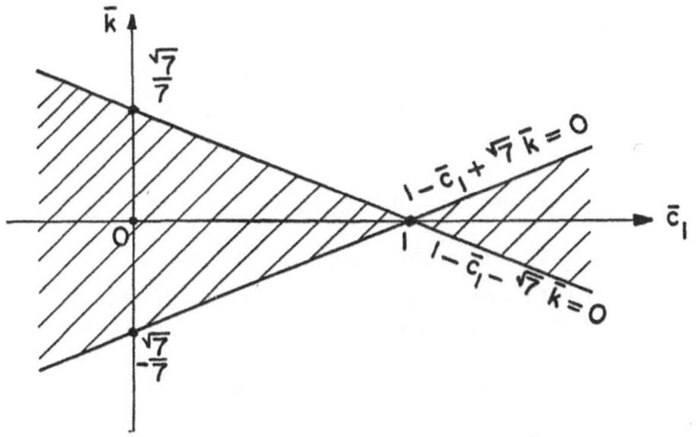


FIG. 2. Graphic representation of the restriction (50a) for stability. Accepted values of $\bar{c}_1$ and $\bar{k}$ must represent points in the shaded region of $\bar{c}_1$, $\bar{k}$-plane.

## 10. Solution of the System (42)

The real roots of the system (42) may be at most nine. By using the number $A$, given by (44a), we reduce to the system (44), of which the real roots are acceptable. By regarding the equations (44) as equations in $x$, whose coefficients are functions of $y$, we get the vanishing of the eliminant of this system, say the Sylvester's eliminant, which is the condition for common roots of the system. The system (42) can be written as

$$\begin{aligned} (\mu + \tfrac{1}{4}By)x + (\lambda - A)y &= 0, \\ \tfrac{1}{8}Bx^2 - (\lambda - A)x + (\mu - \tfrac{1}{8}By)y &= 0, \end{aligned} \tag{52}$$

and its Sylvester's eliminant is

$$\begin{vmatrix} \mu + \tfrac{1}{4}By & (\lambda - A)y & 0 \\ 0 & \mu + \tfrac{1}{4}By & (\lambda - A)y \\ \tfrac{1}{8}B & -(\lambda - A) & (\mu - \tfrac{1}{8}By)y \end{vmatrix} = 0, \tag{53}$$

which, for nonzero roots $y$, leads to the cubic:

$$y^3 - 3(4/B)^2[(\lambda - A)^2 + \mu^2]y - 2(4\mu/B)^3 - 2(4/B)^3\mu(\lambda - A)^2 = 0. \tag{54}$$

If we know the coefficients of the differential equation (39), we find the values $\lambda$ and $\mu$ according to (42a). We can then obtain from (46) the amplitude of the subharmonics, which are at most two. On the circumference with radius $\zeta$ there are one or three singular points, of which the ordinates $y$ are the real roots of the cubic (54), when for their abscissas $x$ we apply the Pythagorean theorem. Therefore, the singular points $(x, y)$ are at most seven, including the origin, which in every case is a singular point, as we shall see in Section 16.

## 11. Requirements for Stability of the Solution $(x, y)$

For the stability of $(x, y)$, according to Section 6, we need the partial derivatives of the functions $A_0(x, y)$ and $C_0(x, y)$, given by (41), with respect to $x$ and $y$. These derivatives are

$$\begin{aligned}
\partial A_0/\partial x &= a_1 = \tfrac{1}{2}[-k + \tfrac{3}{2}c_3xy - \tfrac{3}{16}c_3By],\\
\partial A_0/\partial y &= a_2 = \tfrac{1}{2}[-c_1 + \tfrac{3}{4}c_3(x^2 + 3y^2 + \tfrac{1}{32}B^2) - \tfrac{3}{16}c_3Bx],\\
\partial C_0/\partial x &= b_1 = \tfrac{1}{2}[c_1 - \tfrac{3}{4}c_3(3x^2 + y^2 + \tfrac{1}{32}B^2) - \tfrac{3}{16}c_3Bx],\\
\partial C_0/\partial y &= b_2 = \tfrac{1}{2}[-k - \tfrac{3}{2}c_3xy + \tfrac{3}{16}c_3By].
\end{aligned} \tag{55}$$

If in the inequality (24), which is the condition for convergence of the solution (5) and then of (40), we take $t_0 = 0$, $\epsilon = 1$, we thus establish the restriction for the $T = \max t$:

$$T < \frac{\zeta}{4M}. \tag{56}$$

Under this restriction, and by the use of (2), the partial derivatives (55) can be written as

$$\begin{aligned}
a_1 &= \frac{1}{2}\bar{c}_3\left\{-\frac{\bar{k}}{\bar{c}_3} + \frac{3}{2}xy - \frac{3}{16}By\right\},\\
a_2 &= \frac{1}{2}\bar{c}_3\left\{-\frac{1-\bar{c}_1}{\bar{c}_3} + \frac{3}{4}\left(x^2 + 3y^2 + \frac{B^2}{32}\right) - \frac{3}{16}Bx\right\},\\
b_1 &= \frac{1}{2}\bar{c}_3\left\{\frac{1-\bar{c}_1}{\bar{c}_3} - \frac{3}{4}\left(3x^2 + y^2 + \frac{B^2}{32}\right) - \frac{3}{16}Bx\right\},\\
b_2 &= \frac{1}{2}\bar{c}_3\left\{-\frac{\bar{k}}{\bar{c}_3} - \frac{3}{2}xy + \frac{3}{16}By\right\}.
\end{aligned} \tag{57}$$

## 12. Illustrated Example

With the above we have everything we need for treating a numerical example. For the components $x$ and $y$ of the amplitude of the subharmonics of order $\tfrac{1}{3}$ of the differential equation (39), we do not need, as we can see from (42) and (42a), the coefficient $\bar{c}_2$ of the elastic force, since $\bar{c}_3 \neq 0$. Such a thing happens for the components of any order of the subharmonics, except in the case of subharmonics of order $\frac{1}{2}$, as we can see from the equations (38a) and (38b).

Let us take

$$B = 4, \qquad \bar{k} = 3/16, \qquad \bar{c}_1 = 1\tfrac{3}{4}, \qquad \bar{c}_3 = -\tfrac{1}{2}.$$

For this example we have

$$\frac{1 - \bar{c}_1}{\bar{c}_3} = \frac{3}{2}, \qquad \frac{\bar{k}}{\bar{c}_3} = -\frac{3}{8}, \qquad \lambda = 2, \qquad \mu = -\frac{1}{2},$$

and these values, and $B = 4$, are in accordance with the restrictions (50a, b).

We have

$$\zeta_1^2 \cong 2, \qquad \zeta_1 \cong 1.414, \qquad \zeta_2^2 \cong 1.25, \qquad \zeta_2 \cong 1.118,$$

$$A_1 = \zeta_1^2 + \tfrac{1}{2} \cong 2.5, \qquad A_2 = \zeta_2^2 + \tfrac{1}{2} \cong 1.75,$$

$$A_1^2 \cong 6.25, \qquad A_2^2 \cong 3.06.$$

The cubic (54) gives

$$\begin{aligned} y_1^3 - \tfrac{3}{2}y_1 + \tfrac{1}{2} &= 0, \\ y_2^3 - \tfrac{15}{16}y_2 + \tfrac{5}{16} &= 0. \end{aligned} \tag{58}$$

These cubics have three real unique roots. The roots of the first are

$$y_{11} = 0.9999, \qquad y_{12} = -1.3649, \qquad y_{13} = 1.3645; \tag{59a}$$

those of the second

$$y_{21} = 0.7032, \qquad y_{22} = -1.1034, \qquad y_{23} = 1.0431. \tag{60a}$$

The corresponding abscissas are

$$x_{11} = 0.9999, \qquad x_{12} = 0.3648, \qquad x_{13} = -0.3648 \tag{59b}$$

$$x_{21} = 0.8672, \qquad x_{22} = 0.1755, \qquad x_{23} = -0.3991 \tag{60b}$$

Thus, the origin and the above points (59a, b) and (60a, b) are the singular points in our numerical example.

The kind of singularity of these points can be computed from the corresponding characteristic roots, according to (36a) and (57), by using the coordinates of the above points. The results of the computation are

TABLE I

THE SINGULAR POINTS AND THEIR CLASSES FOR THE NUMERICAL EXAMPLE OF SECTION 12.

| | Point | | Class |
|---|---|---|---|
| 0 | $x = 0$ | $y = 0$ | Stable spiral point |
| I | $x_{11} = 0.9999$ | $y_{11} = 0.9999$ | Stable spiral point |
| II | $x_{12} = 0.3648$ | $y_{12} = -1.3645$ | Stable spiral point |
| III | $x_{13} = -0.3648$ | $y_{13} = 1.3645$ | Saddle point |
| IV | $x_{21} = 0.8675$ | $y_{21} = 0.7032$ | Saddle point |
| V | $x_{22} = 0.1755$ | $y_{22} = -1.1034$ | Saddle point |
| VI | $x_{23} = -0.3991$ | $y_{23} = 1.0431$ | Saddle point |

given in Table I. The singularities I, II, III are on the circumference with radius 1.414; the others, IV, V, VI, on the circumference with radius 1.118.

## 13. Nonexistence of Poincaré's Cycles

In order to give a sketch of the behavior of the integral curves and the singularities in the $x$, $y$-plane of the above example, it is useful to know the existence or nonexistence of Poincaré's cycles. "Bendixson's criterion" can be applied here. Bendixson's criterion states that: "If the expression $\partial A_0/\partial x + \partial C_0/\partial y$ does not change its sign within a domain in the $x$, $y$-plane, no Poincaré cycle can exist in the domain." Application of this criterion gives, by using (57),

$$\partial A_0/\partial x + \partial C_0/\partial y = -\bar{k},$$

which is valid in the whole $x$, $y$-plane. Then no Poincaré cycles can exist in the whole $x$, $y$-plane. Should $\bar{k}$ not equal zero none of the singularities can be a "center." However, we may have centers if $\bar{k} = 0$, when "closed integral curves" appear, which have nothing to do with the Poincaré cycles.

## 14. Sketch Corresponding to Our Example

By applying the method of isoclives to the equation

$$\frac{dy}{dx} = \frac{C_0(x, y)}{A_0(x, y)}, \tag{61}$$

we can sketch the regions and integral curves of our example of Section 12, as is shown in Fig. 3. The solid lines are the boundaries from the

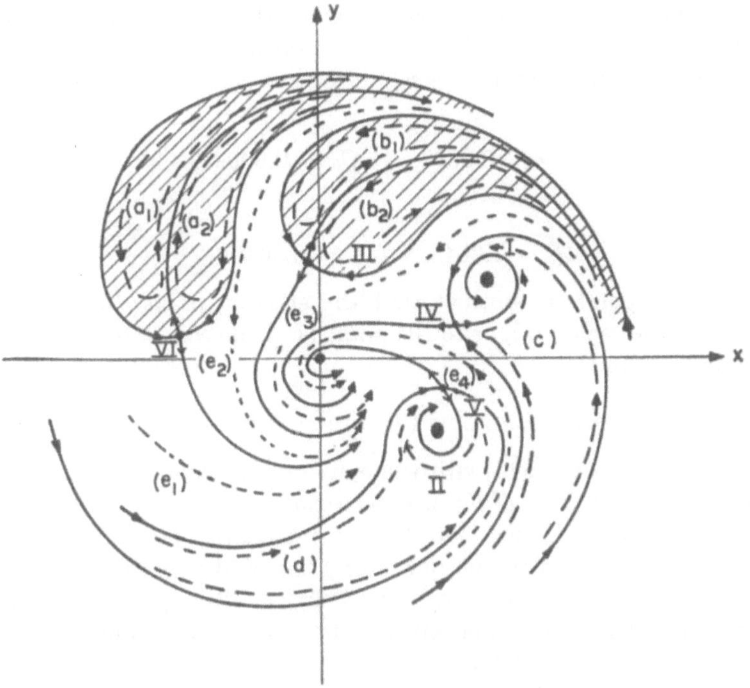


FIG. 3. The regions and the singularities in the $x$, $y$-plane. The solid lines through the saddle points III, IV, V, VI are the boundaries of the regions. The shaded regions $(a_1)$, $(a_2)$ and $(b_1)$, $(b_2)$ do not correspond to any solution. The points I, II and the origin are stable spirals and correspond to the regions $(c)$, $(d)$ and $(e_1, e_2, e_3, e_4)$, respectively. The dotted lines are integral curves.

saddle points, the dotted lines the integral curves. The regions $(a_1)$, $(a_2)$, $(b_1)$, $(b_2)$ correspond to no solution of our example. The regions $(e_1)$, $(e_2)$, $(e_3)$, $(e_4)$ correspond to the stable harmonic solution

$$Q = -\tfrac{1}{2} \sin 3t.$$

The region $(c)$ corresponds to the stable solution

$$Q = 0.9999 \sin t - 0.9999 \cos t - \tfrac{1}{2} \sin 3t,$$

and the region (d) to the stable solution

$$Q = 0.3648 \sin t - 1.3645 \cos t - \tfrac{1}{2} \sin 3t.$$

By "corresponds" we mean that if the starting point is a point of a region, the final point of this region gives the stable solution, which is appropriate to the starting point, and all the starting points of the region give the same final point. According to the remarks of Section 8, if the initial conditions are

$$Q_0 = -0.9999, \qquad \dot{Q}_0 = 0.9999 - 3/2,$$

or

$$Q_0 = 1.3645, \qquad \dot{Q}_0 = 0.3648 - 3/2,$$

which correspond to the fact that the starting point coincides to the point I of the region ($c$), or to the point II of the region ($d$), no transient phenomena can exist.

15. The Converse Problem (Magiros, 1957c)

In this section we deal with the converse problem; that is, *We try to establish the coefficients of the original equation (39) such that for the subharmonic of order 1/3 the amplitude is prescribed.* The solution of this problem corresponds to the determination of the numbers $\lambda$ and $\mu$ in the system (42) and then, with the help of (42a), of the ratios of the coefficients. This determination is to be made in terms of $B$ and $\zeta$, by using the additional equation

$$x^2 + y^2 - \zeta^2 = 0. \tag{62}$$

The determination of $\lambda$ and $\mu$ in this way is impossible because of the form of Eqs. (42). From Eqs. (42) and (62) we can obtain $\lambda$ and $\mu$ in terms of $x, y, \zeta$, and $B$. Then the determination of $\lambda$ and $\mu$ requires us to know $B$ and two of $x$, $y$, $\zeta$. But, if we know $B$ and $x$, $y$, the numbers $\lambda$ and $\mu$ are known by solving the system (42) in $\lambda$ and $\mu$, (62) being a restriction between $x, y, \zeta$. Now, by knowing $\lambda$ and $\mu$, we know the ratios $(1 - \bar{c}_1)/\bar{c}_3$, $\bar{k}/\bar{c}_3$. Thus any two coefficients from $\bar{c}_1$, $\bar{c}_3$, $\bar{k}$ can be determined in terms of a third one, which can get arbitrary values. Since the stability can be treated as in the previous example, the converse problem is finished.

To illustrate the above, let us take two cases with the same amplitude: (a) $x = 1, y = 1$ and (b) $x = 0.3648, y = -1.3645$.

(a) $x = 1, y = 1, B = 4$. According to (40) the corresponding solution is

$$Q = \sin t - \cos t - 1/2 \sin 3t. \tag{63}$$

From (42) and (42a) we find that

$$\frac{1 - \bar{c}_1}{\bar{c}_3} = -\frac{3}{8}, \qquad \frac{\bar{k}}{\bar{c}_3} = \frac{3}{2}. \tag{63a}$$

The test for the stability of the singularity (1, 1) shows that this point is a saddle point. Thus the solution (63) of Eq. (39), when the coefficients satisfy the relations (63a), is not acceptable.

(b) $x = 0.3648$, $y = -1.3645$, $B = 4$. The corresponding ratios of the coefficients are

$$\frac{1 - \bar{c}_1}{c_3} = \frac{3}{2}, \qquad \frac{\bar{k}}{\bar{c}_3} = -\frac{3}{8}, \tag{64a}$$

and the solution is

$$Q = 0.3648 \sin t + 1.3645 \cos t - \tfrac{1}{2} \sin 3t. \tag{64}$$

This is acceptable, since the singularity $x = 0.3648$, $y = -1.3645$ is a stable spiral point.

## 16. The Singularity at the Origin in the General Case (Magiros, 1957c)

From the system (42) we see that the origin $x = 0$, $y = 0$ is, for any case of the coefficients of (39), a singularity of (39). In other words, the function $-\frac{1}{8}B \sin 3t$ is always a solution of (39), the harmonic solution, either stable or unstable. For the stability of this harmonic solution we have, first, from (55):

$$\begin{aligned} a_1 &= -\frac{1}{2}\bar{k} \\ a_2 &= -\frac{1}{2}(1 - \bar{c}_1) + \frac{3}{16^2}\bar{c}_3 B^2, \\ b_1 &= \frac{1}{2}(1 - \bar{c}_1) - \frac{3}{16^2}\bar{c}_3 B^2 \\ b_2 &= -\tfrac{1}{2}\bar{k}, \end{aligned} \tag{65}$$

when, from (36a) and (65), we obtain

$$\rho_{1,2} = -\frac{1}{2}\bar{k} \pm i\left[\frac{1}{2}(1 - \bar{c}_1) - 3\bar{c}_3\frac{B^2}{16^2}\right]. \tag{66}$$

By using the definitions of the singularities of Section 6 we get the following:

(a) If the imaginary part of (66) is not zero, that is, if

$$\frac{1 - \bar{c}_1}{\bar{c}_3} \neq 6 \cdot \frac{B^2}{16^2}, \tag{67a}$$

then (i) for $\bar{k} > 0$, the origin is a stable spiral point; (ii) for $\bar{k} < 0$, the origin is an unstable spiral point; (iii) for $\bar{k} = 0$, the origin is either a center or a spiral.

(b) If the imaginary part of (66) is zero, that is, if

$$\frac{1 - \bar{c}_1}{\bar{c}_3} = 6 \cdot \frac{B^2}{16^2}, \tag{67b}$$

then (i) for $\bar{k} > 0$, the origin is a nodal stable point; (ii) for $\bar{k} < 0$, the origin is an unstable nodal point; (iii) for $\bar{k} = 0$, the origin is not a simple singularity of the kind we know in Section 6, since the characteristic roots are zero.

The result of this section is that *the origin is a stable singularity when* $\bar{k} > 0$, *and of spiral type under the condition* (67*a*), *or of nodal type under the condition* (67b). This is shown in Fig. 4. In Fig. 4, the left-hand of

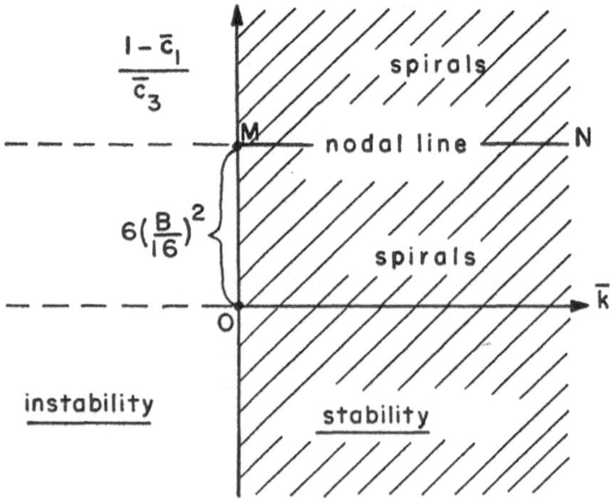


FIG. 4. The kinds of singularity of the origin of Eq. (39). For $\bar{k} > 0$ the origin is stable and either of spiral or of nodal kind. For $\bar{k} < 0$ the origin is an unstable singularity.

the $(1 - \bar{c}_1)/\bar{c}_3$, $\bar{k}$-plane corresponds to instability of the harmonic solution. The right-hand of this plane corresponds to stability, with all spiral points, except the points of the nodal line $MN$, of which all points are nodal points, except the point $M$ which is a singularity but not of simple type. The points of the $[(1 - \bar{c}_1)/\bar{c}_3]$-axis may be either centers or spirals, except $M$. The distance $OM$ has a maximum and a minimum due to the restriction of $B$, given by (50b). For $B = 0$, the $\bar{k}$-axis is the nodal line.

## APPENDIX

The following are concerned with the singular points, that is, the constant solutions, of the system (14). The functions $F_1$ and $F_2$ of (14) satisfy the conditions of Fourier's theorem (17), so:

$$\begin{aligned} \dot{u}_1 &= \epsilon(A_0 + A_1 \cos t + B_1 \sin t + \cdots \\ &\qquad + A_m \cos mt + B_m \sin mt + \cdots) \\ \dot{u}_2 &= \epsilon(C_0 + C_1 \cos t + D_1 \sin t + \cdots \\ &\qquad + C_m \cos mt + D_m \sin mt + \cdots) \end{aligned} \tag{68}$$

where the coefficients $A$, $B$, $C$, $D$ depend on $u_1$ and $u_2$.

Take the system:

$$\begin{aligned} \dot{u}_1 &= \epsilon A_0(u_1, u_2), \\ \dot{u}_2 &= \epsilon C_0(u_1, u_2), \end{aligned} \tag{68a}$$

where the time $t$ does not enter explicitly. Each of the above systems accepts a unique solution $\{u_1, u_2\}$ which assumes given values

$$\{u_1(t_0) = u_{10}, u_2(t_0) = u_{20}\}$$

at $t = t_0$, provided that the functions in their right-hand sides satisfy a Lipschitz condition, when their arguments are restricted to lie in the domain

$$\Delta: \quad |u_1 - u_{10}| \leqq k_1, \quad |u_2 - u_{20}| \leqq k_2, \qquad |t - t_0| < T. \tag{69}$$

Apply Picard's method of successive approximations for the calculation of the solution $\{\bar{u}_1, \bar{u}_2\}$ of the system (68a), by taking as zeroth approximation arbitrary initial conditions $\{\bar{u}_{10}, \bar{u}_{20}\}$. The successive approximations, which converge to the unique solution $\{\bar{u}_1, \bar{u}_2\}$ of (68a), are

$$\bar{u}_1^{(1)} = \bar{u}_{10} + \epsilon \int_{t_0}^{t} A_0(\bar{u}_{10}, \bar{u}_{20})\, dt,$$

$$\bar{u}_2^{(1)} = \bar{u}_{20} + \epsilon \int_{t_0}^{t} C_0(\bar{u}_{10}, \bar{u}_{20})\, dt,$$

$$\cdots\cdots\cdots\cdots \tag{70a}$$

$$\bar{u}_1^{(n)} = \bar{u}_{10} + \epsilon \int_{t_0}^{t} A_0(\bar{u}_1^{(n-1)}, \bar{u}_2^{(n-1)})\, dt,$$

$$\bar{u}_2^{(n)} = \bar{u}_{20} + \epsilon \int_{t_0}^{t} C_0(\bar{u}_1^{(n-1)}, \bar{u}_2^{(n-1)})\, dt.$$

Apply also the above method for the calculation of the solution $\{u_1, u_2\}$ of the system (68), by taking as zeroth solution arbitrary initial conditions $\{u_1(t_0) = u_{10}, u_2(t_0) = u_{20}\}$. The successive approximations are

$$\begin{aligned}
u_1^{(1)} &= u_{10} + \epsilon \int_{t_0}^{t} A_0(u_{10}, u_{20})\, dt \\
&\qquad + \epsilon \int_{t_0}^{t} A_1(u_{10}, u_{20}) \cos t\, dt + \cdots \\
u_2^{(1)} &= u_{20} + \epsilon \int_{t_0}^{t} C_0(u_{10}, u_{20})\, dt \\
&\qquad + \epsilon \int_{t_0}^{t} C_1(u_{10}, u_{20}) \cos t\, dt + \cdots \\
u_1^{(n)} &= u_{10} + \epsilon \int_{t_0}^{t} A_0(u_1^{(n-1)}, u_2^{(n-1)})\, dt \\
&\qquad + \epsilon \int_{t_0}^{t} A_1(u_1^{(n-1)}, u_2^{(n-1)}) \cos t\, dt + \cdots \\
u_2^{(n)} &= u_{20} + \epsilon \int_{t_0}^{t} C_0(u_1^{(n-1)}, u_2^{(n-1)})\, dt \\
&\qquad + \epsilon \int_{t_0}^{t} C_1(u_1^{(n-1)}, u_2^{(n-1)}) \cos dt + \cdots
\end{aligned} \tag{70}$$

which converge to the unique solution $\{u_1, u_2\}$ of the system (68).

Let us take the same initial conditions

$$u_{10} = \bar{u}_{10}, \qquad u_{20} = \bar{u}_{20}, \tag{71}$$

of the above approximation (70) and (70a), and subtract. We get

$$\begin{aligned} u_1^{(n)} - \bar{u}_1^{(n)} &= \epsilon \int_{t_0}^{t} \{A_0(u_1^{(n-1)}, u_2^{(n-1)}) - A_0(\bar{u}_1^{(n-1)}, \bar{u}_2^{(n-1)})\}\, dt \\ &\quad + \epsilon \int_{t_0}^{t} A_1(u^{(n-1)}, u_2^{(n-1)}) \cos t\, dt \\ &\quad + \epsilon \int_{t_0}^{t} B_1(u_1^{(n-1)}, u_2^{(n-1)}) \sin t\, dt + \cdots \\ u_2^{(n)} - \bar{u}_2^{(n)} &= \epsilon \int_{t_0}^{t} \{C_0(u_1^{(n-1)}, u_2^{(n-1)}) - C_0(\bar{u}_1^{(n-1)}, \bar{u}_2^{(n-1)})\}\, dt \\ &\quad + \epsilon \int_{t_0}^{t} C_1(u_1^{(n-1)}, u_2^{(n-1)}) \cos t\, dt \\ &\quad + \epsilon \int_{t_0}^{t} D_1(u_1^{(n-1)}, u_2^{(n-1)}) \cos t\, dt + \cdots \end{aligned} \tag{72}$$

The integrals in (72) are bounded and the right-hand sides contain $\epsilon$ or a common factor, so the approximations, and consequently their limits, are, for small $\epsilon$, of order $\epsilon$; that is,

$$| u_1 - \bar{u}_1 | = O(\epsilon), \qquad | u_2 - \bar{u}_2 | = O(\epsilon). \tag{73}$$

In (73) $\{u_1, u_2\}$ and $\{\bar{u}_1, \bar{u}_2\}$ are solutions of (68) and (68a), respectively. Thus any solution of the system (68a) can be considered as an approximation of the solution of the system (68) of the first order $\epsilon$.

The solutions of the system (14), i.e., of the system (68), are constant if $\epsilon F_1$ and $\epsilon F_2$ tend to zero, or, since the time $t$ enters explicitly in $F_1$ and $F_2$, if $\epsilon \to 0$. But in the system (68a) the time $t$ does not enter explicitly, so we can get constant solutions of (68a), if $\epsilon$ is not necessarily zero, by taking appropriate initial conditions in the approximations (70a), namely, the initial conditions $\{\bar{u}_{10}, \bar{u}_{20}\}$ which satisfy the conditions

$$A_0(\bar{u}_{10}, \bar{u}_{20}) = 0, \qquad C_0(\bar{u}_{10}, \bar{u}_{20}) = 0, \tag{74}$$

when the integrals in (70a) are zero and the solution $\{\bar{u}_1, \bar{u}_2\}$ of (68a) is the constant $\{\bar{u}_{10}, \bar{u}_{20}\}$ for any $\epsilon$, including $\epsilon = 0$. A constant solution

$\{\bar{u}_1, \bar{u}_2\}$ of the approximate systems (68a), which satisfies the condition (74), is considered as an approximate constant solution $\{u_1, u_2\}$ of the exact system (68), or (14), of the first order in $\epsilon$.

The technique of substituting the system (68), where the time $t$ enters explicitly, by the approximate system (68a), where the time does not enter explicitly, consists essentially of replacing a function by its "mean value" over an interval (Fatou, 1929), which is called "moving average" or "sliding mean" (Van der Pol and Bremmer, 1955). Since the moving average is generally smoother than the original function, the above technique, which is known as the "averaging principle" (Kryloff and Bogoliuboff, 1947), offers advantages in the study of the original system. In practice, this technique brings good results, for example, in economics or in electrical problems such as the photoelectric reproduction of sound, television images (Van der Pol and Bremmer, 1955).

RECEIVED: November 27, 1957.

## REFERENCES

ANDRONOW, A., AND WITT, A. (1930). Zur Theorie des Mitnehmens von B. van der Pol. *Arch. Elektrotechn.* **24**, 99.

BENDIXSON, I. (1901). Sur les courbes définies par les équations différentielles. *Acta Math.* **24**, 1–88.

FATOU, P. (1929). Sur le mouvement d'un système soumis à des forces à courte période. *Bull. Soc. math. France* pp. 98–139.

FRIEDRICHS, C. (1953). Fundamentals of Poincaré's theory. "Proceedings of the Symposium on Nonlinear Circuit Analysis," Vol. II, pp. 56–67. Polytechnic Institute of Brooklyn, New York.

HADAMARD, J. (1933). The latest work of Poincaré. *Rice Inst. Pamphlet* **20**(1), 9–28.

HAYACHI, C. (1953). "Forced Oscillations in Nonlinear Systems," p. 23. Nippon Printing and Publishing Company, Osaka, Japan.

KRYLOFF, N. AND BOGOLIUBOFF, N. (1947). Introduction of nonlinear mechanics. *Ann. Math. Studies* **11**, 12.

MAGIROS, D. (1957a). "Subharmonics of any order in case of nonlinear restoring force." *Prakt. Athens Acad. Sci.* **32**, 77–85.

MAGIROS, D. (1957b). "Subharmonics of order one third in case of cubic restoring force." *Prakt. Athens Acad. Sci.* **32**, 101–108.

MAGIROS, D. (1957c). "Remarks on a problem of subharmonics; its inverse problem." *Prakt. Athens Acad. Sci.* **32**, 143–146.

MAGIROS, D. (1957d). "On the singularities of a system of differential equations where the time enters explicitly." *Prakt. Athens Acad. Sci.* **32**, 448–451.

MANDELSTAM, L. AND PAPALEXI, N. (1935). Über einige nichstationäre schwingungsvorgänge." *Tech. Phys. U.S.S.R.* **1**, 415–428.

PETROVSKY, I. (1954). Partial Differential Equations, p. 21. Interscience, New York.

POINCARÉ, H. (1890). Sur le problème des trois corps et les équations différentielles. *Acta Math.* **13,** 88.

POINCARÉ, H. (1892). Sur les courbes définies par une équation différentielle, Vol. 1 of "Oeuvres." Gauthier-Villars, Paris,

STOKER, J. J. (1955). "On the stability on mechanical systems," *Communs. Pure and Appl. Math.* **7,** 132–142.

TSIEN, H. (1956). The "Poincaré-Lighthill-Kuo" method. *Advances in Appl. Mech.* **4,** 281–349.

VAN DER POL, B. (1927). Forced oscillations in a circuit with non-linear resistance. *Phil. Mag.* [7] **3,** 65–80.

VAN DER POL, B. AND BREMMER, H. (1955). "Operational Calculus," 2nd ed., Chapter XIV, §2. Cambridge Univ. Press, London and New York.

VON KÁRMÁN, T. (1940). The engineer grapples with nonlinear problems. *Bull. Am. Math. Soc.* **46,** 615–683.

WHITTAKER, E. T., AND WATSON, G. N. (1952). "A course of Modern Analysis," 4th ed., p. 164. MacMillan, New York.

# On a Problem of Nonlinear Mechanics

Demetrios G. Magiros

*Republic Aviation Corporation, Farmingdale, New York*

The behavior of a forced oscillatory system which is linearly damped but nonlinear in the restoring force is investigated according to the author's previous papers (Magiros, 1957, 1958). It is shown under which conditions the system may contain subharmonics of order $\frac{1}{2}$. The amplitudes of the subharmonics and their components, and the bounds for the amplitude of the external force, are given in terms of the coefficients of the differential equation of the system, which are not necessarily very small, as well as the regions in the $(c_1/c_3, I)$-plane, where we have subharmonics with two, one, or neither amplitudes. Also discussed are the stability of the subharmonics, the free vibrations of the system, and the case when one of the coefficients of the nonlinear terms is zero.

## INTRODUCTION

The investigation of the oscillations of a nonlinear system with frequency half of that of the external force—subharmonics of order $\frac{1}{2}$—is of current interest (Mandelstam, 1935; Melikjan, 1935; Reuter, 1949). A physical example, to mention just one, is the loudspeaker. A sinusoidal current in the coil causes the loudspeaker diaphragm to vibrate axially about a central position. These vibrations may, under certain circumstances, be with frequency half of that of the driving current.

In this paper we study the subharmonics of order $\frac{1}{2}$ when the system is linearly damped and the nonlinear restoring force of cubic type, that is when the system is governed by a differential equation of the form (1), where the coefficients are not necessarily very small. The amplitudes of the subharmonics and their components are found in terms of the coefficients of the differential equation. The regions in the $(c_1/c_3, I)$-plane, where we have subharmonics with two, one, or neither amplitudes, are also discussed. $I$ is a polynomial in the coefficients of the differential equation. Bounds for the amplitude of the external force are given. The stability of the subharmonics, the free vibrations of the system, and the



Reprinted from *Information and Control* **2** (1959), 297–309.

case where one of the coefficients of the nonlinear terms is zero are also treated.

## I. THE AMPLITUDE OF THE SUBHARMONICS

We ask for oscillations with frequency half of that of the excitation of the differential equation

$$\ddot{Q} + \bar{k}\dot{Q} + \bar{c}_1 Q + \bar{c}_2 Q^2 + \bar{c}_3 Q^3 = B \sin 2t. \tag{1}$$

By using

$$\bar{k} = \epsilon k, \qquad 1 - \bar{c}_1 = \epsilon c_1, \qquad \bar{c}_2 = \epsilon c_2, \qquad \bar{c}_3 = \epsilon c_3, \qquad \epsilon > 0, \tag{2}$$

Eq. (1) can be written as

$$\ddot{Q} + Q = \epsilon f(Q,\dot{Q}) + B \sin 2t, \tag{3}$$

the solution of which, in case $\epsilon = 0$, is given by

$$Q = x \sin t - y \cos t - \frac{B}{3} \sin 2t, \tag{4}$$

where $x$ and $y$ are arbitrary constants which depend on the initial conditions of the differential equation. $x$ and $y$ are the components of the amplitude $r$ of the subharmonics. If $\epsilon \neq 0$, (4) also can give the solution of (3) or (1), but $x$ and $y$ are appropriate functions of $t$, say $u_1(t)$ and $u_2(t)$ respectively, which, as $\epsilon \to 0$, must have as limits $x$ and $y$.

By applying Eqs. (38a) and (38b) of Magiros (1958), we can find that $x$ and $y$ satisfy the equations

$$\begin{aligned} A_0(x,y) &\equiv \tfrac{1}{2}[\tfrac{1}{3}c_2 B - k]x + \tfrac{1}{2}[\tfrac{1}{6}c_3 B^2 - c_1]y + \tfrac{3}{8}c_3 y(x^2 + y^2) = 0, \\ C_0(x,y) &= \tfrac{1}{2}[c_1 - \tfrac{1}{6}c_3 B^2]x - \tfrac{1}{2}[k + \tfrac{1}{3}c_2 B]y - \tfrac{3}{8}c_3 x(x^2 + y^2) = 0. \end{aligned} \tag{5}$$

For $c_3 \neq 0$, Eqs. (5) can be written as

$$(\mu - \tfrac{1}{3}\nu B)x + (\lambda - A)y = 0, \qquad (\lambda - A)x - (\mu + \tfrac{1}{3}\nu B)y = 0 \tag{6}$$

with

$$\begin{aligned} \lambda &= \frac{4}{3}\frac{c_1}{c_3} = \frac{4}{3}\frac{1 - \bar{c}_1}{\bar{c}_3}, \qquad \mu = \frac{4}{3}\frac{k}{c_3} = \frac{4}{3}\frac{\bar{k}}{\bar{c}_3}, \\ \nu &= \frac{4}{3}\frac{c_2}{c_3} = \frac{4}{3}\frac{\bar{c}_2}{\bar{c}_3}, \qquad A = r^2 + \frac{2}{9}B^2, r^2 = x^2 + y^2. \end{aligned} \tag{6a}$$

The condition for nonzero amplitude $r$ of the subharmonics, that is for nonzero roots $x$ and $y$ of the system (6), is the vanishing of the determinant of the coefficients of the system (6), which gives

$$r_{1,2}^2 = \lambda - \tfrac{2}{9}B^2 \mp \sqrt{\tfrac{1}{9}\nu^2 B^2 - \mu^2}. \tag{7}$$

## II. THE REALITY OF THE AMPLITUDE $r$

The reality of $r$, as given by (7), implies the following conditions to be satisfied: (a) the expression under the radical must be nonnegative and, (b) the right-hand side of (7) must be positive.

The condition (a) gives the restrictions

$$| B | \geqq 3 | \mu/\nu | = 3 | k/c_2 |, \tag{8a}$$

$$(B - 3k/c_2)(B + 3k/c_2) \geqq 0. \tag{8b}$$

The restriction (8$a$) gives bounds for $B$, and the shaded regions in Fig. 1 are the permissible ones, according to (8$b$).

The condition (b) gives the following restrictions:

(A) The existence of nonzero $r_1$ implies that

$$\sqrt{\tfrac{1}{9}\nu^2 B^2 - \mu^2} < \lambda - \tfrac{2}{9}B^2, \tag{9}$$

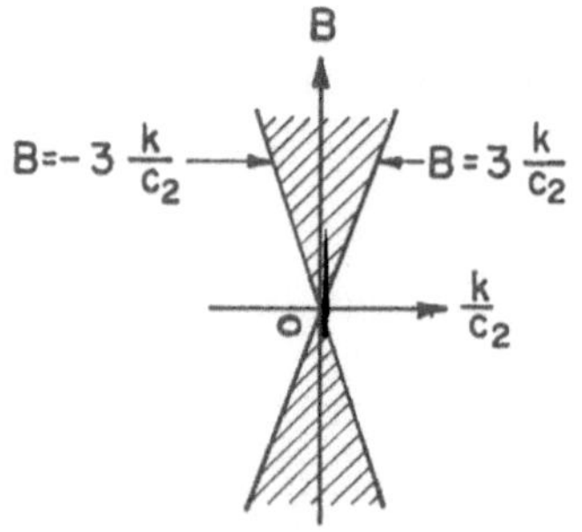


FIG. 1.

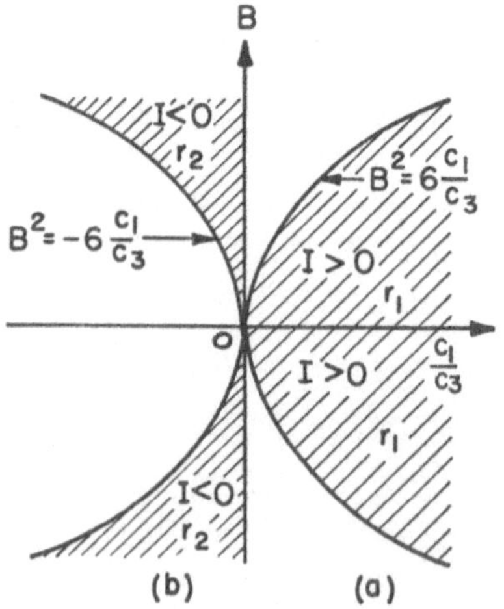


FIG. 2.

FIG. 1. The shaded regions in the ($k/c_2$ , $B$)-plane are the appropriate ones for the reality of the amplitude $r$ of the subharmonics.

FIG. 2. The shaded region ($a$) in the ($c_1/c_3$ , $B$)-plane, for $I > 0$, corresponds to the existence of the amplitude $r_1$ of the subharmonics; the shaded regions ($b$), for $I < 0$, to the existence of the amplitude $r_2$ .

then

$$\lambda - \tfrac{2}{9}B^2 > 0 \quad \text{and} \quad \lambda > 0,$$

or

$$B^2 < \frac{6c_1}{c_3}, \qquad \frac{c_1}{c_3} > 0 \tag{10}$$

Squaring (9) we can get $I > 0$ if

$$I = \tfrac{4}{81}B^4 - \tfrac{1}{9}(\nu^2 + 4\lambda)B^2 + \lambda^2 + \mu^2, \tag{11}$$

and $r_1$ exists only for points of the shaded region Fig. 2(a) with $I > 0$.

(B) The existence of nonzero $r_2$ implies that

$$\sqrt{\tfrac{1}{9}\nu^2B^2 - \mu^2} > \tfrac{2}{9}B^2 - \lambda \tag{12}$$

and we have three cases for the existence of $r_2$ :

(i) For $\lambda < 0$, we have $\frac{2}{9}B^2 - \lambda > 0$, and squaring (12) we get $I < 0$. Then $r_2$ always exists for points of the shaded region in Fig. 2(b) with $I < 0$.

(ii) For $\lambda > 0$, if $| \frac{2}{9}B^2 - \lambda |^2 < \frac{1}{9}\nu^2B^2 - \mu^2$, we have $I < 0$.

(iii) For $\lambda > 0$, if $| \frac{2}{9}B^2 - \lambda |^2 > \frac{1}{9}\nu^2B^2 - \mu^2$, we have $I > 0$.

The above results can be summarized in Fig. 3.

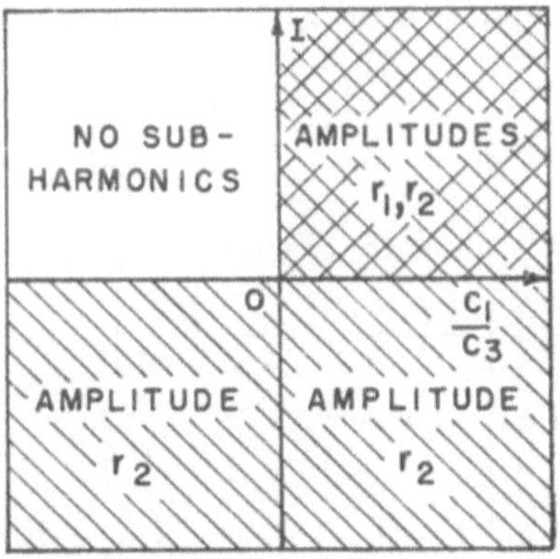


FIG. 3. Only points of the first quadrant of the $(c_1/c_3, I)$-plane can give subharmonics with two different nonzero amplitudes $r_1$, $r_2$. Points of the third and fourth quadrant give subharmonics with one amplitude $r_2$, and points of the second quadrant correspond to *no* subharmonics.

## III. THE COMPONENTS OF THE AMPLITUDES $r$

The system (6) gives

$$\frac{y}{x} = \frac{\mu - \frac{1}{3}\nu B}{A - \lambda} = \frac{\lambda - A}{\mu + \frac{1}{3}\nu B} = \theta \tag{13}$$

when the components $x$ and $y$ are given as solutions of the system

$$y = \theta x, \qquad x^2 + y^2 = r^2 \tag{14}$$

Then

$$\begin{aligned} x_1 &= \frac{r}{\sqrt{1+\theta^2}}, & y_1 &= \frac{r\theta}{\sqrt{1+\theta^2}}, \\ x_2 &= -\frac{r}{\sqrt{1+\theta^2}}, & y_2 &= -\frac{r\theta}{\sqrt{1+\theta^2}}. \end{aligned} \tag{15}$$

These solutions correspond to two symmetric points with respect to the origin in the $x,y$-plane. Since we have two, in general, values of $r$, and $\theta$ depends on $r$, (15) give four points, then the singularities of Eq. (1) are, in general, five, the origin included.

## IV. THE STABILITY OF THE SUBHARMONICS

The coordinates $(x,y)$ of the five singular points of Eq. (1) are given by the solutions of the system: $A_0(x,y) = 0$, $C_0(x,y) = 0$; then the origin and the points (15) are the singular points. Their kind depends on the numbers

$$\rho_{1,2} = \tfrac{1}{2}\{a_1 + b_2 \mp \sqrt{(a_1 - b_2)^2 + 4a_2b_1}\}, \tag{16}$$

where

$$a_1 = \frac{\partial A_0}{\partial x}, \qquad a_2 = \frac{\partial A_0}{\partial y}, \qquad b_1 = \frac{\partial C_0}{\partial x}, \qquad b_2 = \frac{\partial C_0}{\partial y}.$$

For $c_3 \neq 0$, the $a_1$, $a_2$, $b_1$, $b_2$ are given by

$$\begin{aligned} a_1 &= \frac{1}{2}\frac{\bar{c}_3}{\epsilon}\left[-\frac{\bar{k}}{\bar{c}_3} + \frac{3}{2}xy + \frac{B}{3}\frac{\bar{c}_2}{\bar{c}_3}\right], \\ a_2 &= \frac{1}{2}\frac{\bar{c}_3}{\epsilon}\left[-\frac{1-\bar{c}_1}{\bar{c}_3} + \frac{9}{4}y^2 + \frac{3}{4}x^2 + \frac{B^2}{6}\right], \\ b_1 &= \frac{1}{2}\frac{\bar{c}_3}{\epsilon}\left[\frac{1-\bar{c}_1}{\bar{c}_3} - \frac{9}{4}x^2 - \frac{3}{4}y^2 - \frac{B^2}{6}\right], \\ b_2 &= \frac{1}{2}\frac{\bar{c}_3}{\epsilon}\left[-\frac{\bar{k}}{\bar{c}_3} - \frac{3}{2}xy - \frac{B}{3}\frac{\bar{c}_2}{\bar{c}_3}\right]. \end{aligned} \tag{17}$$

The tangential direction of the integral curves at the neighborhood of a nodal or a saddle singular point is characterized by the numbers

$$\mu_{1,2} = \frac{1}{2a_2}\left[b_2 - a_1 \mp \sqrt{(a_1 - b_2)^2 + 4a_2b_1}\right]. \tag{18}$$

For a "stable singular point," a point on the integral curve near the singular point, tends, with the increase of time, to that singular point; otherwise the singular point is "unstable." For a "stable nodal point," the values of $\mu_1$, $\mu_2$ must be real and of different signs. For a "spiral point," when $\rho_{1,2}$, $\mu_{1,2}$ are complex numbers, the negative real part of $\rho$ gives information about its "stability." The integral curves around a "stable spiral point" are of the "clockwise direction," if $a_2 > 0$ (Hayashi, 1953). Let us illustrate the above by the following example.

## V. NUMERICAL EXAMPLE

Take $B = 1, \bar{k} = 0.1, \bar{c}_1 = 1\frac{3}{4}, \bar{c}_2 = \frac{3}{8}, \bar{c}_3 = -\frac{1}{2}$. Notice first that $c_1/c_3 = (1 - \bar{c}_1)/\bar{c}_3 = \frac{3}{2}, I = 3.1205$. Then, according to Fig. 3, we expect two subharmonics with different amplitudes. The result of the computation is shown in Table II below.

## VI. FREE NONLINEAR VIBRATIONS

The previous forced vibrations are dependent on, or governed by, the characteristics of the system and the external force. For the free nonlinear vibrations we must put in (1) $B = 0$. In this case the amplitude of the free vibrations of the system is given by:

$$r_{1,2}^2 = \lambda \mp \sqrt{-\mu^2} \tag{19}$$

The reality of $r$ implies $\mu = 0, \lambda > 0$, that is $\bar{k} = 0, c_1/c_3 = (1 - \bar{c}_1)/\bar{c}_3 > 0$; then the free vibrations of (1) occur in the system when we have either

$$\begin{aligned} &\text{(i)} \quad k = 0, \quad c_1 > 0, \quad c_3 > 0, \quad \text{or} \\ &\text{(ii)} \quad k = 0, \quad c_1 < 0, \quad c_3 < 0. \end{aligned} \tag{20}$$

TABLE I

| | | | |
|---|---|---|---|
| $\lambda = 2$ | $\mu = -\frac{4}{15}$ | $\nu = -1$ | |
| $r_1^2 = 1.57778$ | $r_2^2 = 1.97778$ | $r_1 = 1.25610$ | $r_2 = 1.40634$ |
| $\theta_1 = 0.3333$ | $\theta_1^2 = 0.1111$ | $\sqrt{1+\theta_1^2} = 1.05409$ | |
| $\theta_2 = 0.3333$ | $\theta_2^2 = 0.1111$ | $\sqrt{1+\theta_2^2} = 1.05409$ | |

TABLE II
THE SINGULAR POINTS AND THEIR CLASSES FOR THE EXAMPLE IN SECTION V

| Point | | | |
|---|---|---|---|
| 0* | $x_0 = 0$ | $y_0 = 0$ | $\rho_{1,2} = \frac{1}{4\epsilon}(0.1 + i\, 0.65484)$ |
| $I^\dagger$ | $x_1 = 1.19164$ | $y_1 = -0.39721$ | $\rho_{1,2} = \frac{1}{4\epsilon}(-0.1 + i\, 1.82791)$ |
| $II^\dagger$ | $x_2 = -1.19164$ | $y_2 = 0.39721$ | $\rho_{1,2} = \frac{1}{4\epsilon}(-0.1 + i\, 1.82791)$ |
| $III^\dagger$ | $x_3 = 1.33417$ | $y_3 = 0.44471$ | $\rho_{1,2} = \frac{1}{4\epsilon}(-0.1 + i\, 2.40058)$ |
| $IV^\dagger$ | $x_4 = -1.33417$ | $y_4 = -0.44472$ | $\rho_{1,2} = \frac{1}{4\epsilon}(-0.1 - i\, 2.40058)$ |

* Unstable spiral with clockwise direction.
† Stable spiral with counterclockwise direction.

or, in terms of the coefficients of (1),

$$\begin{aligned} &\text{(i)} \qquad \bar{k} = 0, \quad 1 > \bar{c}_1, \quad \bar{c}_3 > 0. \\ &\text{(ii)} \qquad \bar{k} = 0, \quad 1 < \bar{c}_1, \quad \bar{c}_3 < 0. \end{aligned} \tag{21}$$

In Fig. 4 we have the inequalities (21) graphically. The amplitude $r$ of the free vibrations becomes

$$r = 2\sqrt{(1 - \bar{c}_1)/3\bar{c}_3}\,. \tag{22}$$

From (15) we see that $\theta$ is of undetermined form; then the components $x$ and $y$ of the amplitude $r$ are also undetermined, but such that

$$x^2 + y^2 = \frac{4}{3}\frac{1 - \bar{c}_1}{\bar{c}_3}. \tag{23}$$

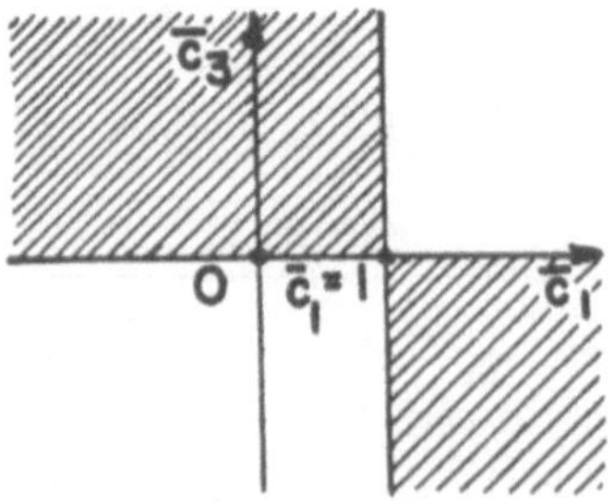


FIG. 4. The shaded regions in $\bar{c}_1$, $\bar{c}_2$-plane correspond to free undamped nonlinear vibrations.

The solution (4) in this case is

$$Q = x \sin t - y \cos t. \tag{24}$$

If the initial conditions are, for $t = 0$, $Q_0$ and $\dot{Q}_0$, then

$$Q_0 = -y, \qquad \dot{Q}_0 = x, \tag{25}$$

the solution becomes

$$Q = \dot{Q}_0 \sin t + Q_0 \cos t, \tag{26}$$

the restriction of the initial conditions $Q_0$, $\dot{Q}_0$ in the $(Q_0, \dot{Q}_0)$-plane being of the cyclic type:

$$Q_0^2 + \dot{Q}_0^2 = \frac{4}{3}\frac{1 - \bar{c}_1}{\bar{c}_3}. \tag{27}$$

The results from the above are the following:

(a) the necessary condition for the existence of free oscillations is the system to be undamped, $\bar{k} = 0$.

(b) the free oscillations do not depend on the coefficient $\bar{c}_2$;

(c) the amplitude $r$ of the free vibrations is given by (22), where the coefficients $\bar{c}_1$ and $\bar{c}_3$ must be such that $r$ is positive,

(d) the components $x$ and $y$ of the amplitude $r$ may be arbitrary but they must obey the restriction (23).

(e) the initial conditions $Q$, $\dot{Q}_0$ must be of the cyclic restriction (27); then if one of them is prescribed the other must have a value given by (27).

## VII. DUFFING'S EQUATION

In case $\bar{c}_2 = 0$, we have the Duffing's equation (Stoker, 1950). In this case the formula (7) gives

$$r^2 = \frac{4}{3}\frac{1 - \bar{c}_1}{\bar{c}_3} - \frac{2}{9}B^2 \mp \sqrt{-(4\bar{k}/3\bar{c}_3)^2} \tag{28}$$

For the existence of the subharmonics we must have

$$\bar{k} = 0, B^2 < 6\,\frac{1 - \bar{c}_1}{\bar{c}_3} \tag{29}$$

which, for positive values of $(1 - \bar{c}_1)/\bar{c}_3$, that is for values of $\bar{c}_1$, $\bar{c}_2$ in the shaded region of Fig. 4, gives bounds for $B$:

$$|B| < \left(6\,\frac{1 - \bar{c}_1}{\bar{c}_3}\right)^{1/2}. \tag{30}$$

The subharmonics have the single amplitude

$$r = \left(\frac{4}{3}\frac{1-\bar{c}_1}{\bar{c}_3} - \frac{2}{9}B^2\right)^{1/2}. \tag{31}$$

VIII. THE CASE $\bar{c}_2 \neq 0$

For the nonlinearity of the system either one or both the coefficients $\bar{c}_2$ and $\bar{c}_3$ must be nonzero. In the foregoing we took $\bar{c}_3 \neq 0$, when $\bar{c}_2$ might be negligible or zero. Let us consider now the case $\bar{c}_2 \neq 0$ when $\bar{c}_3$ might be negligible or zero.

If

$$\frac{c_1}{c_2} = \frac{1-\bar{c}_1}{\bar{c}_2} = l, \qquad \frac{k}{c_2} = \frac{\bar{k}}{\bar{c}_2} = m, \qquad \frac{c_3}{c_2} = \frac{\bar{c}_3}{\bar{c}_2} = n, \qquad A = r^2 + \frac{2}{9}B^2, \qquad r^2 = x^2 + y^2 \tag{32}$$

the system (5) can be written as:

$$\begin{aligned} (\tfrac{1}{3}B - m)x + (-l + \tfrac{3}{4}nA)y &= 0, \\ (l - \tfrac{3}{4}nA)x - (\tfrac{1}{3}B + m)y &= 0. \end{aligned} \tag{33}$$

The vanishing of the determinant of the coefficients of the system (33), which is the requirement for the existence of nonzero roots, gives

$$A^2 - \frac{8}{3}\frac{l}{n}A + \frac{16}{9n^2}\left(l^2 - m^2 + \frac{B^2}{9}\right) = 0, \tag{34}$$

We can obtain from this, by solving for $A$, the amplitudes $r$ of the subharmonics:

$$r_{1,2}^2 = -\frac{2}{9}B^2 + \frac{4}{3n}\left(l \mp \sqrt{(B/3)^2 - m^2}\right). \tag{35}$$

The requirement for the reality of $r$ gives the same restriction as that established in Section 3. Here, instead of the expression (11) for $I$, we have

$$I = \frac{1}{36}B^4 - \frac{1}{3}\left(\frac{l}{n} + \frac{1}{3n^2}\right)B^2 + \left(\frac{l}{n}\right)^2 + \left(\frac{m}{n}\right)^2. \tag{36}$$

The components of the amplitude $r$ of (35) are given by (15), where the quantity $\theta$, calculated from (33), is the same as that given by (13).

As a special case of the above let us treat the case when $n = \bar{c}_3/\bar{c}_2$ is negligible and especially when $n \to 0$ either from positive or from negative values. We write (35) in the form

$$r_{1,2}^2 = \frac{4/3[l \mp \sqrt{(B/3)^2 - m^2}] - \frac{2}{9}B^2 n}{n}. \tag{37}$$

We have two cases:

(A) If $n$ goes to zero from positive values, the reality of $r$ implies that $l$ and $m$ vary in such a way that $[l \mp \sqrt{(B/3)^2 - m^2}]$, being positive and larger than $B^2 n$, goes to zero.

$$l \mp \sqrt{(B/3)^2 - m^2} > 0,$$

or

$$\frac{1 - \bar{c}_1}{\bar{c}_2} > \pm \sqrt{(B/3)^2 - (\bar{k}/\bar{c}_2)^2};$$

and

(i) for $(1 - \bar{c}_1)/\bar{c}_2 \geqq 0$, that is, for $1 \geqq \bar{c}_1$ and $\bar{c}_2 < 0$, or $1 \leqq \bar{c}_1$ and $\bar{c}_2 > 0$, we get

$$\left(\frac{1 - \bar{c}_1}{\bar{c}_2}\right)^2 + \left(\frac{\bar{k}}{\bar{c}_2}\right)^2 > \left(\frac{B}{3}\right)^2; \tag{38a}$$

(ii) for $(1 - \bar{c}_1)/\bar{c}_2 \leqq 0$, that is, for $1 \geqq \bar{c}_1$ and $\bar{c}_2 < 0$, or $1 \leqq \bar{c}_1$ and $\bar{c}_2 > 0$ we have

$$\left(\frac{1 - \bar{c}_1}{\bar{c}_2}\right)^2 + \left(\frac{\bar{k}}{\bar{c}_2}\right)^2 < \left(\frac{B}{3}\right)^2. \tag{38b}$$

Then as $\bar{c}_3 \to 0$, subharmonics may exist in the region $(a_1)$, Fig. 5a, if $\bar{c}_2 > 0$, $1 \geqq \bar{c}_1$, or $\bar{c}_2 < 0$, $1 \leqq \bar{c}_1$; as well as in **the region $(b_1)$ Fig. 5b, if $\bar{c}_2 > 0$, $1 \leqq \bar{c}_1$, or $\bar{c}_2 < 0$, $1 \geqq \bar{c}_1$.**

(B) If $n$ goes to zero from negative values, the reality of $r$ implies that $l \mp \sqrt{(B/3)^2 - m^2} < 0$, then, in the same manner, we can find the region $(a_2)$ and $(b_2)$ of Fig. 6a and Fig. 6b where we may have subharmonics. In the undamped case, $\bar{k} = 0$, the subharmonics correspond to the appropriate segment of $[(1 - \bar{c}_1)/\bar{c}_2]$-axis in the shaded regions.

In the case $\bar{c}_3 = 0$, by combining appropriately the above figures, we have that subharmonics may exist outside the circle if $(1 - \bar{c}_1)/\bar{c}_2 > 0$, and inside if $(1 - \bar{c}_1)/\bar{c}_2 < 0$, Fig. 7(a), 7(b).

Let us call "resonance" the situation where the amplitude $r$ of the

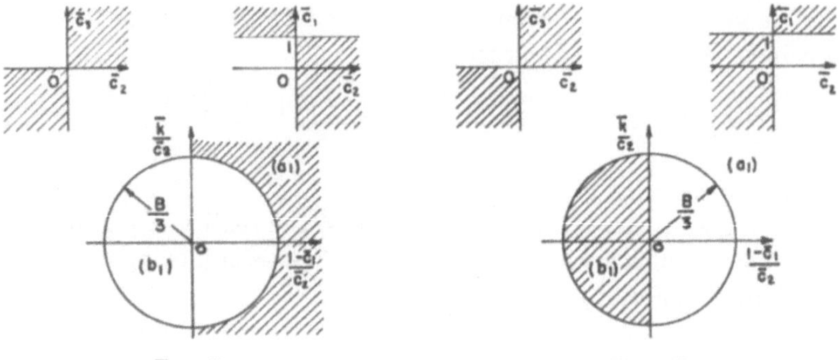


Fig. 5a　　　　Fig. 5b

Fig. 5. The shaded regions $(a_1)$ and $(b_1)$ outside and inside the circle respectively correspond to $\bar{c}_3/\bar{c}_2 > 0$. For the region $(a_1)$ we have $(1 - \bar{c}_1)/\bar{c}_2 > 0$, and for region $(b_1)$: $(1 - \bar{c}_1)/\bar{c}_2 < 0$.

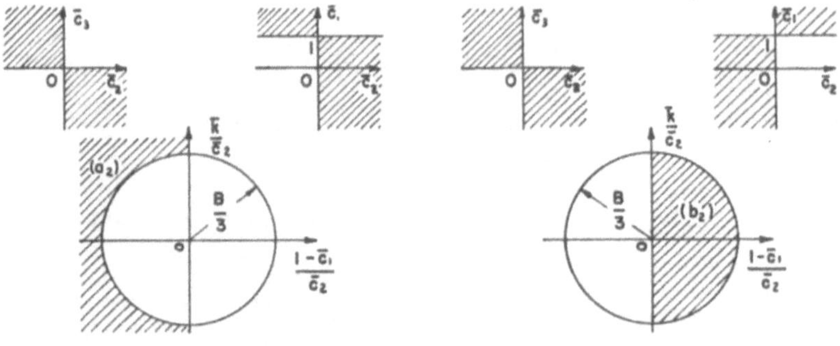


Fig. 6a　　　　Fig. 6b

Fig. 6. The shaded regions $(a_2)$ and $(b_2)$ outside and inside the circle respectively correspond to $\bar{c}_3/\bar{c}_2 < 0$. For the region $(a_2)$ we have $(1 - \bar{c}_1)/\bar{c}_2 > 0$, and for the region $(b_2)$: $(1 - \bar{c}_1)/\bar{c}_2 < 0$.

subharmonics is not finite. The resonance here is characterized by the ratio

$$\frac{l \pm \sqrt{(B/3)^2 - m^2}}{n}.$$

This ratio is not finite, then resonance appears, if $l \neq 0$, $|m| \to B/3$, $n = 0$; that is, if, for $\bar{c}_2 \neq 0$, $\bar{c}_1 \neq 1$, $|\bar{k}/\bar{c}_2| \neq B/3$, $\bar{c}_3 = 0$. Also we have resonance if $l \to 0$, $|m| \to B/3$, $n \to 0$ in such a way that $n$ goes to zero

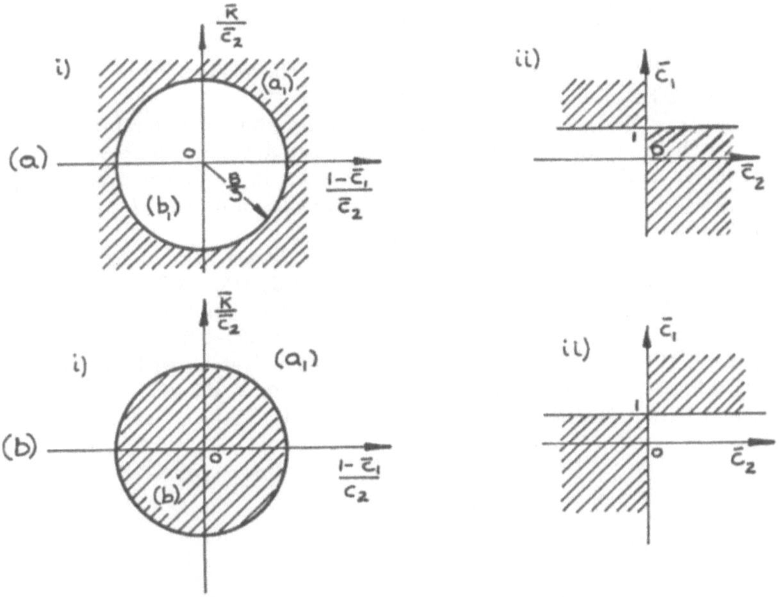


FIG. 7. The shaded region outside the circle, under the condition of the shaded region ($a$, $ii$), and the shaded region inside the circle, under the condition of the shaded region ($b$, $ii$), correspond to subharmonics.

faster than the expression $l \pm \sqrt{(B/3)^2 - m^2}$ does. If the above ratio is finite and bigger than the number $B^2/6$, subharmonics exist.

G. Reuter (1949) studied the subharmonics of order $\frac{1}{2}$ in a special case, the case when the damping coefficient $\bar{k}$ is small and positive, the nonlinear coefficient $\bar{c}_2$ small, the linear coefficient $\bar{c}_1$ near to 1, and the amplitude $B$ of the external force not too large. His results about the the existence of the subharmonics in this special case agree with the above results, and this fact gives a direct verification and a natural check of the validity of the results given above in our general case.

RECEIVED: May, 1959.

REFERENCES

HAYASHI, C. (1953). "Forced Oscillations in Nonlinear Systems," pp. 84–88, 143–147. Nippon Printing and Publishing Company, Osaka, Japan.

MAGIROS, D. (1957a). Subharmonics of any order in case of nonlinear restoring force. Prakt. Athens Acad. Sci. **32,** 77–85.

MAGIROS, D. (1957b). Subharmonics of order one third in case of cubic restoring force. *Prakt. Athens Acad. Sci.* **32,** 101–108.

MAGIROS, D. (1957c). Remarks on a problem of subharmonics; its inverse problem. *Prakt. Athens Acad. Sci.* **32,** 143–146.

MAGIROS, D. (1957d). On the singularities of a system of differential equations where the time enters explicity. *Prakt. Athens Acad. Sci.* **32,** 448–451.

MAGIROS, D. (1958). Subharmonics of any order in nonlinear systems of one degree of freedom: Application to subharmonics of order $\frac{1}{3}$. *Inform. and Control* **1,** 198–227.

MANDELSTAM,. L, AND PAPALEXI, N. (1935). "Über einige nichtstationäre Schwingungsvorgänge". *Tech. Phys. U.S.S.R.* **1,** 415–428.

MELIKJAN, A. (1935). Über das anwachsen der Amplitude bei Resonanzerscheinungen 2-ter Art," *Tech. Phys. U.S.S.R.* 429–448.

REUTER, G. (1949). "Subharmonics in a nonlinear system with unsymmetrical restoring force." *Quart. J. Mech. and Appl. Math.* **2,** 198–207.

STOKER, J. J. (1950). "Nonlinear Vibrations." Interscience, New York.

# *A METHOD FOR DEFINING PRINCIPAL MODES OF NONLINEAR SYSTEMS UTILIZING INFINITE DETERMINANTS*

By Demetrios G. Magiros

REPUBLIC AVIATION CORPORATION, FARMINGDALE, N. Y.

*Communicated by L. Brillouin, October 3, 1960*

1. The concept of "principal modes" plays the predominant role in the analysis of oscillatory systems, no matter what field the systems occur in. For the calculations of principal modes of systems, either linear or nonlinear, two appropriate definitions may be used, namely, the "proportionality definition" of principal modes and their definition as solutions of special "initial value problems."

It is the purpose of this note to provide a procedure for the calculation of principal

Reprinted from the *Proceedings of the National Academy of Sciences* **46** (1960), 1608–1611.

modes of nonlinear systems utilizing "infinite determinants." For a "two-mass-three-spring" free nonlinear system, R. Rosenberg[1] carried out the calculation of principal modes by using the "potential function" of the corresponding conservative system. We restrict ourselves, without loss of generality, to the above "dual-mode" nonlinear system for reasons of comparison with the results obtained by using other methods. The motion of this system is governed by two second-order ordinary differential equations one of which is nonlinear. When the solutions of these equations are assumed to have the form of complex exponential series, the calculation of the frequency and the coefficients leads to recursion formulae, which give rise to "infinite determinants."

2. In this note, the following "proportionality definition" of "principal modes" of an oscillatory nonlinear system is used. By using the terminology of mechanical systems, the "principal modes" are defined as those oscillations for which the amplitudes of the fundamental and the corresponding harmonics of the displacements of any two of the oscillating masses have, separately, a constant ratio at all times. For such motions, the masses all oscillate about their equilibrium positions with the same frequency, the "principal frequency" of the system. If $x_i$, $i = 1, 2, \ldots s$ are the displacements of the masses $m_i$, $i = 1, 2, \ldots s$ from their equilibrium positions and $x_{i_n}$, $x_{i'_n}$ the corresponding amplitudes of the $n$th harmonic of the displacements $x_i$, $x_i'$ of the two masses $m_i$, $m_{i'}$, and if there exist constants $c_n$ such that the conditions,

$$x_{i_n}t \,/x_{i'_n}(t) \equiv c_n,\ i \neq i',\ i, i' = 1, 2, \ldots s\ , c_n \neq 0, \infty \tag{1}$$

are satisfied, then these motions are, by definition, the "principal modes" of the system.

3. If $m_1$ and $m_2$ are the oscillating masses, $K_2$ and $K_3$ the spring constants of the coupling and the second anchor springs, $K_1 = K_1 + \mu x^2$ that of the first anchor spring, $\mu$ a constant which characterizes the nonlinearity, and x and y the displacements of $m_1$ and $m_2$ from their equilibrium positions, the equations of motion of this system are

$$\ddot{x} + \omega_1{}^2 x + \lambda_1 x^3 - \lambda_2 y = 0$$
$$\ddot{y} + \omega_2{}^2 y - \lambda_3 x = 0 \tag{2}$$

where

$$\omega_1{}^2 = \frac{K_1 + K_2}{m_1},\ \omega_2{}^2 = \frac{K_2 + K_3}{m_2},\ \lambda_1 = \frac{\mu}{m_1},\ \lambda_2 = \frac{K_2}{m_1},\ \lambda_3 = \frac{K_2}{m_2}. \tag{2a}$$

Assume solution of (2) to be represented by the exponential series,

$$x(t) = \sum_{n=-\infty}^{\infty} \alpha_n e^{in\omega t},\quad y(t) = \sum_{n=-\infty}^{\infty} A_n e^{in\omega t}. \tag{3}$$

For the reality of the solution, one takes $\bar{\alpha}_n = \alpha_{-n}$, $\bar{A}_n = A_{-n}$, where $\bar{\alpha}_n$ and $\bar{A}_n$ are conjugates of $\alpha_n$ and $A_n$. Also assume that

$$\sum_{n=-\infty}^{\infty} |\alpha_n| < \infty,\ \sum_{n=-\infty}^{\infty} |A_n| < \infty,\ \sum_{u=-\infty}^{\infty} n^2|\alpha_n| < \infty,\ \sum_{n=-\infty}^{\infty} n^2|A_n| < \infty.$$

The first two of these inequalities guarantee the convergence of the series (3), while the second ones the existence of the second derivatives $\ddot{x}$, $\ddot{y}$. Inserting (3) into (2), if one puts $x^3$ into the form,

$$x^3 = \sum_n \sum_{\rho_1} \sum_{\rho_2} \alpha_{\rho_1}\alpha_{\rho_2}\alpha_{n-\rho_1-\rho_2}e^{in\omega t},$$

one can get the system,

$$\left(\omega_1^2 - n^2\omega^2 - \frac{\lambda_2\lambda_3}{\omega_2^2 - n^2\omega^2}\right)\alpha_n + \lambda_1 \sum_{\rho_1}\sum_{\rho_2} \alpha_{\rho_1}\alpha_{\rho_2}\alpha_{n-\rho_1-\rho_2} = 0 \tag{4i}$$

$$A_n/\alpha_n = \frac{\lambda_3}{\omega_2^2 - n^2\omega^2} \tag{4ii}$$

where $n$, $\rho_1$, $\rho_2$ are integers. This system consists of infinitely many nonlinear ($\lambda_1 \neq 0$) homogeneous equations for the infinitely many unknown coefficients $\alpha_n$ and $A_n$.

If $\alpha_1$ is the dominant coefficient of the first of the series (3), the approximate value of the double series in (4i) is

$$c_1\alpha_1^2\alpha_{n-2} + c_2|\alpha_1|^2\alpha_n + c_3\bar{\alpha}_1^2\alpha_{n+2}$$

with an error $o(\alpha_1^2)$. $c_1$, $c_2$, $c_3$ are appropriately calculated integers. Then one can get from (4i) the following recursion formula:

$$p_n\alpha_{n-2} + \alpha_n + q_n\alpha_{n+2} = 0 \tag{5}$$

where

$$p_n = \frac{c_1\lambda_1\alpha_1^2}{K_n},\ q_n = \frac{c_3\lambda_1\bar{\alpha}_1^2}{K_n},\ K_n = \omega_1^2 + c_2\lambda_1|\alpha_1|^2 - n^2\omega^2 - \frac{\lambda_2\lambda_3}{\omega_2^2 - n^2\omega^2}. \tag{5a}$$

For non-zero $\alpha_n$'s, the following necessary and sufficient condition in infinite determinant form must be satisfied:

$$\Delta \equiv \left\|\begin{matrix} \cdots & \cdots & \cdots & \cdots & \cdots & \cdots & \cdots & \cdots & \cdots \\ \cdots & 1 & 0 & q_{-3} & 0 & 0 & 0 & 0 & \cdots \\ \cdots & 0 & 1 & 0 & q_{-2} & 0 & 0 & 0 & \cdots \\ \cdots & p_{-1} & 0 & 1 & 0 & q_{-1} & 0 & 0 & \cdots \\ \cdots & 0 & p_0 & 0 & 1 & 0 & q_0 & 0 & \cdots \\ \cdots & 0 & 0 & p_1 & 0 & 1 & 0 & q_1 & \cdots \\ \cdots & 0 & 0 & 0 & p_2 & 0 & 1 & 0 & \cdots \\ \cdots & 0 & 0 & 0 & 0 & p_3 & 0 & 1 & \cdots \\ \cdots & \cdots & \cdots & \cdots & \cdots & \cdots & \cdots & \cdots & \cdots \end{matrix}\right\| = 0 \tag{6}$$

The "principal frequency" $\omega$ can be determined as a root of the "frequency equation" (6).

To calculate the coefficients $\alpha_n$, in the case $\alpha_1$ dominant, we see that by using the notation,

$$u_n = \frac{\alpha_n}{\alpha_{n-2}}, \tag{7}$$

the recursion formula (5) gives

$$u_n = -\frac{p_n}{1 + q_n u_{n+2}}$$

which leads to the infinite continued fraction,

$$\alpha_n/\alpha_{n-2} = -\cfrac{p_n}{1 - \cfrac{p_{n+2}q_n}{1 - \cfrac{p_{n+4}q_{n+2}}{1 - \dots}}} \tag{8}$$

The infinite continued fraction (8) can be expressed by means of infinite determinants, and the coefficients $\alpha_n$ can be calculated in terms of the arbitrary $\alpha_0$ and $\alpha_1$

Having now calculated the principal frequency $\omega$ and the coefficients $\alpha_n$, the solution (3), by using (4ii), reads:

$$x(t) = \sum_{n=-\infty}^{\infty} \alpha_n e^{in\omega t}, \quad y(t) = \sum_{n=-\infty}^{\infty} \frac{\lambda_3}{\omega_2^2 - n^2\omega^2} \alpha_n e^{in\omega t} \tag{9}$$

The principal frequency $\omega$ is the same for both oscillators, and since the coefficients of (9) satisfy the condition (1), then (9) gives the "principal modes" of our nonlinear system.

The method followed is based on the use of exponential series with complex coefficients leading to infinite determinants when one calculates either the frequency or the coefficients of the series. It is similar to the discussion given by L. Brillouin for Hill and Mathieu functions.[2] Next will be discussed the infinite determinants involved in the present note and the solution in its final form given. It may be mentioned that Hill, in his "Lunar Theory," brought into notice the infinite determinants, and Poincaré first gave conditions of their convergence.

The author is indebted to Dr. G. Nomicos for a comment on a point of the note.

[1] Rosenberg, R., ASME, Appl. Mech. Div. paper No. 59-A-93.

[2] Brillouin, L., *Wave Propagation in Periodic Structures*, 2nd ed. (New York: Dover Publications, Inc., 1953).

# A METHOD FOR DEFINING PRINCIPAL MODES OF NONLINEAR SYSTEMS UTILIZING INFINITE DETERMINANTS, II

By Demetrios G. Magiros

MISSILE AND SPACE VEHICLE DEPARTMENT, GENERAL ELECTRIC COMPANY, PHILADELPHIA

*Communicated by Leon Brillouin, April 17, 1961*

1. (*a*) In a recent paper,[1] the writer has discussed the "principal modes" of a "dual-mode" nonlinear system utilizing "infinite determinants." However, this paper did not give the calculation of the infinite determinants involved and the solution in its final form; also, it did not give the "principal modes" defined as solutions of special "initial-value problems." This is the object of the present paper.

(*b*) We can state for reference that in a "dual-mode" nonlinear system, examined in the previous paper, the "principal modes" are given by

$$x(t) = \sum_{n=-\infty}^{\infty} \alpha_n e^{in\omega t}, \qquad \psi(t) = \sum_{n=-\infty}^{\infty} \frac{\lambda_3}{\omega_2^2 - n^2\omega^2} \alpha_n e^{in\omega t}, \tag{1}$$

and the "recursion formula," for the calculation of the "principal frequency" $\omega$ and the coefficients $\alpha_n$, is

$$p_n\alpha_{n-2} + \alpha_n + q_n\alpha_{n+2} = 0 \tag{2}$$

where

$$p_n = \frac{3\lambda_1\alpha_1^2}{k_n + 6\lambda_1|\alpha_1|^2}, \quad q_n = \frac{3\lambda_1\bar{\alpha}_1^2}{k_n + 6\lambda_1|\alpha_1|^2}, \quad k_n = \omega_1^2 - n^2\omega^2 - \frac{\lambda_2\lambda_3}{\omega_2^2 - n^2\omega^2}. \tag{2a}$$

The factors 3, 6, 3 of the expressions (2*a*) are the coefficients of the trinomial ($A_1$) of the Appendix, which gives the dominant sum of the double series of the formula:

$$k_n\alpha_n + \lambda_1 \sum_{p_1}\sum_{p_2} \alpha_{p_1}\alpha_{p_2}\alpha_{n-p_1-p_2} = 0. \tag{3}$$

For the convergence of the series (1), it is necessary that $(n^2\omega^2 - \omega_2^2)$ be neither zero nor very small, i.e., $\omega$ must not be either a submultiple of $\omega_2$ or very close to a submultiple of $\omega_2$.

2. The "frequency equation" for the calculation of the "principal frequency" $\omega$ comes from the condition of nonzero $\alpha_n$'s of the infinite system (2). This condition in a determinatal form is

$$\Delta(n, \infty) = 0, \tag{4}$$

and the determinant $\Delta(n, \infty)$, for arbitrary integer $n$, is the following one-sided infinite determinant:

$$\Delta(n, \infty) = \lim_{\substack{n = \text{fixed integer} \\ m = 1, 2, 3, \ldots \\ m \to \infty}} \Delta(n, m) \equiv \begin{Vmatrix} 1 & 0 & q_n & 0 & \cdots & & & 0 \\ 0 & 1 & 0 & q_{n+1} & 0 & \cdots & & 0 \\ p_{n+2} & 0 & 1 & 0 & q_{n+2} & 0 & \cdots & 0 \\ \cdots & & & & & & & \cdots \\ 0 & \cdots & & & 0 & 1 & & q_{n+m-2} \\ 0 & \cdots & & & p_{n+m-1} & 0 & 1 & 0 \\ 0 & \cdots & & & 0 & p_{n+m} & 0 & 1 \end{Vmatrix} \tag{5}$$



Reprinted from the *Proceedings of the National Academy of Sciences* 47 (1961), 883–887.

Consider a weak nonlinearity, i.e., $\lambda_1 |x|^3 \ll \omega_1^2 |x|$. Then, since $\alpha_1$ is the dominant coefficient of the series (3), max $x^2 = 2\alpha_1^2$ when

$$\lambda_1 \ll \frac{K_1 + K_2}{2m_1\alpha_1^2}. \tag{6a}$$

For such a small $\lambda_1$, we can get

$$p_n = \lambda_1 \frac{3\alpha_1^2}{k_n} - \lambda_1^2 \frac{18\alpha_1^2|\alpha_1|^2}{k_n^2} + \ldots, \quad q_n = \lambda_1 \frac{3\bar{\alpha}_1^2}{k_n} - \lambda_1^2 \frac{18\bar{\alpha}_1^2|\alpha_1|^2}{k_n^2} + \ldots \tag{6}$$

Thus, the elements of the determinant (5) not in the main diagonal have the small $\lambda_1$ as a common factor, and we can get an expansion of the determinant in powers of $\lambda_1$. Following the appropriate way to get this expansion,[2] by taking the first terms of (6), we obtain

$$\Delta(n, \infty) = 1 - 9\lambda_1^2|\alpha_1|^4 \sum_{m=0}^{\infty} \frac{1}{k_{n+m}\cdot k_{n+m+2}} + 0(\lambda_1^3) \tag{7}$$

For any $\omega$, which is not a zero of the denominators of the elements of the series in (7), this series converges. Combining (4) and (7), if we confine ourselves to the first term of the series in (7), and putting $n = 1$, the frequency equation becomes approximately the following equation, quartic in $\omega^2$:

$$\left(\omega_1^2 - \omega^2 - \frac{\lambda_2\lambda_3}{\omega_2^2 - \omega^2}\right)\left(\omega_1^2 - 9\omega^2 - \frac{\lambda_2\lambda_3}{\omega_2^2 - 9\omega^2}\right) - 9\lambda_1^2|\alpha_1|^4 = 0, \tag{8}$$

which gives the principal frequency $\omega$ of order $0\ (\lambda_1^2)$. For values of the principal frequency of order higher than $0\ (\lambda_1^2)$, we find in the expansion of $\Delta(n, \infty)$ terms of order higher than $0\ (\lambda_1^2)$ and continue in the same way to find the corresponding algebraic equation.

3. The calculation of the coefficients $\alpha_n$ is based on the following formula of the previous paper:[1]

$$\alpha_n/\alpha_{n-2} = -p_n \cdot \cfrac{1}{1 + \cfrac{X_{n,1}}{1 + \cfrac{X_{n,2}}{1 + \ldots}}} \tag{9}$$

with

$$X_{n,m} = -p_{n+4m} \cdot q_{n+2(m-1)}$$

$$n = \text{any fixed integer},\ m = 1,2,3, \ldots \tag{9a}$$

By using the product of $m$ linear transformations of the form

$$X_1(v) = 1, \quad X_m = \frac{1}{1 + X_m v}, \ m = 2,3,4, \ldots$$

$$X_p \neq 0,\ p = 2, 3, \ldots m; \quad X_p = 0,\ p > m, \tag{10}$$

one can prove that the $m^{\text{th}}$ approximant of the continued fraction of (9) may be equal to the ratio of two determinants of which the denominator is

$$\bar{\Delta}(n,m) = \begin{Vmatrix} 1 & -1 & 0 \ldots 0 & 0 & 0 \\ -p_{n+2}q_n & 1 & -1 \ldots 0 & 0 & 0 \\ 0 & -p_{n+4}\cdot q_{n+2} & 1 \ldots 0 & 0 & 0 \\ \cdots & \cdots & \cdots & \cdots & \cdots \\ 0 & 0 & 0 \ldots -p_{n+2(m-1)}\cdot q_{n+2(m-2)} & 1 & -1 \\ 0 & 0 & 0 \ldots 0 & -p_{n+2m}\cdot q_{n+2(m-1)} & 1 \end{Vmatrix} \tag{11}$$

and the numerator, $\bar{\Delta}(n + 2,m)$, is obtained from (11) by omitting its first row and first column. For fixed $n$, and $m \to \infty$, we can write

$$\alpha_n/\alpha_{n-2} = -p_n \frac{\bar{\Delta}(n + 2, \infty)}{\bar{\Delta}(n, \infty)} \tag{12}$$

where the infinite determinants are of von Koch's type.

The induction procedure applied to (12) gives for the coefficients with even index:

$$\alpha_{2n} = (-1)^n \alpha_0 p_2 p_4 \ldots p_{2n} \cdot \frac{\bar{\Delta}(2n + 2, \infty)}{\bar{\Delta}(2, \infty)}, \quad n = 1,2,3, \ldots \tag{13a}$$

For coefficients with odd index, starting from

$$\alpha_3 = -\lambda_1 \frac{\alpha_1^3}{k_3}, \tag{13b}$$

which can be found by using (3) and the Appendix, we have

$$\alpha_{2n+1} = (-1)^{n+1}\lambda_1 \frac{\alpha_1^3}{k_3} \cdot p_5 p_7 \ldots p_{2n+1} \frac{\bar{\Delta}(2n + 3, \infty)}{\bar{\Delta}(5, \infty)}, \quad n = 2,3, \ldots, \tag{13c}$$

$\alpha_0$ and $\alpha_1$ are arbitrary coefficients. By using the property $\alpha_{-n} = \bar{\alpha}_n$ and taking $\alpha_1$ real, the solution (1) can be written as follows:

$$\chi(t) = \alpha_0 + 2\alpha_1 \cos \omega t - 2\alpha_3 \cos 3\omega t + 2 \sum_n \alpha_N \cos N\omega t,$$

$$\psi(t) = \frac{\lambda_3}{\omega_2^2}\alpha_0 + \frac{2\lambda_3\alpha_1}{\omega_2^2 - \omega^2}\cos \omega t - \frac{2\lambda_3\alpha_3}{\omega_2^2 - 2^2\omega^2}\cos 3\omega t + 2\sum_n \frac{\lambda_3\alpha_n}{\omega_2^2 - N^2\omega^2}\cos N\omega t, \tag{14}$$

where $\alpha_3$, $\alpha_{N=2n}$, and $\alpha_{N=2n+1}$ are given by (13$b$), (13$a$), and (13$c$), respectively. Formulas (14) give the solution in its final form, and by using (13$a$), (13$b$), and (13$c$), we can calculate as many coefficients of the solution (14) as we want, in terms of powers of $\lambda_1$. By expanding the determinants of (13$a$) and (13$c$) in powers of the small $\lambda_1$,[2] we find that their ratios are equal to $1 + 0\,(\lambda_1^2)$, and, as we can easily see, the calculation of the coefficients, in terms of powers of $\lambda_1$, needs only the unit as value of these ratios. The first two terms of (14) are independent of $\lambda_1$. For terms of (14) of order $O\,(\lambda_1)$, we take the first term of $p_n$ of (6) and, combining it with (13$a$) for $n = 1$, we get the 2nd harmonic terms of the series (14), when the terms of order $O\,(\lambda_1)$ are

$$-2\lambda_1\left\{\frac{3\alpha_0\alpha_1^2}{k_2}\cos 2\omega t + \frac{\alpha_1^3}{k_3}\cos 3\omega t\right\}, \text{ of } \chi(t).$$

$$-2\lambda_1\left\{\frac{3\lambda_3\alpha_0\alpha_1^2}{k_2(\omega_2^2 - 2^2\omega^2)}\cos 2\omega t + \frac{\lambda_3\alpha_1^3}{k_3(\omega_2^2 - 3^2\omega^2)}\cos 3\omega t\right\}, \text{ of } \psi(t). \qquad (14a)$$

If we take the terms of $p_n$ of (6) up to the order $O(\lambda_1^2)$ and combine them with (13*a*) for $n = 2$, we obtain the 4th harmonic terms of order $O(\lambda_1^2)$:

$$18\lambda_1^2\frac{\alpha_0\alpha_1^4}{k_2k_4}\cos 4\omega t, \quad 18\lambda_1^2\frac{\alpha_0\alpha_1^4\lambda_3}{k_2k_4(\omega_2^2 - 4^2\omega^2)}\cos 4\omega t \qquad (14b)$$

of $\chi(t)$ and $\psi(t)$, respectively. Following the same way, we can calculate the higher harmonics in terms of higher powers of $\lambda_1$. The solution (14), constructed as indicated, must be convergent, and its coefficient of the fundamental term must be much larger than any other coefficient. These requirements imply that the following conditions are satisfied:

$$\alpha_0 << \alpha_1, \quad \lambda_1 << \min\left\{\frac{K_1 + K_2}{2m_1\alpha_1^2}, \frac{k_2.}{2\alpha_0\alpha_1}, \frac{k_3}{\alpha_1^2}\right\}. \qquad (15)$$

For the second condition, the inequality (6*a*) was taken into account. The 4th, 5th, ... harmonic terms are of order $O(\lambda_1^2)$, $O(\lambda_1^3)$, ..., and the convergence is guaranteed.

4. Another approach for the determination of principal modes is based on the manner in which the system is set into motion. This is equivalent to considering the principal modes as solutions of special initial-value problems. The solution (14), found in the previous section, gives an indication about the appropriate initial conditions. The initial displacements are, according to (14), the sum of the coefficients, and the initial velocities are zero.

Consider the differential equations of the "dual-mode" nonlinear system subject to the two sets of initial conditions:

$$\chi(0) = \chi_0, \quad \psi(0) = \psi_0, \quad \dot{\chi}(0) = \dot{\psi}(0) = 0 \qquad (16a)$$

$$\chi(0) = \chi_0, \quad \psi(0) = -\psi_0, \quad \dot{\chi}(0) = \dot{\psi}(0) = 0. \qquad (16b)$$

If the initial displacements $\chi_0$ and $\psi_0$ are appropriately related, the initial velocities being zero, then each set of the initial conditions (16*a*) or (16*b*) gives rise to special vibration modes, which are, by definition, the principal modes of the system. If we assume the series

$$\chi(t) = \sum_{n=-\infty}^{\infty}\alpha_n e^{in\omega t}, \quad \psi(t) = \sum_{n=-\infty}^{\infty}A_n e^{in\omega t} \qquad (17)$$

as solution of the differential equations of the system, by repeating the previous procedure, we will terminate to get the solution in the form (14), which gives the appropriate relationship between the initial displacements of the two oscillators. An approximation of this solution for very small $\lambda_1$, and $\alpha_1$ as dominant coefficient, is:

$$\chi(t) = 2\alpha_1\cos\omega t, \quad \psi(t) = 2\frac{\lambda_3}{\omega_2^2 - \omega^2}\alpha_1\cos\omega t, \qquad (18)$$

where the initial conditions are

$$\chi_0 = 2\alpha_1, \quad \psi_0 = \frac{\lambda_3}{\omega_2^2 - \omega^2}\chi_0, \quad \dot{\chi}_0 = \dot{\psi}_0 = 0. \tag{18a}$$

Formula (18*a*) give the initial conditions required for the solution to be of "principal modes" type.

*Appendix.*—If in the elements of the sequence $\{\alpha_n\}$, $n = 0, \pm 1, \pm 2, \ldots$, where $\alpha_{-n} = \bar{\alpha}_n$, the element $\alpha_1$, and then $\alpha_{-1}$, is dominant, it is easily seen that the dominant sum of the double series of (3) is given by

$$3\alpha_1^2\alpha_{n-2} + 6|\alpha_1|^2\alpha_n + 3\bar{\alpha}_1^2\alpha_{n+2}, \tag{$A_1$}$$

where $n$ is any integer except $n = \pm 1, \pm 3$. For these exceptions, the dominant terms of the double series are

$$3\bar{\alpha}_1\alpha_1^2 \text{ for } n = 1, \quad 3\alpha_1\bar{\alpha}_1^2 \text{ for } n = -1, \quad \alpha_1^3 \text{ for } n = 3, \bar{\alpha}_1^3 \text{ for } n = -3. \tag{$A_2$}$$

[1] Magiros, D. G., these PROCEEDINGS, **46,** 1608 (1960).

[2] Brillouin, L., *Wave Propagation in Periodic Structures,* 2nd ed. (New York: Dover Publications, Inc., 1953), pp. 34, 35.

# Method for Defining Principal Modes of Nonlinear Systems Utilizing Infinite Determinants

DEMETRIOS G. MAGIROS
*General Electric Company, Missile and Space Vehicle Department, Philadelphia, Pennsylvania*
(Received April 21, 1961)

A method for calculation of "principal modes" of linear or nonlinear systems is discussed. The physical definition of "principal modes" is formulated mathematically in two ways. The trial solution of the differential equation of the motion of the system is taken in an appropriate structure. The calculation of principal modes leads to infinite determinants of Hill's and von Koch's type, which are analyzed. The above method yields the possibility of getting the "principal modes" in the form of a series, all the coefficients of which can be calculated.

## 1. INTRODUCTION

THE concept of "principal modes" plays the predominant role in the analysis of the oscillatory systems, no matter what field the systems occur in.

The principal modes of linear systems are, by definition the fundamental set of solutions of which a linear combination gives the general solution of the linear differential equations, which govern the motion of the linear system. This means that any kind of oscillations in linear systems can be discussed in terms of some special modes of oscillation of the system, the "principal modes" of the system.

This definition of "principal modes" is meaningless in nonlinear systems, since the "principle of superposition" does not hold in those systems.

The study of the principal modes of systems, either linear or nonlinear, may be made by using two definitions, namely the "proportionality definition" of principal modes and their definition as solutions of "initial value problems" of special type. Calculations, based on these definitions, are shown for a nonlinear "dual-mode" system. If the solution of the differential equations of this system is taken as an exponential series with complex coefficients, the calculation of the frequency $\omega$ and the coefficients of the series leads to a recursion formula, which gives rise to "infinite determinants" of special type. The analysis of the infinite determinants involved, and the solution in its final form is discussed. The nonunit elements of the determinants contain the coefficient of the nonlinearity as a common factor, and, for a weak nonlinearity, we can get an expansion of these determinants in powers of the coefficient, then an appropriate approximation of them. Thus the "frequency equation," in an infinite determinantal form, is reduced to a quartic in $\omega^2$, and the ratio of the determinants of the coefficients of the series to unit.

The solution given is in accordance with both definitions of "principal modes," and imposes a relation between the initial displacements of the masses, and this relation and the condition for the initial velocities to be zero distinguished the special initial value problems appropriate for the "principal modes."

A brief discussion of the present paper has been published in the Proceedings of National Academy of Sciences.[1a,b]

## 2. THE DEFINITION OF PRINCIPAL MODES, ITS APPLICATION TO A FREE NONLINEAR SYSTEM OF TWO DEGREES OF FREEDOM, AND THE RECURSION FORMULA FOR THE SOLUTION

By using the terminology of mechanical systems with $s$ degrees of freedom, the principal modes of oscillations of the system are defined as those oscillations of the system for which the nonzero amplitudes of the fundamental and the corresponding harmonics of the displacements of any two of the oscillating masses have, separately, a constant ratio. For such motions, the masses all oscillate about their equilibrium positions, where they pass at the same time. Their common frequency is the "principal frequency" of the system.

If $x_i$, $i=1, 2, \cdots s$ are the displacements of the masses $m_i$, $i=1, 2, \cdots s$ from their equilibrium positions, and $x_{in}$, $x_{i'n}$ the corresponding amplitudes of the $n$th harmonic of the displacements $x_i$, $x_{i'}$, of the two masses $m_i$, $m_{i'}$, and if there exist constants $c_n$ such that the conditions

$$x_{in}/x_{i'n}=c_n, \quad i\neq i'; \quad i, i'=1, 2, \cdots s, \quad c_n\neq 0, \infty \qquad (1)$$

are satisfied, then these motions are, by definition, the "principal modes" of the system.

We restrict ourselves, without loss of generality, to a two-degrees-of-freedom nonlinear system; namely, we get—as a mechanical model—the "two-masses-three-springs" system with one of the anchor springs nonlinear and such that the corresponding restoring force is an odd-cubic function of the distance, or—as an electrical model—the "two-inductances-three-capacitances" system with one capacitance variable and the others constant, Fig. 1 (a), (b).

If $m_1$ and $m_2$ are the oscillating masses, $\bar{K}_1=K_1+\mu x^2$, $K_3$ and $K_2$ the stiffnesses of the first and second anchor springs and of the coupling, $\mu$ a constant which characterizes the nonlinearity, and $x$ and $y$ the displacements of $m_1$ and $m_2$ from their equilibrium positions, the equa-

[1] (a) D. G. Magiros, Proc. Natl. Acad. Sci. U. S., Dec. (1960). (b) D. G. Magiros, Proc. Natl. Acad. Sci. U. S., June (1961).



Reprinted from the *Journal of Mathematical Physics* 2 (1961), 869–875.

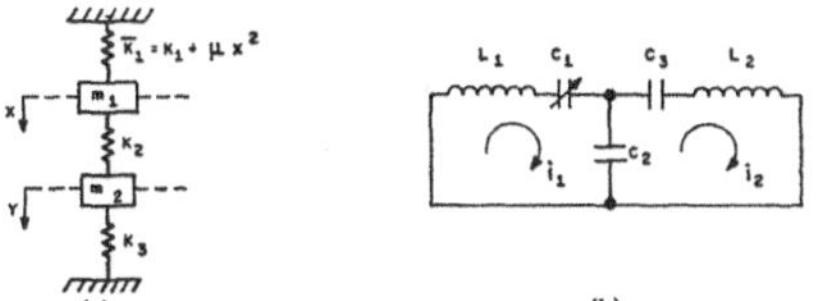

(a) (b)

FIG. 1. (a) The mechanical model. (b) The electrical model.

tions of motion of this sytem are:

$$\ddot{x}+\omega_1^2x+\lambda_1x^3-\lambda_2y=0,$$
$$\ddot{y}+\omega_2^2y-\lambda_3x=0, \tag{2}$$

where

$$\omega_1^2=\frac{K_1+K_2}{m_1},\quad \omega_2^2=\frac{K_2+K_3}{m_2},$$

$$\lambda_1=\frac{\mu}{m_1},\quad \lambda_2=\frac{K_2}{m_1},\quad \lambda_3=\frac{K_2}{m_2}. \tag{2a}$$

We proceed to find the solutions $x(t)$ and $y(t)$ of (2) with one fundamental frequency $\omega$ for both oscillators by imposing a certain organic structure for the functions $x(t)$ and $y(t)$. Assume solutions of (2) of the form

$$x(t)=\sum_{n=-\infty}^{\infty}\alpha_n e^{in\omega t},\quad y(t)=\sum_{n=-\infty}^{\infty}A_n e^{in\omega t}. \tag{3}$$

The coefficients $\alpha_n$ and $A_n$ are complex, then they include the phase angle.

For the reality of the solution, one takes $\bar{\alpha}_n=\alpha_{-n}$, $\bar{A}_n=A_{-n}$, where $\bar{\alpha}_n$ and $\bar{A}_n$ are conjugates of $\alpha_n$ and $A_n$. Also assume that

$$\sum_{n=-\infty}^{\infty}|\alpha_n|<\infty,\quad \sum_{n=-\infty}^{\infty}|A_n|<\infty,$$

$$\sum_{n=-\infty}^{\infty}n^2|\alpha_n|<\infty,\quad \sum_{n=-\infty}^{\infty}n^2|A_n|<\infty.$$

The first two of these inequalities guarantee the convergence of the series (3), while the second ones the existence of the second derivatives $\ddot{x}$, $\ddot{y}$.

By using the first of (3), $x^3$ is given by

$$x^3=\sum_{p_1=-\infty}^{\infty}\sum_{p_2=-\infty}^{\infty}\sum_{p_3=-\infty}^{\infty}\alpha_{p_1}\alpha_{p_2}\alpha_{p_3}e^{i(p_1+p_2+p_3)\omega t},$$

which can be written as follows:

$$x^3=\sum_n\sum_{p_1}\sum_{p_2}\alpha_{p_1}\alpha_{p_2}\alpha_{n-p_1-p_2}e^{in\omega t}. \tag{3a}$$

Inserting (3) and (3a) into (2), the following system results:

$$\left(\omega_1^2-n^2\omega^2-\frac{\lambda_2\lambda_3}{\omega_2^2-n^2\omega^2}\right)\alpha_n+\lambda_1\sum_{p_1}\sum_{p_2}\alpha_{p_1}\alpha_{p_2}\alpha_{n-p_1-p_2}=0, \tag{4a}$$

$$A_n/\alpha_n=\lambda_3/(\omega_2^2-n^2\omega^2), \tag{4b}$$

where $n$, $p_1$, $p_2$ are integers. This nonlinear system consists of infinitely many nonlinear ($\lambda_1\neq0$) homogeneous equations for the infinitely many unknown coefficients $\alpha_n$ and $A_n$.

The equation (4b) expresses the definition of principal modes applied to the above dual-mode system, and by (4b) the calculation of $A_n$ is deduced from $\alpha_n$, then the calculation of principal modes of the system is deduced from $\alpha_n$, by using (4a). If $\alpha_1$ is the dominant coefficient of the sequence $\{\alpha_n\}$, then the approximate value of the double series of (4a), see Appendix I, is given by

$$3\alpha_1^2\alpha_{n-2}+6|\alpha_1|^2\alpha_n+3\bar{\alpha}_1^2\alpha_{n+2}, \tag{5}$$

with an error $o(\alpha_1^2)$.

Inserting the expression (5) into the place of the double series of (4a), we can get the following recursion formula:

$$p_n\alpha_{n-2}+\alpha_n+q_n\alpha_{n+2}=0, \tag{6}$$

where

$$p_n=\frac{3\lambda_1\alpha_1^2}{k_n+6\lambda_1|\alpha_1|^2},\quad q_n=\frac{3\lambda_1\bar{\alpha}_1^2}{k_n+6\lambda_1|\alpha_1|^2},$$
$$k_n=\omega_1^2-n^2\omega^2-\frac{\lambda_2\lambda_3}{\omega_2^2-n^2\omega^2}. \tag{6a}$$

It is the recursion formula (6) which will be used for the calculation of the "principal frequency" $\omega$ and the coefficients $\alpha_n$ of the solution (3). We notice here that for the convergence of the series (3) it is necessary, according to (4b), that $(\omega_2^2-n^2\omega^2)$ is neither zero nor very small, i.e., $\omega$ must not be either a submultiple $\omega_2$ or very close to a submultiple of $\omega_2$.

### 3. CALCULATION OF THE PRINCIPAL FREQUENCY $\omega$

The recursion formula (6) gives infinitely many homogeneous equations for the infinitely many unknowns $\alpha_n$. The corresponding infinite matrix of the coefficients of these equations is

$$\begin{bmatrix} \cdots & \cdots & \cdots & \cdots & \cdots & \cdots & \cdots & \cdots & \cdots \\ \cdots 0 & p_{-1} & 0 & 1 & 0 & q_{-1} & 0 & \cdots & \\ \cdots & 0 & p_0 & 0 & 1 & 0 & q_0 & 0 \cdots & \\ \cdots & & 0 & p_1 & 0 & 1 & 0 & q_1 & 0\cdots \\ \cdots & \cdots & \cdots & \cdots & \cdots & \cdots & \cdots & \cdots & \cdots \end{bmatrix}.$$

For nonzero unknowns $\alpha_n$, the corresponding infinite determinant must be zero. This doubly infinite determinant, by taking $n$ arbitrary integer, becomes one-sided infinite determinant, and we can write

$$\Delta(n,\infty)=0, \tag{7}$$

where the infinite determinant $\Delta(n,\infty)$ is given by the limit

$$\lim_{\substack{n=\text{fixed integer}\\ m=0,1,2,3,\cdots\\ m\to\infty}}\Delta(n,m)=\left\|\begin{array}{ccccccccc} 1 & 0 & q_n & 0 & 0 & \cdots & \cdots & 0 & 0\\ 0 & 1 & 0 & q_{n+1} & 0 & \cdots & \cdots & 0 & 0\\ p_{n+2} & 0 & 1 & 0 & q_{n+2} & \cdots & \cdots & 0 & 0\\ \cdots & \cdots & \cdots & \cdots & \cdots & \cdots & \cdots & \cdots & \cdots\\ 0 & 0 & \cdots & \cdots & \cdots & 0 & 1 & 0 & q_{n+m-2}\\ 0 & 0 & \cdots & \cdots & \cdots & p_{n+m-1} & 0 & 1 & 0\\ 0 & 0 & \cdots & \cdots & \cdots & 0 & p_{n+m} & 0 & 1\end{array}\right\|. \qquad (8)$$

Consider a weak nonlinearity, i.e., $\lambda_1|x|^3 \ll \omega_1^2|x|$, or $\lambda_1 x^2 \ll (K_1+K_2)/m_1$. From the first of (3) we get

$$\max x^2=\left[\sum_{n=-\infty}^{\infty}\alpha_n\right]^2=\sum_{p_1=-\infty}^{\infty}\sum_{p_2=-\infty}^{\infty}\alpha_{p_1}\alpha_{p_2},$$

and since $\alpha_1$ is considered as dominant element of that of the sequence $\{\alpha_n\}$, when $\max x^2=2\alpha_1^2$, then

$$\lambda_1 \ll (K_1+K_2)/(2m_1\alpha_1^2). \qquad (9a)$$

For a weak nonlinearity we can write

$$\rho_n=\frac{3\lambda_1\alpha_1^2}{k_n+6\lambda_1|\alpha_1|^2}=\frac{3\lambda_1\alpha_1^2}{k_n}\times\left\{1-\lambda_1\frac{6|\alpha_1|^2}{k_n}+\lambda_1^2\left(\frac{6|\alpha_1|^2}{k_n}\right)^2-\cdots\right\}$$

$$=\lambda_1\frac{3\alpha_1^2}{k_n}-\lambda_1^2\frac{18\alpha_1^2|\alpha_1|^2}{k_n^2}+\cdots, \qquad (9)$$

$$q_n=\lambda_1\frac{3\bar{\alpha}_1^2}{k_n}-\lambda_1^2\frac{18\bar{\alpha}_1^2|\alpha_1|^2}{k_n^2}+\cdots.$$

All the elements not in the main diagonal of the determinant (8) have the small coeffici[illegible]t [illegible] a common factor. Then, by applying formula [illegible] of Appendix II, and taking the first terms of $\rho_n$ and $q_n$ from formulas (9), the determinant (8) can be written approximately as

$$\Delta(n,\infty)=1-9\lambda_1^2|\alpha_1|^4\sum_{m=0}^{\infty}\frac{1}{k_{n+m}k_{n+m+2}}+0(\lambda_1^3). \qquad (10)$$

$k_{\bar{n}}$, given by (6a), is an even function of $\bar{n}$. To examine the convergence of the series in (10) we confine ourselves to non-negative integers $\bar{n}$. Since $\{|k_{\bar{n}}|\}$ is a sequence with positive terms monotonically increasing with $\bar{n}$, and $|k_{\bar{n}}|\to\infty$, as $\bar{n}\to\infty$, the series in (10) converges. Convergence requirements of the series in (10) necessitates the $\omega^2$ must not be a zero of $k_{\bar{n}}(\omega^2)$; hence

$$\omega^2\mp\omega_{0\pm}^2=\frac{1}{2\bar{n}^2}\{\omega_1^2+\omega_2^2\pm[(\omega_1^2-\omega_2^2)^2+4\lambda_2\lambda_3]^{\frac{1}{2}}\}. \qquad (11)$$

The principal frequency $\omega$ can be determined from (7) by using (10).

Then, if we confine ourselves to the first term of the series in (10) and put $n=1$, the principal frequency $\omega$ is approximately a root of

$$k_1k_3=9\lambda_1^2|\alpha_1|^4,$$

or root of

$$\left(\omega_1^2-\omega^2-\frac{\lambda_2\lambda_3}{\omega_2^2-\omega^2}\right)\left(\omega_1^2-9\omega^2-\frac{\lambda_2\lambda_3}{\omega_2^2-9\omega^2}\right)-9\lambda_1^2|\alpha_1|^4=0, \qquad (12)$$

which is quartic in $\omega^2$. Formula (12) gives the principal frequency $\omega$ of order $O(\lambda_1^2)$. For values of the principal frequency of order higher than $O(\lambda_1^2)$, we find in the expansion of $\Delta(n,\infty)$ terms of order higher than $O(\lambda_1^2)$,[2] and continue in the same way to find the corresponding algebraic equation.

## 4. CALCULATION OF THE COEFFICIENTS

To calculate the coefficients $\alpha_n$, we use in the recursion formula (6) the notation

$$u_n=\alpha_n/\alpha_{n-2}, \qquad (13)$$

when we can get

$$u_n=-\frac{\rho_n}{1+q_nu_{n+2}},$$

which leads to the infinite continued fraction

$$\frac{\alpha_n}{\alpha_{n-2}}=-\cfrac{\rho_n}{1-\cfrac{\rho_{n+2}q_n}{1-\cfrac{\rho_{n+4}q_{n+2}}{1-\cdots}}}, \qquad (14)$$

where the $\rho$'s and $q$'s are given by (9). The formula (14) is written as

$$\alpha_n/\alpha_{n-2}=-\rho_nZ_{n,\infty}, \qquad (15)$$

[2] W. Magnus, "Infinite determinants in the theory of Mathieu's and Hill's equations," Research Report No. BR-1, Mathematical Research Group, Washington Square College of Arts and Science, New York University, 1953.

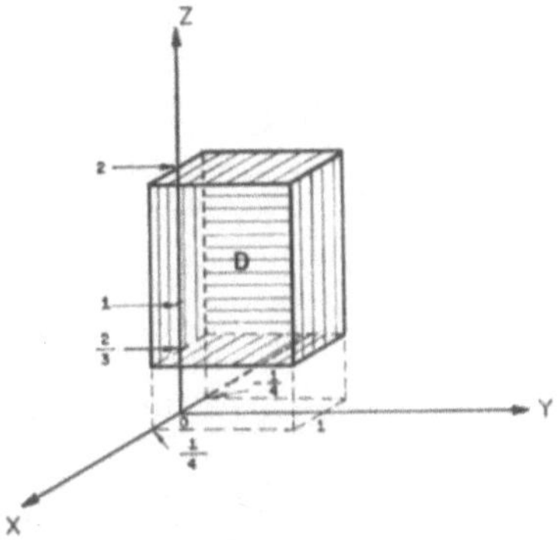


FIG. 2. The domain D in the $X$, $Y$, $Z$ space is the appropriate one for the convergence of the continued fraction (16).

where

$$Z_{n,\infty}=\cfrac{1}{1+\cfrac{X_{n,1}}{1+\cfrac{X_{n,2}}{1+\cdots}}}. \tag{16}$$

$$X_{n,m}=-\rho_{n+2m}q_{n+2(m-1)}, \tag{16a}$$

$n$= fixed integer, $m=1,2,3,\cdots$.

Since the elements $X_{n,m}$ of (16) are functions of the real variable $\omega$, the regions $E$, $Y$, $V$, defined in Appendix III, are segments of lines, and according to von Koch's[3] and Worpitzky's theorems, as stated in Appendix III,

$$\begin{aligned} &E:\ -\tfrac{1}{4}\leq X_{n,m}\leq\tfrac{1}{4},\\ &Y:\ 0<Y_{n,m}=\sum_{m=2}|X_{n,m}|<1, \\ &V:\ \tfrac{2}{3}\leq Z_{n,m}\leq 2. \end{aligned} \tag{17}$$

Consider a Cartesian coordinate system in space, and let us take the region element on the $X$ axis, the series region on the $Y$ axis, and the value region on the $Z$ axis. The inequalities (17) correspond then to the interior of the orthogonal parallelepiped D, Fig. 2, which is an open domain.

The domain

$$D:\ -\tfrac{1}{4}\leq X\leq\tfrac{1}{4},\quad 0<Y<1,\quad \tfrac{2}{3}\leq Z\leq 2$$

is the appropriate one for the continued fraction (16).

To evaluate the continued fraction (16) we apply the theory of Appendix III. The denominator of the ratio, which gives the $m$th approximant $Z_m$ of the continued fraction (16) is

$$B_{n,m-1}=\left\|\begin{matrix} 1 & -1 & 0 & \cdots & 0 & 0 & 0\\ -\rho_{n+2}q_n & 1 & -1 & \cdots & 0 & 0 & 0\\ 0 & -\rho_{n+4}q_{n+2} & 1 & \cdots & 0 & 0 & 0\\ \cdots & \cdots & \cdots & \cdots & \cdots & \cdots & \cdots\\ 0 & 0\cdots & & \cdots & -\rho_{n+2(m-1)}q_{n+2(m-2)} & 1 & -1\\ 0 & 0\cdots & & \cdots & 0 & -\rho_{n+2m}q_{n+2(m-1)} & 1 \end{matrix}\right\|. \tag{18}$$

The numerator $A_{n,m-1}$ of the ratio can be obtained from (18) if we omit its first row and first column. Taking the limit as $m\rightarrow\infty$, we obtain

$$\frac{\alpha_n}{\alpha_{n-2}}=-\rho_n\frac{\bar{\Delta}(n+2,\infty)}{\bar{\Delta}(n,\infty)}, \tag{19}$$

where

$$\bar{\Delta}(n,\infty)=\lim_{m\to\infty}B_{n,m-1},\qquad \bar{\Delta}(n+2,\infty)=\lim_{m\to\infty}A_{n,m-1}. \tag{19a}$$

The determinants of (19) are of von Koch's type and they converge by von Koch's rule, that is, when the series $\sum_n|\rho_nq_{n-2}|$ converges, which happens here, as was pointed out at the discussion of the convergence of the series of (10).

Since $\alpha_n=\bar{\alpha}_n$, we may restrict ourselves to non-negative integers for the calculation of the coefficients $\alpha_n$. Formula (19) suggests starting with $n=2$; then $\alpha_0$ and $\alpha_1$ are arbitrary. The induction procedure applied to (19) for the coefficients with even index gives

$$\alpha_{2n}=(-1)^n\alpha_0\rho_2\rho_4\cdots\rho_{2n}\frac{\bar{\Delta}(2n+2,\infty)}{\bar{\Delta}(2,\infty)},\qquad n=1,2,3\cdots. \tag{20}$$

For the determination of the coefficients with odd index, we first calculate the coefficient $\alpha_3$. This, according to Appendix I, is an exception.

If, according to Appendix I, we take $\alpha_1^3$ instead of the double series of the formula (4a), and we apply this formula for $n=3$, there results

$$\alpha_3=-(\lambda_1\alpha_1^3/k_3), \tag{21}$$

where

$$k_3=\omega_1^2-9\omega^2-[\lambda_2\lambda_3/(\omega_2^2-9\omega^2)]. \tag{21a}$$

[3] H. von Koch, Compt. rend., 120, 144 (1895).

Now, by applying the induction procedure to (19) starting from the coefficient $\alpha_5$ and using the value of $\alpha_3$ given by (21), one can get

$$\alpha_{2n+1}=(-1)^{n+1}\frac{\lambda_1\alpha_1^3}{k_3}\rho_5\rho_7\cdots\rho_{2n+1}\frac{\bar{\Delta}(2n+3,\infty)}{\bar{\Delta}(5,\infty)},\qquad n=2,3,4\cdots. \quad (22)$$

By applying formula (B) of Appendix II to the determinants of (20) and (22), there is obtained

$$\frac{\bar{\Delta}(2n+2,\infty)}{\bar{\Delta}(2,\infty)}=1-9\lambda_1^2|\alpha_1|^4\sum_{m=0}^{\infty}\left(\frac{1}{k_{2n+2+m}k_{2n+4+m}}-\frac{1}{k_{2+m}k_{4+m}}\right)+O(\lambda_1^3),$$

$$\frac{\bar{\Delta}(2n+5,\infty)}{\bar{\Delta}(5,\infty)}=1-9\lambda_1^2|\alpha_1|^4\sum_{m=0}^{\infty}\left(\frac{1}{k_{2n+5+m}k_{2n+7+m}}-\frac{1}{k_{5+m}k_{7+m}}\right)+O(\lambda_1^3). \quad (23)$$

The formulas (21), (20), and (22) give the coefficients of the first of the series (3) for any positive $n$, with arbitrary $\alpha_0$ and $\alpha_1$. For the determination of the coefficients $\alpha_{-n}$ we use the property $\alpha_{-n}=\bar{\alpha}_n$. The coefficient $\alpha_0$ is real, and $\alpha_1$ in general complex, $\alpha_1=|\alpha_1|e_i^{\varphi_1}$. If we take $\alpha_1$ real, the solution (3) can be written as follows:

$$x(t)=\alpha_0+2\alpha_1\cos\omega t-2\alpha_3\cos 3\omega t+2\sum_n\alpha_N\cos N\omega t,$$

$$y(t)=\frac{\lambda_3}{\omega_2^2}\alpha_0+\frac{2\lambda_3}{\omega_2^2-\omega^2}\alpha_1\cos\omega t-\frac{2\lambda_3}{\omega_2^2-3^2\omega^2}\alpha_3\cos 3\omega t+2\sum_n\frac{\lambda_3}{\omega_2^2-N^2\omega^2}\alpha_N\cos N\omega t, \quad (24)$$

where the coefficients $\alpha_3$, $\alpha_{N=2n}$, $\alpha_{N=2n+1}$ are given by the formulas (21), (20), (22), respectively. The formulas (24) give the solution in its final form, and the formulas (20)–(23) can be used for the calculation of as many coefficients of the solution (24) as we want in terms of powers of $\lambda_1$.

We can easily see that for the calculation of the coefficients, in terms of powers of $\lambda_1$, does not need but only the unit as value of the ratios of the determinants (23). The first two terms of the solution (24) are independent of $\lambda_1$. For terms of order $O(\lambda_1)$, we take the first term of $\rho_n$ of (9) and combining it with (20) for $n=1$ we get the 2nd harmonic, which, with (21), gives

$$x(t)=\alpha_0+2\alpha_1\cos\omega t-2\lambda_1\left\{\frac{3\alpha_0\alpha_1^2}{k_2}\cos 2\omega t+\frac{\alpha_1^3}{k_3}\cos 3\omega t\right\}+O(\lambda_1^2), \quad (25)$$

$$y(t)=\frac{\lambda_3}{\omega_2^2}\alpha_0+\frac{2\lambda_3}{\omega_2^2-\omega^2}\alpha_1\cos\omega t-2\lambda_1\left\{\frac{3\lambda_3\alpha_0\alpha_1^2}{k_2(\omega_2^2-2^2\omega^2)}\cos 2\omega t+\frac{\lambda_3\alpha_1^3}{k_3(\omega_2^2-3^2\omega^2)}\cos 3\omega t\right\}+O(\lambda_1^2). \quad (25)$$

If we take the terms of $\rho_n$ of (9) up to the order $O(\lambda_1^2)$, and combine them with (22) for $n=2$ we obtain the 4th harmonic terms of order $O(\lambda_1^2)$:

$$18\lambda_1^2\frac{\alpha_0\alpha_1^4}{k_2k_4}\cos 4\omega t,\quad 18\lambda_1^2\frac{\alpha_0\alpha_1^4\lambda_3}{k_2k_4(\omega_2^2-4^2\omega^2)}\cos 4\omega t, \quad (25a)$$

of $x(t)$ and $y(t)$, respectively. The above procedure indicates how we can get higher harmonics in terms of higher powers of $\lambda_1$. The solution (24), constructed as indicated above, must be convergent and its coefficient of the fundamental term must be much larger than any other coefficient. These requirements imply that the following conditions are satisfied:

$$\alpha_0\ll\alpha_1,\quad \lambda_1\ll\min\left\{\frac{K_1+K_2}{2m_1\alpha_1^2},\frac{k_2}{2\alpha_0\alpha_1},\frac{k_3}{\alpha_1^2}\right\}. \quad (26)$$

For the second condition, the inequality (6a) was taken into account. The 4th, 5th, $\cdots$ harmonic terms are of order $O(\lambda_1^2)$, $O(\lambda_1^3)$, $\cdots$, and the convergence is guaranteed.

Since $\alpha_1$ is much larger compared to $\alpha_0$, the solution in the linear case is approximately

$$x(t)=2\alpha_1\cos\omega t,\qquad x(t)=2\frac{\lambda_3}{\omega_2^2-\omega^2}\alpha_1\cos\omega t, \quad (25b)$$

and the motions of the oscillators are "in phase" for $\omega_2<\omega$, and "180° out of phase" for $\omega_2<\omega$.

## 5. THE PRINCIPAL MODES AS SOLUTIONS OF INITIAL VALUE PROBLEMS

Another approach for the determination of principal modes may be based on the manner in which the system is set into motion. This is equivalent to considering the principal modes as solutions of special initial value problems.

The differential equations of the "dual-mode" system are considered subject to the restriction that the masses are displaced from their equilibrium positions either both up, or one up and the other down, by amounts $x_0$ and $y_0$, respectively, and released without velocity; i.e.,

$$x(0)=x_0,\quad y(0)=y_0,\quad \dot{x}(0)=\dot{y}(0)=0, \tag{27a}$$

$$x(0)=x_0,\quad y(0)=-y_0,\quad \dot{x}(0)=\dot{y}(0)=0. \tag{27b}$$

If the initial displacements $x_0$ and $y_0$ are appropriately related, then each one of these initial conditions gave rise to special vibration modes, which are, by definition, the "principal modes" of the system. To calculate the principal modes of our system utilizing infinite determinants and using the above definition, assume a solution in the form of complex exponential series (3), as in the previous case. The calculation of the principal frequency and the coefficients of the series has been completed throughout the preceding sections and the solution is found to be in the form given by (25). An approximation of this solution is given by (26), associated with the initial conditions

$$x_0=2\alpha_1,\quad y_0=\frac{\lambda_3 x_0}{\omega_2^2-\omega^2}=\frac{2\alpha_1\lambda_3}{\omega_2^2-\omega^2}. \tag{28}$$

Formulas (28) give the relations of the initial displacements required for the solution to be of "principal modes" type. These sinusoidal motions are "in phase" for the initial conditions (27a) if $\omega<\omega_2$.

Both definitions of principal modes lead to the same solution; they have the same physical interpretation and they are equivalent.

The discussion here is based on two definitions of principal modes and the final solution found by analyzing the infinite determinants involved. It may be mentioned that G. W. Hill, in his *Lunar Theory* brought into notice the infinite determinants, and H. Poincaré first gave conditions for their convergence.


## ACKNOWLEDGMENT

The author is greatly indebted to Professor L. Brillouin for his unflagging interest and advice.


## APPENDIX I. THE DOMINANT SUM OF A DOUBLE SERIES

Suppose in the sequence $\{\alpha_n\}$, $n=0, \pm1, \pm2, \cdots$ the complex elements have the property $\alpha_{-n}=\bar{\alpha}_n$. If $\alpha_1$ and $\alpha_{-1}$ are dominant elements in the sequence, then it is easily seen that the dominant sum of the double series of (4a) is given by

$$3\alpha_1^2\alpha_{n-2}+6|\alpha_1|^2\alpha_n+3\bar{\alpha}_1^2\alpha_{n+2}, \tag{A1}$$

where $n$ is any integer except $n=\pm1, \pm3$. For these exceptions the dominant terms of the double series are

$$3\bar{\alpha}_1\alpha_1^2 \text{ for } n=1,\quad 3\alpha_1\bar{\alpha}_1^2 \text{ for } n=-1,$$
$$\alpha_1^3 \text{ for } n=3,\quad \bar{\alpha}_1^3 \text{ for } n=-3. \tag{A2}$$

## APPENDIX II. APPROXIMATE VALUE OF AN INFINITE DETERMINANT

If in an infinite convergent determinant,

$$\Delta=\|B_{m,n}\|_{-\infty}^{+\infty},$$

the elements in the main diagonal are equal to unity, and all the elements not in the main diagonal have a small common factor,[4] say $\epsilon$, i.e., if $B_{m,m}=1$, $B_{m,n}=\epsilon\bar{B}_{m,n}$, $m\neq n$, we may get an expression of the determinant in powers of $\epsilon$. The first term in this expression is independent of $\epsilon$; it is the product of all the elements in the main diagonal, that is 1. The next terms in the expansion are in $\epsilon^2$ and are obtained by replacing the product $\prod_m B_{m,m}$ the elements in the main diagonal $(m,m)$ and $(n,n)$ by the elements not in the main diagonal $(m,n)$ and $(n,m)$. These terms have a minus sign, according to the laws of determinants, and they are $-\epsilon^2\sum_m\sum_n \bar{B}_{m,n}\bar{B}_{n,m}$; the determinant $\Delta$ can be written in the form

$$\Delta=1-\epsilon^2\sum_m\sum_n \bar{B}_{m,n}\bar{B}_{n,m}+0(\epsilon^3). \tag{B}$$

## APPENDIX III. A CONTINUED FRACTION AS A RATIO OF TWO INFINITE DETERMINANTS

Given the continued fraction

$$Z_\infty=\cfrac{1}{1+\cfrac{X_2}{1+\cfrac{X_3}{1+\cdots}}}, \tag{C1}$$

where the complex elements $X$ are subject to specified conditions, its $m$th approximant $Z_m$, obtained by stopping with the $m$th partial quotient, can be estimated If the elements $X$ of the sequence $\{X_m\}$ of (C1) have arbitrary values in a region, the "region element" $E$, then the correspondent series $\sum_{m=2}^{\rho}|X_m|$, $\rho=2, 3, 4, \cdots$ has its values in the "series region" $Y$, and the approximants $Z_m$ have all their values in the value region $V$. The following theorems give relationships between the above regions.[5]

### *H. von Koch's Theorem*

"If

$$Y:\quad \sum_{m=2}^{\rho}|X_m|<1,\quad \rho=2,3,4,\cdots \tag{C2}$$

then: the continued fraction (C1) converges."

### *Worpitzky's Theorem*

"If

$$E:\quad |X_m|\leq\tfrac{1}{4},\quad m=2,3,4\cdots \tag{C3}$$

then

$$V:\quad |Z_m-\tfrac{4}{3}|\leq\tfrac{2}{3}." \tag{C4}$$

[4] L. Brillouin, *Wave Propagation in Periodic Structures* (Dover Publications, New York, 1953), 2nd ed., pp. 34, 35.

[5] H. Wall, *Analytic Theory of Continued Fractions* (D. Van Nostrand Company, Inc., Princeton, New Jersey, 1948), pp. 26, 42, and 51.

The $m$th approximant of (C1), $Z_m$, can have the form of a ratio of two determinants. To show that, one associates the continued fraction (C1) a sequence of linear transformations

$$X_1(v)=1,\quad X_m=1/(1+X_mv),\quad m=2,3,4,\cdots.$$

If:

$$X_\rho\neq 0,\quad \rho=2,3,\cdots,m;\quad X_\rho=0,\quad \rho>m,$$

then the product of $m$ of the above transformations is

$$Z_m=X_1X_2\cdots X_m(v)=\cfrac{1}{1+\cfrac{X_2}{1+\cfrac{\cdots}{\cdots+\cfrac{X_{m-1}}{1+X_mv}}}}=\frac{X_mvA_{m-2}+A_{m-1}}{X_mvB_{m-2}+B_{m-1}},\tag{C5}$$

where the $A$'s and $B$'s may be calculated by means of the recursion formulas

$$A_\rho=A_{\rho-1}+X_\rho A_{\rho-2},\quad B_\rho=B_{\rho-1}+X_\rho B_{\rho-2},\quad \rho=1,2,3,\cdots.\tag{C6}$$

For the above we require the initial values

$$A_{-1}=1,\quad A_0=0,\quad B_{-1}=0,\quad B_0=1,\quad X_1=1.$$

The $m$th approximant of (C1), $Z_m$, is given by (C5) if $v=0$, then it is equal to the ratio $A_{m-1}/B_{m-1}$.

The recursion formulas (C6) give two systems of homogeneous linear equations, one in the variables $A$, the other in the variables $B$. These systems give rise to two determinants, which give the values of the $A$'s and $B$'s. The $B$'s are given by the determinant

$$B_{m-1}=\begin{Vmatrix} 0 & -1 & 0 & 0 & \cdots & & & \cdots & 0\\ X_2 & 1 & -1 & 0 & \cdots & & & \cdots & 0\\ 0 & X_3 & 1 & -1 & \cdots & & & \cdots & 0\\ \cdots & \cdots & \cdots & \cdots & \cdots & \cdots & \cdots & \cdots & \cdots\\ 0 & & & & \cdots & X_{m-1} & 1 & & -1\\ 0 & & & & \cdots & 0 & X_m & & 1 \end{Vmatrix},\tag{C7}$$

$$m=2,3,\cdots.$$

The determinant for the $A$'s can be obtained from the above determinant by omitting its first row and its first column. These determinants are different from zero.

The value of the continued fraction (C1) is given, by definition, by the limit $\lim_{\rho\to\infty}(A_\rho/B_\rho)$.

ΠΡΑΚΤΙΚΑ
ΤΗΣ ΑΚΑΔΗΜΙΑΣ ΑΘΗΝΩΝ

ΕΤΟΣ 1963 : ΤΟΜΟΣ 38ος

ΑΝΑΤΥΠΟΝ
ΣΕΛ. 33 - 36

# On the Convergence of Series Related to Principal Modes of Nonlinear Systems*.

*by* ***Demetrios G. Magiros*****.

'Ανεκοινώθη ὑπὸ τοῦ 'Ακαδημαϊκοῦ κ. 'Ιωάνν. Ξανθάκη.

## I. INTRODUCTION

a. In previous papers [1], where the principal modes of a «dual mode» nonlinear system have been discussed, the solution is found in the form of series. The object of the present announcement is to give a brief discussion of the convergence of these series. Details of the discussion will appear in a forthcoming paper. The convergence is based on the «Abel's test» of convergence.

b. We state for reference that the solution found is the following series:

$$\text{(a).}\quad x(t)=a_0+2a_1\cos\omega t+\frac{2\lambda_1 a_1^3}{k_3}\cos 3\omega t+2\sum_N a_N \cos N\omega t$$

$$\text{(b).}\quad \psi(t)=\frac{\lambda_3 a_0}{\omega_2^2}+\frac{2\lambda_3 a_1}{\omega_2^2-\omega^2}\cos\omega t+\frac{2\lambda_1\lambda_3 a_1^3}{k_3(\omega_2^2-3^2\omega^2)}\cos 3\omega t+2\sum_N \frac{\lambda_3 a_N}{\omega_2^2-N^2\omega^2}\cos N\omega t \qquad (1)$$

where:

$$\text{(a).}\quad a_{N=2n}=(-1)^n a_0 p_2 p_4 \dots p_{2n}\,,\quad n=1,2,3,\dots$$

$$\text{(b).}\quad a_{N=2n+1}=(-1)^{n+1}\frac{\lambda_1 a_1^3}{k_3}p_5 p_7 \dots p_{2n+1}\,,\quad n=2,3,4,\dots$$

$$\text{(c).}\quad p_N=\frac{3\lambda_1 a_1^2}{k_N}\left[1-\frac{6\lambda_1|a_1|^2}{k_N}+\left(\frac{6\lambda_1|a_1|^2}{k_N}\right)^2-\dots\right] \qquad (2)$$

$$\text{(d).}\quad k_N=-N^2\omega^2+\omega_1^2+\frac{\lambda_2\lambda_3}{N^2\omega^2-\omega_2^2}$$

---

* **Of the Valley Forge Space Technology Center of the General Electric Comp. M.S.D. Philadelphia, Pa.**

** ΔΗΜ. ΜΑΓΕΙΡΟΥ, 'Επὶ τῆς συγκλίσεως σειρῶν τῶν πρωταρχικῶν ταλαντώσεων μὴ γραμμικῶν συστημάτων.

Reprinted from the *Proceedings of the Athens Academy of Sciences* 38 (1963), 33–36.

The values $\omega$ which are either submultiples of $\omega_2$ or zeros of the k's are the sigularities of this solution.

### 2. THE CONVERGENCE OF THE SERIES

The convergence of the series (1) is deduced from that of the series $\sum_N a_N$. The proof of convergence of $\sum_N a_N$ can be done in two steps, namely by using either the first term of the series (2c) (first step), or all its terms (second step).

The formula (2d), for large values of N, shows that $|k_N|$ is of order $N^2$ and that the sign of $k_N$ is negative. Then a value $N_1$ of N can be found such that $k_N$ remains negative for any $N > N_1$ and $|k_N|$ increases with N.

*First Step*

By taking the integers $\overline{N}$, $\overline{n}$, n, $\sigma$ such that $\overline{N} = 2\overline{n} > N_1$, $n = \overline{n} + \sigma$, where $N_1$, $\overline{N}$, $\overline{n}$ are fixed, the coefficients $a_N$ can be written as:

$$a_{N=2n} = C_{2\overline{n}} \cdot (3\lambda_1 a_1^2)^\sigma \cdot \frac{1}{k_2(\overline{n}+1) \cdots k_2(\overline{n}+\sigma)}$$

$$a_{N=2n+1} = C_{2\overline{n}+1} \cdot (3\lambda_1 a_1^2)^\sigma \cdot \frac{1}{k_2(\overline{n}+1)+1 \cdots k_2(\overline{n}+\sigma)+1} \tag{3}$$

Where $C_{2\overline{n}}$ and $C_{2\overline{n}+1}$ are constants, and the k's positive. The series $\sum_N a_N$ can be split as follows:

$$\sum_N a_N = \sum_{N=0}^{N_1-1} a_N + C_{2\overline{n}} \sum_{\sigma=1}^{\infty} (3\lambda_1 a_1^2)^\sigma \frac{1}{K_2(\overline{n}+1) \cdots K_2(\overline{n}+\sigma)} + C_{2\overline{n}+1} \sum_{\sigma=1}^{\infty} (3\lambda_1 a_1^2)^\sigma \frac{1}{k_2(\overline{n}+1)+1 \cdots K_2(\overline{n}+\sigma)+1} \tag{4}$$

By imposing the restriction $\lambda_1 < \frac{1}{3a_1^2}$, the Abel's test for the convergence [2b] of the infinite series of the right-hand member of (4) can be applied, since:

(a) the geometrical series $\sum_{\sigma=1}^{\infty} (3\lambda_1 a_1^2)^\sigma$ is convergent, and

(b) the sequences of the products of the inverse of k's are monotonic decreasing and bounded sequences.

*Second Step*

The coefficients $a_N$ in this case can be expressed as follows:

$$a_{N=2n} = (-1)^n a_0 \frac{(3\lambda_1|a_1|^2)^n}{k_2 \dots k_{2n}} \left[1 - \frac{6\lambda_1|a_1|^2}{k_2} + \left(\frac{6\lambda_1|a_1|^2}{k_2}\right)^2 - \dots\right] \dots \left[1 - \frac{6\lambda_1|a_1|^2}{k_{2n}} + \left(\frac{6\lambda_1|a_1|^2}{k_{2n}}\right)^2 - \dots\right] \tag{5}$$

$$a_{N=2n+1} = (-1)^{n+1} \frac{a_1}{3} \frac{(3\lambda_1|a_1|^2)^n}{k_3 \dots k_{2n+1}} \left[1 - \frac{6\lambda_1 a_1^2}{k_3} + \left(\frac{6\lambda_1 a_1^2}{k_3}\right)^2 - \dots\right] \dots \left[1 - \frac{6\lambda_1 a_1^2}{k_{2n+1}} + \left(\frac{6\lambda_1 a_1}{k_{2n+1}}\right)^2 - \dots\right]$$

Then the series $\sum_N a_N$ can be written as:

$$\sum_N a_N = \sum_{N=2n} A_{2n} \prod_{2n} + \sum_{N=2n+1} A_{2n+1} \prod_{2n+1} \tag{6}$$

where A's are the factors of the right--hand members of (5) outside the brackets, and Π's the products of the brackets, namely:

$$\prod_r^N \left[1 - \frac{6\lambda_1 a_1^2}{k_r} + \left(\frac{6\lambda_1 a_1^2}{k_r}\right)^2 - \dots\right]$$

which for r either even or odd is either $\prod_{2n}$ or $\prod_{2n+1}$, respectively.

The series $\sum A_{2n}$ and $\sum A_{n+1}$ are convergent, according to the preceding step. In addition the Π's of (6) are sequences with monotone and bounded terms, because $\prod_r^N$ can be wiritten as:

$$\prod_r^N \left[1 - \frac{6\lambda_1 a_1^2}{k_r} + \left(\frac{6\lambda_1 a_1^2}{k_r}\right)^2 - \dots\right] = \prod_r^{N1-1} \left[1 - \frac{6\lambda_1|a_1|^2}{k_r} + \left(\frac{6\lambda_1|a_1|^2}{k_r}\right)^2 - \dots\right] \prod_{N=N1}^N \left[1 + \frac{6\lambda_1|a_1|^2}{k_r} + \left(\frac{6\lambda_1|a_1|^2}{k_r}\right)^2 + \dots\right] \tag{7}$$

where $\prod_r^{N1-1}$ is a finite fixed number, and $\prod_{N=N1}^N$ is convergent as $N \to \infty$, since the series $\sum_{r=N1}^{\infty} \left(\frac{6\lambda_1|a_1|^2}{k_r}\right)^2$ is convergent. [2α]

The above proves that the series (1a) is convergent. For the proof of the convergence of the series (1b), we see that the Abel's test of convergence can be applied to the series.

$$\sum_{N=0}^{\infty} \frac{\lambda_3}{N^2\omega^2-\omega_2^2} a_N .$$

The nature of the singularities of the series (1) and some subjects related to these singularities will be discussed in another paper.

## REFERENCES


[1]. D. G. MAGIROS: (a) Nat. Ac. Sc. Vol. 46, 1608 (1960); Vol. 47, 883 (1961); (b) J. Math Phys. Vol 2, No 6, pp. 869-875, (Nov-Dec. 1961).

[2]. K. KNOPP: «Infinite Sequences and Series», Dover Publications, Inc. New York (1956), pg. (a) 94, (b) 137.


## ΠΕΡΙΛΗΨΙΣ


Εἰς προηγουμένας ἐργασίας μας ἔχει εὑρεθῆ ὑπὸ μορφὴν σειρῶν ἡ λύσις προβλήματος τῶν πρωταρχικῶν ταλαντώσεων μὴ γραμμικῶν συστημάτων. Ἐνταῦθα δίδεται σύντομος ἐξέτασις τῆς συγκλίσεως τῶν σειρῶν αὐτῶν. Κατὰ τὴν πορείαν διὰ τὴν ἔρευναν τῆς συγκλίσεως γίνεται χρῆσις τοῦ θεωρήματος τοῦ Abel περὶ συγκλίσεως.

Ἡ πλήρης ἐξέτασις τῆς συγκλίσεως, καθὼς καὶ ἡ φύσις τῶν ἀνωμάλων σημείων τῆς λύσεως θὰ ἐκτεθοῦν εἰς ἐργασίαν, ἡ ὁποία συντόμως δημοσιεύεται ἀλλαχοῦ.

ΠΡΑΚΤΙΚΑ
ΤΗΣ ΑΚΑΔΗΜΙΑΣ ΑΘΗΝΩΝ

ΕΤΟΣ 1960: ΤΟΜΟΣ 35ΟΣ

ΑΝΑΤΥΠΟΝ
ΣΕΛ. 96 - 103

## The motion of a projectile around the earth under the influence of the earth's gravitational atraction and a thrust*,

*by* ***Dem. G. Magiros*** **.

'Ανεκοινώθη ὑπὸ τοῦ κ. 'Ιωάνν. Ξανθάκη.

*Abstract.*

In this paper the motion of a projectile around the earth under the influence of the gravitational attraction of the earth and a thrust is discussed. The orbit of the projectile and its velocity along the orbit are found in two cases, namely when the thrust is suddenly applied to the projectile,

* Republic Aviation Corp., U.S.A.

** ΔΗΜ. ΜΑΓΕΙΡΟΥ, 'Η κίνησις βλήματος πέριξ τῆς γῆς ὑπὸ τὴν ἐπίδρασιν τῆς ἑλκτικῆς δυνάμεως τῆς γῆς καὶ μιᾶς ὠστικῆς δυνάμεως.

Reprinted from the *Proceedings of the Athens Academy of Sciences* **35** (1960), 96–103.

and when it is gradually applied. The types of the Keplerian orbit, when the thrust is suddenly removed, are also discussed.

*Introduction.*

The following problem is discussed:

*A projectile is moving around the earth in a Keplerian orbit, when a thrust is applied. Find the motion of the projectile during the action of the thrust.*

This problem is solved for a general thrust vector, either suddenly or gradually applied to the projectile and acting continuously for non-infinitesimal time. The problem is specialized in the case of constant thrust vector, case which is identical with the problem of Quantum Mechanics in connection with «Stark Effect». Conditions are given in connection of the types of the Keplerian orbit, when the thrust is suddenly removed from the projectile.

I. *Mathematical formulation of the problem.*

If $\underline{T}$ is the thrust per unit mass, acting for time $\tau$, the motion of the projectile during the time $\tau$ is governed by the differential equation:

$$\ddot{\underline{r}} = -\frac{G}{r^3}\underline{r} + \underline{T} \tag{1}$$

where r is the magnitude of $\underline{r}$, G a constant equal to $\frac{k(M+m)}{m}$, M and m are the masses of the earth and the projectile, respectively; k the constant of gravitation.

The time $\tau$ is split according to:

$$\tau = t_0 + t_1, \tag{2}$$

where $t_0$ is infinitesimal, and we take as initial conditions to the differential Eq. (1) the following conditions:

$$\underline{r}(t)_{t=t_0} = \underline{r}_0 \ , \ \dot{\underline{r}}(t)_{t=t_0} = \dot{\underline{r}}_0 + \underline{I}_0, \tag{3}$$

where $\underline{r}_0$ and $\dot{\underline{r}}_0$ are the position vector and velocity vector, respectively, at the point of the original orbit where the thrust is applied to the projectile; and $\underline{I}_0$ is the impulse per unit mass of the thrust $\underline{T}$ acting on the projectile in the infinitesimal time $t_0$. Such a selection of initial conditions leads to a solution which is very helpful to practical problems.

II. *Solution of the problem.*

By changing the time t according to:

$$\bar{t} = t - t_o \tag{2.1}$$

the dif. Eq. (1) keeps its form, and the initial conditions (3) in the new scale time are:

$$\underline{r}(\bar{t})_{\bar{t}=o} = \underline{r}_o, \qquad \underline{\dot{r}}(\bar{t})_{\bar{t}=o} = \underline{\dot{r}}_o + \underline{I}_o \tag{3.1}$$

A solution of the system (1) and (3.1) of the form:

$$\underline{r}(\bar{t}) = a(\bar{t})\,\underline{r}_o + b(\bar{t})\,(\underline{\dot{r}}_o + \underline{I}_o) + c(\bar{t})\,\underline{T}_o \tag{4}$$

is going to be established by calculating the functions $a(\bar{t})$, $b(\bar{t})$, $c(\bar{t})$ appropriately. $\underline{T}_o$ is the thrust at $\bar{t}=o$.

For $\bar{t}=o$ the function of Eq. (4) gives

$$a(o) = 1, \quad b(o) = o, \quad c(o) = o \tag{4.1}$$

and its derivative:

$$\dot{a}(o) = o, \quad \dot{b}(o) = 1, \quad \dot{c}(o) = o \tag{4.2}$$

The function of Eq. (4) must satisfy identically the differential Eq. (1); then:

$$\ddot{a}(\bar{t})\,\underline{r}_o + \ddot{b}(\bar{t})\,(\underline{\dot{r}}_o + \underline{I}_o) + \ddot{c}(\bar{t})\,\underline{T}_o \equiv$$

$$-\frac{G}{r^3}\,a(\bar{t})\,\underline{r}_o - \frac{G}{r^3}\,b(\bar{t})\,(\underline{\dot{r}}_o + \underline{I}_o) - \frac{G}{r^3}\,c(\bar{t})\,\underline{T}_o + \underline{T}(\bar{t}). \tag{5}$$

If the projections of $\underline{T}(\bar{t})$ on the constant vectors $\underline{r}_o$, $\underline{\dot{r}}_o + \underline{I}_o$ and $\underline{T}_o$ are, by omitting constant factors, $T_1(\bar{t})$, $T_2(\bar{t})$, and $T_3(\bar{t})$, respectively from the identity (5) we can get:

$$\ddot{a}(\bar{t}) + \frac{G}{r^3}\,a(\bar{t}) = T_1(\bar{t}), \tag{6.1}$$

$$\ddot{b}(\bar{t}) + \frac{G}{r^3}\,b(\bar{t}) = T_2(\bar{t}) \tag{6.2}$$

$$\ddot{c}(\bar{t}) + \frac{G}{r^3}\,c(\bar{t}) = T_3(\bar{t}) \tag{6.3}$$

The determination of $a(\bar{t})$, $b(\bar{t})$ and $c(\bar{t})$ from Eqs. (6), subject to the initial conditions (4.1) and (4.2), requires r, $T_1$, and $T_2$, and $T_3$ to be known functions of time $\bar{t}$. For a prescribed thrust $\underline{T}$, the functions $T_1(\bar{t})$, $T_2(\bar{t})$ and $T_3(\bar{t})$ are known, but r is unknown, then we can not solve the Eqs. (6) for $a(\bar{t})$, $b(\bar{t})$, and $c(\bar{t})$.

In spite of that, approximations of these functions can be found in the following way.

We restrict ourselves without loss of generality to the case of any thrust constant in magnitude and direction. For such a thrust the Eqs. (6) may be replaced by:

$$\ddot{a}(\bar{t}) + \frac{G}{r^3} a(\bar{t}) = 0 \tag{7.1}$$

$$\ddot{b}(\bar{t}) + \frac{G}{r^3} b(\bar{t}) = 0 \tag{7.2}$$

$$\ddot{c}(\bar{t}) + \frac{G}{r^3} c(\bar{t}) = 1. \tag{7.3}$$

Let us assume for $a(\bar{t})$, $b(\bar{t})$, and $c(\bar{t})$ the Meclaurin's expensions:

$$a(\bar{t}) = a(0) + \dot{a}(0)\bar{t} + \ddot{a}(0)\frac{\bar{t}^2}{2} + \cdots a^{(n)}(0)\frac{\bar{t}^n}{n!} + \cdots \tag{8.1}$$

$$b(\bar{t}) = b(0) + \dot{b}(0)\bar{t} + \ddot{b}(0)\frac{\bar{t}^2}{2} + \ldots b^{(n)}(0)\frac{\bar{t}^n}{n!} + \ldots \tag{8.2}$$

$$c(\bar{t}) = c(0) + \dot{c}(0)t + \ddot{c}(0)\frac{\bar{t}^2}{2} + \ldots c^{(n)}(0)\frac{\bar{t}^n}{n!} + \ldots \tag{8.3}$$

We can calculate as many coefficients of these series as we want, by using Eqs. (7), (4.1), (4.2) and (1). The first two coefficients are known from the conditions (4.1) and (4.2). The third coefficients, found from Eqs. (7), are:

$$\ddot{a}(0) = -\frac{G}{r_0^3}, \quad \ddot{b}(0) = 0, \quad \ddot{c}(0) = 1 \tag{4.3}$$

The fourth coefficients can be obtained from Eqs. (7) by differentiation, and then using conditions (4.1), (4.2) and (3.1). We can get:

$$\dddot{a}(\bar{t}) = -G(r\dot{a} - 3a\dot{r})/r^4, \tag{4.4}$$

and similar formulae for $\dddot{b}(\bar{t})$ and $\dddot{c}(\bar{t})$, then:

$$\dddot{a}(0) = 3G(\dot{r}_0 + I_0)/r_0^4, \quad \dddot{b}(0) = -G/r_0^3, \quad \dddot{c}(0) = 0 \tag{4.5}$$

For the fifth coefficients we use the formula (1). From the Eq. (4.4) we get:

$$\ddddot{a}(\bar{t}) = -G\left\{\frac{\ddot{a}}{r^3} - 3\frac{2\dot{a}\dot{r} + a\ddot{r}}{r^4} + 12a\frac{\dot{r}^2}{r^5}\right\} \tag{4.6}$$

and similar formulae for $\ddddot{b}(\bar{t})$ and $\ddddot{c}(\bar{t})$, which for $\bar{t}=0$ give conditions where all the quantities of their right-hand members are known, the value

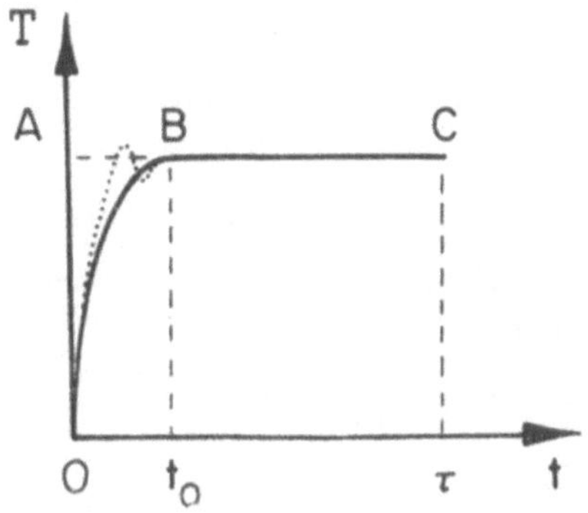


*Fig. 1. The solid line is in accordance with the relation (13), the dotted line with practical problems where $t_0$=0.02 seconds, approximately.*

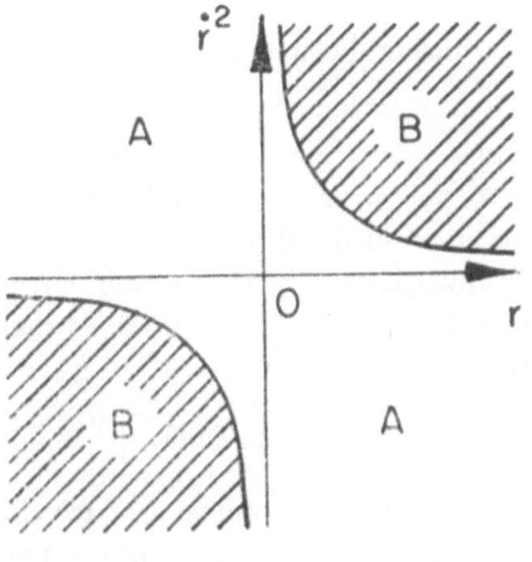


*Fig. 2. The points of the rectangular hyperbola $r.\dot{r}^2 = 2\mu$ in the $r.\dot{r}^2$—plane give a parabolic orbit; the points of the regions A give an elliptic orbit, and that of B a hyberbolic.*

$\ddot{r}_o$ being given by the Eq. (1). For the other coefficients we follow the same procedure.

If we restrict ourselves to the approximations up to the order $\bar{t}^2$, we have:

$$a(\bar{t}) = 1 - \frac{G}{2r_o^3}\bar{t}^2,\ b(\bar{t}) = \bar{t},\ c(\bar{t}) = \frac{1}{2}\bar{t}^2 \tag{9}$$

then an approximation of the solution (4), if we come back to the original scale time by using (2,1), is:

$$\underline{r}(t) = \left\{1 - \frac{G}{2r_o^3}(t - t_o)^2\right\}\underline{r}_o + (t - t_o)(\dot{\underline{r}}_o + \underline{I}_o) + \frac{1}{2}(t - t_o)^2\underline{T}_o, \tag{10}$$

from which we get:

$$\dot{\underline{r}}(t) = -\frac{G}{r_o^3}(t - t_o)\underline{r}_o + (\dot{\underline{r}}_o + \underline{I}_o) + (t - t_o)\underline{T}_o. \tag{11}$$

III. *The solution when the thrust is either suddenly or gradually applied.*

The formulae (10) and (11) are valid in $t_o \leq t \leq \tau$, and the thrust $\underline{T}_o$ and its impulse $\underline{I}_o$ are the values of the thrust and the impulse at $t = t_o$. The impulse $\underline{I}_o$ during the time $t_o$ is, by definition, given by:

$$\underline{I}_o = \int_0^{t_o} \underline{T}(t)\,dt. \tag{12.1}$$

If the thrust $\underline{T}(t)$ is, during the time $t_o$, a constant, $\underline{T}_o = T_o\underline{n}$ say, then:

$$\underline{I}_o = t_o\underline{T}_o, \tag{12.2}$$

and we speak about «*sudden application*» of the thrust. If the thrust is changed in $0 \leq t \leq t_o$, following any law, being zero at $t = 0$, we speak about «*gradual application*» of the thrust.

If the thrust is, say, constant in its direction but its magnitude varies accarding to parabolic law during the infinitesimal time $t_o$, being constant at the remainder time, i.e.

$$\underline{T}(t) = T(t)\underline{n},\quad \underline{n} = \text{constant},\quad T(t) = \begin{cases} at^{1/2}, & 0 \leq t \leq t_o \\ at_o^{1/2}, & t_o \leq t \leq \tau \end{cases} \tag{13}$$

where $\underline{n}$ is the unit vector, and the number a, which characterizes the parameter of the parabola, is a large positive constant, the impulse is:

$$\underline{I}_o = \frac{2}{3}t_o\underline{T}_o,\quad \underline{T}_o = at_o^{1/2}\underline{n}. \tag{12.3}$$

The above is an example of «gradual application», which can approximate the practical problems. In Fig. 1, the solid line is the graph of the thrust (13), while the dotted line is the thrust according to practical problems, $t_0$ being approximately equal to: 0.02 seconds.

The cases (12.2) and (12.3) can be written as:

$$\underline{I}_0 = m t_0 \underline{T}_0, \quad \underline{T}_0 = a t_0^{1/2} \underline{n} \tag{12.4}$$

where m=1 corresponds to sudden application, and $m = 2/3$ to gradual application, then the formulae (10) and (11) can be written as:

$$\underline{r}(t) = \left\{1 - \frac{G}{2r_0^3}(t-t_0)^2\right\}\underline{r}_0 + (t-t_0)\dot{\underline{r}}_0 + \left\{m t_0 (t-t_0) + \frac{1}{2}(t-t_0)^2\right\}\underline{T}_0, \tag{10.1}$$

$$\dot{\underline{r}}(t) = -\frac{G}{r_0^3}(t-t_0)\underline{r}_0 + \dot{r}_0 + \left\{m t_0 + (t-t_0)\right\}\underline{T}_0. \tag{11.1}$$

These formulae describe the motion of the projectile at time $t_0 \leq t \leq \tau$; $m=1$ corresponds to a sudden application of the constant thrust $\underline{T} = a t_0^{1/2} \underline{n}$, $m = 2/3$ corresponds to a gradual application of a thrust according to (13).

IV. *The type of the Keplerian orbit if the thrust is suddenly removed.*

The position vector $\underline{r}(t)$ and the velocity vector $\dot{\underline{r}}(t)$, given above, determine the motion of the projectile on the non-Keplerian arc of its orbit during the action of the thrust. When the thrust is suddenly removed at time $\tau$, the values $\underline{r}(\tau)$ and $\dot{\underline{r}}(\tau)$ determine completely the Keplerian orbit of the projectile after time $\tau$.

The type of this Keplerian orbit depends upon the sign of the right-hand member of the relation:

$$\frac{1}{A_1} = \frac{2}{r} - \frac{\dot{r}^2}{\mu} \tag{14}$$

where $A_1$ is the semi-major axis of the orbit, $\mu = k(M+m)$, and if this member is bigger, equal to or less than zero, we have an ellipse, parabola or hyperbola, respectively, and this leads to the restrictions:

$$r \cdot \dot{r}^2 \lesseqgtr 2\mu \tag{15}$$

for an ellipse, parabola or hyperbola, respectively. In fig. 2 the graph of (15) in the $r, \dot{r}^2$—plane is given. The regious A correspond to an ellipse; the regious B to a hyperbola, and their boundary, the curve $r\dot{r}^2 = 2\mu$, to a parabola.

The auther is indebted to Mr. S. Pines for fruitful discussion on some points of the paper, and to Dr. H. Wolf for suggesting the problem.

ΠΕΡΙΛΗΨΙΣ

Ἐνταῦθα σπουδάζεται ἡ κίνησις ὀχήματος πέριξ τῆς γῆς ὑπὸ τὴν ἐπίδρασιν τῆς ἑλκτικῆς δυνάμεως τῆς γῆς καὶ μιᾶς ὠστικῆς δυνάμεως. Ἡ τροχιὰ καὶ ἡ ταχύτης τοῦ ὀχήματος εὑρίσκονται εἰς δύο περιπτώσεις, ὅταν δηλαδὴ ἡ ὠστικὴ δύναμις ἐφαρμόζεται ἐπὶ τοῦ ὀχήματος εἴτε ἀκαριαίως εἴτε βραδέως. Δίδονται ἐπίσης καὶ αἱ συνθῆκαι ὑπὸ τὰς ὁποίας διακρίνομεν τὰ εἴδη τῆς Κεπλερείου τροχιᾶς, ὅταν ἡ ὠστικὴ δύναμις παύσῃ ἀκαριαίως νὰ δρᾷ.

ΑΝΑΤΥΠΟΝ
ΣΕΛ. 191 - 202

# The Keplerian orbit of a projectile around the earth, after the thrust is suddenly removed**.

*By* ***Dem. G. Magiros*** *.

Ἀνεκοινώθη ὑπὸ τοῦ κ. Ἰωάνν. Ξανθάκη.

*Introduction.*

In the following we discuss the elements of the Keplerian orbit of a projectile around the earth, after the thrust is suddenly removed, in the cases of sudden or gradual application of the thrust, if the thrust acts continuously either for infinitesimal time $t_o$ or for non-infinitesimal time $\tau$. Formulae are given for the elements of the Keplerian orbit in terms of the elements of the Keplerian orbit either the original or that which corresponds to time $t_o$. For the calculation of the elements of the Keplerian orbit when the thrust is removed, the position vector and the velocity vector at that time must be known. These vectors are given in a suitable form in a previous paper [1], «paper I», contained in the present volume. We treat first the case of infinitesimal time, then the case of non-infinitesimal time, if the thrust in both cases is suddenly or gradually applied. The numbers $\varepsilon$ throughout the paper, if multiplied by 100, give the percentage of increment of the corresponding element.

I. *The case of infinitesimal time.*

If the thrust, acting for infinitesimal time $t_o$, ceases at the point $M_o$

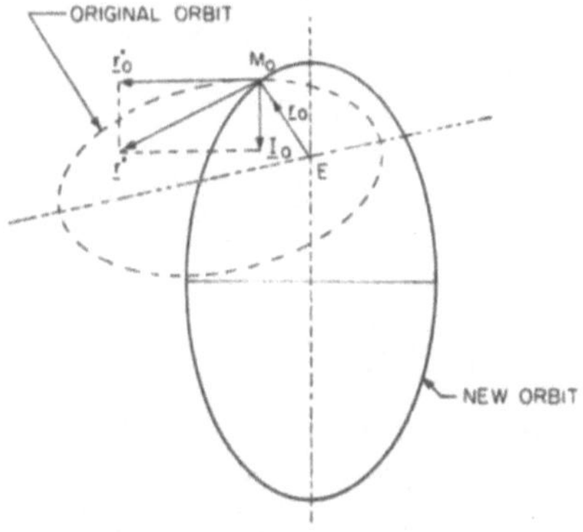


*Fig. 1.—The velocity $\dot{\underline{r}}$, tangent to the new orbit at the point $M_o$, is the resultant of the initial velocity $\dot{\underline{r}}_o$ and the thrust $\underline{I}_o$. The original and the new orbits have the same focus E, the earth.*

say, of the original orbit, Fig. 1, then the position vector and the velocity

* ΔΗΜ. ΜΑΓΕΙΡΟΥ, Τὰ στοιχεῖα τῆς Κεπλερίου τροχιᾶς ὀχήματος πέριξ τῆς γῆς, ὅταν ἀποτόμως παύσῃ ἡ ἐπ' αὐτοῦ ἐνεργοῦσα ὠστικὴ δύναμις.

** Republic Aviation Corp., U.S.A.

Reprinted from the *Proceedings of the Athens Academy of Sciences* **35** (1960), 191–202.

vector at $M_o$ are given, according to the formulae (10) and (11) or (10.1) and (11.1), of the «paper I», by:

$$\underline{r}(t_o)=\underline{r}_o,\quad \dot{\underline{r}}(t_o)=\dot{\underline{r}}_o+\underline{I}_o. \tag{1}$$

After the remarks on the impulse $\underline{I}_o$ made in Chapter III of «Paper I», we proceed to calculate the elements of the orbit, after the thrust is suddenly removed at $t=t_o$, these elements being designated by the subscript I.

a. *The semi-major axis $a_I$.*

For the original and the new semi-major axis, a and $a_I$ respectively, we have [2].

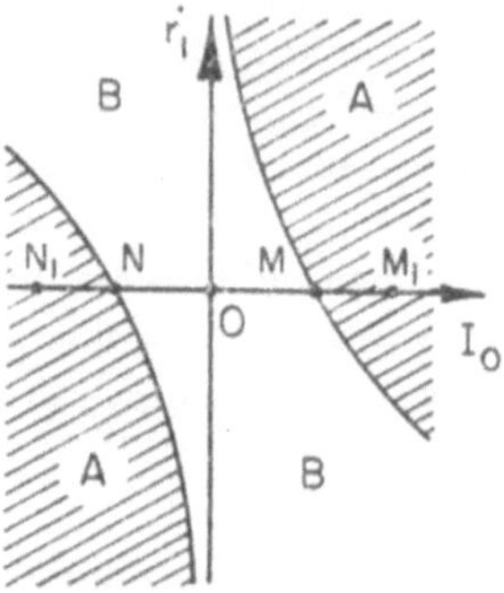


*Fig. 2.—The points of the regions A give hyperbolic orbits, those of the region B elliptic orbits, and their boundary, the curve: $I_o^2+2I_o\dot{r}_1-\frac{\mu}{\alpha}=0$, parabolic orbits.*

$$\frac{1}{\alpha}=\frac{2}{r_o}-\frac{(\dot{\underline{r}}_o)^2}{\mu} \tag{2.1}$$

$$\frac{1}{\alpha_I}=\frac{2}{r_o}-\frac{(\dot{\underline{r}}_o+\underline{I}_o)^2}{\mu} \tag{2.2}$$

which give:

$$\frac{1}{\alpha_I}=\frac{1}{\alpha}-\frac{1}{\mu}\,(2\,\dot{r}_o I_o\cos\vartheta+I_o^2) \tag{2.3}$$

then:

$$\frac{1}{\alpha_I}=\frac{1}{\alpha}\,\psi \tag{2.4}$$

with

$$\psi=1-\frac{\alpha}{\mu}\,(2\,\dot{r}_o I_o\cos\vartheta+I_o^2). \tag{2.5}$$

We also can have:

$$\frac{1}{\alpha_I}=\frac{1}{\alpha}\,(1+\varepsilon),\quad \varepsilon=\psi-1. \tag{2.6}$$

In the above $r_o$, $\dot{r}_o$, $I_o$ are magnitudes of $\underline{r}_o$, $\dot{\underline{r}}_o$, $\underline{I}_o$; and $\vartheta$ the angle of $\dot{\underline{r}}_o$ and $\underline{I}_o$.

The conditions for the kind of the new orbit are:

$$I_o^2+2\,I_o\,\dot{r}_1-\frac{\mu}{\alpha}\lesseqgtr 0 \tag{3}$$

for elliptic, parabolic and hyperbolic ones, respectively, $\dot{r}_1$ is the projection of $\dot{\underline{r}}_o$ along $\underline{I}_o$, $\dot{r}_1=\dot{r}_o\cos\vartheta$. The graph of the conditions (3) is given in Fig. 2.

b. *The angular momentum vector $\underline{H}_I$ and the angle between the original and the new orbits.*

The original and the new angular momentum vectors, $\underline{H}$ and $\underline{H}_I$, are, by definition, given by the vector products:

$$\underline{H} = \underline{r}_o \times \dot{\underline{r}}_o, \quad \underline{H}_I = \underline{r}_o \times (\dot{\underline{r}}_o + \underline{I}_o). \tag{4.1}$$

If $\Delta\underline{H}$ is the increment vector, we can write:

$$\underline{H}_I = \underline{H} + \Delta\underline{H}, \quad \Delta\underline{H} = \underline{r}_o \times \underline{I}_o.$$

The length of $\underline{H}$ and $\Delta\underline{H}$ are:

$$H = r_o \dot{r}_o \sin \varphi_1, \quad \Delta H = r_o I_o \sin \varphi,$$

$\varphi_1$ being the angle of $\underline{r}_o$ and $\dot{\underline{r}}_o$, $\varphi$ that of $\underline{r}_o$ and $\underline{I}_o$, Fig. 3a. The vectors $\underline{H}$ and $\Delta\underline{H}$ are perpendicular to the $\underline{r}_o$, $\dot{\underline{r}}_o$ -plane and to the $\underline{r}_o$, $\underline{I}_o$ -plane, res-

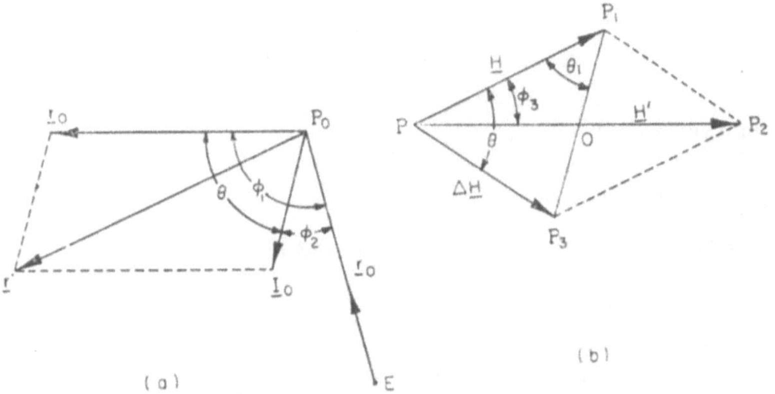


*Fig. 3.—a) The vectors $\underline{r}_o$, $\dot{\underline{r}}_o$, $\underline{I}_o$ define a trihedral angle with vertex $P_o$. b) The new angular momentum $\underline{H}'$ is the resultant of the original angular momentum $\underline{H}$ and its increment $\Delta\underline{H}$.*

pectively. If $\vartheta$ is the angle of these planes, $\vartheta$ must be the angle of $\underline{H}$ and $\Delta\underline{H}$. The new angular momentum vector $\underline{H}$ Is parallel to the $\underline{H}$, $\Delta\underline{H}$ -plane. Fig. 3b helps for calculation of the length $H_I$ of $\underline{H}_I$ and its angle $\varphi_3$ with $\underline{H}$. From the triangle $P_1 P P_3$ we get:

$$(P_1 P_3) = \varrho = r_o \{ (\dot{r}_o \sin \varphi_1)^2 + (I_o \sin \varphi_2)^2 - 2 \dot{r}_o I_o \sin \varphi_1 \sin \varphi_2 \cos \vartheta \}^{1/2},$$

$$\sin \vartheta_1 = (\Delta H / P) \sin \vartheta.$$

From the triangle $P_1PO$ we get:

$$(PO) = \left\{ H^2 + \left(\frac{\varrho}{2}\right)^2 + H\varrho \cos \vartheta_1 \right\}^{1/2}, \quad \sin \varphi_3 = ((P_1O)/(PO)) \sin \vartheta_1.$$

Then

$$H_I = 2\,(PO) = 2\left\{H^2 + \left(\frac{\varrho}{2}\right)^2 - H\varrho \cos \vartheta_1\right\}^{1/2} \tag{4.2}$$

$$\sin \varphi_s = \frac{1}{2}\left\{\frac{1}{4} + \left(\frac{H}{\varrho}\right)^5 + \frac{H}{\varrho}\cos \vartheta\right\}^{-1/2} . \sin \vartheta_1 . \tag{4.3}$$

c. *The period $P_I$, and the mean angular motion $n_I$.*

We have:

$$P = \frac{2\pi}{K\sqrt{\mu}}\,\alpha^{3/2}, \quad n = \frac{2\pi}{P} = K\sqrt{\mu}\left(\frac{1}{\alpha}\right)^{3/2}$$

$$P_I = \frac{2\pi}{K\sqrt{\mu}}\,\alpha_I^{3/2}, \quad n_I = \frac{2\pi}{P_I} = K\sqrt{\mu}\left(\frac{1}{\alpha_I}\right)^{3/2},$$

then

$$P_I = P\,(1 + \varepsilon_1), \quad n_I = n\,(1 + \bar{\varepsilon}_1) \tag{5.1}$$

with

$$\varepsilon_1 = \psi^{-3/2} - 1, \quad \bar{\varepsilon}_1 = \psi^{3/2} - 1. \tag{5.2}$$

$\psi$ is given by (2.5).

d. *The eccentric anomaly $E_o$, and the eccentricity $e_I$.*

From the formulae for the original orbit [3]:

$$e \sin E_o = \frac{\underline{r_o}\, \underline{\dot{r}_o}}{\sqrt{\mu\alpha}} \tag{6.1}$$

$$e \cos E_o = \frac{\alpha - r_o}{\alpha} \tag{6.2}$$

we get:

$$\tan E_o = \sqrt{\frac{\alpha}{\mu}} \cdot \frac{\underline{r_o}\, \underline{\dot{r}_o}}{\alpha - r_o}, \tag{6.3}$$

$$e^2 = \frac{(\underline{r_o}\, \underline{\dot{r}_o})^2}{\mu\alpha} + \frac{(\alpha - r_o)^2}{\alpha^2}. \tag{6.4}$$

Applying (6.3) we have for the eccentric anomaly $E_{oI}$:

$$\tan E_{oI} = \sqrt{\frac{\alpha}{\mu}} \cdot r_o \cdot \frac{\sqrt{\psi}}{\alpha - r_o\psi}\,(\dot{r}_o \cos \varphi_1 + I_o \cos \varphi_2), \tag{6.5}$$

$\varphi_1$ the angle of $\underline{r_o}$ and $\underline{\dot{r}_o}$, $\varphi_2$ that of $\underline{r_o}$ and $\underline{I_o}$.

Therefore:

$$\tan E_{oI} = (1 + \varepsilon_2) \tan E_o \tag{6.6}$$

with

$$\varepsilon_2 = (\alpha - r_o) \frac{\sqrt{\psi}}{\alpha - r_o\psi} \left(1 + \frac{\dot{I}_o}{r_o} \frac{\cos\varphi_2}{\cos\varphi_1}\right) - 1. \tag{6.7}$$

For the new eccentricity $e_I$ we get:

$$e_I = \frac{1}{\alpha} X^{1/2} \tag{7.1}$$

with:

$$X = \frac{\alpha}{\mu} \psi r_o^2 (\dot{r}_o \cos\varphi_1 + I_o \cos\varphi_2)^2 + (\alpha - r_o\psi)^2, \tag{7.2}$$

then:

$$e_I = e(1 + \varepsilon_3) \tag{7.3}$$

with:

$$\varepsilon_3 = \sqrt{\mu X} \left\{\alpha (r_o \dot{r}_o \cos\varphi_1)^2 + \mu(\alpha - r_o)^2\right\}^{-1/2} - 1. \tag{7,4}$$

e. *The «Perigee» $q_I$, «parameter» or «latus rectum» $p_I$ and «true anomaly $V_{oI}$.*

For the perigee we have:

$$q = \alpha(1 - e), \quad q_I = \alpha_I (1 - e_I) = \frac{\alpha}{\psi} \left(1 - \frac{1}{\alpha} X^{1/2}\right),$$

then:

$$q_I = (1 + \varepsilon_4) q \tag{8.1}$$

$$\varepsilon_4 = \frac{\alpha - X^{1/2}}{\alpha(1 - e)\psi} - 1. \tag{8.2}$$

For the parameter:

$$p = \alpha(1 - e^2), \quad p_I = \alpha_I (1 - e_I^2)$$

then:

$$p_I = (1 + \varepsilon_5) p, \tag{9.1}$$

$$\varepsilon_5 = \frac{\alpha^2 - X}{\alpha^2 (1 - e^2)\psi} - 1. \tag{9.2}$$

For the true anomaly $V_{oI}$ we use the formula [4].

$$\cos V_o = \frac{\cos E_o - e}{1 - e \cos E_o}$$

which, with the help of (6.4), can be written as:

$$\cos V_o = \frac{\alpha - \alpha e - r_o}{r_o e}$$

when:

$$\cos V_{oI} = \frac{\alpha_I - \alpha_I e_I^2 - r_o}{r_o e_I} = \frac{\alpha(\alpha - r_o\psi) - X}{r_o \psi X^{1/5}},$$

then:

$$\cos V_{oI} = (1 + \varepsilon_6) \cos V_o \tag{10.1}$$

$$\varepsilon_6 = \frac{e \{ \alpha(\alpha - r_o\psi) - X \}}{\psi X^{1/2} (\alpha - \alpha e^2 - r_o} - 1. \tag{10.2}$$

f. *Orientation cosines:* $\underline{P}_I$, $\underline{Q}_I$, $\underline{W}_I$.

The orientation cosines $\underline{P}$, $\underline{Q}$, $\underline{W}$ are unit vector; $\underline{P}$ the «perigee vector» from the earth towards the perigee; $\underline{Q}$ vector directed along the lactus rectum; $\underline{W} = \underline{P} \times \underline{Q}$. The $\underline{P}$ and $\underline{Q}$ are given by: [5]

$$\underline{P} = \frac{\cos E_o}{r_o} \underline{r}_o - \sqrt{\frac{\alpha}{\mu}} \sin E_o. \dot{\underline{r}}_o \tag{11,1}$$

$$\underline{Q} = \frac{1}{\sqrt{1-e^2}} \left\{ \frac{\sin E_o}{r_o} \underline{r}_o + \sqrt{\frac{\alpha}{\mu}} (\cos E_o - e). \dot{\underline{r}}_o \right\}. \tag{11.2}$$

For the new $\underline{P}_I$ and $\underline{Q}_I$, by taking into account (1) (2.5), (6.1), (6.2), (7.1), (7.2), we get:

$$\underline{P}_I = \frac{\alpha - r_o\psi}{r_o X^{1/2}} \underline{r}_o - \frac{\alpha}{\mu X^{1/2}} r_o \dot{r}_o \cos\varphi_1 (\dot{\underline{r}}_o + \underline{I}_o), \tag{12.1}$$

$$\underline{Q}_I = \frac{\alpha}{\sqrt{\alpha^2 - X}} \left\{ \sqrt{\frac{\alpha\psi}{\mu X}} \cos\varphi_1 . r_o . \underline{r}_o + \frac{1}{\sqrt{\alpha\mu\psi X}} \Big( \alpha(\alpha - r_o\psi) - X \Big) \Big( \dot{\underline{r}}_o + \underline{I}_o \Big) \right. \tag{12.2}$$

We can omit the subscript o from the formulae (12.1) and (12.2) since the orientation cosines do not vary with time, then they are independent of the position of the point $M_o$ on the orbit.

g. *Orientation angles:* $i_I$, $\omega_I$, $\Omega_I$.

The «inclination» i is the angle between the plane of the orbit and that of the equator or of the ecliptic. The «argument of perigee» $\omega$ is the angle between the nodal line (to the direction of the ascending node) and the semi-major axis of the orbit (to the direction of the perigee). The «longitude of node» $\Omega$ is the angle (on the equator) of the nodal line to the ascending node and the intersection of equator - ecliptic.

For the new inclination $i_I$ we notice that the inclinations are measured from the plane of equator and the angle $\varphi_3$ of the original and the new orbits from the plane of the original orbit. If i, $i_I$, $\varphi_3$ are positive in the same rotation as shown in Fig. 4, we can get the relationship:

$$i_I = i \pm \varphi_s \tag{13}$$

$\varphi_s$ being given by the formula (4.3).

Now, for the new angles $\omega_I$ and $\Omega_I$, take the orthogonal xyz-system as it is shown in Fig. 4. If $P_{Ix}$, $P_{Iy}$, $P_{Iz}$ are the components of the new perigee vector $\underline{P_I}$ along the axes of this system, we can get the following formula: [3]

$$\begin{aligned} P_{Ix} &= \cos\omega_I . \cos\Omega_I - \sin\omega_I . \sin\Omega_I . \cos i_I, \\ P_{Iy} &= \cos\omega_I \ \sin\Omega_I + \sin\omega_I \ \cos\Omega_I \ \cos i_I, \\ P_{Iz} &= \sin\omega_I . \sin i_I . \end{aligned} \tag{14}$$

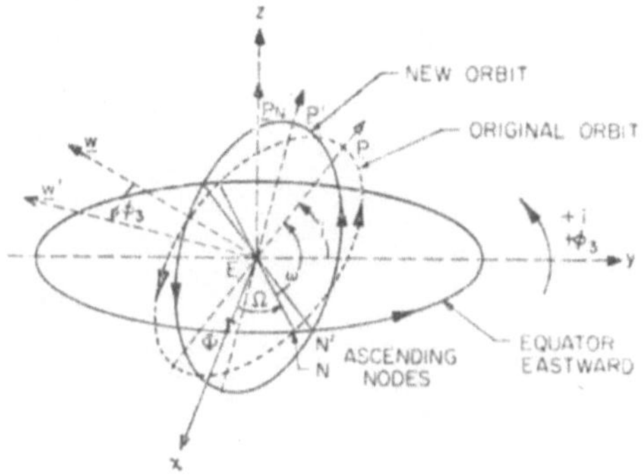


*Fig. 4.—The location of the original and the new orbits with respect to the equator. The arrows in the orbits show the direction of the projectile.*

We have the same formulae for the components $P_x$, $P_y$, $P_z$ of the original perigee $\underline{P}$ by omitting the subscript I from (14).

By using the foamulae (12.1) and (13), the only unknowons in the system (14) are the angles $\omega_I$ and $\Omega_I$. The last equation of (14) gives the new argument of perigee $\omega_I$:

$$\sin\omega_I = P_{Iz}/\sin i_I . \tag{15}$$

Inserting the known value of $\omega_I$ into the first equation of (14), we can determine the new longitude of node $\Omega_I$:

$$\cos\Omega_I = \frac{1}{\varrho^2 + \cos^2\omega_I} \left\{ P_{Ix} \cos\omega_I \pm \varrho \left(\varrho^2 + \cos^2\omega_I - P^2_{Ix}\right)^{1/2} \right\}, \tag{16.1}$$

with:

$$\varrho = \sin \omega_I . \cos i_I . \tag{16.2}$$

The second equation of (14) must be satisfied by the values found above, and this gives indication for the selection of plus or minus sign of the formula (16.1).

II. *The case of non-infinitesimal time.*

We consider in this section the case of a thrust of special type suddenly or gradually applied to the projectile and suddenly removed either after infinitesimal time $t_o$, case I, or after time $\tau$, case II.

The formulae for the elements of the Keplerian orbit in case II are given in terms of that of the case I. The procedure can be used as a model to treat the calculation when other types of thrust are given.

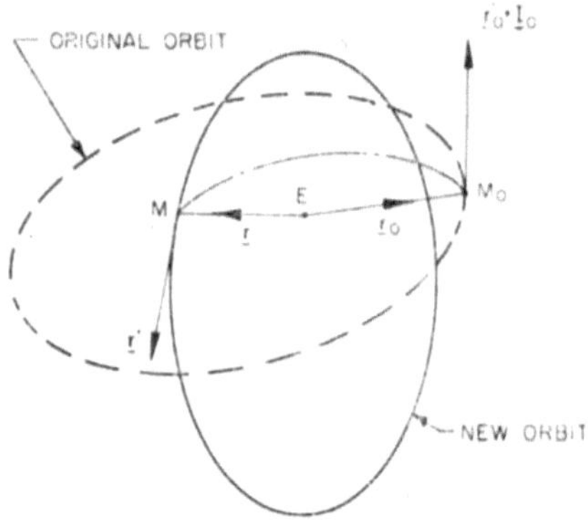


*Fig 5.— The dotted arc $M_o$ $M$ shows the part of the orbit during the action of the thrust.*

Take the direction of the thrust parallel to the initial velocity $\dot{\underline{r}}_o$, that is $\underline{T}(t) = \lambda(t) \dot{\underline{r}}_o$, $\lambda$ being the factor of proportionality, and its magnitude constant for sudden application, or according to the law shown in Fig. 1 «Paper I», for gradual application.

The impulse $\underline{I}_o$ in the formulae (10) and (11) of «Paper I» is given by $\underline{I}_o = \frac{2}{3} \alpha t_o^{3/2} \dot{\underline{r}}_o$ in case of a gradual application with parabolic law in

$o \leqq t \leqq t_o$; and by $\underline{I}_o = \alpha t_o{}^{3/2}\, \dot{\underline{r}}_o$ in case of a sudden application of the constant thrust $\underline{T} = \alpha t_o{}^{1/2}\, \dot{\underline{r}}_o$. If we write: where $m=1$ for sudden application, and $m=2/3$ for gradual application of parabolic type, the formulae (10) and (11) of «paper I» can be written as:

$$\underline{r}(t) = \left\{ 1 - \frac{G}{2r_o{}^3}(t-t_o)^2 \right\} \underline{r}_o + (t-t_o)\left\{1 + mt_o\lambda + \frac{1}{2}(t-t_o)\lambda\right\}\dot{\underline{r}}_o. \quad (17)$$

$$\dot{\underline{r}}(t) = -\frac{G}{r_o{}^3}(t-t_o)\,\underline{r}_o + \left\{1 + mt_o\lambda + (t-t_o)\lambda\right\}\dot{\underline{r}}_o, \quad (18)$$

where $\lambda = \alpha t_o{}^{1/2}$. These formulae describe the motion of the projectile at time $t_o \leqq t \leqq \tau$, when the arc of the orbit is represented by the broken arc of Fig. 5.

a. *The semi-major axes $a_I$, $a_{II}$.*

From the formulae (2) we have:

$$\frac{1}{\alpha_I} = \frac{2\mu - (1+mt_o\lambda)^2 r_o \dot{r}_o{}^2}{\mu r_o}, \quad \frac{1}{\alpha_{II}} = \frac{2\mu - r_o \dot{r}^2}{\mu r}$$

then we can write:

$$\frac{1}{\alpha_{II}} = \frac{1}{\alpha_I}(1+\varepsilon_1) \quad (18.1)$$

with:

$$\varepsilon_1 = \frac{r_o}{r}\,\frac{2\mu - r\dot{r}_o}{2\mu - r_o\dot{r}_o{}^2(1+mt_o\lambda)^2} - 1 \quad (19.2)$$

$r$ and $\dot{r}$ are the magnitudes of $\underline{r}$ and $\dot{\underline{r}}$, given by (17) and (18).

b. *The angular momentum vectors $\underline{H}_I$, $\underline{H}_{II}$.*

For the angular momentum vectors $\underline{H}$, $\underline{H}_I$, $\underline{H}_{II}$ we have:

$$\underline{H} = \underline{r}_o \times \dot{\underline{r}}_o, \quad \underline{H}_I = (1+mt_o\lambda)^2\, \underline{r}_o \times \dot{\underline{r}}_o, \quad \underline{H}_{II} = \underline{r} \times \dot{\underline{r}},$$

then for their magnitudes:

$$H = \dot{r}_o r_o \sin\varphi_1, \quad H_I = (1+mt_o\lambda)^2\, r_o \dot{r}_o \sin\varphi_1, \quad H_{II} = r\dot{r}\sin\varphi,$$

$\varphi_1$ being the angle of $r_o$, $\dot{r}_o$ and $\varphi$ that of $r$, $\dot{r}$.

We can write:

$$H_2 = H_1 (1+\varepsilon_8) \tag{20.1}$$

with:

$$\varepsilon_8 = \frac{r \dot{r} \sin \varphi}{H(1+mt_0\lambda) \sin \varphi_1} - 1 \tag{20.2}$$

The plane of the original and the new orbits are the same, then we have the same direction for the original and the new angular momentum vectors, and the inclination remains constant.

c. *The periode $P_I$, $P_{II}$ and the mean angular motions $n_I$, $n_{II}$.*

For the new periods we have:

$$P_{II} = \frac{2\pi}{K\sqrt{\mu}} \alpha_{II}^{3/2}, \quad P_I = \frac{2\pi}{K\sqrt{\mu}} \alpha_I^{3/2}$$

then:

$$P_{II} = P_I \left(\frac{\alpha_{II}}{\alpha_I}\right)^{3/2} = P_I (1+\varepsilon_7)^{3/2} = P_I \left(1 - \frac{3}{2}\varepsilon_7\right) \tag{21}$$

for small $\varepsilon_7$, which is given by: (19.2).
In the same way for the new mean angular motion we can get:

$$n_{II} = n_I \left(1 + \frac{3}{2}\varepsilon_7\right). \tag{22}$$

d. *The eccentric anomalies $E_{oI}$, $E_{oII}$, and the eccentricities $e_I$, $e_{II}$.*

By using (6.3) we get:

$$\tan E_{oI} = \sqrt{\frac{\alpha_I}{\mu}} \; \frac{r_o \dot{r}_o (1+mt_o \lambda) \cos \varphi_1}{\alpha_I - r_o}$$

$$\tan E_{oII} = \sqrt{\frac{\alpha_{II}}{\mu}} \; \frac{r \dot{r} \cos \varphi}{\alpha_{II} - r} = \sqrt{\frac{\alpha_I (1+\varepsilon_7)^{-1}}{\mu}} \cdot \frac{r \dot{r} \cos \varphi}{\alpha_I (1+\varepsilon_7)^{-1} - r},$$

then:

$$\tan E_{oII} = (1+\varepsilon_9) \tan E_{oI} \tag{23.1}$$

with:

$$\varepsilon_9 = \left(1 - \frac{1}{2}\varepsilon_7\right) \frac{r \dot{r} \cos \varphi}{r_o \dot{r}_o (1+mt_o \lambda) \cos \varphi_1} \cdot \frac{\alpha_I - r_o}{\alpha_I (1-\varepsilon_7) - r} - 1. \tag{23.2}$$

For the new eccentricities, by applying (6.4) we have:

$$e_I^2 = \frac{\{r_o \dot{r}_o (1+mt_o \lambda) \cos \varphi_1\}^2}{\mu \alpha_I} + \frac{(\alpha_I - r_o)}{\alpha_I^2}$$

$$e_{II}^2 = \frac{(r \dot{r} \cos \varphi)^2}{\mu (1-\varepsilon_7) \alpha_I} + \frac{\{\alpha_I (1-\varepsilon_7) - r\}}{(1-2\varepsilon_7) \alpha_I^2},$$

then:

$$e_{II}^2 = e_I^2 (1 + \varepsilon_{10}) \tag{24.1}$$

with:

$$\varepsilon_{10} = \frac{(1-\varepsilon_7)\,\alpha_I (r\dot{r}\cos\varphi)^2 + \mu\,[(1-\varepsilon_7)\,\alpha_I - r]^2}{(1-2\varepsilon_7)\,[\,\{r_o \dot{r}_o (1+m t_o \lambda)\cos\varphi_1\}^2 \alpha_I + \mu(\alpha_I - r_o)^2} - 1 \tag{24.2}$$

e. *The perigees $q_I$, $q_{II}$, and the parameter $p_I$, $p_{II}$.*

For the perigees we have:

$$q_I = \alpha_I (1 - e_I),\quad q_{II} = \alpha_{II}(1 - e_{II}) = \alpha_I (1 - \varepsilon_7) \left| 1 - e_I \left(1 + \frac{1}{2}\varepsilon_{10}\right) \right|,$$

then:

$$q_{II} = (1 - \varepsilon_7)\, q_I - \frac{1}{2}\varepsilon_{10}\, \alpha_I\, e_I; \tag{25}$$

and for the parameters:

$$p_I = \alpha_I (1 - e_I^2),\quad p_{II} = \alpha_{II}(1 - e_{II}^2) = \alpha_I (1 - \varepsilon_7) \left| 1 - e_I^2 (1 + e_{10}) \right|,$$

when, for small $\varepsilon_7$ and $\varepsilon_{10}$, we can write:

$$p_{II} = (1 - \varepsilon_7)\, p_I - \varepsilon_{10}\, \alpha_I\, e_I^2. \tag{26}$$

## ΠΕΡΙΛΗΨΙΣ

Ἡ παροῦσα ἐργασία ἀποτελεῖ συμπλήρωμα καὶ συνέχειαν προηγουμένης ἐργασίας περιεχομένης εἰς τὸν παρόντα τόμον τῶν Πρακτικῶν σελ. 96 – 103. Εἰς τὴν προηγουμένην μελέτην μελετᾶται ἡ κίνησις ὀχήματος πέριξ τῆς γῆς ὑπὸ τὴν ἐπίδρασιν τῆς ἑλκτικῆς δυνάμεως τῆς γῆς καὶ μιᾶς ὠστικῆς δυνάμεως. Εἰς τὴν παροῦσαν ἐξετάζονται τὰ στοιχεῖα τῆς Κεπλερείου τροχιᾶς τοῦ ὀχήματος, ὅταν ἡ ὠστικὴ δύναμις παύσῃ ἀποτόμως νὰ ἐνεργῇ ἐπὶ τοῦ ὀχήματος. Δίδονται τύποι συνδέοντες τὰ στοιχεῖα τῆς Κεπλερείου τροχιᾶς πρὸς τὰ στοιχεῖα τῆς Κεπλερείου τροχιᾶς εἴτε τῆς ἀρχικῆς (ὅταν ἤρχισε ἐνεργοῦσα ἡ ὠστικὴ δύναμις), εἴτε τῆς ἀντιστοιχούσης εἰς τὸν χρόνον $t_o$. Οἱ ἀριθμοὶ ε εἰς τοὺς διδομένους τύπους, ἑκατονταπλασιαζόμενοι, δίδουν τὸ ποσοστὸν ἐπὶ τοῖς ἑκατὸν τῆς αὐξήσεως τῶν ἀντιστοιχούντων στοιχείων.

## REFERENCES

[1] D. G. Magiros, «The motion of a projectile arount the earth under the influence of the earth's gravitational attraction and a thrust». this volume, Pag 96-103.

[2] W. Smart, «Celestial Mechanics», (1953). Longmans, Green and Company, p. 21, paragraph 2.12.

[3] S. Herrick, «Formulas, Constants, Definitions, Notation for Geocentric and Heliocentric Orbits», Systems Laboratories Corporation, Spacenautics Division Report SN 1.

[4] C. Hilton and S. Herrick, «A Technical Note Concerning Coordinate Systems for Linear Vehicle Orbits». (1958), p. 13.

[5] J. Feyk and H. Karrenberg, «Equation Relating to the Trajectory of a Lunar Vehicle», (1958). Notes Systems Corporation of America.

ΠΡΑΚΤΙΚΑ
ΤΗΣ ΑΚΑΔΗΜΙΑΣ ΑΘΗΝΩΝ

ΕΤΟΣ 1963: ΤΟΜΟΣ 38ος

ΑΝΑΤΥΠΟΝ
ΣΕΛ. 36 - 39

# On the convergence of the solution of a special two - body problem *.

*by* ***Demetrios G. Magiros (**)***.

'Ανεκοινώθη ὑπὸ τοῦ 'Ακαδημαϊκοῦ κ. 'Ιωάνν. Ξανθάκη.

## 1. INTRODUCTION

a. In previous papers [1], where the motion of a projectile a Newtonian center during the action of general thrust vector was investigated, a series solution for this problem was constructed. The purpose of the present note is to give a brief discussion of the determination of the time in-

(*) Of the Valley Forge Space Technology Center, of the General Electric Co., M.S.D. Philadelphia, Pa.

(**) ΔΗΜ. Γ. ΜΑΓΕΙΡΟΥ, 'Επὶ τῆς συγκλίσεως τῆς λύσεως ἑνὸς εἰδικοῦ προβλήματος τῶν δύο σωμάτων.

Reprinted from the *Proceedings of the Athens Academy of Sciences* **38** (1963), 36–39.

terval for which the solution found is valid. Details of the present note and related subjects will appear elsewhere.

b. We state for reference that the differential equation and the initial conditions of the problem in vector form are:

$$\ddot{\underline{r}}(\tau) = -\frac{\mu}{r^3(\tau)}\,\underline{r}(\tau) + \underline{T}(\tau)$$

$$\underline{r}(o) = \underline{r}_0,\quad \dot{\underline{r}}(o) = \dot{\underline{r}}_0 + \underline{I}_0 \tag{1}$$

valid in the region D: $|\underline{r}(\tau)| < M_1$, $|\dot{\underline{r}}(\tau)| < M_2$ for any value of time $\tau$ in $D_1$: $o \leqq \tau \leqq \tau'$. $\mu$ is a constant, $\underline{T}$ the thrust, $\underline{I}_0$ the impulse of the thrust for very small time, $\underline{r}$ and $\dot{\underline{r}}$ the displacement and velocity vectors.

If the reference coordinate system is: $(P;\ \underline{r}_0^*,\ \underline{s}_0^*,\ \underline{T}_0^*)$, where P is the position of the projectile when the thrust starts, $\underline{r}_0^*$, $\underline{s}_0^*$, $\underline{T}_0^*$ the unit vectors along $\underline{r}_0$, $\underline{r}_0 + \underline{I}_0$, $\underline{I}_0$, respectively, a solution of the form

$$\underline{r}(\tau) = \alpha_1(\tau)\,\underline{r}_0^* + \alpha_2(\tau)\,\underline{s}_0^* + \alpha_3(\tau)\,\underline{T}_0^* \tag{2}$$

can be determined by calculating the scalar functions $\alpha_i(\tau)$, $i = 1,2,3$, in Mac Laurin's expansions at $\tau = o$.

$$\alpha_i(\tau) = \sum_{n=o}^{\infty} \frac{\alpha_i^{(n)}(o)}{n!}\,\tau^n \quad ;\ i = 1,2,3 \tag{3}$$

The functions $\alpha_i(\tau)$ satisfy the conditions

$$\begin{aligned} &\ddot{\alpha}_i + \frac{\mu}{r^3}\,\alpha_i = T_i \quad ;\ i = 1,2,3 \\ &\alpha_1(o) = r_0,\ \alpha_2(o) = \alpha_3(o) = 0 \\ &\dot{\alpha}_1(o) = \dot{\alpha}_3(o) = 0,\ \dot{\alpha}_2(o) = s_0 \end{aligned} \tag{4}$$

where $s_0 = |\dot{\underline{r}}_0 + \underline{I}_0|$ and $T_1$, $T_2$, $T_3$, projections of $\underline{T}$ on the $\underline{r}^*_0$, $\underline{s}_0^*$, $\underline{T}_0^*$ – axis. By using (1), (2), (4) we can determine the coefficients of the series (3).

### 2. THE RADIUS OF CONVERGENCE OF THE SERIES (3).

The radius of convergence of the series (3) is the reciprocal of the upper limit: [2]

$$\overline{\lim_{n\to\infty}} \left\{ |\alpha_i^{(n)}(o)| \,/\, n! \right\}^{1/n},\ i = 1,2,3 \tag{5}$$

The $n^{th}$ derivative of $\alpha_i(\tau)$, found from the equation (4), is:

$$\alpha_i^{(n)}(\tau) = -\mu\left[\frac{\alpha_i(\tau)}{r^3(\tau)}\right]^{(n-2)} + T_i^{(n-2)}(\tau) \tag{6}$$

The general term $\left(\alpha_i^{(n)}(\tau)\,/\,n!\right)$, if we rake into account the formula (e) of the Appendix, becomes:

$$\frac{\alpha_i^{(n)}(\tau)}{n!} = -\sum_{m=0}^{n-2}\frac{\alpha_i^{(m)}(\tau)\left[V_1 r^{n-m-3} + V_2 r^{n-m-4} + \cdots + V_{n-m-2}\right]}{(n-1)n\, m!\,(n-m-2)!\, r^{n-m+1}} + \frac{T_i^{(n-2)}(\tau)}{n!} \tag{7}$$

where V's are polynomicals in the derivatives of r up to the order $(n-m-2)$ with coefficients smaller than $(n-m)!$

To find the limit of the $n^{th}$ root of the absolute value of the right-hand member of (7) for $\tau=0$ as $n\to\infty$, we consider each term of it as positive, and then take the $n^{th}$ root of each such term. Their sum Sn must satisfy the relation:

$$\left[\,|\alpha_i^{(n)}(0)|\,/\,n!\,\right]^{1/n} \leqq Sn \tag{8}$$

Since $r_0 \neq 0$ and all derivatives of r and $T_i$ are bounded at $\tau=0$, the sum $S_n \to 0$ as $n\to\infty$, then the limit of the left-hand member of (8) is zero, and, as a result, the radius of convergence is $\infty$, and the series (3) converge for any value of $\tau'$.

APPENDIX:

The $n^{th}$ derivative of the product of the functions $\sigma(\tau)$, $\varphi(\tau)$ is given by:

$$[\sigma\varphi]^{(n)} = \sum_{m=0}^{n}\frac{n!}{m!\,(n-m)!}\,\sigma^{(m)}\,\varphi^{(n-m)} \tag{a}$$

and that of:

$$\varphi(\tau) = \frac{1}{r^3(\tau)} \tag{b}$$

by:

$$\varphi^{(n)} = \frac{P(r)}{r^{n+3}} \tag{c}$$

with:

$$P(r) = V_1 r^{n-1} + V_2 r^{n-2} + \cdots + V_n \tag{d}$$

where the V's are polynomials in the derivatives of r up to the order $n^{th}$ with coefficients smaller than $(n+2)!$

Combining (a), (b), (c) we can get:

$$[\sigma/r^3]^{(n)} = \sum_{m=0}^{n}\frac{n!}{m!\,(n-m)!}\,\sigma^{(m)}\,\frac{P(r)}{r^{n-m+3}} \tag{e}$$

where:

$$P(r) = V_1 r^{n-m-1} + V_2 r^{n-m-2} \dots + V_{n-m} \quad \text{(f)}$$

The V's polynomials have coefficients smaller than $(n-m+2)!$

## REFERENCES


[1] D. G. Magiros: (a). Praktika of Athens Academy of Sciences, Vol. 35 (1960), 96 - 103, Athens, Greece. (b). «The motion of a Projectile under the influence of the Attractive Force of a Newtonian Center and a Thrust». General Electric Co., MSD, Technical Information Series, No. 625SD221, Nov. 15, 1962.

[2] K. Knopp: «Theory and Applications of Infinite Series», § 18, Section 94, 154 - 155 Hafner Publishing Co., New York.


## ΠΕΡΙΛΗΨΙΣ


Εἰς προηγουμένας ἐργασίας κατεσκευάσθη ἡ λύσις τοῦ προβλήματος τῆς κινήσεως ὀχήματος ὑπὸ τὴν ἐπίδρασιν Νευτωνίου ἑλκτικοῦ κέντρου καὶ μιᾶς γενικῆς ὠστικῆς δυνάμεως. Ἐνταῦθα ἐκτίθεται ἐν συντομίᾳ τὸ ζήτημα τῆς συγκλίσεως τῆς λύσεως αὐτῆς. Λεπτομέρειαι, καθὼς καὶ ἄλλα συναφῆ ζητήματα, θὰ ἐκτεθοῦν εἰς ἐργασίαν, ἡ ὁποία θὰ δημοσιευθῇ ἀλλαχοῦ.

# The Impulsive Force Required to Effectuate a New Orbit Through a Given Point in Space*

by

**Demetrios G. Magiros**[1]

ABSTRACT

A method for treatment of a special two-body problem is discussed in this paper. The problem is: "To calculate the impulse or impulsive force required to effectuate a new Keplerian orbit around a center through a given point $T$ in space." The solution of this problem, according to the present method, is based on the solution of an auxiliary problem, treated here geometrically, on the "projection property" of the admissible impulse, and trigonometric calculations, which make the admissible impulse an appropriate one. Three groups of problems are discussed: one when the point $T$ is on the plane of the original orbit, and two when $T$ is out of this plane. In the case when the $(\dot{\mathbf{r}}_0, \mathbf{I})$-plane is not perpendicular to the line $CM$, use is made of projections on a plane perpendicular to $CM$. The calculation of the impulsive force in terms of the impulse is also given. Illustrated problems for each group complete the paper.

SYMBOLS

| | |
|---|---|
| $\mathbf{r}_0$, $r_0$ | initial position vector, its magnitude |
| $\dot{\mathbf{r}}_0$, $\dot{r}_0$ | initial velocity vector, its magnitude |
| $\mathbf{F}$, $F$ | impulsive force vector, its magnitude |
| $\mathbf{I}$, $I$ | impulse vector, its magnitude |
| $\mathbf{I}_r$, $I_r$ | resultant of $\mathbf{I}$ and $\dot{\mathbf{r}}_0$, its magnitude |
| $t_0$ | time of duration of $\mathbf{I}$ |
| $\mu$ | constant |
| $C$ | center of attraction (first focus) |
| $E$, $E_3$ | second focus |
| $a$, $c$ | semi-major axis, semi- inter-focal distance of ellipse |
| $\tau$, $E$ | period and total energy related to ellipse |
| $\sigma_1$, $\sigma_2$ | arcs of circumference $(C, l)$ |

* As General Electric Company, MSD, Technical Report 63SD256, April, 1963, appeared in slightly different form; also presented and discussed at the XIVth International Astronautical Congress, Paris, France, 1963.

[1] Consultant, General Electric Company, Valley Forge Space Technology Center, Philadelphia, Pa.



Reprinted from the *Journal of the Franklin Institute* **276** (1963), 475–489.

| | |
|---|---|
| $l, l', l'', L$ | distances defined in the text |
| $M, T, T_1, T_2, T_3, P, P_1, P_2, P', P_1', P_2', N$ | points defined in the text |
| $M_1M_2, MM_3$ | lines defined in the text |
| $\varphi, \varphi_1, \varphi_2, \varphi', \varphi_1', \varphi_2', \psi_1, \psi_1', \psi_2$ | angles defined in the text |
| $\vartheta, \vartheta', \gamma, \tau, \tau_1, \tau_2, \tau_3, \tau_4$ | |
| $\sphericalangle$ | angle |
| $\delta$ | Dirac $\delta$-function |

## INTRODUCTION

In this paper a special two-body problem is treated. For the solution of the problem, without reference to general theory, an elementary method is proposed which may be used as a preliminary discussion for practical problems of current interest.

A vehicle of constant mass is in an elliptical orbit around a Newtownian center $C$ and an impulsive force $\mathbf{F}$ is applied at a place $M$ of the vehicle at time $t = 0$, when the position and velocity vectors are $\mathbf{r}_0$ and $\dot{\mathbf{r}}_0$. If the impulsive force is removed after acting for a short time $t_0$, the vehicle is placed in a new Keplerian orbit. The problem is: *To calculate the impulsive force required to effectuate a new orbit around the center through a given point T in space* (1).[2]

In **Part A** of the paper an auxiliary problem is treated geometrically. In **Part B**, the appropriate impulse is calculated by using the auxiliary problem. By the acceptance of the so-called "projection property" of the impulse, the impulse becomes an admissible one when, by trigonometric calculations, we get the appropriate impulse. In **Part C** the impulse force is calculated by using the corresponding impulse. A variety of problems in a single plane and in space is illustrated

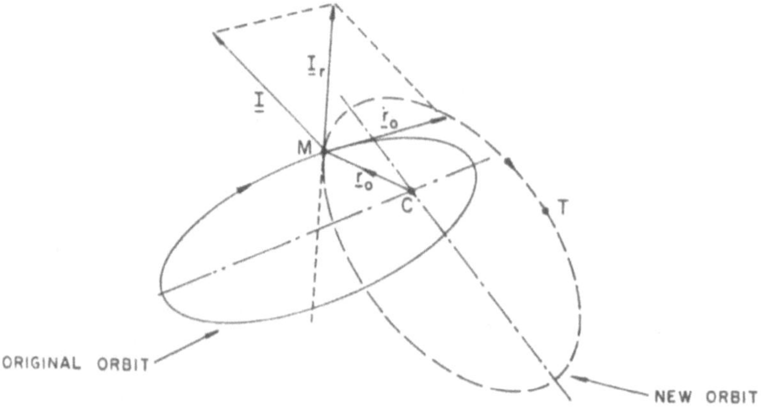


FIG. 1. The original and the new orbits of the vehicle $\mathbf{I}_r$ is the resultant of $\dot{\mathbf{r}}_0$ and $\mathbf{I}$.

[2] The boldface numbers in parentheses refer to the references appended to this paper.

PART A. THE AUXILIARY PROBLEM

The position and velocity vectors, when the force is removed, may be $\mathbf{r}_0$ and $\mathbf{I}_r = \dot{\mathbf{r}}_0 + \mathbf{I}$, respectively. $\mathbf{I}$ is the impulse of the impulsive force $\mathbf{F}$ during the time $t_0$ given by:

$$\mathbf{I} = \int_0^{t_0} \mathbf{F}(t)dt. \tag{A.1}$$

The plane of the new orbit must contain the vector $\mathbf{I}_r$, which must be tangent to the new orbit at the point $M$, Fig. 1. Then it is suggested physically to find the solution of the following auxiliary problem:

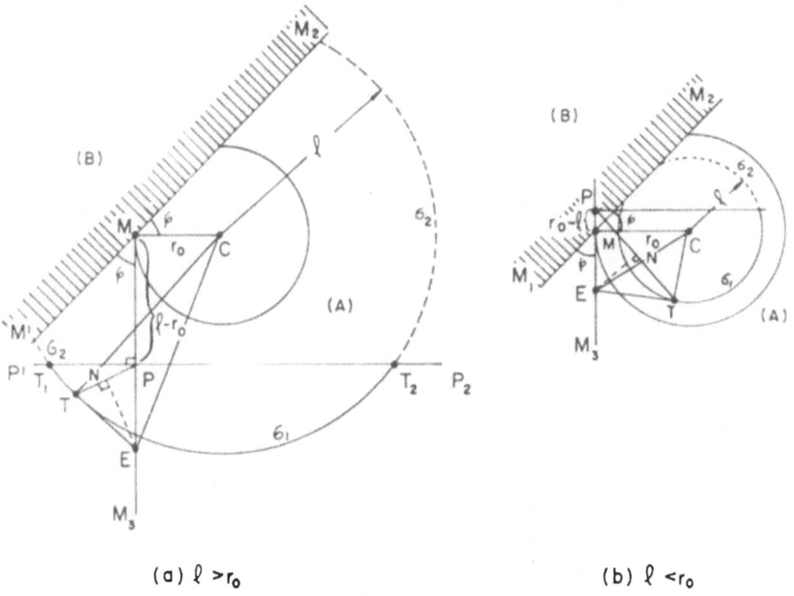


FIG. 2. The construction of the ellipse.

"*Construct an ellipse of which are known the focus C, the points M and T, and the tangent $M_1M_2$ through the point M.*"

The solution to the problem stated in the Introduction can be obtained by the solution of this auxiliary problem.

The ellipse of the auxiliary problem can be constructed by a unique construction of its second focus, when, based on this construction, one can calculate the major axis and the interfocal distance of the ellipse.

1. *Construction of the ellipse.*

For the construction of the ellipse, four conditions, independent of one another, are given. These conditions are such that the following property of an ellipse can be applied: "*The normal to an ellipse at any point of the ellipse bisects the angle included by the focal radii at that point.*" Application of this property in the problem at hand suggests the construction of the line $MM_3$, Fig. 2, such that $\sphericalangle M_1MM_3 = \sphericalangle CMM_3 = \varphi$. The second focus $E$ must be a point of the line $MM_3$ and, by using this line, the focus $E$ can be uniquely constructed as follows. Take first $TC = l$, $MC = r_0$, and draw the circle $(C, l)$ with the center $C$ and radius $l$. For the construction of $E$ two cases can be distinguished.

(a) *The point $T$ is not inside the circle $(C, l)$, $l \geqq r_0$.*

We take a point $P$ on the line $MM_3$, Fig. 2a, such that $MP = l - r_0$, draw the segment $TP$ and construct the perpendicular to its midpoint $N$. The intersection point $E$ with the line $MM_3$ is the second focus of the ellipse, for:

$$TE + TC = TE + l$$

$$ME + MC = MP + PE + r_0 = (l - r_0) + TE + r_0 = TE + l.$$

(b) *The point $T$ is inside the circle $(C, l)$, $l < r_0$.*

We take a point $P$ on the extension of $M_3M$, Fig. 2b, such that $MP = r_0 - l$, draw the segment $TP$ and the perpendicular to its midpoint, when the intersection point $E$ is the second focus, for:

$$TE + TC = TE + l$$

$$ME + MC = PE - PM + r_0 = TE - (r_0 - l) + r_0 = TE + l.$$

2. *Calculation of the major axis and interfocal distance of the ellipse.*

The previous construction of the ellipse gives a procedure for the calculation of the major axis 2a and the interfocal distance 2c of the ellipse by using triangles of the construction. As shown in Fig. 3, we have $2a = ME + MC$, $2c = EC$. The points $M$, $C$, $T$ we consider fixed, then $r_0$, $l$, $\gamma$, $\varphi$ are known. We calculate from the triangle $MCK$ the length $MK$ and the $\sphericalangle MKC$, from triangle $KTP$ the lengths $KP$ and $TP$ and the $\sphericalangle KPT$, and from the triangle $TPE$ the length $PE$, when the calculation of the major axis 2a results. The triangle $CEK$ can be used for the calculation of the interfocal distance $CE$.

3. *Discussion of the construction.*

The tangent $M_1M_2$ divides the $CM_1M_2$-plane into parts (A) and (B), when the ellipse and the second focus must lie on the same part with the given focus $C$, namely on the part (A), Fig. 2a & b.

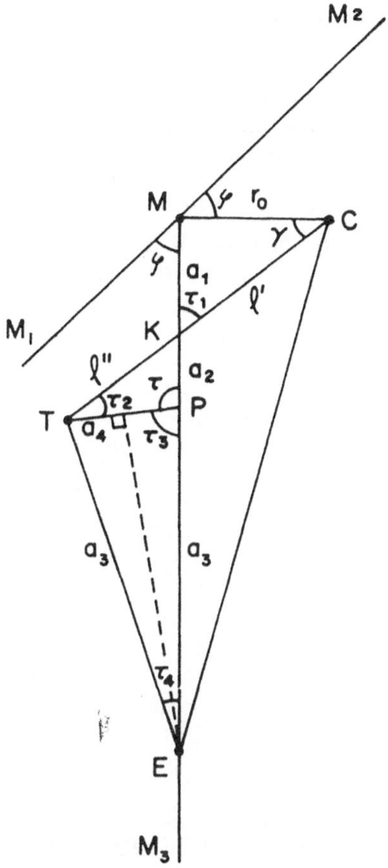


FIG. 3. Triangles for calculation of the major axis $2_a$ and the interfocal distance $2_c$ of the ellipse.

The line $P_1P_2$ perpendicular to $MM_3$ at $P$, which is fixed point for any place of the point $T$ on the fixed circle $(C, l)$, has important properties in connection with the above construction. $P_1P_2$ divides the circumference $(C, l)$ into arcs $\sigma_1$ and $\sigma_2$, solid and dashed, respectively, in Fig. 2a & b, and we have the following cases, if $l \geqq r_0$.

(a) In case $T$ is on $P_1P_2$, that is $T \equiv T_1$ or $T \equiv T_2$, the perpendicular to the midpoint $N$ of the segment $PT$ intersects $MM_3$ at infinity, $PE = \infty$, and the ellipse is a parabola.

(b) In case $T$ is point of the arcs $\sigma_2$, the point $E$ appears above $P$ and in general above $M$ in the part (B), when the construction is not valid, the places of $T$ on the arcs $\sigma_2$ are not acceptable, and we have a domain of non-permissible places of $T$.

(c) In case $T$ is point of the arc $\sigma_1$, the point $E$ is in part (A) of the plane, the distance $PE$ is finite, the construction is valid, the places of $T$ on the arc $\sigma_1$ are acceptable, and we have ellipses.

We can get appropriate remarks if $l < r_0$.

4. *Special important case of the construction,* $\varphi = \pi/2$.

The case $\varphi = \pi/2$, that is when $M$ is a vertex of the ellipse and $CM$ contains the second focus $E$, is of special importance. We examine this case.

(a) *Case* $l \geqq r_0$.

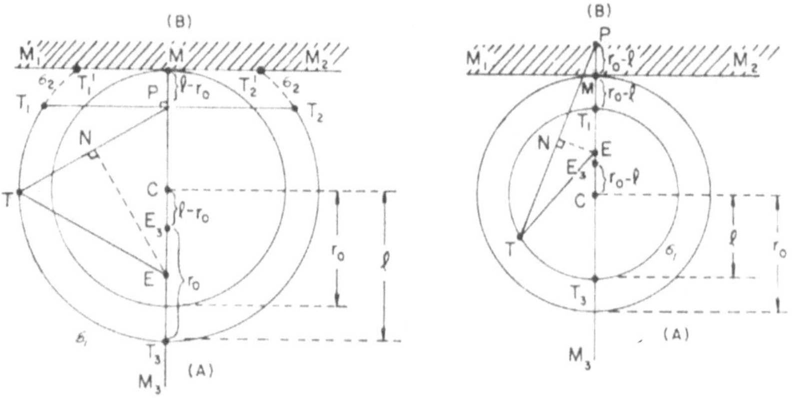


(a) CASE $l \geq r_0$ (b) CASE $l < r_0$

FIG. 4. Construction when $MC$ perpendicular to $M_1M_2$.

The line $T_1PT_2$, Fig. 4a, which gives the arcs $\sigma_1$ and $\sigma_2$ of the circumference $(C, l)$ is parallel to the tangent $M_1M_2$. If $L$ is the distance of $T$ from the tangent $M_1M_2$, we can have:

(i) The condition $L < l - r_0$ corresponds to non-permissible places of $T$.
(ii) The condition $L = l - r_0$ corresponds to parabolas ($T \equiv T_1$ or $T \equiv T_2$).
(iii) The condition $L > l - r_0$ corresponds to ellipses.
(iv) When $T \equiv T_3$, then $E \equiv E_3$, $2a = l + r_0$, $2c = l - r_0$.
(v) The condition $l = r_0$ corresponds to a circle.
(vi) The focus $E$ of any ellipse is not above $E_3$, then for all ellipses the vertex $M$ is pericenter.

(b) *Case* $l < r_0$.

From Fig. 4b we have:

(i) The condition $L = r_0 - l$ corresponds to non-permissible places of $T(T \equiv T_1$ only)
(ii) The condition $L > r_0 - l$ corresponds to ellipses.
(iii) When $T \equiv T_3$, then $E \equiv E_3$, $2a = l + r_0$, $2c = r_0 - l$.
(iv) The focus $E$ of any ellipse is not below $E_3$, then for all ellipses the vertex $M$ is apocenter.

The graph of a and c versus $l$, based on $a = \frac{1}{2}(l + r_0)$, $c = \frac{1}{2}|l - r_0|$, is given in Fig. 5.

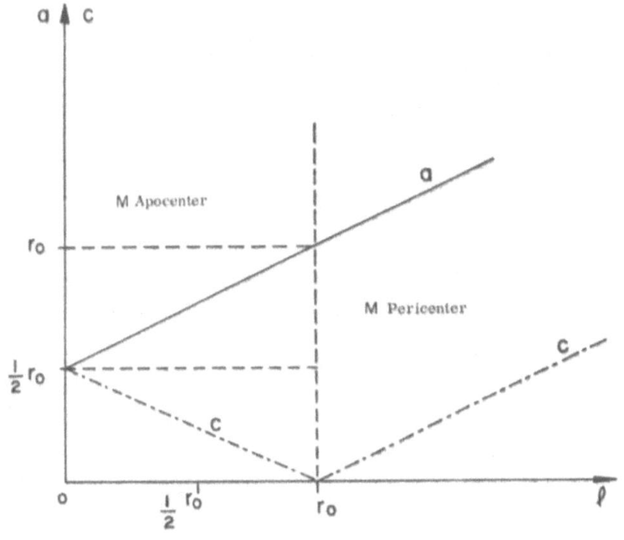


FIG. 5. The "$a$ line" and "$c$ line" versus the distance $l$.

## PART B. THE APPROPRIATE IMPULSE

In the previous section we have discussed the orbit through the point $T$, the new orbit, in its plane. In this section we will discuss the impulse compatible with the new orbit, the appropriate impulse.

The central point $C$, the main point $M$ and the target point $T$ are taken fixed, when $l$, $\mathbf{r}_0$, $\dot{\mathbf{r}}_0$, $\gamma$, $\varphi$, Fig. 6, are known.

The point $T$ may be either on the plane of the original orbit or out of it, and we consider the "single plane problems" and "space problems," separately.

1. *The single plane problems.*

When the target point $T$ is on the plane of the original orbit, the impulse **I**, the angles $\vartheta$ and $\varphi$ and the new orbit must lie on this plane. The angles $\vartheta$, $\varphi_1$, $\varphi$, $\varphi_2$, $\psi_1$, $\psi_2$ are defined as shown in Fig. 6. We have:

$$\varphi = \varphi_1 + \psi_1, \qquad \vartheta = \psi_1 + \psi_2 \tag{B.1}$$

and from triangle $MPP_1$:

$$I_r = \{\dot{r}_0^2 + I^2 + 2\dot{r}_0 I \cos\vartheta\}^{1/2} \tag{B.2}$$

$$\tan\psi_1 = \frac{\sin\vartheta}{\cos\vartheta + (\dot{r}_0/I)}. \tag{B.3}$$

By using the above formulae and the construction of the new orbit according to the previous section, we can solve a variety of problems, the problems of "Group A."

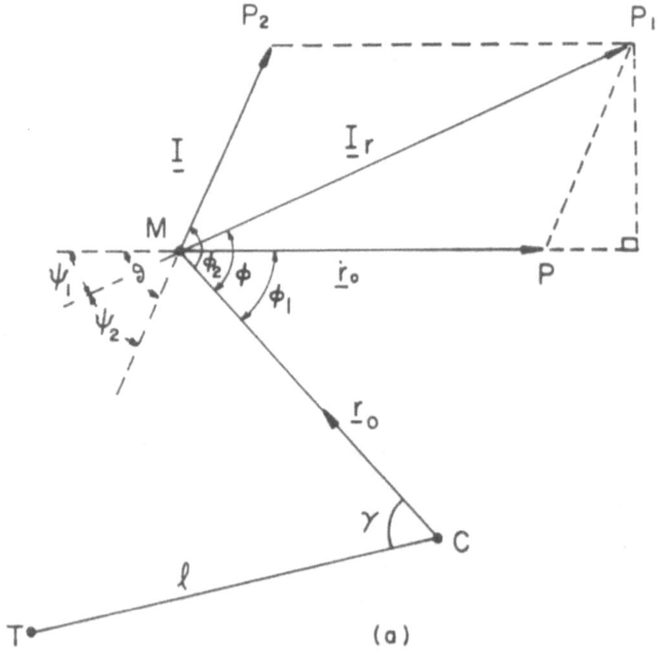


FIG. 6. The resultant $\mathbf{I}_r$ must be on the $(\mathbf{r}_0, T)$-plane.

Some of these problems are illustrated here.

*Problem 1: "Given an impulse-vector, determine the new orbit through the target point, if this point is in a permissible place."*

*Solution:* The problem accepts a unique solution. Since $\dot{\mathbf{r}}_0$, $I$, $\vartheta$ are known, tan $\psi_1$ can be calculated from Eq. B.3, when from the two angles $\psi_1$ we select the appropriate one to the directions of the known vectors $\dot{\mathbf{r}}_0$ and $\mathbf{I}$. The angle $\varphi_1$ is known, the angle $\psi_1$ is now calculated, when from Eq. B.1 we have the angle $\varphi$. The appropriate ellipse can now be constructed according to the instructions of the previous section and the construction is valid—that is, the solution is physically accepted—if the target point is in a permissible place.

*Problem 2: "Given the direction of an impulse-vector and the angle $\psi_1$, determine the magnitude of the appropriate impulse and the orbit through the target point."*

*Solution:* The problem accepts a unique solution. Since $\psi_1$, $\varphi_1$ are known, $\varphi$ can be found, when the orbit through the target point can be determined. Since $\psi_1$, $\vartheta$, $\dot{r}_0$ are known, we calculate the magnitude $I$ of the appropriate impulse $\mathbf{I}$ by using Eq. B.3.

*Problem 3: "Given the magnitude $I$ of an impulse-vector and the angle $\psi_1$, determine the direction of the appropriate impulse-vector and the corresponding orbit through the target point."*

*Solution:* The problem has a non-unique solution. Since $\psi_1$ and $\varphi_1$ are known, $\varphi$ can be found and then the orbit can be constructed. For the direction of the impulse, we calculate sin $\vartheta$ by using Eq. B.3, when two angles $\vartheta$ are accepted, then two directions of the impulse-vector.

*Problem 4: "Discuss the orbit through the target point and related subjects under the following restrictions:*

(i) *The impulse is tangential to the original orbit;*
(ii) *The point $M$ is vertex of the original orbit;*
(iii) *The point $T$ is on the line $MC$.*

*Solution:* This problem is related to the special case of construction, paragraph 4 of the previous section, Fig. 4.

(i) Since the impulse is tangential, the directions of the impulse will be either $\vartheta = 0$ or $\vartheta = \pi$, then, for prescribed magnitude $I$ of the impulse, we have extremes to the magnitude $I_r$ of the resultant vector $\mathbf{I}_r = \dot{\mathbf{r}}_0 + \mathbf{I}$ at $M$, namely max $I_r = \dot{r}_0 + I$ when $\vartheta = 0$, min $I_r = \dot{r}_0 - I$ when $\vartheta = \pi$.

(ii) Since $M$ is vertex of the original orbit and the impulse tangential at $M$, then $M$ must necessarily be vertex of the new orbit.

(iii) Since $T$ is on $MC$, the only permissible place of $T$ is $T \equiv T_3$, Fig. 4, and we have:

$$a = \tfrac{1}{2}(l + r_0), \qquad c = \tfrac{1}{2}|l - r_0|.$$

As $T_3$ moves on $MC$, $l$ is changed and we have the following cases:

As $l = r_0$, the ellipse becomes the circle $(C, r_0)$; as $l > r_0$, $M$ is pericenter; and as $l < r_0$, $M$ is apocenter.

(iv) The family of the ellipses, corresponding to $l > r_0$ has an ellipse with minimum major axis; and the family of the ellipses, corresponding to $l < r_0$, has an ellipse with maximum major axis. For both families, and for these distinct extremal ellipses, the point $T$ is the point $T \equiv T_3$ of line $CM$.

This remark, associated with the formula (2):

$$\tau = 2\pi\sqrt{a^3/\mu} = \tfrac{1}{2}\sqrt{\pi\mu/E^3}$$

where $\tau$ is the periodic time of the ellipse in the Newtonian two-body field, and $E$ the total energy, gives the following result:

"From all ellipses of the family corresponding to $l > r_0$ (or $l < r_0$), the ellipse of the present problem ($T \equiv T_3$) has the minimum (maximum) period, and the maximum (minimum) total energy."

2. *The space problems.*

In case the point $T$ is not on the plane of the original orbit, we have the following planes, Fig. 6.

| | |
|---|---|
| The $(\mathbf{r}_0, \dot{\mathbf{r}}_0)$-plane, the original orbit plane, | "First-plane" |
| The $(\mathbf{r}_0, T)$-plane, the new orbit plane, | "Second-plane" |
| The $(\mathbf{r}_0, \mathbf{I}_r)$-plane, plane corresponding to resultant $\mathbf{I}_r$, | "Third-plane" |
| The $(\mathbf{r}_0, \mathbf{I})$-plane, | "Fourth-plane" |
| The $(\dot{\mathbf{r}}_0, \mathbf{I})$-plane, which contains $\mathbf{I}_r$. | "Fifth-plane" |

All the above planes are identical with the plane of the original orbit if the target point $T$ is in this plane.

(a) The impulse $\mathbf{I}$ must be such that the resultant $\mathbf{I}_r$ necessarily lies on the new orbit plane, that is the above "second" and "third" planes must coincide. In the general case when the above "fifth-plane" is not perpendicular to $CM$, by using a plane per-

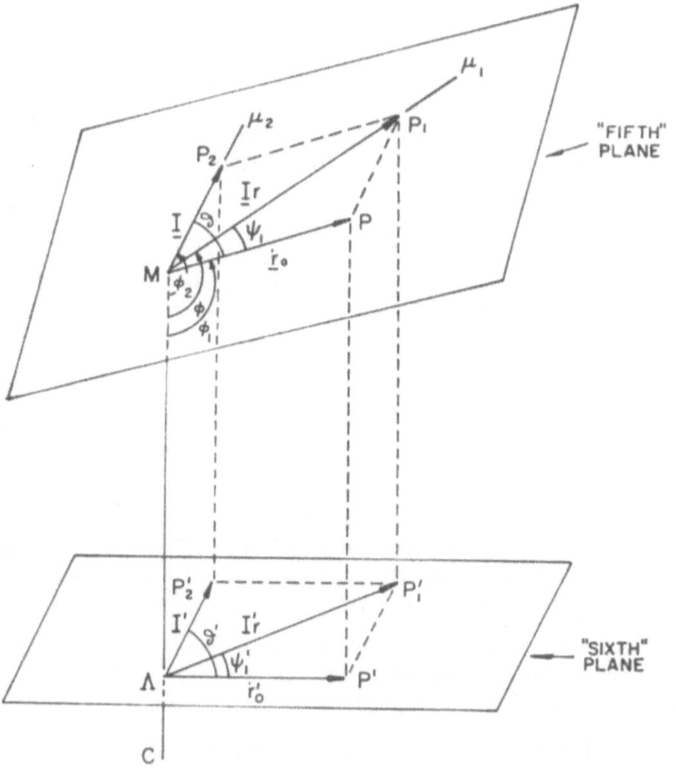


FIG. 7. The "fifth"-plane nonperpendicular to $MC$ contains $MPP_1P_2$. The "sixth"-plane perpendicular to $MC$ contains the projection of $MPP_1P_2$.

pendicular to $CM$, the "sixth-plane," the above restriction of the impulse can be stated as follows:

"*Projection property of the impulse:*" "*The orthogonal projection of the angle* ($\dot{\mathbf{r}}_0$, $\mathbf{I}_r$) *on the sixth plane must be equal to dihedral angle of the planes of the original and the new orbits.*"

The impulse which obeys this "projection property" is a "permissible" impulse and has the "possibility" of making the impulse required. Trigonometric calculations can make the permissible impulse an appropriate one.
If we define the angles $\varphi_1$, $\varphi$, $\varphi_2$, $\psi_1$, $\vartheta$ as shown in Fig. 7, and the angles $\varphi_1'$, $\varphi'$, $\varphi_2'$ as the inclination angles of $\dot{\mathbf{r}}_0$, $\mathbf{I}_r$, $\mathbf{I}$, respectively,

to the "sixth-plane," we can have:

$$\varphi_1 = \frac{\pi}{2} + \varphi_1', \qquad \varphi = \frac{\pi}{2} + \varphi', \qquad \varphi_2 = \frac{\pi}{2} + \varphi_2'. \quad \text{(B.4)}$$

The orthogonal projection of the parallelogram $MPP_1P_2$ of the "fifth-plane" on the "sixth-plane" is a new parallelogram $\Lambda P'P_1'P_2'$, for which we can obtain:

$$\Lambda P' = \dot{r}_0 \cos \varphi_1', \qquad \Lambda P_1' = I_r \cos \varphi',$$
$$\Lambda P_2' = I \cos \varphi_2', \qquad \vartheta' = \text{proj.}\, \vartheta, \qquad \psi_1' = \text{proj.}\, \psi_1 \quad \text{(B.5)}$$

$$I_r' = \{(\dot{r}_0 \cos \varphi_1')^2 + (I \cos \varphi_2')^2 + 2\dot{r}_0 I \cos \varphi_1' \cos \varphi_2' \cos \vartheta'\}^{1/2} \quad \text{(B.6)}$$

$$\tan \psi_1' = \frac{\sin \vartheta'}{\cos \vartheta' + \dfrac{\dot{r}_0}{I} \dfrac{\cos \varphi_1'}{\cos \varphi_2'}}. \quad \text{(B.7)}$$

The angle $\psi_1'$ must be, according to the above restriction, equal to the dihedral angle of the original and new orbital planes.

A variety of problems, the problems of "*Group B*" can be solved by using the above formulae and the construction of the ellipse of the previous section. We give just two:

*Problem 1: "Given the line of the action of an impulse, find the appropriate magnitude of the impulse required, and the new orbit through the target point, if it is in a permissible place."*

*Solution:* The line of action of the impulse and that of $\dot{\mathbf{r}}_0$ determine the "fifth-plane" where the angle $\vartheta$ is given, Fig. 7. The "second-plane" which contains the target point, intersects the "fifth-plane" into the line $M\mu_1$, which must necessarily be the action line of the resultant $\mathbf{I}_r$ and tangent to the new orbit. The "first," "second," and "fifth" planes are known in this problem and from $MP$, $\psi_1$, $\vartheta$ we can construct the parallelogram $MPP_1P_2$. The construction leads to calculation of the magnitudes $MP_1 = I_r$, $MP_2 = I$. For the construction now of the new orbit we need to know the angle $\varphi$, and, since $\varphi = \frac{\pi}{2} + \varphi'$, we calculate $\varphi'$. For the calculation of $\varphi'$, we use the orthogonal projection of the parallelogram $MPP_1P_2$ on the "sixth-plane," namely the parallelogram $\Lambda P'P_1'P_2'$. In the formula Eq. B.7 we know $\varphi_1'$, $\varphi_2'$, $\psi_1'$, $\dot{r}_0$, $I$, then we calculate $\cos \vartheta'$, when from formula Eq. B.6 we calculate $I_r'$. The angle $\varphi'$ needed can be determined from $\cos \varphi' = I_r'/I_r$.

*Problem 2: "Given the magnitude I of the impulse and the line of action of the resultant* $\mathbf{I}_r$*, find the appropriate line of the action of the impulse re-*

*quired, and the new orbit through the target point, if it is in a permissible place.*"

*Solution:* The line $M\mu_1$ of action of the unknown resultant $\mathbf{I}_r$ must be on the "second-plane," and the "fifth-plane" is the $(MP, M\mu_1)$-plane. The angle $\psi_1$ is known. We construct the parallelogram $MPP_1P_2$ on the "fifth-plane." The circle $(P, I)$ in the "fifth-plane" intersects $M\mu_1$ at $P_1$, and the line of action of the impulse is the line $M\mu_2$ parallel to $PP_1$. Again, for the construction of the new orbit, we calculate $\varphi'$.

(b) Investigating the properties of the point $M$ as a vertex of the original and/or the new orbit in connection with an impulse perpendicular or oblique to $CM$, another group of problems, the "Group C," may arise.
Properties of the inclination angles $\varphi_1'$, $\varphi'$, $\varphi_2'$ of the coplanar vectors $\dot{\mathbf{r}}_0$, $\mathbf{I}_r$, $\mathbf{I}$ on the "sixth-plane" are used to distinguish this group of problems. We use the following statements of solid geometry:

(i) "A line is, by definition, perpendicular to a plane when it is perpendicular to every line on the plane which meets it;"
(ii) "If a line is perpendicular to each of two intersecting lines at their point of intersection, it is perpendicular to the plane determined by the two lines;"
(iii) "A line oblique to a plane is perpendicular to only one line in the plane through the point of intersection;"

We can have the following properties of $\varphi_1'$, $\varphi'$, $\varphi_2'$, based on the above statements of solid geometry:

(i) "These angles may be altogether either zero or non-zero;"
(ii) "If one of them is zero, the other two must be either both zero or both non-zero;"
(iii) "If one of them is non-zero, the other two must be either both non-zero or one zero and the other non-zero, but never both zero;"
(iv) "If two of them are zero, the third one must be zero; if two of them are non-zero, the third one may be either zero or non-zero."

By using these properties of $\varphi_1'$, $\varphi'$, $\varphi_2'$, and that the meaning of $\varphi_1' = 0$ (or $\varphi' = 0$) is that the point $M$ is a vertex of the original (or new) orbit, and $\varphi_2' = 0$ is that the impulse is perpendicular to $CM$, we may have some of the problems of "Group C:"

*Problem 1:* "*If the impulse is perpendicular to CM, then:*

(i) *M being a vertex of the original (new) orbit implies that M is a vertex of the new (original) orbit;*
(ii) *M being a non-vertex of the original (new) orbit implies that M is non-vertex of the new (original) orbit;*"

*Problem 2:* "*If M is a vertex of the original (new) orbit, then:*

(i) *M being a vertex of the new (original) orbit, implies that the impulse is perpendicular to CM;*
(ii) *M being a non-vertex of the new (original) orbit implies that the impulse is oblique to CM.*"

*Problem 3:* "*If the impulse is oblique to CM and M non-vertex of the original (new) orbit, then M may or may not be a vertex of the new (original) orbit.*"

*Problem 4:* "*If the impulse is perpendicular to CM and M a vertex of the original (new) orbit, then M is a vertex of the new (original) orbit.*"

*Problem 5:* "*If M is non-vertex of the original and the new orbits, then the impulse may be either perpendicular or oblique to CM.*"

### PART C. THE APPROPRIATE IMPULSIVE FORCE

The determination of the appropriate impulsive force, based on the integral definition Eq. A.1 of the impulse, in terms of the impulse, is a problem of special singular integral equations, subject to modern mathematical theories.

The impulsive forces are, by nature, not forces of ordinary type. Their convenient idealization is the improper "Dirac $\delta$-function," of which the impulsive forces represent a simple physical interpretation. We can approximate the impulsive force by appropriate assumptions. If for example, we assume that it is constant during its short duration $t_0$, we can get from the formula Eq. A.1 that: $F = I/t_0$.

This formula gives an approximate transition from the impulse required to the corresponding impulsive force, valid approximately for any direction of the impulse, which direction is the same with that of the impulsive force.

* * * *

The author dedicates this paper to B. H. Caldwell and W. R. Becraft of General Electric Company, Missile and Space Division, for their encouragement and help in connection with this research work.

**REFERENCES**

(1) Transfer between elliptical orbits has been investigated by D. Lawden, P. Longe, G. Smith, R. Plimmer, J. Horner, etc. We may have another procedure for the solution of the present problem, without reference to the general theory, by using the D. Lawden article: "Impulse Transfer Between Elliptical Orbits," which is Chapter 11 of G. Leitmann's book: "Optimization Techniques with Applications to Aerospace Systems,', New York, Academic Press, 1962.

(2) J. Synge and B. Griffith, "Principles of Mechanics," 3rd Ed., New York, McGraw-Hill, 1959, Page 165.

# *Motion in a Newtonian Forced Field Modified by a General Force*

*by* DEMETRIOS G. MAGIROS

*General Electric Company*

*Valley Forge Space Technology Center, Philadelphia, Pa.*

ABSTRACT: *The motion of a man-made celestial body under the influence of an attractive center according to the inverse square Newton's law and a general force is discussed. A solution of special form, referred to a specific inertial coordinate system, is established. For the time-dependent coefficients of the solution MacLaurin expansions are taken. It is proven that the formal solution constructed satisfies all the requirements needed for really being a solution of the physical problem; besides its region of convergence is found. The problem is specialized in the case of a constant force. The case of suddenly or gradually applied forces is also examined.*

## *1. Introduction*

A variety of current problems, especially from the point of view of engineering, are related to the following problem:

*"A man-made celestial body is moving under the influence of a 'centripetal force' obeying the inverse square Newton's law toward an attractive center, when an impulsive force is applied, it acts for an interval of time, then is removed. Find the motion of the body during the action of the force."*

A mathematical discussion of this particular two-body problem is given in t reference paper (1). A solution of special form, referred to a specific inert coordinate system, is established. The time-dependent coefficients of this solution are taken in MacLaurin expansions. The solution is not only a formal one, but is actually accepted as a solution of the above physical problem since it satisfies all the appropriate requirements, besides the solution is valid for any interval of time, however large. In discussing the convergence of the series, by using the Leibnitz formula appropriately applied to this problem as shown in the Appendix, a specific form for their general term is given when the root test is successfully applied. This is discussed in Sections 3 and 4.

As an application, the problem is specialized in the case of a constant force, when the solution is simplified and taken in an approximation of the error found. A direct physical application of this is given in Section 5.[1]

The solution contains the impulse of the force for small time $t_o$, and this

[1] From the author's discussions with John Immel, Guidance and Advance Requirements Operation, G. E., Valley Forge Space Tech. Center.



Reprinted from the *Journal of the Franklin Institute* 278 (1964), 407–416.

*Demetrios G. Magiros*

impulse differs in the case when the force is either suddenly or gradually applied to the body. Further remarks on this subject are given in Section 6.

## 2. *Mathematical Formulation of the Problem*

The differential equation in vector form:

$$\ddot{\mathbf{r}}(t) = -\frac{\mu}{r^3(t)}\mathbf{r}(t) + \mathbf{T}(t), \qquad 0 \leq t \leq t' \tag{1}$$

governs the motion of the body during the action of the force. $\mathbf{T}$ is the force, $r$ the magnitude of the position vector $\mathbf{r}$, the constant $\mu = K(M + m)$, $M$ and $m$ the masses of the attractive center and the body, respectively, $K$ the constant of gravity.

The time $t'$ of the action of the force is split according to

$$t' = t_o + \tau', \tag{2}$$

where $t_o$ is very small time.

Initial conditions attached to Eq. 1 are:

$$\mathbf{r}(t_o) = \mathbf{r}_o, \qquad \dot{\mathbf{r}}(t_o) = \dot{\mathbf{r}}_o + \mathbf{I}_o, \tag{3}$$

where $\mathbf{r}_o$ and $\dot{\mathbf{r}}_o$ are the position vector and velocity vector of the body at time $t = 0$; $\mathbf{I}_o$, the impulse of the force $\mathbf{T}$ during the time $t_o$, is defined by the formula

$$\mathbf{I}_o = \int_o^{t_o} \mathbf{T}(t)dt. \tag{4}$$

Small terms of the order of the increments $\delta r$ and $\delta\dot{r}$ of the vectors $\mathbf{r}$ and $\dot{\mathbf{r}}$, due to the action of an attractive center during the small time $t_o$, are omitted in Eq. 3.

The introduction of Eqs. 2 and 3 leads to a solution which is helpful in practical problems, as we can easily see.

By changing the time scale to

$$\tau = t - t_o \tag{5}$$

the system Eqs. 1 and 3 become

$$\ddot{\mathbf{r}}(\tau) = -\frac{\mu}{r^3(\tau)}\mathbf{r}(\tau) + \mathbf{T}(\tau), \qquad 0 \leq \tau \leq \tau' \tag{6}$$

$$\mathbf{r}(0) = \mathbf{r}_o, \qquad \dot{\mathbf{r}}(0) = \dot{\mathbf{r}}_o + \mathbf{I}_o. \tag{7}$$

Equation 6 is nonlinear because of the appearance of $r^3$. The time $\tau'$ of the action of the force is not necessarily considered small, as we cannot take the "average value" of $r$ during the action of the force and work with linearized equations.

Every function $\mathbf{r}(\tau)$ which satisfies Eq. 6 must be twice differentiable, when this function and its first derivative $\dot{\mathbf{r}}(\tau)$ must necessarily be continuous functions and bounded for any $\tau$ in the interval $D$: $0 \leq \tau \leq \tau'$. Thus, if the constants $M_1$ and $M_2$ are the least upper bounds of $\mathbf{r}(\tau)$ and $\dot{\mathbf{r}}(\tau)$, respectively, then any solution $\mathbf{r}(\tau)$ of Eq. 6 as well as its derivative $\dot{\mathbf{r}}(\tau)$ are continuous in the domain:

$$D_1: \ |\mathbf{r}(\tau)| < M_1, \qquad |\dot{\mathbf{r}}(\tau)| < M_2$$

for all $\tau$ in $D$.

### 3. *A Solution of the Problem*

Let us select a reference coordinate system $(P; \mathbf{r}_o^*, \mathbf{s}_o^*, \mathbf{T}_o^*)$ defined as follows: $P$ is the position of the body when the force starts acting; $\mathbf{r}_o^*$, $\mathbf{s}_o^*$, $\mathbf{T}_o^*$ are unit vectors of $\mathbf{r}_o$, $\dot{\mathbf{r}}_o + \mathbf{I}_o$, $\mathbf{I}_o$, respectively, see Fig. 1. We establish in the following a solution of the system Eqs. 6 and 7 valid in the domain $D_1$ for any $\tau$ in $D$, of the form:

$$\mathbf{r}(\tau) = a_1(\tau)\mathbf{r}_0^* + a_2(\tau)\mathbf{s}_o^* + a_3(\tau)\mathbf{T}_o^* \tag{8}$$

by an appropriate calculation of the scalar functions $a_i(\tau)$, $i = 1, 2, 3$. This function $\mathbf{r}(\tau)$, as solution of Eq. 6 must be differentiable twice, and we take this function differentiable as many times as needed. Then, the functions $a_i(\tau)$ of Eq. 8 are differentiable, continuous and bounded in the domain $D_1$, for any $\tau$ in $D$.

We can now insert the function $\mathbf{r}(\tau)$ of Eq. 8 into Eq. 6 when, if $T_1(\tau)$, $T_2(\tau)$, $T_3(\tau)$ are the projections of the force $\mathbf{T}(\tau)$ on the axis of the coordinate system selected, Eqs. 6 and 7 yield

$$\left.\begin{array}{llll}
\text{(i)} & \ddot{a}_1(\tau) + \dfrac{\mu}{r^3(\tau)} a_1(\tau) = T_1(\tau), & a_1(0) = r_o, & \dot{a}_1(0) = 0 \\[2ex]
\text{(ii)} & \ddot{a}_2(\tau) + \dfrac{\mu}{r^3(\tau)} a_2(\tau) = T_2(\tau), & a_2(0) = 0, & \dot{a}_2(0) = s_o \\[2ex]
\text{(iii)} & \ddot{a}_3(\tau) + \dfrac{\mu}{r^3(\tau)} a_3(\tau) = T_3(\tau), & a_3(0) = 0, & \dot{a}_3(0) = 0
\end{array}\right\}. \tag{9}$$

By using the conditions Eqs. 9, satisfied by the functions $a_i(\tau)$, we can uniquely determine $a_i(\tau)$. To do that, let us take Eq. 9(i).

By changing the notation according to

$$a_1 = x_1, \qquad \dot{a}_1 = x_2, \tag{10}$$

Eq. 9(i) can be replaced by the equivalent normal system

$$\left.\begin{array}{l}
\dot{x}_1(\tau) = x_2(\tau) \\[1ex]
\dot{x}_2(\tau) = -\dfrac{\mu}{r^3(\tau)} x_1(\tau) + T_1(\tau)
\end{array}\right\} \tag{11}$$

valid in the domain $D \times D_1$: $0 \leq \tau \leq \tau'$, $|x_1| < M_1$, $|x_2| < M_2$ with initial conditions $x_{10} = r_0$, $x_{20} = 0$. The right-hand members of Eq. 11:

$$f_1 \equiv x_2, \qquad f_2 \equiv -\frac{\mu}{r^3} x_1 + T_1, \tag{12}$$

*Demetrios G. Magiros*

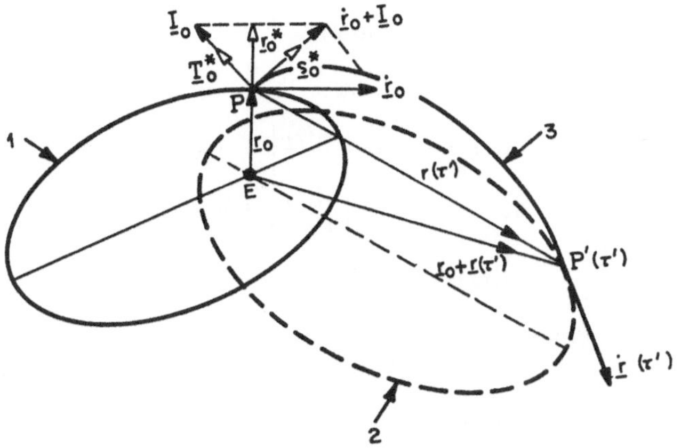


FIG. 1.
1. Original orbit; 2. New orbit; 3. Orbit during the action of the thrust.

if $T_1$ is taken as a differentiable function of $\tau$ in $D$, and the value $r = 0$ is excluded, are single-valued, continuous and bounded functions of all their arguments simultaneously, and satisfy a Lipschitz condition on $x_1$, $x_2$ in $D$ for any value $\tau$ in $D$. Then, (2) solution of Eq. 11 will exist and be uniquely defined by given values $x_{10}$, $x_{20}$ of $x_1$, $x_2$ in $D_1$, for any value $\tau_o$ of $\tau$ in $D$, say $x_{10} = r_o$, $x_{20} = 0$ for $\tau_0 = 0$. Therefore, Eq. 9i has only one solution $a_1(\tau)$.

By the same reasoning and using the other equations of (9), the functions $a_2(\tau)$, $a_3(\tau)$ can be uniquely defined.

For a prescribed force $\mathbf{T}(\tau)$, its components $T_i(\tau)$, $i = 1, 2, 3$ are known, but as $r(\tau)$ is unknown, it looks impossible to calculate $a_i(\tau)$ from Eq. 9. In spite of that, approximations of $a_i(\tau)$ can be found by considering their MacLaurin expansions at $\tau = 0$.

Due to the properties of $a_i(\tau)$ pointed out above, MacLaurin expansions of $a_i(\tau)$ at $\tau = 0$ convergent for any value $\tau$ in $D_1$, can be assumed:

$$\left.\begin{aligned} a_1(\tau) &= a_1(0) + \dot{a}_1(0)\tau + \frac{\ddot{a}_1(0)}{2}\tau^2 + \cdots + \frac{a_1^{(n)}(0)}{n!}\tau^n + \cdots \\ a_2(\tau) &= a_2(0) + \dot{a}_2(0)\tau + \frac{\ddot{a}_2(0)}{2}\tau^2 + \cdots + \frac{a_2^{(n)}(0)}{n!}\tau^n + \cdots \\ a_3(\tau) &= a_3(0) + \dot{a}_3(0)\tau + \frac{\ddot{a}_3(0)}{2}\tau^2 + \cdots + \frac{a_3^{(n)}(0)}{n!}\tau^n + \cdots \end{aligned}\right\} \quad (13)$$

We may calculate as many coefficients of these series as we want by successive differentiation of Eq. 9 and using Eqs. 6 and 7.

The first two coefficients are given in Eq. 9. For the third coefficients, by using Eq. 9, we get

$$\ddot{a}_1(0) = -\frac{\mu}{r_o^3} + T_1(0), \qquad \ddot{a}_2(0) = T_2(0), \qquad \ddot{a}_3(0) = T_3(0). \tag{14}$$

For the fourth coefficients, we differentiate Eq. 9, when

$$\left.\begin{aligned} \dddot{a}_1(\tau) &= -\mu \frac{d}{d\tau}\left\{\frac{a_1(\tau)}{r^3(\tau)}\right\} + \dot{T}_1(\tau) \\ \dddot{a}_2(\tau) &= -\mu \frac{d}{d\tau}\left\{\frac{a_2(\tau)}{r^3(\tau)}\right\} + \dot{T}_2(\tau) \\ \dddot{a}_3(\tau) &= -\mu \frac{d}{d\tau}\left\{\frac{a_3(\tau)}{r^3(\tau)}\right\} + \dot{T}_3(\tau) \end{aligned}\right\} \tag{15}$$

and take their values at $\tau = 0$. Applying the procedure of successive differentiation, we can get as many coefficients of the series 13 as we want when the series become known series, then the coefficients $a_i(\tau)$ of the function $\mathbf{r}(\tau)$ given by Eq. 8 are determined, and the solution $\mathbf{r}(\tau)$ of Eqs. 6 and 7 is formally constructed.

### *4. Further Discussion of the Solution*

The solution constructed in the preceding section is a formal solution, which may or may not be an actual solution of the physical problem stated in the introduction.

The "existence" of the solution of the form Eq. 8, its "uniqueness" and "continuous dependence" with the data, which is a kind of stability of the solution, properties sometimes called "Hadamard's postulates," (**3**) decide about the acceptance of the solution physically.

Solution of Eq. 8 should satisfy all the above requirements, if the functions $a_i(\tau)$ exist uniquely and depend continuously on the data.

We prove that the functions $a_i(\tau)$ have these properties.

First, about the function $a_1(\tau)$. The functions $f_1$ and $f_2$ of Eq. 12 are differentiable in $D_1$, for all $\tau$ in $D$, the partial derivatives $\frac{\partial}{\partial x_j} f_i(x_1, x_2, \tau)$ of the functions $f_1$ and $f_2$ with respect to $x_1$ and $x_2$ exist and are continuous in $D \times D_1$ Then (**2**) the partial derivatives $\frac{\partial}{\partial x_{j0}} x_i(x_{10}, x_{20}, \tau)$ of the functions $x_1$ and $x$ with respect to the initial conditions $x_{10}$ and $x_{20}$ exist and are continuous functions of $x_{10}, x_{20}, \tau$, when $x_1$ and $x_2$ themselves depend continuously on the initial conditions given in $D \times D_1$.

This proves that the function $a_1(\tau)$, as a solution of Eq. 9(i), exists, i unique and depends continually on the initial conditions.

In the same way we can prove also that the functions $a_2(\tau)$ and $a_3(\tau)$ have the above properties.

Therefore, as a result, the function $r(\tau)$ of Eq. 8 satisfies the requirements needed for being an actual solution of the series 13.

It remains to determine the domain of the validity of the solution found and for that we can use the correspondent radius of convergence.

*Demetrios G. Magiros*

The radius of convergence of the series 13 is the reciprocal of the upper limit (4):

$$\overline{\lim_{n\to\infty}} [|a_i^{(n)}(0)|/n!]^{1/n}, \qquad i = 1, 2, 3. \tag{16}$$

To determine this limit, we first find the general term of the sequence

$$\{a_i^{(n)}(\tau)/n!\}, \qquad i = 1, 2, 3. \tag{17}$$

By using Eqs. 9, the $n^{\text{th}}$ derivative of $a_i(\tau)$ for any $n \geq 2$, is given by

$$a_i^{(n)}(\tau) = -\mu \left[\frac{a_i(\tau)}{r^3(\tau)}\right]^{(n-2)} + T_i^{(n-2)}(\tau), \qquad i = 1, 2, 3. \tag{18}$$

The general term of the sequence 17, if we take into account in Eq. 18 formula $V$ of the Appendix, can be given by

$$\frac{a_i^{(n)}(\tau)}{n!} = -\sum_{m=0}^{n-2} \frac{a_i^{(n)}(\tau)[V_1 r^{n-m-3} + V_2 r^{n-m-4} + \cdots + V_{n-m-2}]}{(n-1)nm!(n-m-2)!r^{n-m+1}} + \frac{T_i^{(n-2)}(\tau)}{n!}, \tag{19}$$

where $V_1$, $V_2$, $\cdots$ are polynomials in the derivatives of $r$ up to the order $(n-m-2)$ with coefficients smaller than $(n-m)!$

Due to the convergence of the series 13, the ratio $|a_i^{(n)}(0)|/n!$ as $n \to \infty$ must tend to zero, and its $n^{\text{th}}$ root to a limit less than one.

To find the limit of the $n^{\text{th}}$ root of the above ratio, that is, to find the limit of the $n^{\text{th}}$ root of the absolute value of the right-hand member of Eq. 19 for $\tau = 0$ as $n \to \infty$, consider each term of it as positive, and then take the $n^{\text{th}}$ root of each such term. Their sum $S_n$ must satisfy the relation

$$[|a_i^{(n)}(0)|/n!]^{1/n} \leq S_n. \tag{20}$$

Since $r_o \neq 0$ and all derivatives of $r$ and $T_i$ are bounded at $\tau = 0$, the sum $S_n$ tends to zero as $n \to \infty$, then the limit of the left-hand member of Eq. 20 is zero as $n \to \infty$, and, as a result, the radius of convergence is infinite, and the series 13 converge for any value of $\tau'$.

In case $r_o$ is large compared to other quantities appearing in the coefficients of the series 13, and the time $\tau'$ of the action of the force is not large, the series 13 converge rapidly, then we can get in this case a good approximation of the coefficients of the solution Eq. 8, if we confine ourselves to a few terms of the series 13.

### *5. The Case of a Constant Force*

In the previous discussion a solution of a specific form of the problem at hand is established. In the following an application is given.

Consider the case of a force constant in magnitude and direction with respect to the inertial coordinate system taken. In this case $T_1$, $T_2$, $T_3$ of

Eq. 9 are constants and the calculation of the coefficients of the series 13 is simplified. More specifically, take the force identical with the vector $\mathbf{T}_o^*$. In this specific case, $T_1 = T_2 = 0$, $T_3 = 1$, and the coefficients of the series 13 are

$$\left.\begin{array}{llll} a_1(0) = r_o, & \dot{a}_1(0) = 0, & \ddot{a}_1(0) = -\dfrac{\mu}{r_o^2}, & \dddot{a}_1(0) = \dfrac{3\mu s_o}{r_o^3}, \cdots \\ a_2(0) = 0, & \dot{a}_2(0) = s_o, & \ddot{a}_2(0) = 0 & \dddot{a}_2(0) = -\dfrac{\mu s_o}{r_o^3}, \cdots \\ a_3(0) = 0, & \dot{a}_3(0) = 0, & \ddot{a}_3(0) = 1 & \dddot{a}_3(0) = 0, \cdots \end{array}\right\}. \quad (21)$$

For large $r_o$ but small $\tau$, we can get a good approximation for $a_1$, $a_2$, $a_3$ if we take a few terms of the corresponding series 13; we can write

$$\begin{aligned} a_1(\tau) &= r_o - \frac{\mu}{2r_o^2}\tau^2 + 0(\tau^3/r_o^4) \\ a_2(\tau) &= s_o\tau + 0(\tau^3/r_o^4) \\ a_3(\tau) &= \tfrac{1}{2}\tau^2 + 0(\tau^4/r_o^5). \end{aligned} \quad (22)$$

Therefore, by using the original time scale, the solution Eq. 8 and its derivative become

$$\left.\begin{aligned} \mathbf{r}(t) &= \left(r_o - \frac{\mu}{2r_o^2}(t-t_o)^2\right)\mathbf{r}_o^* + (t-t_o)s_o\mathbf{s}_o^* + \tfrac{1}{2}(t-t_o)^2\mathbf{T}_o^* \\ \dot{\mathbf{r}}(t) &= -\frac{\mu}{r_o^2}(t-t_o)\mathbf{r}_o^* + s_o\mathbf{s}_o^* + (t-t_o)\mathbf{T}_o^* \end{aligned}\right\}. \quad (23)$$

Equations 23 give the position vector and the velocity vector of the body in this special case during the action of the force.

When the force is removed at time $t'$, say, the values $\mathbf{r}_o + \mathbf{r}(t')$ and $\dot{\mathbf{r}}(t')$ calculated from Eq. 23, can be used for the determination of the elements of the new Keplerian orbit of the body (5) which has the same focus, the attractive center $E$, Fig. 1, as the original Keplerian orbit.

### *A Physical Application*

A direct physical application of the above special case can be given.

Consider a projectile that is moving around the earth in a Keplerian orbit and it is desired to change the initial Keplerian orbit into a new one that intersects the surface of the earth for the purpose of recovering the projectile. The thrust required for the desired change in orbit is produced by a rocket attached to the projectile, and the direction of the thrust vector is maintained in a specified initial direction by rotating the projectile at a high angular velocity $\omega$ about the thrust axis. The gyroscopic action of the rotating projectile resists changes in the direction of the thrust vector; thus, if the rocket provides a constant magnitude of the thrust after time $t_o$, and if no changes occur in the direction of the thrust vector, it can be seen that the conditions $T_1 = T_2 = 0$, $T_3 = 1$ will be realized. Also, because the projectile is rotating prior to ignition of the rocket, $\mathbf{I}_o$ is in the same direction as $\mathbf{T}_o^*$. To further illustrate the physical application of the solution consider a typical situation encountered in practice. In many cases the orbiting object is composed of two parts, one being especially adapted for injection into orbit and attitude control while in orbit, and the other for re-entry into the earth's atmosphere. Figure 2 illustrates the above situation. It should be pointed out that the thrust direction shown is typical, chosen to provide the most favorable re-entry conditions for a given rocket impulse capacity. Also, the re-entering portion of the projectile is designed to be aerodynamically stable so that atmospheric forces align the nose of the projectile with the velocity vector as re-entry occurs. Figure 2 shows (a)

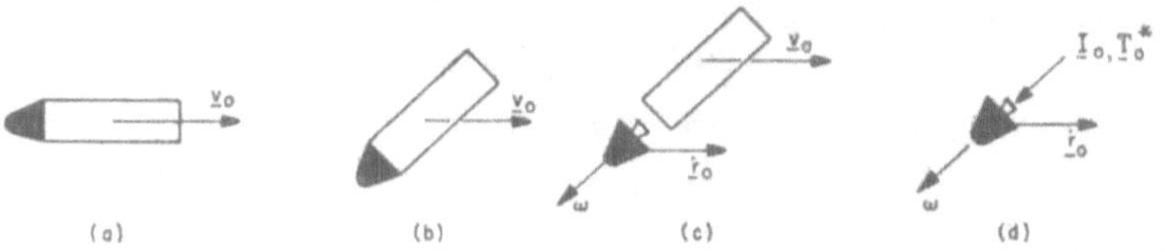


FIG. 2.

projectile in orbit with re-entry portion shaded; (b) projectile in attitude for thrust application; (c) re-entry portion separated and rotating at angular velocity $\omega$; the unshaded portion remains moving around the earth in Keplerian orbit; and (d) enlarged view of the re-entry portion during thrust application.

### 6. *The Impulse of the Force in Two Cases*

The solution found is valid in $t_o \leq t \leq t'$ and contains the impulse $\mathbf{I}_o$, which is given by the formula Eq. 4, it then depends on the function $\mathbf{T}(t)$ in $0 \leq t \leq t_o$. We have two cases.

(a) For a force $\mathbf{T}(t)$ constant in magnitude and direction during time $t_o$ with the same direction as $\mathbf{T}_o^*$, $\mathbf{T} = T_o \mathbf{T}_o^*$, the impulse is

$$\mathbf{I}_o' = t_o T_o \mathbf{T}_o^* \tag{24}$$

and in this case we can speak about a "sudden application" of force. The broken line $00_1 0_2 0_3$ in the $t$, $T$-plane, Fig. 3, corresponds to this situation.

(b) Let us take in this case as the direction of the force in $0 \leq t \leq t_o$ the direction of $\mathbf{T}_o^*$ and its magnitude varied according to a parabolic law during the time $0 \leq t \leq t_o$, being constant at the remainder time, i.e.,

$$\mathbf{T}(t) = T(t)\mathbf{T}_o^*, \qquad T(t) = \begin{cases} a t^{1/2}, & \text{for} \quad 0 \leq t \leq t_0 \\ a t_o^{1/2}, & \text{for} \quad t_0 \leq t \leq t', \end{cases} \tag{25}$$

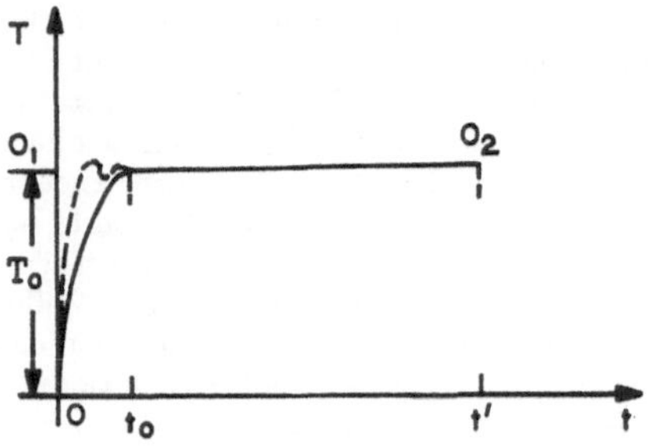


FIG. 3.

where the number $a$, the parameter of the parabola, is a large positive number. The impulse in this case is

$$\mathbf{I}_o'' = \tfrac{2}{3} a t_o^{3/2} \mathbf{T}_o^*$$

or if, as a comparison, we take $T_o = a t_o^{1/2}$, then

$$\mathbf{I}_o'' = \tfrac{2}{3} t_o T_o \mathbf{T}_o^* = \tfrac{2}{3} \mathbf{I}_o'. \tag{26}$$

The above is an example of a "gradual application," which approximates practical problems. In Fig. 3 the solid line is the graph of the magnitude of the force according to formula 25, while the dotted curve is the force according to practical experience, $t_o$ being approximately equal to 0.02 seconds.

### *Appendix*

### *The $n^{\text{th}}$ Derivative of the Function $\sigma(t)/r^3(t)$*

We find upon examination that the $n^{\text{th}}$ derivative of the product of the functions $\sigma(t)$ and $\varphi(t)$ is given by the Leibnitz formula:

$$(\sigma \cdot \varphi)^{(n)} = \sum_{m=0}^{n} \frac{n!}{m!(n-m)!} \sigma^{(m)} \cdot \varphi^{(n-m)} \tag{I}$$

and that, if

$$\varphi(t) = \frac{1}{r^3(t)} \tag{II}$$

its $n^{\text{th}}$ derivative is

$$\varphi^{(n)} = \frac{P(r)}{r^{n+3}}, \tag{III}$$

where

$$P(r) = V_1 r^{n-1} + V_2 r^{n-2} + \cdots + V_n \tag{IV}$$

$V_1, V_2, \cdots$ are polynomials in the derivatives of $r$ up to the order $n^{\text{th}}$ with coefficients smaller than $(n+2)!$ Combining I, II, III we derive

$$\left(\frac{\sigma}{r^3}\right)^{(n)} = \sum_{m=0}^{n} \frac{n!}{m!(n-m)!} \sigma^{(m)} \frac{P(r)}{r^{n-m+3}} \tag{V}$$

where

$$P(r) = V_1 r^{n-m-1} + V_2 r^{n-m-2} + \cdots + V_{n-m} \tag{VI}$$

and the polynomials $V_1$, $V_2$, $\cdots$ have coefficients smaller than $(n - m + 2)!$

* * *

This paper appeared in a slightly different form as a General Electric M. S. D. technical report, No. 625D256. It was also presented and discussed at the XV[th] International Astronautical Congress, Warsaw, Poland, 1964.

## *References*

(1) D. G. Magiros, "On the Convergence of the Solution of a Special Two-Body Problem," Athens, Greece, Practica of Athens Academy of Sciences, Vol. 38, pp. 36–39, 1963.
(2) W. Hurewicz, "Lectures on Ordinary Differential Equations," New York, M.I.T. Technical Press and John Wiley and Sons, 1958.
(3) R. Courant, "Boundary Value Problems in Modern Dynamics," *Proc. Intern. Congress of Mathematicians*, II, 1952.
(4) K. Knopp, "Theory and Applications of Infinite Series" New York, Hafner Publishing Company, Sec. Ed., §18, Sect. 94, pp. 154–155, 1948.
(5) W. Smart, "Celestial Mechanics," New York, Longmans, Green and Company, p. 21, paragraph 2.12, 1953.

# MOTION IN A NEWTONIAN FORCE FIELD MODIFIED BY A GENERAL FORCE. II

by

D. G. MAGIROS*

General Electric Company, Philadelphia, Pennsylvania (U.S.A.)

## 1. Introduction

In a preceding paper published under a similar title,** the motion of an artificial celestial body $P$, which is moving under the influence of a Newtonian force $\underline{N}$ from an attractive center $E$ and a general force $\underline{T}$, was discussed.

The forces $\underline{T}$ may be such that the moving body either collides or not with the attractive center $E$, that is the paths of the body pass or not through the center $E$ and they are either "collision paths" or "non-collision paths", and the forces $\underline{T}$ are either "collision forces" or "non-collision forces", respectively, when one has two distinct cases of the problem, the "collision case" and the "non-collision case".

(a) In the preceding paper the "non-collision case" of the problem was discussed. It is established for this case an actual mathematical solution of specific form in terms of power series in time, found the general term of the series, proved its convergence, determined the region of convergence, and indicated the conditions for a rapid convergence, which permits an approximation of the solution within a given error. The procedure was based on the hypothesis of "non-zero distance $r$" between the attractive center $E$ and the moving body $P$ during its motion. $r = 0$ is a singularity of the series found, and this singularity is associated with important physical significance, which makes the subject very worth studying.

(b) In the present paper the "collision case" of the problem is investigated. Conditions are discussed for the forces to be "collision forces", and calculation is given of the "collision time" $\tau$, that is the finite time needed

* Consulting Mathematician, Valley Forge Space Technology Center.

** MAGIROS, D. G., Motion in a Newtonian force field modified by a general force, *Journal of Franklin Institute* (December 1964).

[349]

Reprinted from the *Journal of the Franklin Institute* **278** (1964), 349–355.

for the moving body $P$ to collide with the attractive center $E$. The series solution of the problem, found in the preceding paper, which is valid for very large time in the "non-collision case", is now valid for the "collision case" for time smaller than the finite collision time $\tau$, which corresponds to the particular collision problem one has.

## 2. Useful Remarks

The following two remarks are useful for the collision problem.

(a) In a Newtonian force field, the attractive force $\underline{N}$ to the body $P$ of distance $r$ from the attractive center $E$ is: $\underline{N} = -(\mu/r^2)\underline{r}^*$, where $\underline{r}^*$ is a unit vector, Figure 1(a), $\mu = K(M+m)$, $K$ the constant of gravity, $M$ and $m$ the masses of the attractive center and the body, respectively.

If $\underline{r}(t)$ is the position vector in this field,

$$\underline{r}(t) = r(t) \cdot \underline{r}^*(t) \tag{1}$$

the corresponding linear velocity $\underline{V}(t)$ is

$$\begin{aligned} \dot{\underline{r}} &= \dot{r}\underline{r}^* + r\dot{\underline{r}}^* \\ &= \dot{r}\underline{r}^* + r\omega\underline{r}_1^* \\ &= \underline{V}(t) = V(t)\cdot\underline{V}^*(t) \end{aligned} \tag{2}$$

where $\omega$ is the angular velocity $\dot{\varphi}$. The vectors $\underline{r}^*$ and $\dot{\underline{r}}_1^*$ are perpendicular.

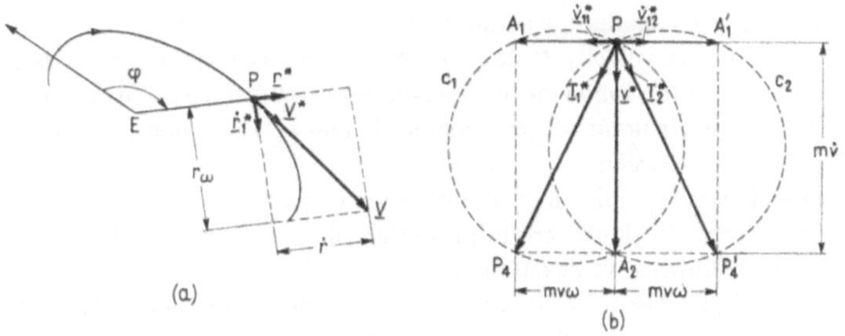

Fig. 1

The speed $V$ of the velocity $\underline{V}$ is given by

$$V = \begin{cases} \sqrt{\dfrac{\mu}{a}} \cdot \sqrt{\dfrac{2a-r}{r}}, & \text{for ellipse } (0 \leqslant e < 1) \\ \sqrt{\dfrac{\mu}{a}} \cdot \sqrt{\dfrac{2a+r}{r}}, & \text{for hyperbola } (e > 1) \\ 2\mu \cdot \dfrac{1}{\sqrt{r}}, & \text{for parabola } (e = 1), \end{cases} \tag{3}$$

where $e$ is the orbital eccentricity and $\alpha$ the semi-major axis of the orbit. The Eqs. (3), if $V_1 = \sqrt{2\mu/r}$ and $V_P$ the speed at $P$, give

$$V_P < V_1, \quad V_P > V_1, \quad V_P = V_1, \tag{4}$$

when $P$ is on an ellipse, hyperbola or parabola, respectively, with the same focus $E$, and $V_P$, speed at $P$, is elliptic, hyperbolic or parabolic speed, respectively.

(b) The general force $\underline{T}$ and its corresponding velocity $\underline{v}$ are such that

$$\begin{aligned} \underline{T}(t) &= T(t)\cdot\underline{T}^*(t) \\ &= m\cdot\dot{\underline{v}}(t) \\ &= m\frac{d}{dt}\{v(t)\cdot\underline{v}^*(t)\} \\ &= m\dot{v}\underline{v}^*+mv\dot{\underline{v}}^* \\ &= m\dot{v}\underline{v}^*+mv\omega\underline{v}_1^* \end{aligned} \tag{5}$$

$\underline{v}^*$ and $\dot{\underline{v}}_1^*$ are perpendicular. Given $\underline{v}^*$ one may have two directions for the perpendicular unit vector $\dot{\underline{v}}_1^*$, namely $\dot{\underline{v}}_{11}^*$ and $\dot{\underline{v}}_{12}^*$, then two forces symmetric to $\underline{v}^*$ and of the same magnitude, Fig. 1b.

## 3. Collision Problem

The body $P$ is under the action of the Newtonian force $\underline{N}$ and the general force $\underline{T}$, and its velocity $\underline{R}$ due to these forces is tangent to the path of the body. For the following we assume that the velocity $\underline{R}$ of the body is considered as resultant of the velocities $\underline{V}$ and $\underline{v}$, associated with the forces $\underline{N}$ and $\underline{T}$, respectively.

One may have collision or non-collision cases of the problem, by restricting the forces $\underline{T}$ to have certain properties of rate of change.

In the following we discuss the:

**Collision Problem:** "*Determine the force $\underline{T}$ which is such that the resultant $\underline{R}$ of the velocities $\underline{V}$ and $\underline{v}$ is through the center $E$ for all time $t$ in $0 \leqslant t < \tau$, $\tau$ being appropriate time.*" Forces $\underline{T}$, which satisfy the requirement of this problem, are "collision forces", and the time $\tau$ is the "*collision time*" of the problem. The series solution of the preceding paper is now valid for time $t < \tau$.

The discussion of the above problem is based on the following two auxiliary problems:

**Auxiliary Problem A:** "*Given the vector $\underline{V}$, determine a vector $\underline{v}$ such that their resultant $\underline{R}$ is through a constant point $E$.*" One has two cases, accordingly as the direction or the magnitude of $\underline{v}$ is given.

(a) Given the direction of $\underline{v}$, that is the angle $\varphi_2$, Fig. 2, from the triangle $PP_1P_2$ of the parallelogram $PP_1P_2P_3$ one has

$$\left.\begin{aligned} &\text{(i)} \quad R = \frac{\sin\varphi_2}{\sin(\varphi_2-\varphi_1)} \cdot V \\ &\text{(ii)} \quad v = \frac{\sin\varphi_1}{\sin(\varphi_2-\varphi_1)} \cdot V \end{aligned}\right\}. \tag{6}$$

The graphical construction shows that the vector $\underline{v}$ must be in the angle $EPP_1'$, then for the above auxiliary problem the direction of $\underline{v}$ is restricted according to

$$0 < \varphi_1 < \varphi_2 < \pi. \tag{7}$$

**Remark.** The restrictions (7) hold in the regular construction. In an extreme case, that is in case either $\varphi_1 = 0$ or $\varphi_2 = 0, \pi$, one of these conditions implies the other.

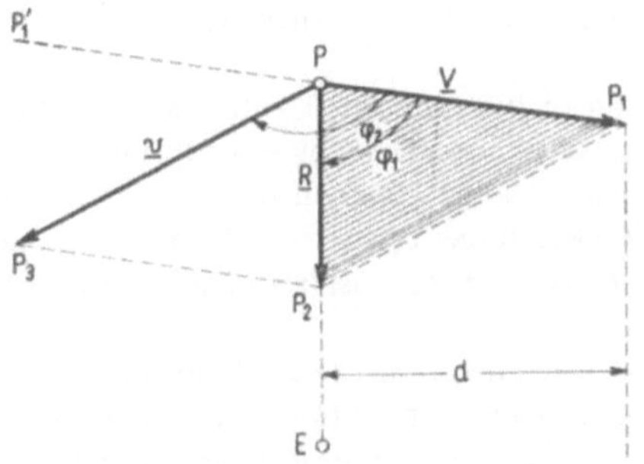


Fig. 2

(b) Given the magnitude of $\underline{v}$, one determines the appropriate direction of $\underline{v}$ as follows. With center the end-point $P_1$ of $\underline{V}$, Fig. 2, and radius the magnitude $v$ of $\underline{v}$ one draws a circumference which intersects the line $PE$ in either one or two points, or does not intersect it. If $P_2$ is such a point of intersection, then the direction of $\underline{v}$ is parallel to $P_1P_2$ from $P$. Therefore, one may have either one or two solutions, or no solution, if

$$d = v, \quad d < v, \quad d > v, \tag{8}$$

respectively. $d$ is the distance of $P_1$ from the line $PE$.

**Auxiliary Problem B:** "*Determine the force $\underline{T}$ in terms of its corresponding velocity $\underline{v}$, if $\underline{v}$ satisfies the auxiliary problem A.*"

Given $v$, the force $\underline{T}$ can be determined by using formulae (5) and Fig. 1b. Either the diagonal $PP_4$ or $PP_4'$ is the force $\underline{T}$ of magnitude

$$T = m(v^2\omega^2 + \dot{v}^2)^{\frac{1}{2}}. \tag{9}$$

In case the velocity $\underline{v}$ is of constant direction, the force $\underline{T}$ is along this direction and its magnitude is: $T = m\dot{v}$.

In case $\underline{v}$ is of constant magnitude, the force is perpendicular to the velocity $\underline{v}$ and its magnitude is: $T = mv\omega$.

Combining the above auxiliary problems, one can have the following problems equivalent to the collision problem:

**Problem A:** "*Given the Newtonian velocity $\underline{V}$ and the direction of the velocity $\underline{v}$, which satisfies the requirement of the collision problem, find the appropriate collision forces $\underline{T}$.*"

**Problem B:** "*Given $\underline{V}$ and the magnitude of $\underline{v}$, which satisfies the requirement of the collision problem, find the appropriate collision forces $\underline{T}$.*"

For the solution of the problem A, if the direction of $\underline{v}$ satisfies the requirement (7) the magnitude $v$ of $\underline{v}$ is given by the formula (6ii), then the component $PA_2$, Fig. 1b, of the force is known. There are two directions for the perpendicular to $\underline{v}^*$, namely $\dot{\underline{v}}_{11}^*$ and $\dot{\underline{v}}_{12}^*$, Fig. 1b, and the second component of the force is either $PA_1$, or $PA_1'$, both of magnitude $mv\omega$, according to auxiliary problem B. Then, the problem A has two solutions.

For the solution of the problem B, one first must check which of the conditions (7) is satisfied by $v$. Then, by using formula (6ii), one determines the angle $\varphi_2$, which gives the direction of $\underline{v}$, when, Fig. 1b, one sees that for each accepted $v$ two forces can be found. Therefore one has:

(i) Case $v < d$, when there exist no collision forces $\underline{T}$ satisfying the collision problem;

(ii) Case $v = d$, when there exist two solutions of the collision problem;

(iii) Case $v > d$, when there exist four solutions of the collision problem.

**Application:** "*Motion in case velocity $\underline{v}$ is parallel to velocity $\underline{V}$.* Combining the above remark of the auxiliary problem A and the requirements of the collision problem, one can distinguish two cases:

(a) The regular case of Keplerian orbit. In this case $\varphi_1 \neq 0$, and $\varphi_2 = \varphi_1$ or $\varphi_2 = \varphi_1 + \pi$, then the forces $\underline{T}$, which produce $\underline{v}$, can not be collision forces, and one has a non-collision case of the problem, when the series solution is valid for large time.

(b) The degenerate case of Keplerian orbit. In this case $\varphi_1 = 0$, and $\varphi_2 = 0$ or $\varphi_2 = \pi$, then the forces $\underline{T}$ are collision forces and one has a collision case of the problem, when the series solution is valid for time smaller than the collision time corresponding to this collision case.

## 4. The Collision Time

The collision time $\tau$ determines the time-interval for convergence of the series solution, $0 \leqslant t < \tau$.

We calculate the collision time in the case of the present collision problem.

In this problem, the motion of the body $P$ is along the line $PE$ towards $E$, and its speed $R$, according to formula (6i) and (3) is known function of the distance $r$ of $P$ from $E$, $R = R(r)$. Then, along the $r$-axis one has

$$\dot{r} = R(r)$$

when

$$t-t_0 = \int_{r_0}^{r} \frac{dr}{R(r)}. \tag{10}$$

The limits $r_0$ and $r$ are values of $r$ at time $t_0$ and $t$, respectively. One can get the "collision time" $\tau$ from formula (10) by taking $t_0 = 0$ and $r = 0$, when

$$\tau = \int_{r_0}^{0} \frac{dr}{R(r)} = -\int_{0}^{r_0} \frac{dr}{R(r)}. \tag{11}$$

The integrand of (11), according to formulae (6i) and (3), is given by

$$-\frac{1}{R(r)} = \begin{cases} \dfrac{\sin(\varphi_1-\varphi_2)}{\sin\varphi_2}\sqrt{\dfrac{a}{\mu}}\sqrt{\dfrac{r}{2a-r}}, & \text{for ellipse} \\ \dfrac{\sin(\varphi_1-\varphi_2)}{\sin\varphi_2}\sqrt{\dfrac{a}{\mu}}\sqrt{\dfrac{r}{2a+r}}, & \text{for hyperbola} \\ \dfrac{\sin(\varphi_1-\varphi_2)}{\sin\varphi_2}\dfrac{\sqrt{r}}{\sqrt{2\mu}}, & \text{for parabola} \end{cases} \tag{12}$$

We remark that: for any place of the body in the line $PE$, a speed $R$ is associated with the body, and this $R$ depends on the distance $r$ of the body from the center, on the number $a$ and on the angles $\varphi_1, \varphi_2$.

If at that place of the body one takes specified values for $a, \varphi_1, \varphi_2$, then $R$, and therefore the integrand of (11) given by (12), is a function of $r$. If we assume $a, \varphi_1, \varphi_2$ as parameters, from (11) and (12) one can get

$$\left.\begin{aligned} &\text{(i)} \quad \tau_e = \sqrt{\frac{a}{\mu}}\cdot\frac{\sin(\varphi_1-\varphi_2)}{\sin\varphi_2}\cdot I_e \\ &\text{(ii)} \quad \tau_h = \sqrt{\frac{a}{\mu}}\cdot\frac{\sin(\varphi_1-\varphi_2)}{\sin\varphi_2}\cdot I_h \\ &\text{(iii)} \quad \tau_p = \frac{1}{\sqrt{2\mu}}\cdot\frac{\sin(\varphi_1-\varphi_2)}{\sin\varphi_2}\cdot I_p \end{aligned}\right\} \tag{13}$$

where

$$\text{(i) } I_e = \int_0^{r_0} \sqrt{\frac{r}{2a-r}}\,dr, \quad \text{(ii) } I_h = \int_0^{r_0} \sqrt{\frac{r}{2a+r}}\,dr, \quad \text{(iii) } I_p = \int_0^{r_0} \sqrt{r}\,dr. \tag{14}$$

The indices $e, h, p$ in (13) and (14) correspond to elliptic, hyperbolic, parabolic Keplerian velocity $\underline{V}$, respectively.

By using the transformation $2a-r = y^2$ in the integral (14i), and $2a+r = y^2$ in the integral (14ii), the calculation of the integrals (14) gives

$$\left.\begin{aligned}
&\text{(i)} \quad I_e = a\pi - \sqrt{r_0(2a-r_0)} - 2a\sin^{-1}\left(\sqrt{\frac{2a-r_0}{2a}}\right)\\
&\text{(ii)} \quad I_h = \sqrt{r_0(2a+r_0)} + 2a\log\frac{\sqrt{2a}}{\sqrt{r_0}+\sqrt{2a+r_0}}\\
&\text{(iii)} \quad I_p = \tfrac{2}{3} r_0\sqrt{r_0}
\end{aligned}\right\} \qquad (15)$$

Inserting (15) into (13) we have the collision time in our case

$$\left.\begin{aligned}
&\text{(i)} \ \tau_e = \sqrt{\frac{a}{\mu}}\,\frac{\sin(\varphi_1-\varphi_2)}{\sin\varphi_2}\times\\
&\qquad\times\left\{a\pi - \sqrt{r_0(2a-r_0)} - 2a\sin^{-1}\left(\sqrt{\frac{2a-r_0}{2a}}\right)\right\}\\
&\text{(ii)} \ \tau_h = \sqrt{\frac{a}{\mu}}\,\frac{\sin(\varphi_1-\varphi_2)}{\sin\varphi_2}\times\\
&\qquad\times\left\{\sqrt{r_0(2a+r_0)} + 2a\log\frac{\sqrt{2a}}{\sqrt{r_0}+\sqrt{2a+r_0}}\right\}\\
&\text{(iii)} \ \tau_p = \frac{1}{3}\sqrt{\frac{2}{\mu}}\,\frac{\sin(\varphi_1-\varphi_2)}{\sin\varphi_2}\, r_0\sqrt{r_0}.
\end{aligned}\right\} \qquad (16)$$

*Remark.* The calculation of the collision time in the present collision problem was based on the fact that the integration line is the segment $PE$, and on the assumption that $a$, $\varphi_1$, $\varphi_2$ are parameters. This assumption needs a special consideration.

# MOTION IN A NEWTONIAN FORCE FIELD MODIFIED BY A GENERAL FORCE III APPLICATION: THE ENTRY PROBLEM

by

D. MAGIROS and G. REEHL

General Electric Company, Valley Forge Space Technology Center, Philadelphia, Pennsylvania (U.S.A.)

In the present paper the entry problem is discussed as an application of papers published by one of the authors during the last years.[1(a)(b)(c)]

The discussion is of mathematical type, but it can help for a treatment of physical entry problems.

## 1. The Problem

The problem is the following: "An artificial celestial body is moving in an elliptical Keplerian orbit around the Earth. A force $\boldsymbol{T}$ is applied to the body, acts for some time, then stops, when the body starts moving in a new Keplerian orbit around the Earth. Conditions are required under which the new Keplerian orbit intersects the surface of the Earth."

## 2. Solution of the Problem

We discuss first three kinds of orbits separately.

### The original orbit

The position vector $\boldsymbol{r}_0$ and the velocity vector $\dot{\boldsymbol{r}}_0 = \boldsymbol{V}_0$ at a point $P_0$, due only to the Newtonian force, determine the Keplerian orbit through $P_0$ with focus the attractive center $E$ of the Earth. If these vectors are known, the distance $r_0 = EP_0$, the true anomaly $\vartheta_0$, the angle $\varphi_0$, and the speed $V_0$, associated with the point $P_0$, will be known.

The Keplerian orbit is an ellipse, parabola or hyperbola if $V_0$ is smaller than, equal to or bigger than $\sqrt{2\mu/r_0}$, respectively.



Reprinted from the *Proceedings of the XVIIth Intl. Astron. Congress*, Madrid (1966), 149–154.

$\mu = k(m_1+m_2)$, $k$ = gravitational constant, $m_1$ and $m_2$ masses of the Earth and the moving body, respectively.

The semi-major axis $a_0$, the eccentricity $e_0$, the perigee distance $p_0$ of the orbit, and the angle $\varphi_0$ can be calculated in terms of the known quantities by using appropriate formulae, which, in the case of an elliptical orbit, are:[2]

$$\left.\begin{aligned} a_0 &= \frac{\mu}{(2\mu/r_0)-V_0^2}, \quad e_0 = \left\{1-\frac{r_0}{a_0^2}(2a_0-r_0)\sin^2\varphi_0\right\}^{1/2} \\ p_0 &= a_0(1-e_0), \quad \sin\varphi_0 = \frac{1+e_0\cos\vartheta_0}{\{1+e_0^2+2e_0\cos\vartheta_0\}^{1/2}} \end{aligned}\right\} \quad (1)$$

**The transfer orbit**

If the time is split according to $t = t_0+\tau$, where $t_0$ is very small, and if the reference coordinate system is $(P_0; \boldsymbol{r}_0^*, \boldsymbol{s}_0^*, \boldsymbol{T}_0^*)$ as indicated in Fig. 1, the conditions at $P_0$ can be:

$$\left.\begin{aligned} &\text{(i)} \quad \boldsymbol{r}(0) = \boldsymbol{r}(t_0) = \boldsymbol{r}_0 = r_0\cdot\boldsymbol{r}_0^* \\ &\text{(ii)} \quad \dot{\boldsymbol{r}}(t_0) = \dot{\boldsymbol{r}}(0)+\boldsymbol{I}_0 = \boldsymbol{V}_0+\boldsymbol{I}_0 = s_0\cdot\boldsymbol{s}_0^* \\ &\text{(iii)} \quad \boldsymbol{I}_0 = \int_0^{t_0}\boldsymbol{T}(t)dt, \quad \text{the impulse.} \end{aligned}\right\} \quad (2)$$

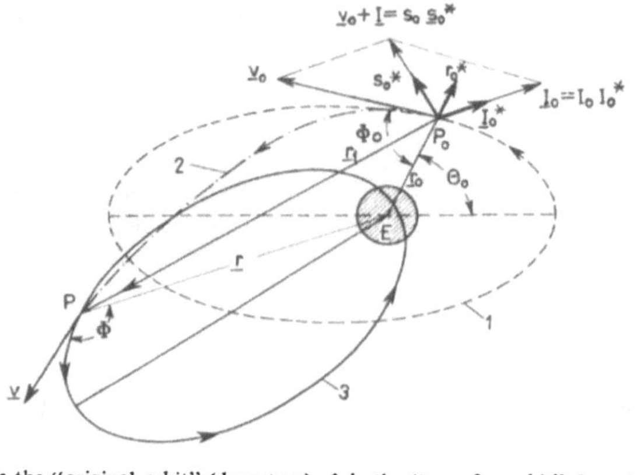


Fig. 1 *1* is the "original orbit" (departure); *2* is the "transfer orbit" (non keplerian); *3* is the "new orbit" (destination). The force $\boldsymbol{T}$ starts at $P_0$ and stops at $P$ with duration $t'$.

The motion of the body during the action of the force on the non-Keplerian transfer orbit is given by the position vector $\boldsymbol{r}(\tau)$ and the velocity $\boldsymbol{r}(\tau)$ according to the formulae:

$$\left.\begin{aligned} \boldsymbol{r}(\tau) &= \boldsymbol{r}_0+\boldsymbol{r}_1(\tau) = \alpha_1(\tau)\boldsymbol{r}_0^*+\alpha_2(\tau)\boldsymbol{s}_0^*+\alpha_3(\tau)\boldsymbol{T}_0^* \\ \dot{\boldsymbol{r}}(\tau) &= \boldsymbol{V}(\tau) = \dot{\alpha}_1(\tau)\boldsymbol{r}_0^*+\dot{\alpha}_2(\tau)\boldsymbol{s}_0^*+\dot{\alpha}_3(\tau)\boldsymbol{T}_0^* \end{aligned}\right\} \quad (3)$$

where:

$$\left.\begin{aligned}
\alpha_1(\tau) &= \alpha_1(0)+\dot{\alpha}_1(0)\tau+\frac{1}{2}\ddot{\alpha}_1(0)\tau^2+ \dots +\frac{1}{n!}\alpha_1^{(n)}(0)\,\tau^n+ \dots \\
\alpha_2(\tau) &= \alpha_2(0)+\dot{\alpha}_2(0)\tau+\frac{1}{2}\ddot{\alpha}_2(0)\tau^2+ \dots +\frac{1}{n!}\alpha_2^{(n)}(0)\,\tau^n+ \dots \\
\alpha_3(\tau) &= \alpha_3(0)+\dot{\alpha}_3(0)\tau+\frac{1}{2}\ddot{\alpha}_3(0)\tau^2+ \dots +\frac{1}{n!}\alpha_3^{(n)}(0)\,\tau^n+ \dots
\end{aligned}\right\} \quad (4)$$

In the series (4), which is proved to be convergent, the first two coefficients are known, while the others can be calculated by using the formulae:

$$\left.\begin{aligned}
\ddot{\alpha}_1(\tau) &= -\frac{\mu}{r^3(\tau)}\,\alpha_1(\tau)+T_1, &\quad \alpha_1(0) &= r_0, &\quad \dot{\alpha}_1(0) &= 0 \\
\ddot{\alpha}_2(\tau) &= -\frac{\mu}{r^3(\tau)}\,\alpha_2(\tau)+T_2, &\quad \alpha_2(0) &= 0, &\quad \dot{\alpha}_2(0) &= s_0 \\
\ddot{\alpha}_3(\tau) &= -\frac{\mu}{r^3(\tau)}\,\alpha_3(\tau)+T_3, &\quad \alpha_3(0) &= 0, &\quad \ddot{\alpha}_3(0) &= 0
\end{aligned}\right\} \quad (4a)$$

$T_1$, $T_2$, $T_3$ are projections of the force $\boldsymbol{T}$ on the coordinate axis.
The three-dimensional entry problem is characterized by the fact that the force $\boldsymbol{T}$ and the unit-vector $\boldsymbol{T}_0^*$ are not on the $(\boldsymbol{V}_0, \boldsymbol{r}_0^*)$-plane of the original Keplerian orbit. Having the impulse $\boldsymbol{I}_0$ at $P_0$ we calculate the vector $\boldsymbol{V}_0+\boldsymbol{I}_0 = s_0 \cdot \boldsymbol{s}_0^*$. During the action of the force $\boldsymbol{T}$, the body is moving on the transfer orbit, when the position vector and the velocity vector of the body are given by the formulae (3) indicated above.

When $\boldsymbol{T}$ stops acting, say at the point $P$, the position vector $\boldsymbol{r} = \boldsymbol{EP}$ and the velocity vector $\dot{\boldsymbol{r}} = \boldsymbol{V}$ at $P$, that is the magnitudes $r = EP$ and $V$ as well as the angles which determine the direction of the vectors $\boldsymbol{r}$ and $\boldsymbol{V}$, are known by the above formulae.

## The destination orbit

The destination orbit is determined by the position vector and the velocity vector at the last point $P$ of the transfer orbit, that is, by using $r$, $V$, $\varphi$ corresponding to $P$ in the $(E, V)$-plane. The destination orbit is an ellipse in case: $\sqrt{2\mu/r} > V$. The destination orbit intersects the surface of the Earth, when the entry problem is realized, if its perigee distance is smaller than or equal to the radius $\varrho$ of the Earth taken spherical, that is if, in case of elliptical destination orbit, the following condition is satisfied:

$$\alpha(1-e) \leqslant \varrho. \quad (5)$$

We notice that $\alpha$ and $e$ are dependent upon known and calculated angles and lengths and upon the time of the duration of the force $\boldsymbol{T}$.

### 3. The Solution of Some Restricted Problems

By imposing restrictions to magnitude and/or to direction of the force $\boldsymbol{T}$ one can have a variety of entry problems. We distinguish here two most important cases, namely the cases of constant force $\boldsymbol{T}$ in three-dimensional and in two-dimensional entry problems.

#### The force $T$ constant in space

Let us take:

$$T_1 = 0 \quad \text{on} \quad \boldsymbol{r}_0^*, \qquad T_2 = 0 \quad \text{on} \quad \boldsymbol{s}_0^*, \qquad T_3 = T \quad \text{on} \quad \boldsymbol{T}_0^* \tag{6}$$

From formulae (4a) one can get:

$$\left.\begin{aligned}
&\alpha_1(0) = r_0,\ \dot{\alpha}_1(0) = 0,\ \ddot{\alpha}_1(0) = -\frac{\mu}{r_0^2},\ \dddot{\alpha}_1(0) = 0,\ \ddddot{\alpha}_1(0) = \frac{\mu^2}{r_0^5},\ \ldots \\
&\alpha_2(0) = 0,\ \dot{\alpha}_2(0) = s_0,\ \ddot{\alpha}_2(0) = 0,\ \dddot{\alpha}_2(0) = -\frac{\mu s_0}{r_0^3},\ \ddddot{\alpha}_2(0) = 0,\ \ldots \\
&\alpha_3(0) = 0,\ \dot{\alpha}_3(0) = 0,\ \ddot{\alpha}_3(0) = T,\ \dddot{\alpha}_3(0) = 0,\ \ddddot{\alpha}_3(0) = -\frac{\mu T}{r_0^3},\ \ldots
\end{aligned}\right\} \tag{6a}$$

For large $r_0$ and small $\tau$, if we restrict ourselves to the first four terms of the expressions (4), the solution (3) becomes:

$$\left.\begin{aligned}
\boldsymbol{r}(\tau) &= \left(r_0 - \frac{\mu\tau^2}{2r_0^2}\right)\boldsymbol{r}_0^* + \left(s_0\tau - \frac{\mu s_0 \tau^3}{6r_0^3}\right)\boldsymbol{s}_0^* + \tfrac{1}{2}T\tau^2\boldsymbol{T}_0^* \\
\dot{\boldsymbol{r}}(\tau) &= -\frac{\mu\tau}{r_0^2}\boldsymbol{r}_0^* + \left(s_0 - \frac{\mu s_0\tau^2}{2r_0^3}\right)\boldsymbol{s}_0^* + T\tau\boldsymbol{T}_0^*
\end{aligned}\right\} \tag{6b}$$

The errors of the coefficients in the solution (6b) are known.

#### The constant force $T$ on the original orbit plane

We take:

$$T_1 = 0 \quad \text{on} \quad \boldsymbol{r}_0^*, \qquad T_2 = 0 \quad \text{on} \quad \boldsymbol{s}_0^*,\ T_3 = T \text{ on } \boldsymbol{T}_0^* = -\boldsymbol{s}_0^* \tag{7}$$

when the solution (6b) becomes:

$$\left.\begin{aligned}
\boldsymbol{r}(\tau) &= \left(r_0 - \frac{\mu\tau^2}{2r_0^2}\right)\boldsymbol{r}_0^* + \left(s_0\tau - \tfrac{1}{2}T\tau^2 - \frac{\mu s_0\tau^3}{6r_0^3}\right)\cdot\boldsymbol{s}_0^* \\
\dot{\boldsymbol{r}}(\tau) &= -\frac{\mu\tau}{r_0^2}\boldsymbol{r}_0^* + \left(s_0 - T\tau - \frac{\mu s_0\tau^2}{2r_0^3}\right)\cdot\boldsymbol{s}_0^*
\end{aligned}\right\} \tag{7a}$$

## 4. Numerical Examples

We now apply equations (7a) to specific numerical examples, taking the coordinate system as indicated in Fig. 2 and the original orbit circular with 100 miles altitude. Thus we have:

radius of spherical Earth: $\varrho = 2.090290 \times 10^7$ ft

radius of circular original orbit: $r_0 = 2.143090 \times 10^7$ ft

the satellite velocity: $\boldsymbol{V}_0 = 2.562762 \times 10^{14} \dfrac{\text{ft}}{\text{sec}} \cdot \boldsymbol{s}_0^*$

the constant coefficient: $\mu = 1.407528 \times 10^{16} \dfrac{\text{ft}^3}{\text{sec}^2}$.

Assume $t_0 = 10^{-4}$ sec and a retro-engine starting at $t = 0$ and acting for time $t'$ giving a constant thrust $\boldsymbol{T} = -10 \dfrac{\text{ft}}{\text{sec}^2} \cdot \boldsymbol{s}_0^*$.

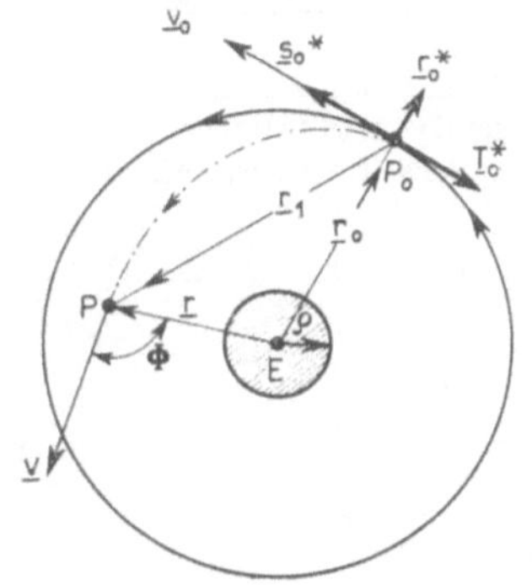


Fig. 2

At time $t_0$ we get:

$$\left.\begin{aligned} \boldsymbol{r}(0) = \boldsymbol{r}(t_0) &= 2.143090 \times 10^7 \cdot \boldsymbol{r}_0^* \\ \dot{\boldsymbol{r}}(t_0) &= 2.562762 \times 10^4 \cdot \boldsymbol{s}_0^* \end{aligned}\right\} \tag{8}$$

(i) For time $t' = 100$ sec, Eqs. (7a) and the values (8) give:

$$\boldsymbol{r}(100) = 2.127767 \times 10^7 \cdot \boldsymbol{r}_0^* + 2.506654 \times 10^5 \cdot \boldsymbol{s}_0^*$$
$$\dot{\boldsymbol{r}}(100) = -3.064617 \times 10^2 \cdot \boldsymbol{r}_0^* + 2.444438 \times 10^4 \cdot \boldsymbol{s}_0^*$$

Initial conditions for the new Keplerian orbit are:

$r = 2.142481 \times 10^7$ ft

$V = 2.463574 \times 10^4 \dfrac{\text{ft}}{\text{sec}}$

$$\varphi = 90° - \tan^{-1} \frac{2.506654 \times 10^5}{2.127767 \times 10^7} + \tan^{-1} \frac{3.064617 \times 10^2}{2.444438 \times 10^4} = 90°25'37.5''$$

when:

$$\alpha = 1.990836\times 10^7 \text{ ft}, \quad e = 0.076534, \quad p = 1.838471\times 10^7 \text{ ft}.$$

and since $p < \varrho = 2.090290\times 10^7$ ft, the entry problem has been achieved in the above case.

(ii) For time $t' = 10$ sec, we get:

$$r(10) = 2.142937\times 10^7 \cdot r_0^* + 2.557701\times 10^5 \cdot s_0^*,$$

$$\dot{r}(10) = -3.064617\times 10^2 \cdot r_0^* + 2.552579\times 10^4 \cdot s_0^*$$

$$\varphi = 90^\circ\, 0'\, 14.5''$$

$$\alpha = 2.126527\times 10^7 \text{ ft}, \quad e = 0.0077698, \quad p = 2.110004\times 10^7 \text{ ft}$$

and since $p > \varrho$ is this case, the entry problem has not been achieved.

## References

[1] MAGIROS. D. G., Proc. International Astronautical Congress: (a) XIV 1963 (Paris), (b) XV 1964 (Warsaw), (c) XVI 1965 (Athens)

[2] ROY. A. E., The Foundations of Astrodynamics, pp. 80, 83, The MacMillan Co., New York (1965)

ΜΑΘΗΜΑΤΙΚΑ.— **The Entry Problem***, *by* ***Demetrios G. Magiros and George Reehl****. Ἀνεκοινώθη ὑπὸ τοῦ Ἀκαδημαϊκοῦ κ. Ἰ. Ξανθάκη.

## I. INTRODUCTION

### (a) The Entry Problem.

*«An artificial celestial body is moving in an elliptical Keplerian orbit around the earth. A force T is applied to the body, acts for some time, then stops, when the body starts moving in a new Keplerian orbit around the earth. Condition are required under which the new Keplerian orbit intersects the surface of the earth».*

The conditions required will be found from the restriction of the perigee distance of the new Keplerian orbit to be smaller than or equal to the radius of the earth, taken spherical.
Use will be made of the results of one of the author's papers in connection with the motion of the body on the non-Keplerian orbit during the action of the general force $T$.

### (b) The Motion of the Body During the Action of the Force.

We state for reference [1(b)] that if the time is split according to: $t = t_o + \tau$, where $t_o$ is very small, the reference coordinate system: $(P_o; r^*_o, S^*_o, T^*_o)$, which is in general non-orthogonal and three-dimensional, is defined as shown in Figure 1, the conditions at $P_o$ are:

$$\left.\begin{array}{ll} \text{(i)} & r(0) = r(t_o) = r_o = \mathrm{r}_o \cdot r^*_o \\ \text{(ii)} & \dot{r}(t_o) = \dot{r}(0) + I_o = V_o + I_o = \mathrm{s}_o \cdot s^*_o \\ \text{(iii)} & I_o = \int_0^{t_o} T(t)\, dt \end{array}\right\} \qquad (1)$$

the motion of the body during the action of the force on the non-Keplerian arc of the orbit is given by the position vector $r(\tau)$ and the velocity $\dot{r}(\tau)$ according to the formulae:

$$r(\tau) = r_o + r_1(\tau) = \mathrm{a}_1(\tau) \cdot r^*_o + \mathrm{a}_2(\tau) \cdot s^*_o + \mathrm{a}_3(\tau)\, T^*_o \qquad (2)$$

$$\dot{r}(\tau) = V(\tau) = \dot{\mathrm{a}}_1(\tau)\, r^*_o + \dot{\mathrm{a}}_2(\tau) \cdot s^*_o + \dot{\mathrm{a}}_3(\tau)\, T^*_o \qquad (3)$$

* DEMETRIOS S. MAGIROS καὶ GEORGE REEHL, Τὸ πρόβλημα τῆς εἰσόδου τεχνητοῦ δορυφόρου.

** General Electric Co, Missile and Space Division, Philadelphia, PA.

Reprinted from the *Proceedings of the Athens Academy of Sciences* 41 (1966), 246–251.

where:

$$\left.\begin{aligned} a_1(\tau) &= a_1(o) + \dot{a}_1(o)\tau + \frac{1}{2}\ddot{a}_1(o)\tau^2 + \dots + \frac{1}{n!}a_1^{(n)}(o)\tau^n + \dots \\ a_2(\tau) &= a_2(o) + \dot{a}_2(o)\tau + \frac{1}{2}\ddot{a}_2(o)\tau^2 + \dots + \frac{1}{n!}a_2^{(n)}(o)\tau^n + \dots \\ a_3(\tau) &= a_3(o) + \dot{a}_3(o)\tau + \frac{1}{2}\ddot{a}_3(o)\tau^2 + \dots + \frac{1}{n!}a_3^{(n)}(o)\tau^n + \dots \end{aligned}\right\} \quad (4)$$

The first two coefficients of these series are given, and the others can be calculated by using:

$$\left.\begin{aligned} \ddot{a}_1(\tau) &= -\frac{\mu}{r^3(\tau)}a_1(\tau) + T_1, & a_1(0) &= r_o, & \dot{a}_1(o) &= 0 \\ \ddot{a}_2(\tau) &= -\frac{\mu}{r^3(\tau)}a_2(\tau) + T_2, & a_2(0) &= o, & \dot{a}_2(o) &= s_o \\ \ddot{a}_3(\tau) &= -\frac{\mu}{r^3(\tau)}a_3(\tau) + T_3, & a_3(o) &= o, & \dot{a}_3(o) &= 0 \end{aligned}\right\} \quad (4a)$$

$T_1$, $T_2$, $T_3$ are projections of the force $T$ on the coordinate axis, $\mu = K(m_1 + m_2)$, K the gravitational constant, and $m_1$, $m_2$ the masses of the earth and the body, respectively.
The convergence of the series (4a) is proved.[1b]

II. THE SOLUTION OF THE ENTRY PROBLEM

By the following procedure we can solve the three-dimensional entry problem.

**(a) The Original Keplerian Orbit.**

The position vector $r_o$ and the velocity vector $\dot{r}_o = V_o$ at a point $P_o$, due to the Newtonian force only, determine the Keplerian orbit through $P_o$ with focus the attractive center E of the earth. If these vectors are known, the distance $r_o = EP_o$, the true anomaly $d_o$, the angle $\varphi_o$, and the speed $V_o$, associated with the point $P_o$, will be known.
The Keplerian orbit is an ellipse, parabola, or hyperbola if $V_o$ is smaller than, equal to, or bigger than $\sqrt{2\mu/r_o}$, respectively.[1(c)]
The semi-major axis $a_o$, the eccentricity $e_o$ and the perigee distance $p_o$ of the orbit can be calculated in terms of the known quantities by using appropriate formulae, which in case of elliptic orbit are:[2]

$$\left.\begin{aligned} a_o &= \frac{\mu}{\dfrac{2\mu}{r_o} - V_o}, & e_o &= \left\{1 - \frac{r_o}{a_o^2}(2a_o - r_o)\sin^2\varphi_o\right\} \\ P_o &= a_o(1 - e_o), & \sin\varphi_o &= \frac{1 + e_o\cos d_o}{(1 + e_o^2 + 2e_o\cos d_o)^{1/2}} \end{aligned}\right\} \quad (5)$$

**(b) The Non - Keplerian Arc of the Orbit.**

The three-dimensional entry problem is characterized by the fact that the force $T$ and the unit vector $T_o^{\bullet}$ are not on the plane ($V_o$, $r_o^{\bullet}$) of the original Keplerian orbit, Figure 1.

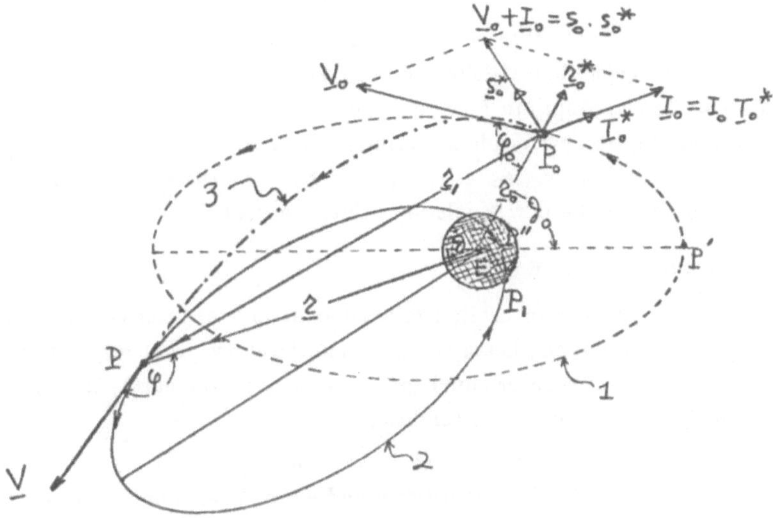


1 : Original orbit (Departure). $P_o$ point where the force starts.
2 : New orbit (Destination). P point where the force stops.
3 : Non - Keplerian arc (Transfer). $P_o$ P during the action of the force.

**Figure 1**

Having the impulse $I_o$ at $P_o$, given by the formula (1,iii) we calculate the vector: $V_o + I_o = s_o \cdot s_o^{\bullet}$.

During the action of the force $T$, the body is moving on the non-Keplerian arc of the orbit, which is a curve in space, and the position vector and the velocity vector of the body are given by the formulae (2) and (3).

When the force $T$ stops acting, say at the point P, formulae (2) and (3) give the position vector $r = EP$ and the velocity vector $\dot{r} = V$ at P, that is the magnitudes r = EP and V, as well as the angles which determine the direction of the vectors $r$ and $V$.
We notice that the plane $(r_1, V)$ in general does not contain the point E.

**(c) The New Keplerian Orbit.**

To determine the new Keplerian orbit, which corresponds to the body at P, Figure 1, where the force $T$ stops, one must know the position vector $r = EP$ and the velocity vector $V$ at P, that is, one must know the magnitudes r and V, and the angle φ on the plane (E, $V$) of the new orbit, which are now known.
The new orbit is an ellipse, parabola or hyperbola, if V is smaller than, equal to, or bigger than $\sqrt{2\mu/r}$, respectively.
By using the appropriate formulae the Keplerian orbit can be calculated. In case of elliptic new Keplerian orbit formulae (5) can be used, if, instead of the quantities $r_o$, $V_o$, $a_o$, $\varphi_o$, $e_o$, $p_o$, $d_o$, the quantities r, V, a, φ, e, p, d, corresponding to the new Keplerian orbit, are used.

**(b) The Condition for Entry.**

The new Keplerian orbit, constructed as above, intersects the surface of the earth, (when one has a realization of the entry problem), if its perigee distance is smaller than or equal to the radius ρ of the earth, that is, in the case of elliptical orbit, if the condition:

$$EP'' = a\,(1 - e) \leqq \rho \qquad (6)$$

is satisfied, Figure 1. The quantities a and e are dependent, as we saw, upon known and calculated angles and lengths, and upon the time $t'$ of the duration of the force $T$.

III. THE SOLUTION IN RESTRICTED PROBLEMS

By restricting the force $T$, that is by imposing restrictions to its magnitude and direction, which are valid during its action on the body, a variety of entry problems may occur, when the above general procedure for their solution is simplified.
We consider two restricted cases of the entry problem of special interest.

**(a) Force *T* Constant.**

If the magnitude and the direction of $T$ are constant, the problem is simplified. As an example, in case $T_1 = 0$, $T_2 = 0$, $T_3 \doteq T$, and large $r_0$ and small $\tau$, formulae (2) and (3) become:[1(b)]

$$r(\tau) = \left(r_0 - \frac{\mu\tau^2}{2r_0^2}\right) r_0^* + \left(s_0\tau - \frac{\mu s_0 \tau^3}{6 r_0^3}\right) s_0^* + \frac{1}{2} T \tau^2 T_0^* \tag{7}$$

$$\dot{r}(\tau) = V = -\frac{\mu\tau}{r_0^2} r_0^* + \left(s_0 - \frac{\mu s_0 \tau^2}{2r_0^3}\right) s_0^* + T\tau T_0^* \tag{8}$$

**(b) Two-dimensional Entry Problem.**

The two-dimensional entry problem occurs in case the force $T$ is on the plane of the original orbit. In this case the non-Keplerian arc and the new orbit are on the plane of the original orbit, and the coordinate system ($P_0$; $r_0^*$, $s_0^*$) is suggested. All calculations of the general procedure of the previous section are simplified, especially in case of large $r_0$ and small $\tau$, when the following formulae can be used:

$$r(\tau) = \left(r_0 - \frac{\mu\tau^2}{2r_0^2}\right) r_0^* + \left(s_0\tau - \frac{T\tau^2}{2} - \frac{\mu s_0 \tau^3}{6 r_0^3}\right) s_0^* \tag{9}$$

$$\dot{r}(\tau) = V = -\frac{\mu\tau}{r_0^2} r_0^* + \left(s_0 - T\tau - \frac{\mu s_0 \tau^2}{2 r_0^3}\right) s_0^* \tag{10}$$

REMARK. The present mathematical discussion of the entry problem can be taken as basic discussion of «physical entry problems».

REFERENCES

1. D.G. Magiros: «Proceedings of International Astronautical Congress», (a) XIV 1963 (Paris), (b) XV 1964 (Warsaw), (c) XVI 1965 (Athens).
2. A.E. Roy: «The foundations of Astrodynamics», The MacMillan Company, New York, (1965), pg. 80, 83.

ΠΕΡΙΛΗΨΙΣ

Τὸ πρόβλημα τῆς εἰσόδου ἑνὸς τεχνητοῦ δορυφόρου ἐξετάζεται εἰς τὴν παροῦσαν ἐργασίαν ὡς ἐφαρμογὴ γενικωτέρας ἐρεύνης δημοσιευθείσης ὑφ' ἑνὸς τῶν συγγραφέων. Ἡ καμπύλη μεταφορᾶς λαμβάνεται συμφώνως πρὸς τὴν ἐργασίαν 1 (b). Ἡ τροχιὰ προορισμοῦ πρέπει νὰ πληροῖ τὴν συνθήκην (6). Δίδονται δύο εἰδικαὶ περιπτώσεις.

*

Ὁ Ἀκαδημαϊκὸς κ. Ἰω. Ξανθάκης κατὰ τὴν ἀνακοίνωσιν τῆς ἀνωτέρω ἐργασίας εἶπε τὰ κάτωθι:

*Εἰς τὴν παροῦσαν ἐργασίαν ἐξετάζεται τὸ «Πρόβλημα τῆς Εἰσόδου» ὑπὸ τὴν ἀκόλουθον μορφήν:*

*«Τεχνητὸς δορυφόρος τῆς Γῆς κινεῖται πέριξ αὐτῆς εἰς ἐλλειπτικὴν τροχιάν. Μία γενικὴ δύναμις ἐφαρμόζεται ἐπὶ τοῦ δορυφόρου, δρᾷ ἐπὶ χρονικόν τι διάστημα καὶ μετὰ παύει δρῶσα, ὁπότε ὁ δορυφόρος ἀρχίζει νὰ κινῆται ἐπὶ νέας Κεπλερείου τροχιᾶς. Ζητοῦνται αἱ συνθῆκαι καὶ οἱ περιορισμοὶ ὑπὸ τοὺς ὁποίους ἡ νέα τροχιὰ δύναται νὰ συναντήσῃ τὴν ἐπιφάνειαν τῆς Γῆς».*

*Ἡ λύσις τοῦ παρόντος προβλήματος ἐπιτυγχάνεται δι' ἐφαρμογῆς τῶν πορισμάτων προηγηθεισῶν ἐργασιῶν ἑνὸς ἐκ τῶν συγγραφέων τῆς παρούσης ἐρεύνης. Οἱ ζητούμενοι περιορισμοὶ προκύπτουν ἐκ τοῦ γεγονότος ὅτι «ἡ ἐκ τοῦ περιγείου ἀπόστασις τοῦ δορυφόρου τῆς νέας Κεπλερείου τροχιᾶς δέον νὰ εἶναι μικροτέρα ἢ ἴση τῆς ἀκτῖνος τῆς Γῆς, θεωρουμένης σφαιρικῆς».*

*Εἶτα ἐξετάζεται τὸ πρόβλημα τοῦτο τῆς εἰσόδου εἰς τὰς μερικὰς περιπτώσεις κατὰ τὰς ὁποίας:*

*α) Ἡ δρῶσα δύναμις ἔχει σταθερὸν μέγεθος καὶ διεύθυνσιν καὶ*

*β) Αὕτη κεῖται ἐπὶ τοῦ ἐπιπέδου τῆς ἀρχικῆς τροχιᾶς.*

## PART III

# DYNAMICAL SYSTEMS ANALYSIS

In this Part, Magiros' important papers on stability analysis, precessional phenomena and separatrices of dynamical systems were selected. In papers 30 and 31 a unified formulation of the three basic stability concepts in the sense of Lyapunov, Poincaré and Lagrange is given, on the basis of an appropriate interpretation of momentary or permanent effects of the perturbations acting on the system. Specialized stability concepts, as well as some relationships between the general stability concepts are indicated. By appropriate geometrical interpretations, the stability concepts are clarified. Illustrative and application examples are also included. Paper 32 is referred to the "in-orbital-plane" angular motion, and its stability characteristics of a spherical shape satellite, acted upon by aerodynamic and gravitational torques. Different stability concepts are employed and conditions are determined for the stability of the equilibrium attitudes. Paper 33 deals with the stability concepts of the class of DE with deviating arguments, and Paper 34 provides some remarks on stability concepts, by which a better understanding of current stability problems can be obtained. Finally paper 35 is a unification of the concepts and results involved in the previous papers. Paper 42 is a brief version of paper 35, published in Russian.

In paper 36 the stability of the precessions defined and studied in a previous paper of Magiros (see [32] in the full list of publications) is treated on the basis of remarks given in paper 31. The main results are: (i) the "regular" and "non-regular periodic" precessions are stable, but the helicoid precession is unstable in the Lagrange sense, (ii) The "regular" and "non-regular periodic" precessions are stable, but not asymptotically stable in the Poincaré sense. The "helicoid" precession is asymptotically stable in the

Poincaré sense, and (iii) The "non-regular periodic" and the "helicoid" precessions are unstable in the Lyapunov sense, but the "regular" precession is either unstable or stable in the Lyapunov sense, depending on the sudden perturbations which affect or do not affect, respectively, the initial angular velocity component along the roll axis. In paper 37 a re-entry problem is solved through the concept of the helicoid precession. This is the problem of determining the orientation of a spin-stabilized axi-symmetric re-entry vehicle. In paper 38 the "helicoid precession concept" is used as a mean to treat the problem of finding the error of the orientation of the angular momentum vector of a re-entry vehicle at the end of spin-up, and in paper 39 the stability of a class of helicoid precessions is studied. It is found that all members of the class are Lyapunov unstable, but Poincaré (orbitaly) stable, or asymptotically stable, or unstable, if the limiting value of the "pitch distance" of the member at hand is either constant, or zero, or infinite, respectively.

Finally, paper 40 presents new results for the determination of separatrices of dynamical systems and their usefulness in the analysis of physical systems. On the basis of the new definition of separatrices, theorems are formulated for the existence, the number, and the determination of separatrices. The basic concepts used are the concepts of "nonlinearities" and "sectors" with apex singular points, of the system.

Paper 41 is also devoted to the study of separatrices of dynamical systems. Using a supplementery definition of separatrices, some theorems are proved concerning separatrices at the neighborhood of elementary and nonelementary critical points in the plane.

# ON STABILITY DEFINITIONS OF DYNAMICAL SYSTEMS

By Demetrios G. Magiros

GENERAL ELECTRIC COMPANY, VALLEY FORGE SPACE TECHNOLOGY CENTER, PHILADELPHIA, PENNSYLVANIA

*Communicated by Leon Brillouin, March 25, 1965*

1. *Introduction.*—The dominant subject in the analysis of dynamical systems is the stability of the insuing motion. Many fields of interest under current research involve dynamical systems. These range from fields such as astrodynamics, meteorology, and control systems to biology, chemistry, medicine, and economics. In each of these fields dynamical systems can be formulated, and their stability can be examined as special cases of stability problems in dynamics. During recent years the stability concepts of dynamical systems have been advanced, either by modifying old ideas or by creating new stability concepts. These advances permit a deeper penetration into the more profound stability problems.

In this note, stability concepts of physically realistic dynamical systems, which are mathematically consistent, are discussed. Based on appropriate interpretation of the effects of perturbations, acting on the system either momentarily or permanently, a unified formulation of the three basic stability concepts in the sense of Liapunov, Poincaré, and Lagrange is given. Specialized stability concepts, introduced within the basic concepts, are indicated by appropriate remarks, and some relationships between the stability concepts are emphasized, and the stability concepts are clarified by geometrical interpretations.

2. *Physical Stability Considerations.*—Physically, we may have three basic stability aspects, depending on stability considerations of a motion on its orbit, of the orbit of a motion, and of the boundedness of the motion and its orbit.

A motion or its orbit is considered as stable, if, by giving a small disturbance to the motion or to its orbit, the disturbed motion or its orbit, initially near to the unperturbed motion or to its orbit, remains near to it for all time. More specifically,

If for small disturbances the effect on the motion or on its orbit is small, we say that the motion or its orbit is in a "stable" situation.

If for small disturbances the effect is considerable, the situation is "unstable."

If for small disturbances the effect tends to disappear, the situation is "asymptotically stable."

If, regardless of the magnitude of the disturbances, the effect tends to disappear, the situation is "asymptotically stable in the large."

Reprinted from the *Proceedings of the National Academy of Sciences* **53** (1965), 1288–1294.

The stability in the sense of Liapunov and Poincaré is based on the above physical stability definitions.

The boundedness of all motions and orbits of a system in connection with bounded disturbances is another physical stability aspect, on which the stability in the sense of Lagrange is based.

These three different stability aspects are of a qualitative type. In the following paragraphs a unified quantitative discussion and a geometrical interpretation of these stability aspects are given.

The disturbances are due to perturbations, which are considered as minor disturbing forces acting on the system either momentarily or constantly.

3. *The Effects of Perturbations.*—The equations of the dynamical system are

$$\dot{x}_i(t) = X_i(t, x_1, \ldots x_n), \qquad i = 1, \ldots n$$

$$X_i(t, 0, \ldots, 0) = 0, \; x_i(t_0) = x_{i0}. \tag{1}$$

Let $x_i(t)$ be a solution of the system (1), that is, a motion with orbit $L$ (Fig. 1). The effect of perturbations is a change of certain quantities dependent on the motion, that is, a change of the motion into the perturbed motion $\bar{x}_i(t)$ with orbit $\bar{L}$.

If the distance $\rho$ between a point $P$ of $L$ and a point $\bar{P}$ of $\bar{L}$ measures the effect of perturbations at the point $P$, the points $P$ and $\bar{P}$ are correspondent points of the unperturbed motion $x_i(t)$ and the perturbed one $\bar{x}_i(t)$.

We may have different kinds of correspondence between points $P$ of $L$ and points $\bar{P}$ of $\bar{L}$, and this correspondence characterizes different stability concepts of the motion.

We distinguish here two such correspondences, most physical, on which two important stability concepts are based, namely, the stability in the sense of Liapunov and Poincaré.

(*a*) The points $P$ and $\bar{P}$ are points of $L$ and $\bar{L}$, respectively, at the "same time" (Fig. 1), either in the phase space or in the parameter space of the system, and the distance $\rho = P\bar{P}$ is "time-dependent," namely, we have

$$\rho = \left\{\sum_{i=1}^{n} [x_i(t) - \bar{x}_i(t)]^2\right\}^{1/2} \quad \text{in the phase space, and}$$

$$\rho = \left\{\sum_{i=1}^{n} [x_i(t,a_j) - x_i(t,\bar{a}_j)]^2\right\}^{1/2}, \quad \text{in the parameter space.} \tag{2}$$

This kind of correspondence is an appropriate one for stability discussion of a motion on its orbit. We call "Liapunov distance" the distance $\rho$ defined with the above correspondence between points $P$ and $\bar{P}$.

(*b*) The points $P$ and $\bar{P}$ on $L$ and $\bar{L}$ correspond to each other in such a way that the distance $\rho = P\bar{P}$ is the minimum from the distances of $\bar{P}$ from all points of $L$:

$$\rho = \min\left\{\sum_{i=1}^{n} (x_i - \bar{x}_i)^2\right\}^{1/2}, \tag{3}$$

When $\bar{P}$ is sometimes on the normal to $L$ at $P$ (Fig. 2), and the distance $\rho$, defined in this way, is "time-independent".

This kind of correspondence is an appropriate one for stability discussion of the

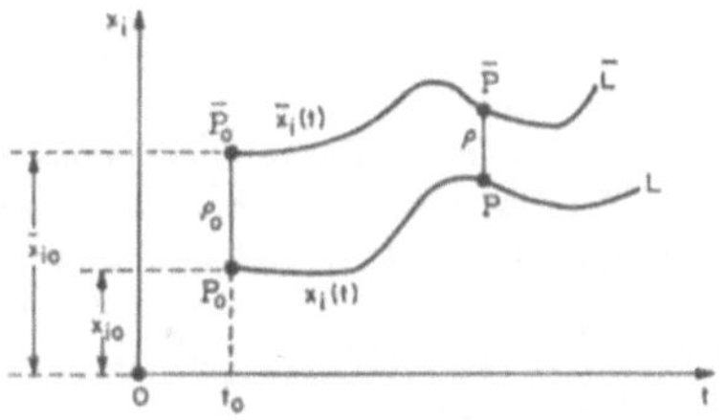

FIG. 1.

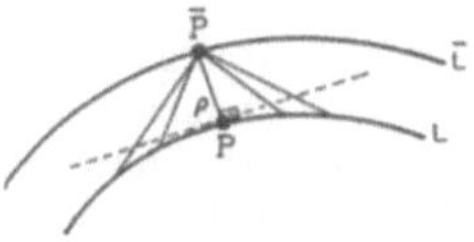

FIG. 2.

orbit of a motion. We call "Poincaré distance" the distance $\rho$ defined in the above manner.

4. *Stability Definitions of a General Trajectory.*—The distance $\rho$ defined in the preceding can be used for finding stability definitions of an analytic type in agreement with the physical stability definitions.

The motion $x_i(t)$ of the system (1) is said to be "stable," if, for any given positive number $\epsilon$, however small, a positive number $\delta$, depending on $\epsilon$ and, in general, on the initial time $t_0$, can be chosen such that, for any perturbed motion $\bar{x}_i(t)$, the inequality

$$\rho_0 < \delta, \tag{4}$$

where $\rho_0$ is the distance initially, implies

$$\rho < \epsilon \tag{5}$$

for all time $t \geq t_0$.

The motion is "unstable," if, given $\epsilon$ as above, for sufficiently small positive $\delta$, the inequality (5) is not satisfied even for at least one perturbed motion.

The motion is "asymptotically stable," if it is stable, and, in addition, a positive number $\delta_1 \geq \delta$ exists such that, starting from any $\rho_0$, which is

$$\rho_0 < \delta_1, \tag{4a}$$

the limiting condition

$$\lim_{t \to \infty} \rho = 0 \tag{6}$$

holds uniformly relative to the quantities $t_0$, $x_{i0}$.[1a]

The motion is "quasi-asymptotically stable," if, from the above stability conditions, only the limiting condition (6) holds.

The motion is "asymptotically stable in the large," or "globally stable," or "completely stable," if it is asymptotically stable and the number $\delta_1$ is very large.

*Remarks:* (*a*) The above definitions are "stability definitions in Liapunov sense," either in the phase space or in the parameter space, if the distances $\rho_0$ and $\rho$ are "Liapunov distances" either in phase space or in parameter space, respectively.

The definitions are "stability definitions in Poincaré sense," or "orbital definitions," if the distances $\rho_0$ and $\rho$ are "Poincaré distances."

(*b*) In case the orbit $L$ shrinks to an equilibrium point of the system, the distances in Liapunŏv and Poincaré sense are identical, and therefore the stability definitions in Liapunov and Poincaré sense are equivalent at the singular points of dynamical systems.

(c) A nonlinear autonomous system, of which no eigenvalue has positive real part, may have, under some appropriate restrictions of the nonlinearities, the origin as an equilibrium point asymptotically stable in the large. Such nonlinear systems are called "absolutely stable," and there are methods to find classes of nonlinearities for absolute stability of the systems. The above kind of stability of the systems plays an important role in modern technology.[2]

(d) The stability definitions in Liapunov sense in periodic motions are narrow definitions, because they classify as unstable motions some motions which can be practically considered as stable. The definitions in Poincaré sense are the appropriate ones in periodic motions.

(e) In case $\delta$ is independent of $t_0$ as in autonomous systems, the definitions are "uniform"; so we may have "uniform stability," "uniform asymptotic stability," etc., in Liapunov or in Poincaré sense. In nonautonomous systems the selection of the initial time $t_0$ is not free; there is in those systems a value $\tau$ of time such that $t_0 \geq \tau$.[3]

(f) The stability conditions (4) and (5), in case of Liapunov stability in parameter space of initial conditions, become conditions of continuity of the solution $x_i\,(t, t_0, x_{10}, \ldots, x_{n0})$ of (1) on the initial conditions. The continuity of the solution on the initial conditions, which is one of the three "Hadamard's postulates" in order that the solution be accepted physically, is then a special stability case in Liapunov sense.[4]

(g) If the stability situation of the motion $x_i(t)$ is invariant in a parameter space $S$ of the system, we speak about a "structural stability" of the system in the parameter space $S$, which is the "domain of structural stability" of the system. This property of the system gives information about the "insensitivity" of the system to perturbations.

5. *Lagrange Stability of a System.*—The boundedness of all motions and orbits of a system is the third basic stability concept of the system.

If $\rho$ is the distance of the origin of the system from any point of any solution $x_i(t, t_0, x_{10}, \ldots, x_{n0})$ of the system, and if, given any finite positive number $\epsilon$, we can find another finite positive number $\delta$ such that condition (4) implies condition (5) for all $t \geq t_0$, the system is "stable in Lagrange sense." The system is "unstable in Lagrange sense," if its response has the tendency to grow without bounds.

*Remarks:* (a) The distance $\rho$ used in stability conditions in Lagrange sense is identical to the distance in Liapunov or in Poincaré sense applied to origin.

(b) The three basic stability concepts are given in the preceding, independently of each other. We justify this by the following examples:

The solution $x = t + c$, where $c$ is arbitrary constant, of the equation $\dot{x} = 1$, is unstable in Lagrange sense, but it is stable in Liapunov sense.

The family of solutions $x = a \sin (at + b)$, where $a$ and $b$ are parameters, of the equation

$$\ddot{x} = -\,{}^1\!/_2\, x\{x^2 + (x^4 + 4\dot{x}^2)^{1/2}\} \tag{7}$$

is stable in Poincaré and Lagrange sense, but unstable, except the zero solution, in Liapunov sense, when the ratio of any two values of $a$ is irrational.

(*c*) Nevertheless, the stability concepts are in many cases connected. For example:[5]

In linear homogeneous systems, $\dot{x} = \sum_{j=1}^{n} a_{ij}(t)x_j$ with $a_{ij}(t)$ continuous in $t \geq t_0$, the Lagrange stability implies the Liapunov stability, and conversely.

In linear inhomogeneous systems, $\dot{x} = \sum_{j=1}^{n} a_{ij}(t)\ x_j + f_i(t)$ with $a_{ij}(t)$ and $f_i(t)$ continuous in $t \geq t_0$, the Lagrange stability implies the Liapunov stability, but for the converse the boundedness of one solution is needed.

In nonlinear systems, connections between the stability concepts either are, in general, hard to find or do not exist.

6. *Stability under Persistent Perturbations.*—If $p_i(t, x_1, \ldots, x_n)$ are persistent perturbations, the equations of the dynamical system are

$$\dot{x}_i(t) = X_i(t, x_1, \ldots, x_n) + p_i(t, x_1, \ldots, x_n), \tag{8}$$

where $p_i$ must be such that (8) has unique solutions corresponding to given initial conditions. The solution $x_i(t)$ of (1) is "stable under persistent perturbations," if, in addition to conditions (4) and (5), some appropriate restrictions for the magnitude of $p_i$ are accepted, and which hold for all accepted $x_i$ and $t \geq t_0$.

In case the magnitude of the perturbations is according to

$$\max|p_i(t, x_1, \ldots, x_n)| < \eta, \tag{9}$$

where $\eta$ is an appropriate positive number, we speak about "total stability" of the solution $x_i(t)$.

If the condition (9) is replaced by

$$\int_0^\infty \max|p_i(t, x_1, \ldots, x_n)|\ dt < \eta, \tag{9a}$$

we speak about "integral stability."

In case of total stability the perturbations must be small, but in case of integral stability they may be large in a small interval of time.

The properties of both these types are possessed by stability under persistent perturbations "bounded in the mean," when we have "stability in the mean." The last type of stability is obtained, if, instead of condition (9), we have

$$\int_t^{t+T} \max|p_i(t, x_1, \ldots, x_n)|\ dt < \eta, \tag{9b}$$

where $\eta$ and $\delta$ of (4) depend on T, in general.

If, in addition to the above, condition (6) holds, we have "stability of asymptotic type."

The above stabilities are in Liapunov or in Poincaré sense, if the distances $\rho_0$ and $\rho$ in (4) and (5) are Liapunov or Poincaré distances.

There are relationships between the above stability under persistent perturbations. For example:[6]

Asymptotic integral stability implies asymptotic stability in the mean, and conversely.

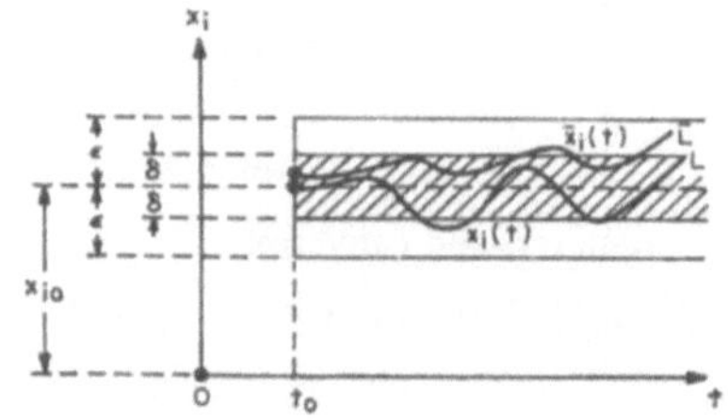

FIG. 3.

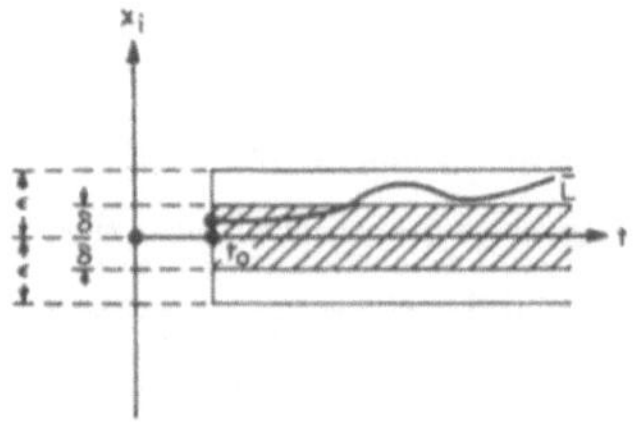

FIG. 4.

Asymptotic integral stability implies total asymptotic stability, and this implies total stability.

Asymptotic stability in the mean implies stability in the mean, and this implies integral stability.

*Remarks:* (*a*) The preceding analytical stability definitions need appropriate modifications, changes, supplements in order to be useful practically, that is, in order to meet the requirements for a "practical stability." These requirements are:[1b, 7] "to know the size of the acceptable deviations of a state, of the region of the appropriate initial conditions, of the acceptable perturbations, and the finite time for which the above happens."

(*b*) The character of the stability, by using any of the previous concepts, is not invariant with a general transformation of the coordinates of the system, that is, the stability depends on the coordinate system with reference to which the variables are to be considered.

7. *Geometric Interpretation of Stability Definitions.*—We clarify the stability concepts by their geometrical interpretation. We call "$r$-tube" around a curve $l$ in $n$-dimensional space the set of all points of which the distance $\rho$ in Poincaré sense from $l$ is smaller than $r$; $r$ is the width and $l$ the central curve of the tube. For finite $r$ the tube is bounded.

The above definitions can appropriately be used for a geometrical interpretation of the stability concepts.

For stability in the Liapunov sense of a general motion $x_i(t)$, we consider the "$\delta$-tube" and "$\epsilon$-tube" around the straight line $x_i = x_{i0}$ which is parallel to $t$-axis through the point $(t_0, x_{i0})$.

A motion $x_i(t, t_0, x_{10}, \ldots, x_{n0})$ of orbit $L$ is stable in Liapunov sense, if given the "$\epsilon$-tube" around the line $x_i = x_{i0}$ (Fig. 3), we can find a "$\delta$-tube" around the line $x_i = x_{i0}$ such that any perturbed motion of orbit $\bar{L}$, starting in the "$\delta$-tube." remains for all $t \geq t_0$ in the "$\epsilon$-tube."

In case of equilibrium points, that is, the origin, the above tubes are around the $t$-axis (Fig. 4).

For the asymptotic stability we need, in addition, another tube, "$\delta_1$-tube," which will be unbounded for asymptotic stability in the large.

For stability of a general motion in Poincaré sense, that is, for "orbital stability" of the motion, the "$\delta$-tube" and "$\epsilon$-tube" around the orbit $L$ can be used. A motion is "orbitally stable" if, given the "$\epsilon$-tube" around $L$, we can find a "$\delta$-tube" around $L$ such that any perturbed orbit $\bar{L}$ starting in the "$\delta$-tube" remains always in the "$\epsilon$-tube" (Fig. 5).

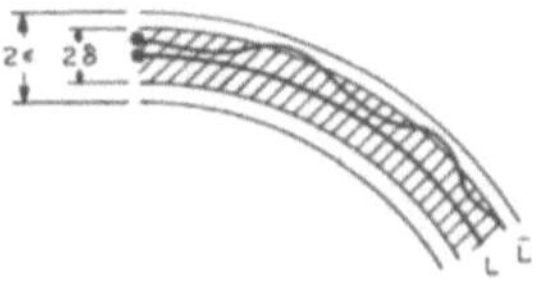


FIG. 5.

For stability of equilibrium points, the orbit $L$ is the $t$-axis, then the tubes are around the $t$-axis (Fig. 4), and the discussion is the same as in Liapunov case. For the periodic motions, the tubes around the closed orbit $L$ are tori with $L$ as central curve.

For Lagrange stability, bounded tubes around the $t$-axis (Fig. 4) can be used.

The system is "Lagrange stable," if for every bounded "$\epsilon$-tube" around the $t$-axis another bounded "$\delta$-tube" around the $t$-axis can be found such that every orbit of the system starting in the "$\delta$-tube" does not leave the "$\epsilon$-tube" for all time $t \geq t_0$.

For stability under persistent perturbations the "$\delta$-tube" and "$\epsilon$-tube" must be around the $t$-axis, but, in addition, an appropriate "$\eta$-tube" around the $t$-axis is needed.

The author wishes to express his thanks to O. Klima, R. Chapman, and J. Weinstein of the General Electric Company, Missile and Space Division, Valley Forge Space Technology Center, the first two for their encouragement and help in connection with this research work, and the third for his assistance in editing this paper.

[1] Zubov, V. I., *Mathematical Methods for the Study of Automatic Control Systems* (New York: The Macmillan Co., 1963): (*a*) chap. I, §1; (*b*) p. 13.

[2] Aizerman, M., and F. Gantmaher, *Absolute Stability of Regulator Systems* (San Francisco: Holden-Day, Inc., 1964).

[3] LaSalle, J., and R. Rath, "Eventual stability," in *Proceedings of the Fourth Joint Automatic Control Conference*, University of Minnesota, Minneapolis (1963), pp. 468–470.

[4] Magiros, D. G., *Physical Problems Discussed Mathematically*, General Electric Co., Valley Forge Space Technology Center, Technical Report No. 64SD286, October 1964.

[5] Cesari, L., *Asymptotic Behavior and Stability Problems in Ordinary Differential Equations* (Berlin: Springer-Verlag, 1959), pp. 7–13.

[6] Vrkŏv, I., "On some stability problems," "Differential equations and their applications," in *Proceedings of a Conference in Prague*, September 1962 (New York: Academic Press, 1963), pp. 217–221.

[7] LaSalle, J., and S. Lefshetz, *Stability by Liapunov's Direct Method* (New York: Academic Press, 1961), p. 121.

# Stability Concepts of Dynamical Systems

DEMETRIOS G. MAGIROS*

*General Electric Company, Missile and Space Division,*
*Philadelphia, Pennsylvania*

## I. INTRODUCTION

Many fields of interest under current research involve dynamical systems. In these fields dynamical systems can be formulated, and their stability examined as a special case of problems of stability in dynamics.

One can distinguish classes of concepts of stability depending on the nature of the dynamical systems, the manner in which the system approaches a given state or deviates from it, the properties of the perturbations of the system, and the space variables selected.

During recent years the concepts of stability of dynamical systems have been advanced, either by modifying old ideas or by creating new ones, and these advances permit a deeper penetration into the more profound problems of stability.

In this paper, of which a first draft is published in the Proceedings of National Academy of Sciences (Magiros, 1965a), we discuss the three basic concepts of stability in the sense of Liapunov, Poincaré, and Lagrange, and some specialized ones. Based on an appropriate interpretation of the effects of perturbations, acting on the systems either momentarily or permanently, a unified formulation of the basic concepts of stability is given. Some relationships between the concepts of stability are emphasized and appropriate examples and geometrical interpretations of the concepts clarify the discussion.

## II. PHYSICAL STABILITY CONSIDERATIONS

Physically, one may have three basic aspects of stability depending on stability considerations of the motion in a given orbit, of the orbit of a given motion and of the boundedness of the motion and its orbit.

* The author wishes to express his thanks to O. Klima and R. Chapman of General Electric Company, Missile and Space Division, for their encouragement and help with this research work.



Reprinted from *Information and Control* 9 (1966), 531–548.

A motion or its orbit is considered stable, if, by giving a small disturbance to the motion or to its orbit, the disturbed motion or its orbit, remains close to the unperturbed one for all time.

More specifically.

a. If for small disturbances the effect on the motion or on its orbit is small, one says that the motion or its orbit is in a "stable" situation.

b. If for small disturbances the effect is considerable, the situation is "unstable."

c. If for small disturbances the effect tends to disappear, the situation is "asymptotically stable."

d. If, regardless of the magnitude of the disturbances, the effect tends to disappear, the situation is "asymptotically stable in the large." The stability in the sense of Liapunov and Poincaré is based on the above physical stability definitions.

The boundedness of motions and orbits of a system in connection with bounded disturbances is another physical aspect of stability, on which the stability in the sense of Lagrange is based.

These three different aspects of stability are of a qualitative type. In the following a unified quantitative discussion of these aspects of stability is given.

## III. THE EFFECTS OF PERTURBATIONS

The disturbances of the systems are due to perturbations, which are considered as minor disturbing forces acting on the system either momentarily or constantly.

The equations of a nonautonomous dynamical system in case of sudden perturbations are:

$$\begin{aligned} \dot{x}_i(t) &= X_i(t, x_1, \cdots, x_n), \qquad x_i(t_0) = x_{i0} \\ X_i(t, 0, \cdots, 0) &\equiv 0; \qquad i = 1, \cdots, n \end{aligned} \tag{1}$$

Let $x_i(t)$ be a solution of the system (1), that is, a motion with orbit $L$, Fig. 1. The effect of perturbations is a change of certain quantities dependent on the motion, that is, a change of the motion into the perturbed motion $\bar{x}_i(t)$ with orbit $\bar{L}$.

If the distance $\rho$ between a point $P$ of $L$ and a point $\bar{P}$ of $\bar{L}$ measures the effect of perturbations at the point $P$, the points $P$ and $\bar{P}$ are corresponding points of the unperturbed motion $x_i(t)$ and the perturbed one $\bar{x}_i(t)$.

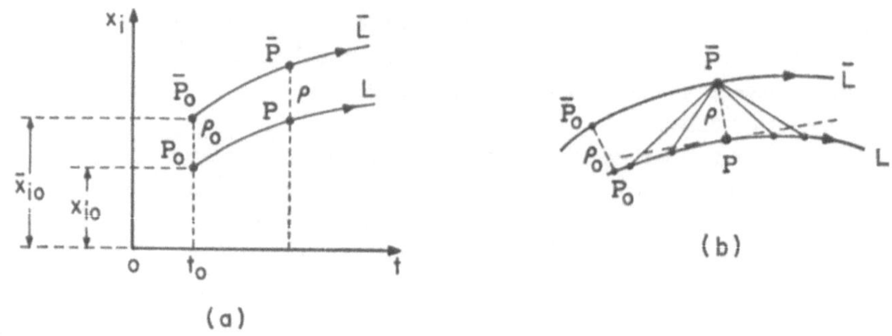


FIG. 1

One may have different kinds of correspondence between points $P$ of $L$ and points $\bar{P}$ of $\bar{L}$, and this correspondence characterizes different stability concepts of the motion.

One can distinguish two such correspondences, which are mostly physical, on which two important concepts of stability are based, namely, the stability in the sense of Liapunov and Poincaré:

a. The points $P$ and $\bar{P}$ are points of $L$ and $\bar{L}$, respectively, at the "same time," Fig. 1(a), either in phase space or in parameter space of the system, and the distance $\rho = P\bar{P}$ is "time-dependent," namely, one has:

$$\rho = \left\{\sum_{i=0}^{n} [x_i(t) - \bar{x}_i(t)]^2\right\}^{1/2}, \quad \text{in the phase space, and}$$

$$\rho = \left\{\sum_{i=0}^{n} [x_i(t, a_j) - x_i(t, \bar{a}_j)]^2\right\}^{1/2}, \quad \text{in the parameter space.}$$

This kind of correspondence is an appropriate one for stability discussion of a motion in its orbit.

We call "Liapunov distance" the distance $\rho$ defined with the above correspondence between points $P$ and $\bar{P}$.

b. The points $P$ and $\bar{P}$ on $L$ and $\bar{L}$ correspond to each other in such a way that the distance $\rho = P\bar{P}$, Fig. 1(b), is the minimum from the distances of $\bar{P}$ from all points of $L$:

$$\rho = \min\left\{\sum_{i=1}^{n} (x_i - \bar{x}_i)^2\right\}^{1/2}$$

The distance $\rho$ defined in the above manner is "time-independent."

This correspondence is an appropriate one for stability discussion of the orbit of a motion.

We call "Poincaré distance" the distance $\rho$ defined in the above manner.

## IV. STABILITY DEFINITIONS OF A GENERAL MOTION

The distance $\rho$ defined in the preceding can be used for finding stability definitions of a general motion of analytical type in agreement with the physical stability definitions.

a. A motion $x_i(t)$ of the system (1) is said to be "stable," if, for any given positive number $\epsilon$ and initial time $t_0 \geqq 0$, a positive number $\delta$, depending on $\epsilon$ and, in general, on $t_0$, can be found such that, for any perturbed motion $\bar{x}_i(t)$, the inequality

$$\rho_0 < \delta \tag{2}$$

where $\rho_0$ is the distance initially, implies, for all time $t \geqq t_0$,

$$\rho < \epsilon \tag{3}$$

b. The motion is "unstable", if, given $\epsilon$ as above, for sufficiently small positive $\delta$ the inequality (3) is not satisfied even for at least one perturbed motion.

c. The motion is "asymptotically stable", if it is stable and in addition a positive number $\delta_1 \geqq \delta$ exists such that, starting from any $\rho_0$, which is:

$$\rho_0 < \delta_1 \tag{2a}$$

the limiting condition:

$$\lim \rho = 0 \tag{4}$$

holds uniformly to the quantities $t_0$, $x_{i0}$ (Žubov, 1963a).

d. The motion is "quasi-asymptotically stable", if, from the above stability conditions, only the limiting condition (4) holds.

e. The motion is "asymptotically stable in the large", if it is asymptotically stable and the number $\delta_1$ is very large.

The above definitions express the stability definitions in a unifying way, since in the above conditions the kind of the distance is not yet specified.

## V. REMARKS

### Remark 1: Stability Definitions in the Liapunov and Poincaré Senses

The above definitions are "stability definitions in the Liapunov sense," either in phase space or in parameter space, if the distances $\rho_0$ and $\rho$ are "Liapunov distances" either in phase space or in parameter space, respectively.

The definitions are "stability definitions in the Poincaré sense," or "orbital definitions," if the distances $\rho_0$ and $\rho$ are "Poincaré distances."

### Remark 2: Equivalence of Liapunov and Poincaré Definitions in Equilibrium Points

In case the orbit $L$ shrinks to an equilibrium point of the system, the distances in the Liapunov and Poincaré senses are identical, and therefore the stability definitions in the Liapunov and Poincaré senses are equivalent at the singular points of dynamical systems.

### Remark 3: Stability Definition in the Lagrange Sense

The Lagrange stability of a system, namely, the stability concept in connection with the boundedness of motions and orbits of the system, is given by the preceding general stability conditions.

If $\rho$ is the distance of the origin of the system from a point of any solution $x_i(t, t_0, x_{i0}, \cdots, x_{n0})$ of the system, and if, given any finite positive number $\epsilon$, one can find another finite positive number $\delta$ such that condition (2) implies condition (3) for all time $t \geqq t_0$, the system is "stable in the Lagrange sense" (Hahn, 1963a).

The system is "unstable in Lagrange sense," if its response can grow without bounds.

The distance $\rho$ used in stability conditions in Lagrange sense is identical with the distance in Liapunov or in Poincaré sense applied to origin.

### Remark 4: Stability at Right and/or at Left

If the above stability definitions hold for $t \geqq t_0$, one can speak about "stability at right"; if they hold for $t \leqq t_0$, one can speak about "stability at left." One may have "stability at right and left" in case the conditions hold for all time (Cesari, 1959a).

REMARK 5: STABILITY DEFINITIONS IN AUTONOMOUS AND NONAUTONOMOUS SYSTEMS

a. If $\delta$ is independent of $t_0$, as in the case of autonomous systems or nonautonomous but periodic ones, the definitions are called "uniform"; so one may have "uniform stability," "uniform asymptotic stability," etc., in the Liapunov or in the Poincaré sense. In nonautonomous systems the selection of the initial time $t_0$ is not free, as, e.g., for stability of equilibrium points there is a value $\tau$ of time such that $\tau \leqq t_0$ (La Salle and Rath, 1963).

b. There is a one-to-one correspondence between the definitions of stability expressed by the previous simple stability conditions where the distances are either Liapunov or Poincaré distances, and the known definitions of stability of invariant sets (Zubov, 1964) applied to equilibrium states or to periodic motions.

REMARK 6: CONTINUOUS DEPENDENCE OF A SOLUTION ON INITIAL CONDITIONS

In case a solution $x(t, x_0)$ of the system is stable in the special parameter region of the initial conditions, for any two sets of values $x_{01}$ and $x_{02}$ of the initial conditions and for appropriate numbers $\epsilon$ and $\delta$, the two inequalities:

$$\rho_0 = \{\sum (x_{01} - x_{02})^2\}^{1/2} < \delta, \qquad \rho = \{\sum [x(t, x_{01}) - x(t, x_{02})]^2\}^{1/2} < \epsilon$$

are compatible.

These two inequalities are the conditions for a continuous dependence of the solution $x(t, x_0)$ on the initial conditions $x_0$ uniformly in $t$, and this property of the solution is a Hadamard's postulate of the solution for being physically acceptable (Magiros, 1965b) .Therefore, "the continuous dependence of a solution on the initial conditions corresponds to a stability situation of the solution in the Liapunov sense in a special parameter space, namely, in the initial conditions space."

REMARK 7: STABILITY OF PERIODIC MOTIONS

The stability definitions in Liapunov sense of periodic motions are narrow definitions, because they classify as unstable motions some motions which can be practically considered as stable.

The definitions in Poincaré sense are the appropriate ones for periodic motions.

REMARK 8: STABILITY OF THE ORIGIN WITH RESPECT TO ONLY CERTAIN COORDINATES (Zubov, 1963b)

If $x_i(t; x_{10}, \cdots, x_{n0})$ is a perturbed solution of a system through the point $(x_{10}, \cdots, x_{n0})$, and the distances $\rho$, $\rho_k$, $\rho_{\bar{k}}$ from the origin are given by the expressions:

$$\rho = \left\{\sum_{i=1}^{k} x_i^2(t, t_0, x_{10}, \cdots, x_{n0})\right\}^{1/2}, \rho_k = \left\{\sum_{i=1}^{k} x_{i0}^2\right\}^{1/2}, \rho_{\bar{k}} = \left\{\sum_{i=k+1}^{n} x_{i0}^2\right\}^{1/2}$$

the origin of the system is "stable with respect to the coordinates $x_1, \cdots, x_k$," $k < n$, if, for every small positive number $\epsilon$, there exist two positive numbers $\delta_1$ and $\delta_2$, $\delta_1 < \epsilon$, such that:

$$\rho_k < \delta_1, \qquad \rho_{\bar{k}} < \delta_2$$

imply, for all $t \geqq t_0$,

$$\rho < \epsilon$$

If, in addition,

$$\lim_{t\to\infty} \rho = 0$$

the origin is "asymptotically stable with respect to the coordinates $x_1, \cdots, x_k$.

*Example:* The system (Zubov, 1963c)

$$\dot{x}_1 = x_1 + x_2, \qquad \dot{x}_2 = -(4x_1 + x_2 + x_3), \qquad \dot{x}_3 = -(2x_1 + x_2 + x_3)$$

has a solution which is "asymptotically stable at the origin" in the third component of the solution, but "asymptotically unsatable at the origin" in the first two components.

REMARK 9: ABSOLUTE STABILITY

A nonlinear autonomous system of which not any eigenvalue has positive real part, may have, under some restrictions of the nonlinearities, the origin as an equilibrium point "asymptotically stable in the large." Such nonlinear systems are called "absolutely stable systems," and there are methods for finding classes of nonlinearities for the absolute stability of the systems (Aizerman and Gantmaher, 1964).

The above kind of stability of the systems plays an important role in modern technology.

REMARK 10: STRUCTURAL STABILITY

If the stability situation of a motion $x_i(t)$ of a system is invariant in a parameter space $S$ of the system, one speaks about a "structural stability" of the system in this space which is the "domain of structural stability" of the system. This property of the system gives information about the "insensitivity" of the system to perturbations (Lefschetz, 1957).

## VI. STABILITY UNDER PERSISTENT PERTURBATIONS

In the preceding, the perturbations were considered momentarily acting to the system, and consequently the perturbations do not appear in the formulation of the equations of the system.

In the case of constantly acting perturbations $p_i(t; x_1, \cdots, x_n)$, the equations of the system must contain the perturbations, for which the equations of the system during the action of persistent perturbations are:

$$\begin{aligned} \dot{x}_i(t) &= X_i(t; x_1, \cdots, x_n) + p_i(t; x_1, \cdots, x_n) \\ x_i(t_0) &= x_{i0}; \qquad i = 1, \cdots, n \end{aligned} \tag{5}$$

where $p_i$ must be such that the system (5) has unique solutions corresponding to the initial conditions.

A solution $x_i(t)$ of the system (1) is "stable under persistent perturbations," if, in addition to conditions (2) and (3), some appropriate restrictions on the magnitude of $p_i$ are accepted and which must hold for all accepted $x_i$ and $t \geqq t_0$.

In case the magnitude of the perturbations is according to (Vrkŏv, 1963):

$$\max_{x_i} | p_i(t; x_1, \cdots, x_n)| < \eta \tag{6}$$

where $\eta$ is an appropriate positive number, one speaks about a "total stability" of the solution $x_i(t)$.

If the condition (6) is replaced by:

$$\int_0^\infty \max_{x_i} | p_i(t; x_1, \cdots, x_n)| \, dt < n \tag{6a}$$

one speaks about "integral stability."

In case of total stability the perturbations must be small in magnitude, but in case of integral stability they may be large in a small interval of time.

These magnitude restrictions of the perturbations of the above stabilities are possessed by stability under perturbations which are "bounded in the mean," and in this case one has "stability in the mean," which corresponds to perturbations satisfying the condition:

$$\int_t^{t+T} \max_{x_i} | p_i(t; x_1, \cdots, x_n)| \, dt < \eta \tag{6b}$$

where $\eta$ of 6($b$) and $\delta$ of (2) depend on $T$ in general.

If, in addition to the above, condition (4) holds, one has "stability of asymptotic type" under persistent perturbations.

The above stability of persistent perturbations are in the Liapunov or Poincaré sense, if the distances $\rho_0$ and $\rho$ in (2) and (3) are Liapunov or Poincaré distances.

## VII. COMPARISON AND CONNECTIONS OF STABILITY CONCEPTS—EXAMPLES

The three basic concepts of stability in case of sudden or persistent perturbations, although expressed by the same mathematical conditions, are independent of one another. It is therefore possible for a solution to have different situations of stability under different definitions of stability. Nevertheless, the concepts of stability are in many cases connected.

In the following we clarify these statements.

### 1. *Liapunov and Poincaré Definitions of Stability Applied to Periodic Motions*

We apply the Liapunov and Poincaré definitions of stability to the following physical problem:

*"Discuss the stability of the motion of a mass in an elliptic orbit under the inverse square Newton's law of attraction of an attractive center."*

The elliptic orbit of the moving mass in this Newtonian field can be determined by knowing the initial conditions, that is, the distance of the moving mass from the attractive center $E$ and its velocity at time $t_0$. Let us take $L$ the orbit by specifying the initial conditions, and $T$ the corresponding period of the motion on $L$. If these initial conditions are changed, then the new initial conditions correspond to a new orbit $\bar{L}$ of the perturbed motion with new period $\bar{T}$.

a. *The motion of the mass on the orbit L is "orbitally stable", but not "asymptotically orbitally stable".*

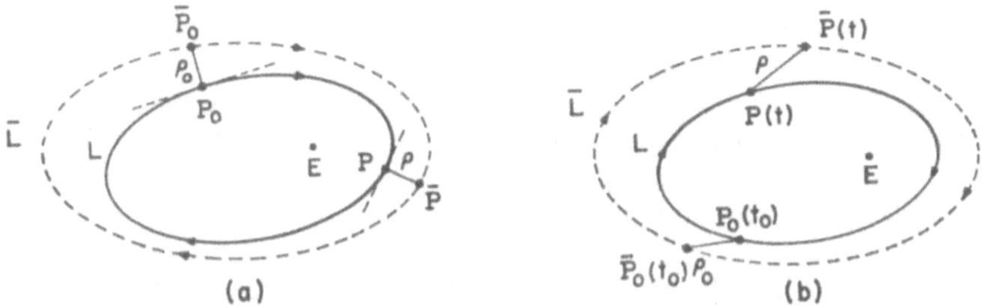


FIG. 2

For any point $P$ on $L$, Fig. 2(a), the corresponding point $\bar{P}$ on $\bar{L}$ must be such that the distance $\rho = P\bar{P}$ is the minimum distance from $\bar{P}$ to $L$.

If one wants this distance $\rho$ to be smaller than a given number $\epsilon > 0$ for all time, one needs for orbital stability to find a $\delta > 0$ such that, if the distance initially is smaller than $\delta$, $\rho_0 < \delta$, one has $\rho < \epsilon$. This is possible. Given a perturbation $p$, the perturbed orbit $\bar{L}$ is known, then the deviation $\rho$, corresponding to any point $P$ of $L$, is known. This deviation for all points of $L$ has a maximum $\rho_M$, and a minimum $\rho_m$; and the smaller the perturbation $p$, the smaller $\rho_M$ and $\rho_m$. Given now a small positive $\epsilon$, if one wants $\rho < \epsilon$ for all points $P$ of $L$, one must use a very small perturbation $p$ such that the corresponding $\rho_M$ is smaller than $\epsilon$, when the appropriate $\delta$ must be smaller than $\rho_m$. Therefore, the motion is "orbitally stable."

Since the distance $\rho$ does not tend to zero as $t$ changes, the motion is not "asymptotically orbitally stable," that is, the ellipse is not a "limit cycle."

b. *The above motion on L is "unstable in Liapunov sense.*

The distance $\rho = P\bar{P}$, Fig. 2(b), in Liapunov definition is the distance of the points $P$ and $\bar{P}$ on $L$ and $\bar{L}$, which are places of the mass at the "same time." The periods $T$ and $\bar{T}$ of the motion on $L$ and $\bar{L}$ are dependent on the length of the corresponding major axes, thus the points $P$ and $\bar{P}$, very close initially but traveling on different ellipses, may find themselves at opposition and thus at great distance from each other in due course of time. Then given small positive $\epsilon$, one can not find a $\delta$ such that $\rho_0 < \delta$ implies $\rho < \epsilon$, and this is instability of the motion in Liapunov sense.

Motions as of the present problem are practically considered as stable, and by the above example one sees that the Liapunov stability definitions are narrow definitions, since they classify as unstable situations some situations which are practically considered as stable.

The Poincaré stability definitions in periodic motions are the appropriate ones in these motions.

We can prove that any periodic motion is unstable in the Liapunov sense if its period is different from the period of the corresponding perturbed motion.

The free vibrations of a simple pendulum are stable in the Liapunov sense if they are linear and unstable if they are nonlinear, because in the case of linearity the isochronous phenomenon exists, and in the case of nonlinearity the isochronous phenomenon is violated.

*2. Examples of Motions Stable or Unstable Under Different Definitions of Stability*

a. The rectilinear motion:

$$x(t) = c_1(t - t_0) + c_2$$

is the general solution of:

$$\ddot{x} = 0, \qquad \dot{x}(t_0) = c_1\,, \qquad x(t_0) = c_2\,.$$

The above motion is stable or unstable depending on the initial conditions and the definitions of stability used. We have the following cases:

(i) $c_1 = 0$, $c_2 = $ *arbitrary*. The motion can be represented in the "$t$, $x$-plane" by lines parallel to "$t$-axis," the Liapunov distance $\rho_L$, the Poincaré distance $\rho_P$ and the Lagrange distance $\rho_{La}$ are, Fig. 3(a), $\rho_L = \rho_P = P\bar{P}$, $\rho_{La} = OP'$, which are constants, then the motion is "stable" in the Liapunov, Poincaré, and Lagrange senses.

(ii) $c_1 \neq 0$ *fixed*, $c_2 = $ *arbitrary*. The motion can be represented by parallel lines with nonzero slope $c_1$, Fig. 3(b), the Liapunov distance $\rho_L = P\bar{P} = $ constant, the Poincaré distance $\rho_P = P\bar{\bar{P}} = $ constant, and the Lagrange distance $\rho_{La} = OP'$ increasing to infinity as $t$, and then $P$, tends to infinity. Then, the motion is "stable" in Liapunov and Poincaré sense, but "unstable" in Lagrange sense.

(iii) $c_1 \neq 0$ *arbitrary*, $c_2$ *either fixed or arbitrary*. The trajectories $L$ and $\bar{L}$ are, in general, as shown in Fig. 3(c). The Liapunov, Poincaré and Lagrange distances are: $\rho_L = P\bar{P}$, $\rho_P = P\bar{\bar{P}}$, $\rho_{La} = OP'$, and all of them increase to infinity with $t$, and then $P$ goes to infinity when the motion is "unstable" in all the three senses.

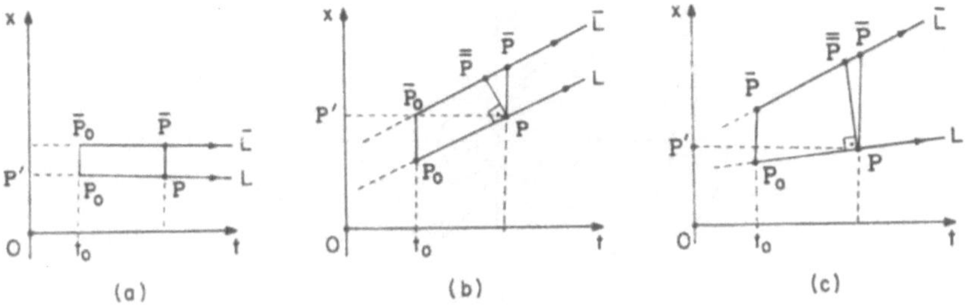


FIG. 3

b. The motion $x(t) = a \sin (at + b)$, where $a$ and $b$ are arbitrary parameters is a solution of the equation:

$$\ddot{x} = -\tfrac{1}{2}x\{x^2 + (x^4 + 4\dot{x}^2)^{1/2}\}$$

it is stable in Lagrange and Poincaré sense, but, with the exception of the origin, it is unstable in Liapunov sense.

(c) The motion $x = c_1 \cos (\mu t + c_2)$, $y = c_1 \sin (\mu t + c_2)$ of the system: $\dot{x} = -\mu y$, $\dot{y} = \mu x$, $\mu \neq 0$ where $c_1$ and $c_2$ are arbitrary constants, is a periodic motion with frequency $\mu$ in the concentric circles $x^2 + y^2 = c_1^2$.

This motion is Liapunov stable in case $\mu$ is a constant, it is Liapunov unstable in case $\mu$ is an arbitrary parameter, say in $\mu_1 \leqq \mu \leqq \mu_2$.

d. The equilibrium point of the system

$$\dot{x} = 2xy, \qquad \dot{y} = y^2 - x^2$$

that is the origin of the system is "quasi-asymptotically stable." The solutions of this system are the members of the one parameter family of curves: $(x - \rho)^2 + y^2 = \rho^2$, then circles through the origin with radius $\rho$ and the center on the $x$-axis, Fig. 4. Starting from any point of the $x$, $y$-plane, the circle through this point ultimately terminates to origin, then the limiting condition (4) is satisfied. All solutions of the system starting from points of the circle $(0, \delta)$ can not be included in the circle $(0, \epsilon)$, then the conditions (2) and (3) are not compatible, even if $\delta$ is small and $\epsilon$ large.

e. The motion $x = ae^{-t} \sin t$, where $a$ is an arbitrary parameter, is "asymptotically stable" in case $t_0 \neq k\pi$, $k$ an integer, and it is "quasi-asymptotically stable" in case $t_0 = k\pi$ (Hahn, 1963b).

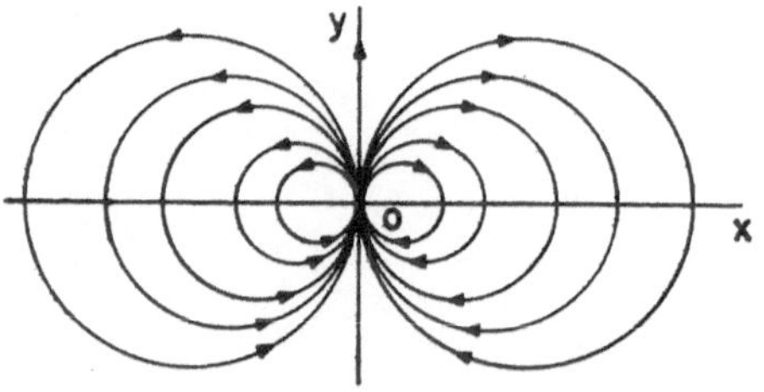


FIG. 4

*3. Connections Between Stability Concepts*

a. In linear homogeneous systems:

$$\dot{x}_i(t) = \sum_{j=1}^{n} d_{ij}(t)x_j\,; \qquad i = 1, \cdots, n$$

where $d_{ij}(t)$ are continuous functions of $t$ in $t \geqq t_0$ :
"*The Lagrange stability of its solutions implies their Liapunov stability*", *and conversely* (Cesari, 1959b).

b. In linear nonhomogeneous systems:

$$\dot{x}_i(t) = \sum_{j=1}^{n} d_{ij}(t)x_j + f_i(t); \qquad i = 1, \cdots, n$$

where $d_{ij}(t)$ and $fi(t)$ are continuous functions of $t$ in $t \geqq t_0$ : "*The Lagrange stability of its solutions implies their Liapunov stability, but for the converse the boundedness of one solution is needed as an additional requirement*" (Cesari, 1959b).

In nonlinear systems, connections between the above concepts of stability either is in general hard to be found or they do not exist.

In case of stabilities of persistent perturbations connections can be introduced by the following statements (Vrkŏv, 1963):

c. "*Asymptotic integral stability implies asymptotic stability in the mean, and conversely*";

d. "*Asymptotic integral stability implies total asymptotic stability, and this implies total stability*";

e. "*Asymptotic stability in the mean implies stability in the mean, and this implies either integral stability or total stability*";

f. "*Integral stability does not imply total stability*";

g. "*Stability in the mean does not imply total asymptotic stability*";

h. "*Total asymptotic stability does not imply integral stability*".

## VIII. GEOMETRICAL INTERPRETATION OF THE STABILITIES

The concepts of stability may be clarified by their geometrical interpretation.

One may define as "$r$-tube" around a curve $l$ in $n$-dimensional space the set of all points of which the distance $\rho$ in the Poincaré sense from $l$ is smaller than $r$. The quantity $r$ is the width and $l$ the central curve of the tube. For finite $r$ the tube is bounded.

These definitions can appropriately be used for a geometrical interpretation of the stability concepts.

A motion $x_i(t, t_0, x_{10}, \cdots, x_{n0})$ of orbit $L$ is stable in Liapunov sense, if given the "$\epsilon$-tube" around the line $x_i = x_{i0}$, Fig. 5(a), one can find a "$\delta$-tube" around the line $x_i = x_{i0}$ such that any perturbed motion of orbit $\bar{L}$, starting in the "$\delta$-tube," remains for all $t \geqq t_0$ in the "$\epsilon$-tube."

In case of equilibrium points, that is, the origin, the above tubes are around the $t$-axis, Fig. 5(b).

For asymptotic stability one needs, in addition, another tube, a "$\delta_1$-tube," which will be unbounded for asymptotic stability in the large.

For stability of a general motion in Poincaré sense, that is, for "orbital stability" of the motion, the "$\delta$-tube" and "$\epsilon$-tube" around the orbit $L$ can be used.

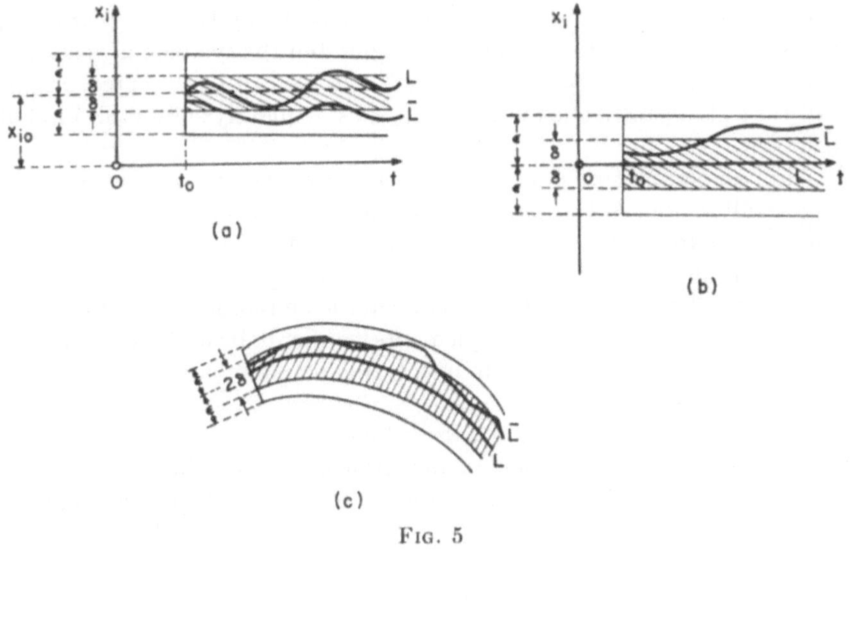


Fig. 5

A motion is "orbitally stable," if given the "$\epsilon$-tube" around $L$, one can find a "$\delta$-tube" around $L$ such that any perturbed orbit $\bar{L}$ starting in the "$\delta$-tube" remains always in the "$\epsilon$-tube", Fig. 5(c).

For stability of equilibrium points, the orbit $L$ is the $t$-axis, then the tubes are around the $t$-axis, Fig. 5(b), and the discussion is the same as in the Liapunov case.

For periodic motions, the tubes around the closed orbit $L$ are tori with $L$ as a central curve.

For Lagrange stability bounded tubes around the $t$-axis, Fig. 5(b) can be used.

The system is "Lagrange stable," if, for every bounded "$\epsilon$-tube" around the $t$-axis another bounded "$\delta$-tube" around the $t$-axis can be found such that every orbit of the system starting in the "$\delta$-tube" does not leave the "$\epsilon$-tube" for all time $t \geqq t_0$.

For stability under persistent perturbations the "$\eta$-tube" around the $t$-axis is needed, in addition. The "$\delta$-tube" and "$\epsilon$-tube" must be around the $t$-axis in the Liapunov case, or around the orbit $L$ in Poincaré case.

## IX. PRACTICAL STABILITY (Zubov, 1963(d); La Salle-Lefschetz, 1961)

All the previous discussion on concepts and definitions of stability, although modified in many points for the purpose of their application to practical problems, are of mathematical type.

One can see that a state of a system may be unstable mathematically, that is, under one of the previous definitions, but the system may oscillate sufficiently near this state and its performance can be accepted practically as a stable one. The motion of missiles frequently shows this kind of behavior.

Also, an equilibrium state of a system may be stable mathematically in a small region, but in practice the perturbations expected may cause the system to go far from the equilibrium state; then the system is practically unstable at the equilibrium state.

The definitions of stability discussed in the preceding need appropriate modifications, changes, and supplements in order to be useful for practical problems, that is, in order to meet the requirements for "practical stability".

For practical stability one needs to know:

(a) the size of the region of deviations, that is, the width of the "$\epsilon$-tube," which gives the "acceptable states" and a "satisfactory operation" of the system;

(b) the size of the region of initial data, that is, the width of the "$\delta$-tube," which gives the permitted size of the initial conditions which can be controlled;

(c) the size of the region of perturbations, that is, the width of the "$\eta$-tube"; and

(d) a finite time $T$ over which the stability is valid, that is, the lengths of the above tubes, in other words, the time $t_0 \leqq t \leqq t_0 + T$, for which the solution $x_i(t)$ satisfies the above requirements.

In determining practical stability, the linearization of the system is not, in general, permitted, since practical stability depends on the non-linearities of the system.

## X. STABILITY AND THE SYSTEM OF COORDINATES

The character of the stability, under any of the previous concepts, is not invariant with a general transformation of the coordinates of the system. In other words, the stability is dependent on the coordinate system with reference to which the variables are to be considered; then for a discussion of stability the selection of the appropriate space variables of the system is suggested.

By means of the following examples, we clarify the dependence of the stability on the coordinate system (Cesari, 1959c).

*Example* 1. The dynamical system:

$$\dot{x} = -y(x^2 + y^2)^{1/2}, \qquad \dot{y} = x(x^2 + y^2)^{1/2} \tag{7}$$

accepts the two parameter family of solutions:

$$x = a \cos (at + b), \qquad y = a \sin (at + b) \tag{8}$$

where $a$ and $b$ are arbitrary.

This family of solutions is stable in Poincaré and Lagrange sense, but it is unstable in Liapunov sense.

By introducing new variables, one can make the solution Liapunov stable.

In case the new variables $r$ and $b$ are introduced by means of the relations:

$$x = r \cos \vartheta, \qquad y = r \sin \vartheta, \qquad \vartheta = at + b \tag{9}$$

the original system (7) is transformed into the new one:

$$\dot{r} = 0, \qquad \dot{b} = 0 \tag{10}$$

for which the solution

$$r = c_1, \qquad b = c_2 \tag{11}$$

is Liapunov stable, where $c_1$ and $c_2$ arbitrary constants.

*Example* 2. The equation of pendulum:

$$\ddot{x} + \sin x = 0 \tag{12}$$

accepts the two parameter family of solutions:

$$x = a \sin \{\varphi(a)t + b\} \tag{13}$$

where $a$ and $b$ are arbitrary, and $\varphi(a)$ can be expressed in terms of elliptic functions.

The solutions (13) are, except for the origin, unstable in Liapunov sense, but by using new coordinates, $r$ and $b$, according to the transformation formulas:

$$x = r \sin \{\varphi(r)t + b\}, \qquad y = r \cos \{\varphi(r)t + b\} \tag{14}$$

the system (12) leads to the system:

$$\dot{r} = 0, \qquad \dot{b} = 0 \tag{15}$$

of which the solutions are Liapunov stable.

*Example* 3. Let us take the nonautonomous system

$$\dot{x}_i(t) = X_i(t, x_1, \cdots, x_n); \qquad i = 1, \cdots, n \tag{16}$$

of which the solution in its implicit form is:

$$\phi_i(t, x_1, \cdots, x_n) = c_i \tag{17}$$

where $c_1, \cdots, c_n$ are constants. By using the transformation:

$$y_i = \phi_i(t, x_1, \cdots, x_n) \tag{18}$$

the original system (16) is reduced to:

$$\dot{y}_i = 0 \tag{19}$$

of which the solution:

$$y_i = c_i \tag{20}$$

is Liapunov stable, but this is not the case, in general for the solution (17) of the system (16).

RECEIVED: October 8, 1965

## REFERENCES

AIZERMAN, M. AND GANTMAHER, F. (1964), "Absolute Stability of Regulator Systems." Holden-Day, San Francisco.

CESARI, L. (1959), "Asymptotic Behavior and Stability Problems in Ordinary Differential Equations," (a) p. 6, (b) pp. 7-8, (c) pp. 12-13. Springer, Berlin.

HAHN, W. (1963), "Theory and Applications of Liapunov's Direct Method," Prentice-Hall, Englewood Cliffs, New Jersey. (a) p. 129, (b) p. 6.

LA SALLE, J. AND RATH, R. (1963), Eventual stability. *Proc. 1963 IFAC.*

LA SALLE, J. AND LEFSCHETZ, S. (1961), "Stability by Liapunov's Direct Method," p. 121. Academic Press, New York.

LEFSCHETZ, S. (1957), "Differential Equations: Geometric Theory," p. 239. Interscience, New York.

MAGIROS, D. (1965a), On stability definitions of dynamical systems. *Proc. Natl. Acad. Sci., U.S.A.* **53**, 1288–1294.

MAGIROS, D. (1965b), Physical problems discussed mathematically. *Bull. Greek Math. Soc., New Ser. II,* **6**, 143–156.

VRKŏV, I. (1963), On some stability problems. *Proc. Conf. Prague, September, 1962,* pp. 217–221 (Academic Press, New York).

ZUBOV, V. (1963), "Mathematical Methods for the Study of Automatic Control Systems," Macmillan, New York. (a) Chap. I, Sect. 1, (b) Sect. 4, (c) Section 8, (d) p. 13.

ZUBOV, V. (1964), "Methods of A. M. Liapunov and Their Application," pp. 22–26. Noordhoff, Groningen, The Netherlands.

# *Attitude Stability of a Spherical Satellite**

*by* D. G. MAGIROS *and* A. J. DENNISON

*General Electric Company*
*Valley Forge Space Technology Center, Philadelphia, Pa.*

ABSTRACT: *The in-orbit-plane angular motion of a spherical satellite with a mass eccentricity is studied, and both aerodynamic and gravitational torques along with large angle motions are considered. The equilibrium attitudes of the motion are investigated and the conditions for stability presented. The stability concepts on which the study is based are delineated in an Appendix.*

## I. Introduction

Problems associated with the angular motion and attitude stability of satellites acted upon by aerodynamic and gravitational torques have received much attention in recent years. Of the many papers publishe ., a few have considered the problem of particular interest here, *viz.*, the attitude stability of a spherical satellite with a mass eccentricity.

D. M. Schrello (**2**), as part of a comprehensive work considering many aerodynamic shapes, has studied the attitude stability of the sphere employing small angle approximations. Nam Tum Po (**7**), while admitting large angular motions, considers only the aerodynamic rate damping effects in studying the attitude stability of a spherical satellite. Finally, the form of the basic differential equation employed in this paper, excluding the dynamic damping term, appears in other works, e.g., Minorsky (**6**) considers the problem of a rotating pendulum and discusses the stability characteristics of that system.

In general, the admission of large angular motion and the inclusion of dynamic damping effects can significantly alter conclusions regarding the stability characteristics of the physical system being studied. Accordingly, this paper considers both gravitational and aerodynamic torques, including the aerodynamic rate damping term, and admits large angular motion in determining equilibrium attitudes and establishing stability conditions.

For the purposes of this paper the motion of the satellite is assumed to be confined to the orbital plane. The authors intend to discuss the out-of-plane rotational problems at a later time.

## II. Development of the Equation Governing the Angular Motion

***Rigid Body Equations of Motion.*** The satellite is considered as a rigid body. In general, the motion of a rigid body consists of two motions in coexistence, a translational motion of its mass center $o_1$ with respect to a fixed orthogonal coordinate system $O_1X_1X_2X_3$, and a rotational motion about axes through

* This is part of the senior author's paper discussed at the 18th International Astronautical Congress, Belgrade, Yugoslavia, Sept. 1967.



Reprinted from the *Journal of the Franklin Institute* **286** (1968), 193–203.

*D. G. Magiros and A. J. Dennison*

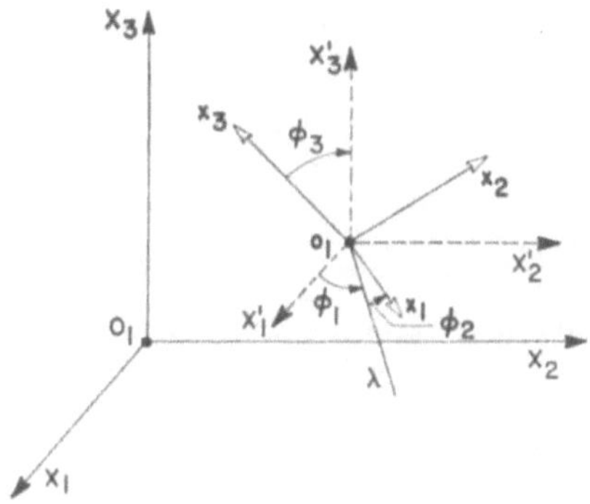


Fig. 1.

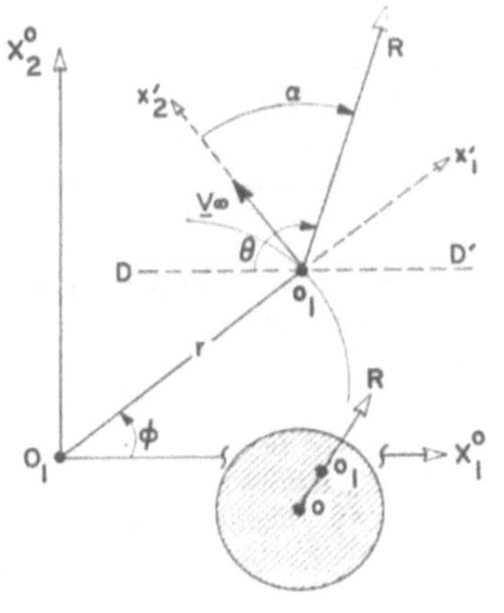


Fig. 2.

the mass center referred to a moving orthogonal coordinate system $o_1x_1x_2x_3$ rigidly fixed in the body, see Fig. 1.

If the resultant **F** of all external forces acting on the body is displaced parallel to itself to act on the mass center, a resultant torque **L** of the external forces results, and if the $x_1$, $x_2$, $x_3$-axes are oriented as the principal axes of inertia of the body, the motion of the body in scalar form is given by:

$$m\dot{v}_i = F_i, \qquad i = 1, 2, 3 \tag{1a}$$
$$I_i\dot{w}_i = L_i, \qquad i = 1, 2, 3 \tag{1b}$$

where $m$ is the constant mass of the body, $v_i$ the translational velocities, $w_i$ the angular velocities, and $I_i$ the principal moments of inertia. Equations 1a refer to the fixed system $O_1X_1X_2X_3$, and Eqs. 1b, the moving system $O_1x_1x_2x_3$.

Equations 1 are assumed to be uncoupled, i.e., the force is independent of the angular velocity and the torque is independent of the translational velocity. Accordingly, the *translational* motion problem and the *rotational* motion problem are considered separately.

***Planar Equations of Motion.*** We assume that the translational motion of the satellite is confined to an inertially-fixed orbit plane and that the motion in this plane consists of a circular orbit with the earth at its center, see Fig. 2. Based on this assumption, Eqs. 1a are not considered further.

Furthermore, it is assumed that the rotational motion of the satellite is confined to the orbital plane and for this case the equation of motion is

$$I_P\ddot{\theta} = L \tag{2}$$

where $I_P$ is the moment of inertia about the axis (of rotation) through the center of mass $o_1$ perpendicular to the orbital plane, $\theta$ the inertial angle, shown in Fig. 2, and $L$ the component of the resultant torque on the axis of rotation. The resultant torque $L$ is taken as the sum of the applied gravitational torque, $L_G$, and the applied aerodynamic torque, $L_A$. Expressions for these applied torques are developed in the next two sections.

***Gravitational Torque.*** For an inverse square central force field the gravitational torque, $L_G$, can be approximated by (**2**)

$$L_G = \tfrac{3}{2}(K/r^3)(I_P - I_R)\sin 2\alpha \tag{3}$$

where $r$ is the radius of the circular orbit of the satellite, $K$ the product of the universal gravitational constant and the mass of the earth, $I_R$ the moment of inertia about the satellite reference axis, and $\alpha$ the angle of attack shown in Fig. 2. Equation 3 is written, for convenience, in the form

$$L_G = I_P G \sin 2\alpha \tag{4}$$

where the constant $G$ is defined by

$$G = \tfrac{3}{2}(K/r^3)[1 - (I_R/I_P)]. \tag{5}$$

***Aerodynamic Torque.*** The aerodynamic torque, $L_A$, for a passive satellite is given by (**2, 3**)

$$L_A = [C_m + C_{m_q}(l_q/2V_\infty)]\tfrac{1}{2}\rho_\infty V_\infty^2 A l \tag{6}$$

where $C_m$ and $C_{m_q}$ are the static and dynamic aerodynamic moment coefficients, respectively; $l$ and $A$ are the aerodynamic reference length and area, respectively; $\rho_\infty$ is the local atmospheric density; $V_\infty$ is the satellite velocity relative to the atmosphere; and $q$ is the total angular rate of the satellite in the orbital plane.

In Eq. 6, $A$, $l$, $\rho_\infty$ and $V_\infty$ can all be taken as constants; $A$ and $l$ are, of course, constants dictated by the size of the satellite. Since the satellite orbit is assumed to be circular, $\rho_\infty$ can be taken as constant. Finally, $V_\infty$ can also be considered as a constant since, for a circular orbit, its deviation from a constant is a second-order effect even when the atmosphere is considered to be rotating.

Owing to the restricted nature of the orbit, the time-rate of change of angle-of-attack, $\dot{\alpha}$, is the prime contributor to the total angular rate, $q$, and, accordingly, it is assumed that

$$q = \dot{\alpha} \tag{7}$$

for the purposes of Eq. 6. The static and dynamic aerodynamic coefficients, $C_m$ and $C_{m_q}$, for a spherical satellite moving near the earth, are given by (**3**)

$$C_m = -k_1 \sin\alpha \tag{8a}$$
$$C_{m_q} = -(k_2 + k_3\cos\alpha) \tag{8b}$$

where $k_1$, $k_2$, $k_3$ are positive constants depending upon the free stream molecular speed ratio which is assumed constant, upon the radius $a$ of the spherical satellite, and upon the distance $x = |oo_1|$, shown in Fig. 2.

Employing Eqs. 7 and 8, Eq. 6 can be put in the following convenient form

$$L_A = -I_P[A_1 \sin\alpha + (A_2 + A_3 \cos\alpha)\dot{\alpha}] \tag{9}$$

where the constants $A_1$, $A_2$ and $A_3$ are defined by

$$A_1 = k_1[(\tfrac{1}{2}\rho_\infty V_\infty^2)Al/I_P] \tag{10a}$$
$$A_2 = k_2[(\tfrac{1}{2}\rho_\infty V_\infty^2)Al/I_P] \cdot (l/2V_\infty) \tag{10b}$$
$$A_3 = k_3[(\tfrac{1}{2}\rho_\infty V_\infty^2)Al/I_P] \cdot (l/2V_\infty) \tag{10c}$$

the magnitudes of which satisfy the inequalities

$$A_1 \gg/ A_3 > A_2 > 0. \tag{11}$$

***Equation Governing Angular Motion.*** From Fig. 2 it is clear that the angle $\theta$ can be expressed in terms of $\alpha$ and $\phi$, the polar angle defining the position of the satellite in its orbit, according to the relation

$$\theta = \tfrac{1}{2}\pi + \alpha - \phi. \tag{12}$$

Since the satellite orbit is assumed to be circular, $\phi$ is a constant and, from Eq. 12

$$\ddot{\theta} = \ddot{\alpha}. \tag{13}$$

Now, substitution of Eqs. 4, 9 and 13 into Eq. 2, gives

$$\ddot{\alpha} + (A_2 + A_3 \cos\alpha)\dot{\alpha} + A_1 \sin\alpha - G \sin 2\alpha = 0 \tag{14}$$

the differential equation describing the in-orbit-plane angular motion of a spherical satellite. Finally, Eq. 14 is put into the equivalent normal form of an autonomous non-linear system

$$\dot{\alpha} = \omega \tag{15a}$$
$$\dot{\omega} = -(A_2 + A_3 \cos\alpha)\omega + (2G \cos\alpha - A_1) \sin\alpha \tag{15b}$$

thereby facilitating the investigation of the satellite's attitude stability.

### *III. Equilibrium Points and their Stability*

***Equilibrium Points of the System (15),*** are defined by the condition $\dot{\alpha} = 0$, $\dot{\omega} = 0$ and the points satisfying this condition are given by

$$\omega = 0, \quad \alpha = m\pi \tag{16a}$$
$$\omega = 0, \quad \alpha = \cos^{-1}(A_1/2G) + 2m\pi \tag{16b}$$

where $m = 0, \pm 1, \pm 2 \cdots$ and the point defined by (16b), of course, exists only on the range $-1 \leq (A_1/2G) \leq +1$.

Since the physical restriction of the angle $\alpha$ is $-\pi < \alpha \leq \pi$ and, since the values $(A_1/2G) = +1$ and $(A_1/2G) = -1$ correspond to $\alpha = 0$ and $\alpha = \pi$, respectively, the equilibrium points studied here are confined to

$$\alpha_1 = 0 \tag{17a}$$
$$\alpha_2 = \pi \tag{17b}$$
$$\alpha_3 = \pm \cos^{-1}(A_1/2G), \quad \text{where} \quad -1 < (A_1/2G) < 1. \tag{17c}$$

We remark here that the stability situation at each of the equilibrium points is a local behavior because of the nonlinearities in the system (15).

***Stability of the Equilibrium Points** (17)*, is investigated here on the basis of the definitions and theorems given in the Appendix.

***Stability of $\alpha_1 = 0$.*** Expanding $\sin\alpha$ and $\cos\alpha$ by use of McLaurin series, the system (15) becomes

$$\dot{\alpha} = \omega \tag{18a}$$
$$\dot{\omega} = (2G - A_1)\alpha - (A_2 + A_3)\omega + F_1(\alpha, \omega) \tag{18b}$$

where $F_1(\alpha, \omega)$ is a known power series starting with terms of the second degree.

It can be shown that $F_1(\alpha, \omega)$ satisfies the requirement of Theorem I so that it is only necessary to consider the linear terms in Eqs. 18 in studying the stability of this equilibrium point. Accordingly, Theorem I, which defines the stability condition in terms of the characteristic roots of the linearized system, is applied in this case.

The characteristic roots of system (18), if linearized, are

$$\lambda_{1,2} = -\tfrac{1}{2}(A_2 + A_3) \pm \tfrac{1}{2}[(A_2 + A_3)^2 + 4(2G - A_1)]^{1/2}. \tag{19}$$

In view of the relative magnitudes of $A_1$, $A_2$ and $A_3$, as shown by (11), the signs of the real parts of $\lambda_{1,2}$ are dictated, for practical cases, by the term $4(2G - A_1)$. This being the case, it is clear that:

(a) If $G < A_1/2$, then both $\lambda_{1,2}$ have negative real parts and the equilibrium point is asymptotically stable.

(b) If $G > A_1/2$, then one of the $\lambda_{1,2}$ has a positive real part and the equilibrium point is unstable.

(c) If $G = A_1/2$, then one of the $\lambda_{1,2}$ has a negative real part and the other has a zero real part. This is a critical case and must be studied by other means.

In order to study the stability of the critical case where $G = A_1/2$, an appropriate Liapunov function $V(\alpha, \omega)$ is developed and Theorem II is applied.

*D. G. Magiros and A. J. Dennison*

Such a function can be determined, according to (4), by

$$V(\alpha, \omega) = \tfrac{1}{2}\omega^2 - \int_0^\alpha \dot{\omega}(\alpha, 0)\, d\alpha \tag{20}$$

which for Eqs. 15, with $G = A_1/2$, gives

$$V(\alpha, \omega) = \tfrac{1}{2}\omega^2 + \tfrac{1}{2}A_1(1 - \cos\alpha)^2 \tag{21}$$

which has the time derivative

$$\dot{V}(\alpha, \omega) = -(A_2 + A_3 \cos\alpha)\omega^2. \tag{22}$$

Clearly, $V(\alpha, \omega)$ is *positive* definite and $\dot{V}(\alpha, \omega)$ is *negative* definite in the neighborhood of the equilibrium point $(\alpha, \omega) = (0, 0)$. Hence, according to Theorem II, this equilibrium point is asymptotically stable when $G = A_1/2$.

In summary, then, it can be stated that the equilibrium point $(\alpha, \omega) = (0, 0)$ is: (a) asymptotically stable if $G \leq A_1/2$; and (b) unstable if $G > A_1/2$.

***Stability of*** $\alpha_2 = \pi$. If in system (15) $\alpha$ is transformed into $x$ according to relation $x = \alpha - \pi$ and if, in addition, sin $x$ and cos $x$ appearing in the resulting expression are expanded about $x = 0$, the following system of equations results:

$$\dot{x} = \omega \tag{23a}$$
$$\dot{\omega} = (A_1 + 2G)x + (A_3 - A_2)\omega + F_2(x, \omega). \tag{23b}$$

Note that the equilibrium point $(\alpha, \omega) = (\pi, c)$ for system (15) corresponds to the equilibrium point $(x, \omega) = (0, 0)$ for system (23).

As before, it can be shown that $F_2(x, \omega)$ satisfies the condition of Theorem I so that only the linear terms in Eqs. 23 need be considered in studying the stability. Also, as with the previous case, the stability conditions can be determined by examining the characteristic roots of the linearized equations.

For this case the characteristic roots are

$$\lambda_{1,2} = \tfrac{1}{2}(A_3 - A_2) \pm \tfrac{1}{2}[(A_3 - A_2)^2 + 4(2G + A_1)]^{1/2}. \tag{24}$$

Examination of Eq. 24, in view of Eq. 11, shows that one of the $\lambda_{1,2}$ always has a positive real part. Hence, according to Theorem I, the equilibrium point $(\alpha, \omega) = (\pi, 0)$ is always unstable.

***Stability of*** $\alpha_3 = \pm\cos^{-1}(A_1/2G)$. As with the previously discussed equilibrium point, if in the system (15) $\alpha$ is transformed into $x$ according to the relation $x = \alpha - \alpha_3$ and if, in addition, the sin $x$ and cos $x$ appearing in the resulting expression are expanded about $x = 0$, the following system of equations results

$$\dot{x} = \omega \tag{25a}$$
$$\dot{\omega} = -2G[1 - (A_1/2G)^2]x - A_3[(A_2/A_3) + (A_1/2G)]\omega + F_3(x, \omega) \tag{25b}$$

Again $F_3(x, \omega)$ satisfies the condition of Theorem I, and this Theorem can be applied. The characteristic roots of the system (25) are

$$\lambda_{1,2} = -\tfrac{1}{2}A_3\left(\frac{A_2}{A_3} + \frac{A_1}{2G}\right) \pm \frac{1}{2}\left[A_3^2\left(\frac{A_2}{A_3} + \frac{A_1}{2G}\right)^2 - 8G\left[1 - \left(\frac{A_1}{2G}\right)^2\right]\right]^{1/2}. \qquad (26)$$

The existence of the equilibrium point $(\alpha, \omega) = (\alpha_3, 0)$ implies that $-1 < (A_1/2G) < 1$, the equal signs being omitted as indicated earlier, so that the term $[1 - (A_1/2G)^2]$ is always positive and only two cases of practical interest result:

(a) If $G < -A_1/2$, then one of the $\lambda_{1,2}$ has a positive real part and the equilibrium point $(\alpha, \omega) = (\alpha_3, 0)$ is unstable.

(b) If $G > A_1/2$, then both of the $\lambda_{1,2}$ have negative real parts and the equilibrium point $(\alpha, \omega) = (\alpha_3, 0)$ is asymptotically stable.

### *Remarks on Dynamic Damping Effects*

The above analysis and conclusions rest heavily on the existence and magnitude of the dynamic damping term, $C_{m_q}$, in the aerodynamic torque as defined by Eq. 6. In practical cases this term is quite small so that the satellite motion considered here is only lightly damped in those situations where asymptotic stability has been previously indicated. Indeed, numerical computations show that the time required to damp a given oscillation to half of its initial magnitude is, at best, of the same order of magnitude as the time required for the orbit to decay to the point where re-entry occurs.

In any case, it can be concluded that the system is stable and, in fact, lightly damped for those cases where asymptotic stability is indicated.

### *Stability and Instability Regions in Coefficient Space*

The results concerning the stability of the equilibrium points $\alpha_1$, $\alpha_2$ and $\alpha_3$, expressed in terms of inequalities in the preceding, can be summarized and clarified by geometrical means, namely, by giving the regions of stability and instability for the equilibrium positions $\alpha_1$, $\alpha_2$ and $\alpha_3$ in a coefficient space.

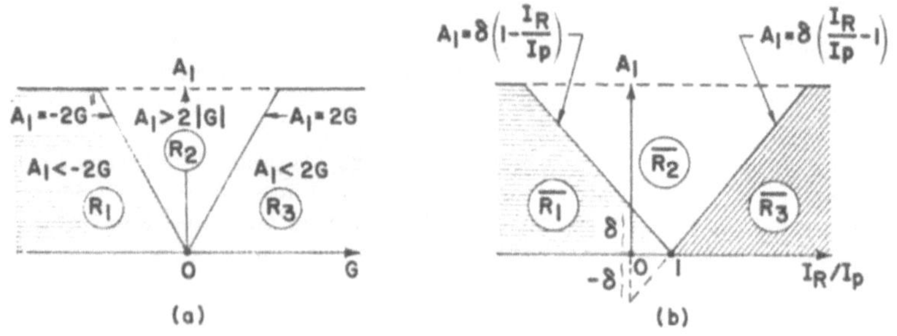


FIG. 3.

*D. G. Magiros and A. J. Dennison*

Figure 3(a) shows the regions of asymptotic stability and instability as well as the non-permissible regions for the equilibrium positions $\alpha_1 = 0$, $\alpha_2 = \pi$ and $\alpha_3 = \pm \cos^{-1} (A_1/2G)$ in the $G$, $A_1$-plane. Table I summarizes the results in terms of the regions shown in Fig. 3(a). The regions $R_1$, $R_2$, $R_3$ of Fig. 3(a) are defined in the parameter space $(G, A_1)$, where $G$ and $A_1$ are given by Eqs 5 and 10a, respectively. By using these formulae, the above regions can be mapped into the regions $\bar{R}_1$, $\bar{R}_2$, $\bar{R}_3$ in the parameter space $[(I_R/I_P), A_1]$, as shown in Fig. 3(b). The lines $A_1 = \pm 2G$ in the $(G, A_1)$-plane can be mapped into the lines $A_1 = \pm\delta[(I_R/I_P) - 1]$ of the $[(I_R/I_P), A_1]$-plane, where $\delta = 3K/r^3$. The quantity $\bar{I}_P$ can be taken either as unity or as a magnification factor. Table II summarizes the results in terms of the regions shown in Fig. 3(b).

TABLE I

| Equilibrium points | Regions $R_1$ | $R_2$ | $R_3$ |
|---|---|---|---|
| $\alpha_1 = 0$ | Asymptotically stable | Asymptotically stable | Unstable |
| $\alpha_2 = \pi$ | Unstable | Unstable | Unstable |
| $\alpha j = \pm \cos^{-1}(A_1/2G)$ | Unstable | Non-permissible | Asymptotically stable |

TABLE II

| Equilibrium points | Regions $\bar{R}_1$ | $\bar{R}_2$ | $\bar{R}_3$ |
|---|---|---|---|
| $\alpha_1 = 0$ | Unstable | Asymptotically stable | Asymptotically stable |
| $\alpha_2 = \pi$ | Unstable | Unstable | Unstable |
| $\alpha_3 = \pm \cos^{-1}\{A_1/\delta[1 - (I_R/I_P)]\}$ | Asymptotically stable | Non-permissible | Unstable |

The above Figures and Tables show that there always exists an asymptotically stable situation, *viz.*, $\alpha_1 = 0$ in the regions $R_1$ and $R_2$ (or $\bar{R}_2$ and $\bar{R}_3$), and $\alpha_3 = \pm \cos^{-1}(A_1/2G)$ in region $R_3$ (or $\bar{R}_1$). Furthermore, when $\alpha_1$ is asymptotically stable all other equilibrium situations are unstable, and when $\alpha_3$ is asymptotically stable, the remaining equilibrium situations are unstable. The equilibrium situation $\alpha_2 = \pi$ is always unstable.

### *Appendix: Definitions and Theorems Relating to Stability*

***Stability Definitions** (5):* Classes of stability concepts can be distinguished according to the nature of the dynamical systems, the manner in which the system approaches a given state or deviates from it, the properties of the perturbations of the system, and the system of coordinates selected.

Physically, there exist three basic stability concepts depending upon considerations of stability of the motion in a given orbit, of the orbit of a given motion, and of the boundedness of the motion and its orbit.

An analytical description of the stability concepts can be achieved by using the Liapunov, Poincaré and Lagrange distances $\rho$, shown in Fig. 4. These concepts interpret the correspondence between a point $P$ of the unperturbed orbit $L$ and a point $\bar{P}$ of the perturbed one $\bar{L}$ in different ways; the distance $\rho$ measures the effect of the perturbations at $P$. The three stability concepts in the sense of Liapunov, Poincaré and Lagrange are of basic type and all others are regarded as special cases of them.

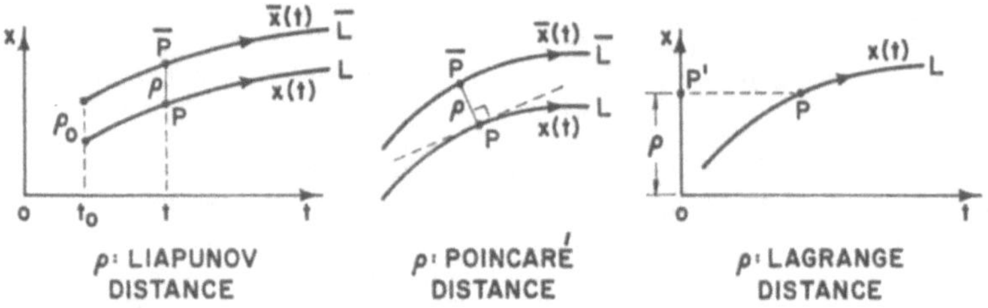


FIG. 4.

All stability concepts of dynamical systems are included in the same stability conditions:

$$\rho_0 < \delta \tag{27a}$$
$$\rho < \epsilon \tag{27b}$$

$$\lim_{t\to\infty} \rho = 0 \tag{27c}$$

$$\| p \| < \eta \tag{27d}$$

by appropriately interpreting the distances $\rho_0$, the initial value of $\rho$, and $\rho$. These conditions express a "unification" of the stability concepts.

In case the perturbations affect the initial conditions only, the first two conditions (27) guarantee the stability of the motion $x(t)$, and the first three conditions the asymptotic stability of the motion. For the case where the whole system is perturbed by persistent perturbations $p(x, t)$, the definition of stability requires a restriction on the perturbations $p$, say, a restriction of its norm $\| p \|$, so that the fourth condition, (27), is needed for the stability definition under persistent perturbations.

The above remarks and unification of the stability concepts lead to important results concerning stability, of which we mention just two needed in the present

*D. G. Magiros and A. J. Dennison*

paper:

(a) For the equilibrium states of a system, the stability concepts in the Liapunov and Poincaré sense are equivalent. For periodic motions, the Liapunov stability concept is narrow compared to the Poincaré stability concept. The isochronism characterizes the use of the Liapunov stability concept in periodic motions.

(b) The Lagrange stability concept, applied either to equilibrium states of to periodic motions in finite distance, always classifies these special motions of a system as stable, so that it is useless.

***Stability Theorems* (*5, 6*):** Consider the nonlinear autonomous system

$$\dot{x}_i = \sum_{i=1}^{n} a_{ij}x_j + X_i(x_1, x_2, \cdots, x_n) \quad \text{with} \quad i = 1, 2, \cdots, n \tag{28}$$

where the $a_{ij}$ are constants and the $X_i(x_1, x_2, \cdots, x_n)$ are power series beginning with terms of at least the second degree.

***Theorem I:*** If the nonlinear terms, $X_i(x_1, x_2, \cdots, x_n)$, in Eqs. 28 are such that

$$\lim_{x_1\to 0 \ldots x_n \to 0} \left[\sum_{j=1}^{n} | X_j(x_1, x_2, \cdots x_n) | / \sum_{j=1}^{n} | x_j |\right] = 0 \tag{29}$$

then the nature of the singularity at the point of equilibrium $x_i = 0$ $(i = 1, \cdots, n)$ may be determined by considering only the linear terms in Eqs. 28.

If the system (28) is such that the condition (29) is satisfied, then the stability of the equilibrium point $x_i = 0$ $(i = 1, \cdots, n)$ is specified by the roots $\lambda_i$ $(i = 1, \cdots, n)$ of the characteristic equation for the linear terms in Eq. 28 as follows:

(a) If all of the distinct roots have negative real parts, then the equilibrium point is asymptotically stable.

(b) If at least one of the non-zero roots has a positive real part then the equilibrium point is unstable.

(c) If none of the roots have positive real parts but some roots have zero real parts, then a so-called critical case exists and the question of stability cannot be resolved on the basis of the characteristic equation roots. Other means must be employed to study this case.

In general, if the nonlinearities in the system (28) are such that they must be considered in studying the stability characteristics of the system, then the following theorem is applicable.

***Theorem II:*** Given the system (28) with the equilibrium point $x_i = 0$ $(i = 1, \cdots, n)$, the equilibrium is

(a) asymptotically *stable*, if it is possible to determine a positive (negative) definite function $V(x_1, x_2, \cdots x_n)$, defined on a domain $D$ including the origin, whose time derivative $\dot{V}$ is negative (positive) definite; and

(b) *unstable*, if it is possible to determine a positive (negative) definite function $V(x_1, x_2, \cdots, x_n)$ whose time derivative $\dot{V}$ is also positive (negative) definite.

### *References*

(1) H. Goldstein, "Classical Mechanics," Reading, Mass., Addison-Wesley Pub. Co., Chap. 5, 1959.

(2) D. M. Schrello, "Passive Aerodynamic Attitude Stabilization of Near-Earth Satellites, Vol. I, Librations due to Combined Aerodynamic and Gravitational Torques," WADD Tech. Rep. 61–133, Wright Air Dev. Div., Wright-Patterson A.F.B., Ohio, July 1961, *Unclassified.*

(3) P. H. Davison, "Passive Aerodynamic Attitude Stabilization of Near-Earth Satellites, Vol. II, Aerodynamic Analysis," WADD Tech. Rep. 61–133, Wright Air Dev. Div., Wright-Patterson A.F.B., Ohio, July 1961, *Unclassified.*

(4) W. Leighton, "On the Construction of Liapunov Functions for Certain Autonomous Nonlinear Differential Equations," Contributions to Differential Equations, New York, Interscience Pub. Co., p. 368, 1963.

(5) D. G. Magiros, "On Stability Definitions of Dynamical Systems," *Proc. Nat. Acad. Sci.*, Vol. 53, No. 6, pp. 1288–1294, June 1965.

(6) N. Minorsky, "Nonlinear Oscillations," Princeton, N. J., D. Van Nostrand Co. Inc., 1962

(7) Nam Tum Po, "On the Rotational Motion of a Spherical Satellite Under the Action of Retarding Aerodynamical Moments," Leningrad Univ., Math., Mech. and Astronomical Ser. No. 19, Vol. 4, pp. 129–134, 1964.

ΕΦΗΡΜΟΣΜΕΝΑ ΜΑΘΗΜΑΤΙΚΑ.— **Stability concepts of solutions of differential equations with deviating arguments,** *by Demetrios G. Magiros* *. Ἀνεκοινώθη ὑπὸ τοῦ Ἀκαδημαϊκοῦ κ. Ἰ. Ξανθάκη.

## 1. Introduction

The majority of physical and social systems can be expressed by relations between quantities under investigation and their rate of change, that is by differential equations.

The duration of the transmission of the action or signal can not in many cases be neglected, and the processes are governed by «differential equations with deviating arguments», as, e.g., by equations with «retarded arguments», which characterize situations with «delay periods» or «aftereffects». Modern needs of automatic regluation, or probability, of biology, of medicine, and of certain other fields, lead to such equations.

The study of the stability properties of the solutions of these equations is a fundamental problem, and this problem can be treated if, and only if, the «concepts of stability» are clarified and selected appropriately.

In this note we formulate stability concepts of the class of differential equations with deviating arguments, and give somme remarks concerning these stability concepts. The stability concepts depend especially upon the nature of the perturbations, the way the perturbations act on the system, their magnitude, the magnitude of their effect, etc. Definition of stability are given in case of sudden and permanent perturbations, and in case of perturbations of the deviating arguments. By the remarks, we clarify questions on the stability concepts.

## 2. Definitions

A system of differential equations with retarded arguments, that is variables containing «delay periods», can be given by:

---

* **ΔΗΜΗΤΡΙΟΥ ΜΑΓΕΙΡΟΥ, Αἱ ἔννοιαι τῆς εὐσταθείας τῶν λύσεων διαφορικῶν ἐξισώσεων μὲ ἀποκλινούσας μεταβλητάς.**

Consulting Scientist, General Electric Co., RESD Philadelphia, Pa., U.S.A.:

Reprinted from the *Proceedings of the Athens Academy of Sciences* 46 (1971), 273–278.

$$\dot{x}_i(t) = f_i[t, x_j(t - \tau_{jk}(t))]$$
$$i, j = 1, 2, \ldots, n; \quad k = 1, 2, \ldots, m, \quad m \leqslant n \tag{1}$$

where the retardations $\tau_{jk}$ are positive.

By «solution» of (1) we mean continuous functions $x_i(t)$ which satisfy (1) on $t \geqslant t_0$, and which, on the initial interval: $E_{t_0}: t_0 - \tau_{jk}(t) \leqslant t \leqslant t_0$ become $x_i(t) = \varphi_i(t)$, and $\varphi_i(t)$ are given continuous functions, called «initial functions» of (1).

By «stationary point» of (1) we mean a constant solution $x_{i_0}$ of (1) on $t \geqslant t_0$, which is also constant on the initial set $E_{t_0}$.

The solution of (1) depends on the given arbitrary functions $\varphi_i(t)$. $\varphi_i(t)$ are extensions of the solutions in the interval $E_{t_0}$, and the solutions are uniquely determined by the initial functions and appropriate properties of $f_i$.

### 3. Stability in case of sudden perturbations

«Sudden perturbations» are perturbations «momentarily» applied to the initial functions $\varphi_i(t)$ attached to (1). Let $x_\varphi$ be the solution of (1) with orbit L corresponding to the initial functions $\varphi_i(t)$ and $x_\psi$ be the solution of (1), after the perturbation, with orbit $\bar{L}$ corresponding to the initial functions $\psi_i(t)$, Fig. (a).

If the points $\bar{P}_{01}$, $\bar{P}_0$, $\bar{P}$ of the perturbed curve $\bar{L}$ correspond to the points $P_{01}$, $P_0$, P of the unperturbed curve L, the following distances can be defined, corresponding to time indicated:

$$\begin{aligned} \varrho_{01} &= P_{01}\bar{P}_{01} = |\psi_i(t) - \varphi_i(t)| && \text{on} \quad E_{t_0}: t_0 - \tau \leqslant t \leqslant t_0 \\ \varrho_0 &= P_0\bar{P}_0 = |\psi_i(t_0) - \varphi_i(t_0)| && \text{at} \quad t = t_0 \\ \varrho &= P\bar{P} = |x_\psi(t) - x_\varphi(t)| && \text{on} \quad t \geqslant t_0 \end{aligned} \tag{2}$$

These distances give the effect of the perturbation at the points $P_{01}$, $P_0$, P of L, Fig. (a).

The solution $x_\varphi$ of (1) is «stable» if, given $\varepsilon > 0$ and $t_0 \geqslant 0$, there exists $\delta = \delta(t_0, \varepsilon) > 0$ such that the inequality $\varrho_{01} < \delta$ implies $\varrho < \varepsilon$.

This solutions is «asymptotically stable», if it is stable and, in addition, $\lim_{t \to \infty} \varrho = 0$.

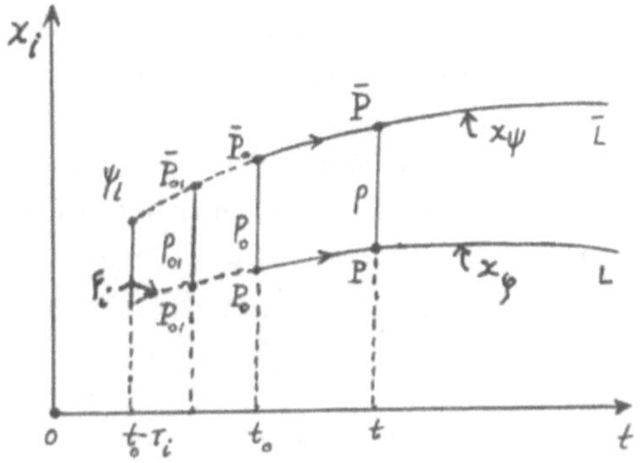

Fig. (a).

These definitions can be accepted for other types of differential equations with deviating arguments, say of neutral types.

## 4. Stability in case of persistent perturbations

If the system (1) is continuously perturbed by the perturbations $R_i$, the perturbed system, corresponding to (1), is :

$$\dot{x}_i(t) = f_i\big(t, x_j(t - \tau_{jk}(t))\big) + R_i\big(t, x_j(t - \tau_{jk}(t))\big) \tag{3}$$

$R_i$ are sufficiently small in absolute value. Given $\varepsilon > 0$ and $t_0 \geqslant 0$, if there exist two positive quantities $\delta_1 = \delta_1(t_0, \varepsilon)$, $\delta_2 = \delta_2(t_0, \varepsilon)$ such that: $\varrho_{01} < \delta_1$ valid on $E_{t_0}$, and $\| R_i \| < \delta_2$ valid on $t \geqslant t_0$, imply $\varrho < \varepsilon$ valid on $t \geqslant t_0$, we say that the solution $x_\varphi$ of (1) is «stable» with respect to persistent perturbations $R_i$.

We remark that many basic theorems on stability can be carried out without essential alteration to the case of differential equations with deviating argument, but, up to now, the stability theory for these equations is essential for stationary equations of the first approximation in noncritical cases.

## 5. Stability in case of perturbations of deviating arguments

In processes with after effects, described by differential equations with deviating arguments, the deviations themselves can not be prescribed exactly, that is deviations themselves may have small disturbances, when the question of stability of the equations with respect to small perturbations of the deviating arguments arises. The important stability problem arises when the perturbations of the deviating arguments have a continuing character, as, e.g., when in some processes with after effects the retardation or delay period is not precisely defined.

Instead of the system (1), we now take the system:

$$\dot{x}_i(t) = f_i\left(t,\ x_j\left(t - \bar{\tau}_{jk}(t)\right)\right) \tag{4}$$

where $\bar{\tau}_{jk}(t)$ are the perturbed deviations. The initial interval is: $E_{\bar{t}_0} : \bar{t}_0 - \bar{\tau}_{jk}(t) \leqslant t \leqslant \bar{t}_0,\ \ t_0 \leqslant \bar{t}_0$.

Now, the solution of (1), $x_\varphi$, defined by $\varphi_i(t)$ on the set $E_{t_0}$, it said to be «stable with respect to perturbations of deviating arguments», if, for $\varepsilon > 0$, $t_0 \geqslant 0$, we can find $\delta_1 > 0$, $\delta_2 > 0$ such that, if $\varrho_{01} < \delta_1$, for t on $E_{t_0}$, and if $/\bar{\tau}_{jk}(t) - \tau_{jk}(t)/ < \delta_2$ for t on $t \geqslant t_0$, then $\varrho < \varepsilon$ for t on $t \geqslant t_0$.

In case $\tau_{jk}$ and $\bar{\tau}_{jk}$ are constant, we have «stability with respect to a continuously perturbed deviating argument».

In case the difference $(\bar{\tau}_{jk} - \tau_{jk})$ is either positive or negative, we have a «one sided perturbation of a deviating argument» (**1**).

### REMARKS (2)

The following remarks may give to the reader an opportunity to think more deeply about the difficulties of the subject.

1. The above stability definitions are in the sense of Liapunov or Poincaré, if $\varrho_{01}$, $\varrho_0$, $\varrho$ are interpreted as Liapunov or Poincaré distances.

The stability definitions in the sense of Liapunov or Poincaré are equivalent in case of equilibrium solutions, but for any other kind of solutions the stability situation may be different, if we employ different stability concepts, and naturally appears the important subject of the selection of the appropriate stability definition for the stability problem at hand.

2. The above classes of stability concepts can give any subclass of stability concepts by appropriate restrictions of the distances, the time, and other quantities. So, e.g., one can speak about «eventually uniform stability», if $\delta_1$ is independent of $t_0$, that is $\delta_1 = \delta_1(\varepsilon)$, and $t_0$ has a minimum $\alpha(\varepsilon)$, that is $\alpha(\varepsilon) \leqslant t_0 \leqslant t$. Nonexistence of $\alpha(\varepsilon)$, that is $\alpha(\varepsilon) \equiv 0$, corresponds to «uniform stability».

3. In case of persistent perturbations, the selection of the kind of the norm of the perturbations, specifies the stability concept. So, one may have «total stability», or «integral stability» or «stability in the mean» under suitable norm of the perturbations (3).

4. The stability, as defined above, is a property of the solution different from its boundedness property, although in some cases there may exist regions where these two properties are equivalent and one implies the other.

5. All the above stability concepts are of mathematical type and the results, theorems or criteria, based on them, may not interpret the reality. In order that these results have a practical usefulness, which must be the ultimate purpose of the investigations, the investigation must be accompanied by some additional requirements, as, e.g., to know the region of the practically permitted deviations of the solutions, that is the «ε-region», the corresponding region of the initial conditions, that is the «$\delta_1$-region», and the «$\delta_2$-region» of the norm of the perturbation $R_i$.

## REFERENCES

1. L. È. Èl'sqol'c: «Qualitative Methods of Mathematical Analysis» American Mathematical Society, Providence, R. I. (1964), pg. 167-199.
2. D. G. Magiros: «Stability Concepts of Dynamical Systems», Journal of Information and Control, Vol. 9, No. 9 (Oct. 1966), pg. 531-548.
3. I. Vrkov: «On Some Stability Problems», Academic Press, Proc. Conf. Prague, (Sept. 1962), pg. 217-221.

★

Ὁ Ἀκαδημαϊκὸς κ. **Ἰω. Ξανθάκης** κατὰ τὴν ἀνακοίνωσιν τῆς ἀνωτέρω ἐργασίας εἶπε τὰ κάτωθι :

«Ἡ ἀνακοίνωσις τοῦ κ. Μαγείρου ἀναφέρεται εἰς τὰς ἐννοίας τῆς εὐσταθείας τῶν λύσεων διαφορικῶν ἐξισώσεων μὲ ἀποκλινούσας μεταβλητάς.

Μία μεγάλη κατηγορία φυσικῶν καὶ κοινωνικῶν φαινομένων ἐκφράζεται μαθηματικῶς διὰ διαφορικῶν ἐξισώσεων μή - γραμμικῶν μὲ ἀποκλινούσας μεταβλητάς. Ἀφ᾽ ἑτέρου, ὡρισμένα προβλήματα αὐτομάτου ἐλέγχου, προβλήματα πιθανοτήτων, ἡ ἀνάπτυξις εἰς τὸν τομέα τῆς βιολογίας καὶ ἰατρικῆς ὁδηγοῦν εἰς ἐξισώσεις τοῦ ἐν λόγῳ τύπου.

Τὸ ζήτημα τῆς εὐσταθείας τῶν λύσεων τῶν ἐξισώσεων αὐτῶν εἶναι θεμελιῶδες. Ἡ σπουδὴ δὲ τῆς εὐσταθείας βασίζεται ἐπὶ διαφόρων ἀντιλήψεων καὶ ὑποθέσεων περὶ εὐσταθείας, αἱ ὁποῖαι ἐξαρτῶνται ἐκ τοῦ τρόπου τῆς δράσεως τῶν «διαταράξεων» καὶ ἐκ τοῦ εἴδους μετρήσεως τοῦ μεγέθους τῶν διαταράξεων.

Εἰς τὴν παροῦσαν ἀνακοίνωσιν ἐκτίθενται αἱ ἔννοιαι εὐσταθείας τῶν λύσεων τῶν ἐν λόγῳ ἐξισώσεων καὶ διατυποῦνται παρατηρήσεις τινὲς ἐπ᾽ αὐτῶν. Διατυποῦνται ἐπίσης οἱ ὁρισμοὶ ὅταν αἱ διαταράξεις εἶναι «αἰφνίδιαι» ἢ «συνεχῶς δρῶσαι», καθὼς καὶ ὅταν, αἱ ἴδιαι αἱ ἀποκλίσεις τῶν μεταβλητῶν ὑπόκεινται εἰς διαταράξεις.

Ἡ εἰσαγωγὴ τῆς ἐννοίας τῆς εὐσταθείας κατὰ Liapounov καὶ Poincaré παρέχει δύο διακεκριμένας γενικὰς κατηγορίας τῶν ἀντιλήψεων εὐσταθείας».

408

ΕΦΗΡΜΟΣΜΕΝΑ ΜΑΘΗΜΑΤΙΚΑ.— **Remarks on stability concepts of solutions of dynamical systems,** *by Demetrios G. Magiros**.

'Ανεκοινώθη ὑπὸ τοῦ 'Ακαδημαϊκοῦ κ. 'Ιω. Ξανθάκη.

## INTRODUCTION

The study of the stability situation of physical and social phenomena, which are modeled as dynamical systems, is based on a variety of stability concepts, and this variety makes the study complicated and the stability results questionable in many cases.

In this paper, we will give a set of remarks on stability concepts, which may permit a better understanding of the difficulties of current stability problems.

The stability concepts may come from sources of different nature.

Examining the stability of a motion in its orbit and of the orbit of a motion, one can distinguish two basic stability concepts, which contain many other specialized concepts as special cases.

The manner in which a state of a system approaches another state, or deviates from it, the way the perturbations act on a system, or the way one measures their norm and their effect on the system, the type of the mathematical model of the system, etc., are sources for stability concepts of different nature.

All these different stability concepts can be «unified» into the same «stability relationships», and this «unification» of the stability concepts brings a natural simplicity in the understanding of subjects concerning stability, and gives rise to new results.

## 1. REMARKS ON PERTURBATIONS

It is necessary to make remarks concerning the purturbations of the systems and their solutions. A variety of stability concepts comes from the manner the purturbations act on the system, the way the norm

---

* **ΔΗΜΗΤΡΙΟΥ Γ. ΜΑΓΕΙΡΟΥ, Παρατηρήσεις ἐπὶ τῶν ἀντιλήψεων εὐσταθείας εἰς δυναμικὰ συστήματα.** Scientific Consultant, General Electric Co., RESD, Philadelphia, Pa. U. S. A.

Reprinted from the *Proceedings of the Athens Academy of Sciences* **49** (1974), 408–416.

of perturbations is taken, and the way their effect on the system is measured.

The perturbations, which can be considered as minor disturbing forces, may act on a system either «momentarily», when only the initial conditions are perturbed, or «permanently», when the system itself is disturbed and the perturbations, during their action, must enter the equations of the system explicitly. The «sudden» and «persistent» perturbations characterize two different classes of stability concepts.

The «norm» of the perturbations can be taken in different ways, and each of them characterizes a special stability concept under persistent perturbations. The perturbations may depend on deviating arguments, when the stability concepts will be related to the deviations of the arguments.

The «effect» of the perturbations is a change of certain quantities pertaining to the original motion and/or to its orbit, and this effect can be visualized by the change of the orbit S of the unperturbed motion $x_i(t)$ into the orbit $\bar{S}$ of the perturbed motion $\bar{x}_i(t)$.

Given a state of a dynamical system, that is a point P on S, if, as a result of the perturbations, $\bar{S}$ is the new orbit, and the point $\bar{P}$ on $\bar{S}$ corresponds to the point P of S, the distance $\varrho = P\bar{P}$ can be taken as the magnitude of the effect of the perturbations at P, Fig. 1.

To a given point P of S, one may make to correspond different points $\bar{P}$ on $\bar{S}$, and each correspondence characterizes a specific stability concept of the motion. Any such correspondence presuposes an assumption, and each assumption comes from a physical reason.

One can distinguish two, the most physical, correspondences between P and $\bar{P}$, when two important stability concepts result, namely, the stability concept in the sence of Liapunov and that in the sense of Poincaré, by using the stability distances $\varrho = P\bar{P}$, called Liapunov and Poincaré distances. In Fig. 2 «Liapunov distances» are shown P and $\bar{P}$ correspond to each other at the «same time». In Fig. 3, P and $\bar{P}$ correspond in such a way that:

$$\varrho = \varrho(P, \bar{S}) = P\bar{P} = \min \left\{ \sum_{i=1}^{n} |\bar{x}_i - x_i|^2 \right\}^{1/2}$$

and $\varrho = P\bar{P}$ is «Poincaré distance».

If the model of the system is expressed by differential equations with deviating arguments, e.g. by differential equations with retarded arguments of retardation $\tau_i(t)$, the «initial functions» $\varphi_i(t)$ of S and $\bar{\varphi}_i(t)$ of $\bar{S}$ define the «initial distance» $\varrho_{01} = P_{01}\bar{P}_{01} = |\bar{\varphi}_i(t) - \varphi_i(t)|$ taken over the interval $t_0 - \tau_i \leqslant t \leqslant t_0$, Fig. 2.

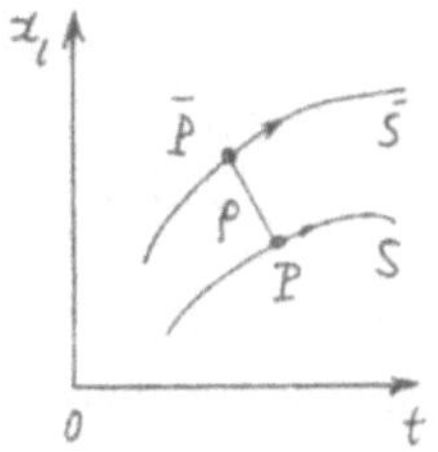

Fig. 1.

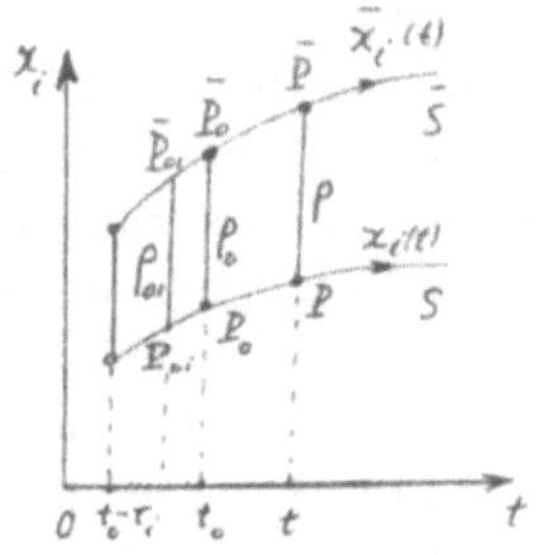

Fig. 2.

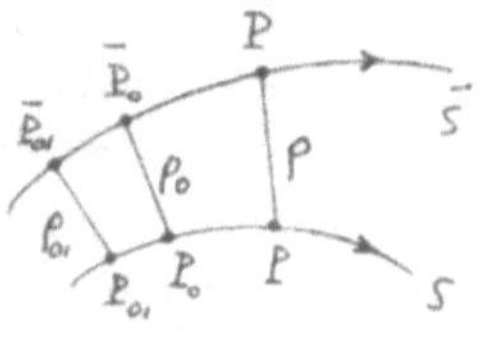

Fig. 3.

The above includes the ordinary differential equations, where $\tau_i = 0$, $\varrho_{01} = \varrho_0 = P_0\bar{P}_0$ at $t = t_0$, Fig. 2.

In case the retardations $\tau_i(t)$ are perturbed, one has a new stability concept characterized by the distance: $\varrho_\tau = |\bar{\tau}_i(t) - \tau_i(t)|$, where $\bar{\tau}_i$ the perturbed regardations (1).

## 2. REMARKS ON THE RELATIONSHIPS OF THE STABILITY CONCEPTS

Following considerations on «physical stability» and using the precending remarks and notations, any stability concept can be described quantitatively by the «stability conditions»:

$$(a): \varrho_{01} < \delta_1, \quad (b): \varrho < \varepsilon, \quad (c): \lim_{t \to \infty} \varrho = 0, \quad (d): \|p_i\| < \delta_2, \quad (e): \varrho_\tau < \delta_3 \qquad (1)$$

which, then, «unify» all the stability concepts. $\varepsilon$, $\delta_1$, $\delta_2$, $\delta_3$ are positive constants, $p_i$ the perturbations, and, in general, $\delta_1$ and $\delta_2$ depend on $t_0$ and $\varepsilon$.

By a suitable combination of these relationships, by an appropriate interpretation of the distances involved, and by some restrictions of some quantities of these relationships, one can express any stability concept.

The following remarks may help to clarify the above statements.

The inequality 1(a) is valid for t in the initial inverval $E_{t_0}: t_0 - \tau_i \leqslant t \leqslant t_0$, while the inequalities 1(b), (d), (e) for t in $t \geqslant t_0$, $t_0 \geqslant 0$.

In case of «sudden perturbations», the inequality 1(d) is meaningless, and in this case 1(a), (b) express the «stability», while 1(a), (b), (c) the «asymptotic stability».

In case of «permanent perturbations», when 1(d) is meaningful, the selection of the kind of the norm of perturbations specifies the stability concept, so, one may have «total stability» or «integral stability», or «stability in the mean», under suitable norm of perturbations.

By restricting $\delta_1$ and $\delta_2$ to depend only on $\varepsilon$ and not on $t_0$, we have the «uniform stabilities», as this happens in periodic systems, or in autonomous systems.

Restriction on $t_0$ to have a minimum, $\min t_0 = \alpha$, implies «eventual stabilities».

If in (1) the distances $\varrho$ are interpreted as Liapunov or Poincaré distances, we have stabilities in the sence of Liapunov or Poincaré.

In case of equilibrium points of a system, when the orbit S shrinks to a point, the distinction between stabilities in the sence of Liapunov and Poincaré is meaningless.

The inequality 1(e) has a meaning in case of perturbed retardations.

In case of periodic motions, when S is a closed curve, the stability concept in Liapunov sense is a narrow concept compared to the stability concept in Poincaré sence. The «isochronism», that is the «constancy of the frequency», characterizes the Liapunov stability, while this notion does not enter in the Poincaré stability. Further we may have that :

. A motion stable in Liapunov sense is also stable in Poincaré sense ;

. A motion unstable in Poincaré sense is also unstable in Liapunov sense ;

. A motion unstable in Liapunov sense may be stable or unstable in Poincaré sense, and

. A motion stable in Poincaré sense may be stable or unstable in Liapunov sense.

The stability is a property of the solution different from its boundedness property, although in some cases there may exist regions where these properties are equivalent and one implies the other. The boundedness of the solution is characterized by the boundedness of $OP = \varrho_1 = |x_i|$, where O the origin of the coordinate system and P point of the orbit; but the stability is characterized by the boundedness of $\varrho = |\bar{x}_i - x_i|$, and it is possible for the orbits $x_i$ and $\bar{x}_i$ to be unbounded as $t \rightarrow \infty$, when the stability distance $\varrho$ gets the form $(\infty - \infty)$, when $\varrho$ will be either infinite, or constant, or zero, and the unbounded solution $x_i$ will be either unstable or stable.

All the stability concepts included in the relationships (1) are of mathematical type, and the results, theorems or criteria, based on them, may not interpret the reality. Also, one and the same phenomenon may be, mathematically speaking, stable or unstable depending on the stability concept employed in the discussion of the stability of the phenomenon when the selection of the stability concept, appropriate for the phenomenon, arises.

The mathematical stability concepts and the stability criteria based on them, represent a possible functionig of the physical system, and in order all these to have a practical usefulness, and to agree with «practical stability», which must be the ultimate purpuse of the stability investigations, appropriate modifications, changes, supplements of the mathematical stability concepts must accompany the investigations.

The notion of «practical stability» is a subject not yet completely studied, but in many cases is characterized by the knowledge of : (2)

. The size of deviation of the state acceptable for a satisfactory operation of the system ;

. The size of permitted initial conditions that can be controlled ;

. The size of permitted perturbations ;

. The finite time T for the stability investigation.

The nolinearities of the system play a decisive role for practical stability.

### 3. AN EXAMPLE

We terminate the discussion on stability remarks by using as an example the investigation of the «stability of some precessional phenomena», by which some of the above statements may be clarified.

The rotational motion of a rigid body around its axis of symmetry is governed by the Euler's (ordinary differential) equations, of which the stability of the solutions has been examined under sudden perturbations, then by employing the first three stability relationships (1).

If $\underline{L} = (L_1, L_2, L_3)$ is the external torque vector acting on the body, $\underline{\omega} = (\omega_1, \omega_2, \omega_3)$ the angular velocity vector, which characterizes the precession of the body, $\underline{\omega}_0 = (\omega_{01}, \omega_{02}, \omega_{03})$ the initial angular velocity vector, and $I_1$, $I_2 = I_3 = I$ the moments of inertia, the precessional motion of the body in the following two cases are given by:

$$\left.\begin{array}{l} \text{(a)}\quad \omega_1 = \omega_{10} = \text{constant},\quad \omega_2 = A\cos Q_1,\quad \omega_3 = A\sin Q_1,\quad Q_1 = \dfrac{(I_1 - I)\,\omega_{10}}{I}t \\ \qquad \text{(in case, } L_1 = L_2 = L_3 = 0;\quad I_1,\ I,\ A = \left(\omega_{02}^2 + \omega_{03}^2\right)^{1/2} \text{ constants)} \\ \text{(b)}\quad \omega_1 = \dfrac{L_1}{I_1}t,\quad \omega_2 = A\cos Q_2,\quad \omega_3 = A\sin Q_2,\quad Q_2 = Q_1 + \dfrac{(I_1 - I)\,L_1}{2\,I_1\,I}t^2 \\ \qquad \text{(in case: } L_1 = \text{constant},\quad L_2 = L_3 = 0) \end{array}\right\} \quad (2)$$

The «regular precession» 2(a) is bounded, while the «helicoid precession» 2(b) is unbounded as $t \to \infty$. The results for their stability situation are the following:

(i) The regular precession 2(a) is «stable» but «not asymptotically stable» in Poincaré sense (orbitally). In Liapunov sense it is «stable» but «not asymptotically stable», if $\omega_{01}$ is not affected by the perturbation; and it is «unstable», if $\omega_{01}$ is affected by the perturbations.

(ii) The helicoid precession 2(b) is «asymptotically stable» in Poincaré sense, but it is «unstable» in Liapunov sense.

(iii) The stability situation of the above example in Poincaré sense is preferred, because in this case it is proved that the requirements for «practical stability» are satisfied.

We remark that the stability of a system, by using any of the previous concepts, depends, in general, on the selection of the major variables of the system, and on the transformation of the variables. This will be subject of a next paper.

## ΠΕΡΙΛΗΨΙΣ

1. Εἰς τὴν παροῦσαν ἐργασίαν δίδονται παρατηρήσεις ἐπὶ τῶν ἀντιλήψεων περὶ εὐσταθείας εἰς φυσικὰ καὶ κοινωνικὰ φαινόμενα, τὰ ὁποῖα μαθηματικοποιοῦνται ὡς δυναμικὰ προβλήματα. Ἡ ποικιλία τῶν ἀντιλήψεων εὐσταθείας κάμνει τὰ προβλήματα εὐσταθείας πολὺ πεπλεγμένα καὶ τ' ἀποτελέσματα τῆς ἐρεύνης βάσει αὐτῶν ὄχι δεκτὰ ἐνίοτε.

2. Ἀναφέρομεν μερικὰς πηγάς, ἀπὸ τὰς ὁποίας δυνάμεθα νὰ ἔχωμεν ποικιλίαν διαφόρου φύσεως ἀντιλήψεων περὶ εὐσταθείας:

—Ὁ τρόπος μὲ τὸν ὁποῖον μία κατάστασις ἑνὸς συστήματος πλησιάζει μίαν ἄλλην κατάστασιν ἢ ἀπομακρύνεται ἀπὸ αὐτήν.

—Ὁ τρόπος μὲ τὸν ὁποῖον προκαλοῦνται καὶ δροῦν αἱ διαταραχαὶ ἑνὸς συστήματος, ὁ τρόπος μὲ τὸν ὁποῖον μετροῦμεν τὴν ἔντασιν τῶν διαταραχῶν, καθὼς καὶ τὴν τῶν ἀποτελεσμάτων των ἐπὶ τοῦ συστήματος.

—Ἡ δομὴ τοῦ μαθηματικοῦ μοντέλου τοῦ συστήματος κλπ.

3. Ὅλαι αἱ ἀντιλήψεις εὐσταθείας, ἂν καὶ διαφόρου φύσεως, δύνανται ν' ἀναχθοῦν εἰς τὰς αὐτὰς μαθηματικὰς «σχέσεις εὐσταθείας», αἱ ὁποῖαι δίδουν μίαν «ἑνοποίησιν» τῶν ἀντιλήψεων εὐσταθείας, αὐτὴ δὲ ἡ ἑνοποίησις ὑποβοηθεῖ τὴν κατανόησιν τῶν ἀντιστοίχων προβλημάτων, δύναται δὲ νὰ ὁδηγήσῃ εἰς νέα συμπεράσματα. Μὲ κατάλληλον συνδυασμὸν τῶν «σχέσεων εὐσταθείας», καὶ κατάλληλον ἑρμηνείαν ἢ περιορισμὸν τῶν ποσοτήτων τῶν σχέσεων αὐτῶν, δύναται νὰ προκύψῃ ὁποιαδήποτε ἀντίληψις εὐσταθείας.

4. Αἱ ἀντιλήψεις εὐσταθείας, ποὺ περικλείονται εἰς τὰς «σχέσεις εὐσταθείας» ὑποδεικνύουν ἐνδεχομένην λειτουργίαν τοῦ συστήματος, τὰ δὲ συμπεράσματα, θεωρήματα ἢ κριτήρια, ποὺ βασίζονται ἐπ' αὐτῶν, ἐνδέχεται νὰ μὴ ἑρμηνεύουν τὴν πραγματικότητα κατὰ ἱκανοποιητικὸν τρόπον, ἢ ἐνδέχεται νὰ ἔχωμεν διὰ τὸ αὐτὸ φαινόμενον διαφόρους καταστάσεις εὐσταθείας, ὁπότε γεννᾶται τὸ πρόβλημα τῆς ἐκλογῆς τῆς καταλλήλου καταστάσεως διὰ τὸ φαινόμενον.

5. Αἱ μαθηματικαὶ ἀντιλήψεις περὶ εὐσταθείας, καθὼς καὶ τὰ βάσει αὐτῶν συμπεράσματα, διὰ νὰ ἑρμηνεύουν τὴν πραγματικότητα κατὰ ἱκανοποιητικὸν τρόπον, πρέπει νὰ συμφωνοῦν μὲ τὰ πορίσματα τῆς «πρακτικῆς εὐσταθείας», πρὸς τοῦτο δὲ χρειάζονται κατάλληλον τροποποίησιν καὶ συμπλήρωσιν. Ἡ πρακτικὴ εὐστάθεια δὲν ἔχει πλήρως σπουδασθῆ, ὅμως κύρια χαρακτηριστικά της δύναται νὰ εἶναι ἡ γνῶσις:

— τοῦ μεγέθους τῆς ἀποκλίσεως δεκτῶν καταστάσεων τοῦ συστήματος πρὸς ἱκανοποιητικὴν λειτουργίαν τοῦ συστήματος,
— τοῦ μεγέθους τῶν ἀρχικῶν συνθηκῶν, αἱ ὁποῖαι δύνανται νὰ ἐλεγχθοῦν,
— τοῦ μεγέθους τῶν ἐπιτρεπομένων διαταραχῶν,
— τοῦ χρόνου, ποὺ μελετῶμεν τὴν εὐστάθειαν.
— Αἱ μὴ γραμμικότητες τοῦ συστήματος παίζουν ἀποφασιστικὸν ρόλον διὰ τὴν πρακτικὴν εὐστάθειαν, καὶ δὲν δύνανται νὰ ἀμεληθοῦν.

## REFERENCES

1. L. Èl'sgol'c, Qualitative methods in mathematical analysis, Amer. Math. Soc. (1964), 167 - 199, Providence, R.I., U.S.A.
2. J. La Salle and S. Lefshetz, Stability by Liapunov direct method with applications, Academic Press. New York, U.S.A. (1961).
3. D. Magiros, (a) Nat. Acad. Sci., Proc. 53, No. 6 (June 1965), U.S.A.
4. ——, (b) J. of Information and Control, 9, No. 9 (Oct. 1966), U.S.A.
5. ——, (c) 5th Intern. Conf. of Nonlinear Oscillations, Proc. 2, pp. 346 - 357, (Sept. 1969), Kiev, U.S.S.R.
6. ——, (d) Athens Acad. Sci., Proc. (May 25, 1972), Athens, Greece.
7. ——, (e) C. R. Acad. Sci., 266 A, (Apr. 8. 1968), 770 - 773. Paris, France.
8. ——, (f) C. R. Acad. Sci., 268 A, (March 1969), 652 - 655 Paris, France.
9. ——, (g) Athens Acad. Sci., Proc. (Dec. 9, 1971) Athens, Greece.

*

Ὁ Ἀκαδημαϊκὸς κ. **Ἰωάννης Ξανθάκης,** παρουσιάζων τὴν ἀνωτέρω ἀνακοίνωσιν, εἶπε τὰ ἑξῆς:

Εἰς τὴν ἐργασίαν ταύτην τοῦ κ. Μαγείρου, τὴν ὁποίαν ἔχω τὴν τιμὴν νὰ παρουσιάσω εἰς τὴν Ἀκαδημίαν, ἐκτίθενται ὡρισμέναι ἐνδιαφέρουσαι παρατηρήσεις ἐπὶ τῶν ἀντιλήψεων περὶ εὐσταθείας εἰς φυσικὰ καὶ κοινωνικὰ φαινόμενα, ποὺ ἐκφράζονται μαθηματικῶς ὡς δυναμικὰ προβλήματα.

Αἱ ἀντιλήψεις περὶ εὐσταθείας προέρχονται ἀπὸ πηγὰς διαφόρου φύσεως. Ἡ μελέτη τῆς σταθερότητος μιᾶς κινήσεως ἐπὶ τῆς τροχιᾶς της ἢ ἡ μελέτη τῆς τροχιᾶς μιᾶς κινήσεως μᾶς παρέχει δύο διακεκριμένας βασικὰς ἀντιλήψεις περὶ εὐσταθείας, αἱ ὁποῖαι περιέχουν πλῆθος ἄλλων εἰδικευμένων ἀντιλήψεων ὡς εἰδικὰς περιπτώσεις. Ὁ τρόπος μὲ τὸν ὁποῖον ἡ κατάστασις ἑνὸς συστήματος πλησιάζει πρὸς μίαν ἄλλην, ἢ παρεκκλίνει ἐξ αὐτῆς, ὁ τρόπος μετρήσεως τῆς ἐντάσεως τῶν διαταραχῶν ἑνὸς συστήματος, ἡ δομὴ τοῦ μαθηματικοῦ μοντέλου ἑνὸς συστήματος καὶ ἄλλα εἶναι πηγαὶ ἀντιλήψεων εὐσταθείας διαφόρου φύσεως. Ἡ ποικιλία αὕτη τῶν ἀντιλήψεων εὐσταθείας κάμνει τὰ προβλήματα λίαν πολύπλοκα καὶ τὰ ἀποτελέσματα τῶν ἐρευνῶν βάσει αὐτῶν ἐνίοτε δὲν εἶναι γενικῶς ἀποδεκτά.

Ὅλαι αἱ ἀντιλήψεις περὶ εὐσταθείας, ἂν καὶ διαφόρου φύσεως, δύνανται νὰ ἀναχθοῦν, κατὰ τὸν κ. Μάγειρον, εἰς τὰς αὐτὰς μαθηματικὰς «σχέσεις εὐσταθείας», αἱ ὁποῖαι παρέχουν μίαν ἑνοποίησιν τῶν διαφόρων περιστάσεων. Ἡ ἑνοποίησις αὕτη ὑποβοηθεῖ εἰς τὴν πληρεστέραν κατανόησιν τῶν ἀντιστοίχων προβλημάτων.

Τὰ συμπεράσματα, θεωρήματα ἢ κριτήρια, τὰ στηριζόμενα εἰς τὰς μαθηματικὰς σχέσεις ἀντιλήψεων εὐσταθείας, εἶναι δυνατὸν νὰ μὴ ἑρμηνεύουν κατὰ ἱκανοποιητικὸν τρόπον τὴν πραγματικότητα, ὁπότε παρίσταται ἀνάγκη τροποποιήσεων ἢ συμπληρώσεων τοῦ μαθηματικοῦ προτύπου, οὕτως ὥστε ἡ μαθηματικὴ διατύπωσις τῆς ἀντιλήψεως εὐσταθείας νὰ πλησιάζῃ, ὅσον τὸ δυνατὸν περισσότερον, πρὸς τὴν λεγομένην «Πρακτικὴν Εὐστάθειαν», ἥτις ὅμως δὲν ἔχει ἀκόμη πλήρως μελετηθῆ.

# STABILITY CONCEPTS OF DYNAMICAL SYSTEMS

By: Demetrios G. Magiros

January, 1980

Reprinted from a *Technical Report, Genl. Electric Co.*, RSD, Philadelphia (Jan. 1980), 1–28.

-2-

TABLE OF CONTENTS

-3-

## INTRODUCTION

Basic problems in many fields of interest under current research in physical, technological and life sciences can be formulated and solved as problems of stability of dynamical systems.
In such problems the investigation faces many difficulties, which make the problems complicated and the results questionable in many cases. There are difficulties in understanding the stability concept, difficulties in its definitions that are reasonable from the physical point of view and consistent from a mathematical point of view, difficulties in developing workable criteria for making a decision regarding stability or instability of the states of the system. During recent years the concepts of stability of dynamical systems and the corresponding stability criteria have been advanced either by modifying old ideas or by creating new ones, and these advances permit a deeper penetration into the more profound problems of stability, very important both in theory and practice.

In the present paper attention is given to the stability concepts.
The variety of the stability concepts is due to sources of different nature, as, e.g., to the type of the mathematical model of the physical system, the nature of the variables of the system, the manner in which a state of the system approaches to, or deviates from, another state, the way the perturbations act on the system, the way the norm of the perturbations and their effect are measured, etc.
The stability concepts are classified into two classes, of which one is characterized by the stability concept in the sense of Liapunov, and the other by the stability concept in the sense of Poincaré.

-4-

The stability concepts are expressed and unified by the same mathematical relationships, and such a unification of the stability concepts brings a natural simplicity in the understanding of problems concerning stability and gives rise to new results. These unification relationships, by specialization or restrictions of the quantities involved, give a variety of special stability concepts and some relations between them.

Appropriate remarks and examples clarify the discussion.

-5-

## 1. PHYSICAL STABILITY CONCEPTS

The notion of stability was originated in physical problems as a characterization of specific situations. A state of a phenomenon or system is said to be "stable" if small disturbances to the state have as an effect small changes to the state. If the effect is considerable, the state is "unstable", and if, for small disturbances of the state, the effect tends to disappear, the state is "asymptotically stable"; and if, regardless of the magnitude of the disturbances, the effect tends to disappear, the state is "asymptotically stable in the large".

These physical stability concepts are of "qualitative" nature, and by using them no stability problems can, in general, be solved. The discussion of stability problems necessitates a "quantitative" knowledge of the stability concepts, for which mathematical relationships are needed between the magnitude of the disturbances of the state of the system and the magnitude of their effect on the state.
The subject in the following is a quantitative discussion of the stability concepts, which is essentially based on remarks on perturbations and their effect.

## 2. SYSTEMS UNDER PERTURBATIONS ACTING MOMENTARILY OR PERMANENTLY.

The disturbances to a state of a system are due to perturbations, which can be considered as "disturbing forces" acting on the system either momentarily or permanently.

The "momentarily" acting perturbations give disturbances to the initial states of the system but they do not appear in the formulation of the equations

-6-

of the motion of the system. We take as the model of the physical system in case of momentarily acting perturbations the $n$-dimensional system of non-autonomous differential equations in its normal form:

$$\left.\begin{aligned} \dot{x}_i(t) &= X_i(t, x_1, \ldots, x_n) \\ x_i(t_0) &= x_{i0}\ , \quad X_i(t, 0, \ldots, 0)\ , \ t_0 \leq t \end{aligned}\right\} \qquad (1)$$

where $x_1, \ldots, x_n$ are the "state variables", and $x_{i0}$ the initial conditions.

The "permanently" acting perturbations give disturbances to the system itself, $X_i$ will be changed, and the equations of the motion of the system must contain these persistent perturbations, and, in this case, we take the model of the system in the form:

$$\dot{x}_i(t) = X_i(t, x_1, \ldots, x_n) + p_i(t, x_1, \ldots, x_n) \qquad (2)$$

where $p_i$ are the persistent perturbations. For the existence of unique solutions of the systems (1) or (2), the functions $X_i$ and $p_i$ must satisfy appropriate conditions. We remark that the distinction of the perturbations into sudden and persistent leads to a classification of the stability concepts into two important categories.

## 3. THE EFFECT OF THE PERTURBATIONS. SOME DEFINITIONS.

The effect of the perturbations on a system is a change of certain quantities pertaining to the system, and this effect can be visualized by the change of the trajectory $S$ of the original (the unperturbed) system into the trajectory $\bar{S}$ of the new (the perturbed) system.

A state of a system is designated by a point of the trajectory, and if a point $M$ of the unperturbed trajectory $S$ corresponds to the point $\bar{M}$ of the perturbed trajectory $\bar{S}$ , the distance $\rho = M\bar{M}$ , Figure 1(a), can be considered as measuring the magnitude of the effect of the perturbations on $M$ .

-7-

This distance $\rho$ of which the unperturbed end $M$ corresponds to the perturbed end $\bar{M}$ , plays a decisive role in the stability concepts and in the formulation of the stability relationships. Let us call it a "stability distance".[7(a)(c)] To a given point $M$ on $S$ one may make to correspond different points $\bar{M}$ on $\bar{S}$ . Any such correspondence presupposes an assumption and characterizes a specific stability concept of the trajectory $S$ . Each assumption must be related to physical reality.

One can distinguish two the most physical, correspondences between $M$ and $\bar{M}$ , when one has two important and basic stability concepts.

(a): One correspondence between $M$ on $S$ and $\bar{M}$ on $\bar{S}$ is when $M$ and $\bar{M}$ are taken at the "same time", Figure 1(b). In this case we call the distance $\rho=\rho_\ell = M\bar{M}$ a "Liapunov distance", which characterizes a basic stability concept, the "stability in Liapunov sense". This distance in the $n$-dimensional Euclidian space is:

$$\rho_\ell=\left\{\sum_{i=1}^{n}[\bar{x}_i(t)-x_i(t)]^2\right\}^{1/2}, \quad \text{or:} \quad \rho_\ell=\left\{\sum_{i=1}^{n}[x_i(t,\bar{a}_j)-x_i(t,a_j)]^2\right\}^{1/2}$$

in the state space and in the parameter space, respectively.

$\bar{x}_i$ and $\bar{a}_j$ are the perturbed state variables and parameters corresponding to the unperturbed ones $x_i$ and $a_j$ respectively.

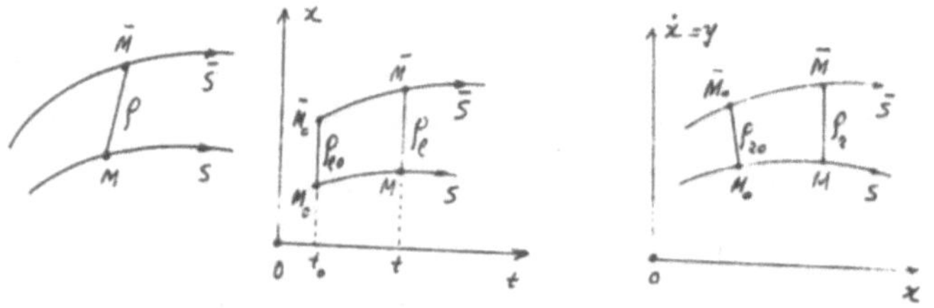


(a) (b) (c)

Figure 1

-8-

(b): In the second correspondence, the distance $\rho = \rho_2 = M\bar{M}$, Figure 1(c), is taken as the "minimum" of the distances from $M$ to the points of $\bar{S}$, that is:

$$\rho = \rho(M, \bar{S}) = \rho_2 = min\left\{ \sum_{i=1}^{n} (\bar{x}_i - x_i)^2 \right\}^{1/2}$$

in the "phase space" ( $x$, $\dot{x} = y$ ), parametrized by the time variable $t$.
The equations in the phase space can be found by eliminating $t$ from the equations of the model, thus reducing them to differential equations containing $x$ and $\dot{x} = y$.
The distance $\rho_2$, called a "Poincaré distance", characterizes a basic special stability concept, the "stability in Poincaré sense", also called an "orbital stability".

4. UNIFICATION OF STABILITY CONCEPTS

A variety of the stability concepts can be quantitatively described by the "stability relationships":

(a): $\rho_0 < \delta_1$ , (b): $\rho < \varepsilon$ , (c): $\lim_{t \to \infty} \rho = 0$ , (d): $\|p\| < \delta_2$ , (e): $\rho_\tau < \delta_3$ (3)

which, then, "unify" the stability concepts. In these relationships $\varepsilon, \delta_1, \delta_2, \delta_3$ are positive constants, $\delta_1$ and $\delta_2$ depend on $\varepsilon$ and on the initial time $t_0$, $t_0 \geq 0$, $\rho$ is the stability distance, and $\|p\|$ designates the norm of the perturbations.
By a suitable combination of these relationships and an appropriate interpretation and restriction of the quantities involved, one can define specific stability concepts.
In the following we analyze the above statement.

-9-

## 5. STABILITY CONCEPTS IN CASE OF MOMENTARILY ACTING PERTURBATIONS.

In case the perturbations act on the system momentarily we have variations only of initial conditions, and we take into account the first three relations of (3).

- Let a solution of the system (1) be $x_i(t)$ and its perturbed $\bar{x}_i(t)$. Given a positive number $\varepsilon$ however small, if it is possible to find a positive number $\delta_1 = \delta_1(t_0, \varepsilon)$ such that $\rho_0 < \delta_1$, where $\rho_0$ is the distance initially, implies, for all $t \geq t_0$, $\rho < \varepsilon$, then $x_i(t)$ is said to be "stable".

- The motion $x_i(t)$ is "unstable", if, for given small number $\varepsilon$, and for sufficiently small number $\delta_1$ for which $\rho_0 < \delta_1$, the inequality $\rho < \varepsilon$ is not satisfied.

- $x_i(t)$ is "asymptotically stable", if it is stable and, in addition, a number $\delta \geq \delta_1$ exists such that starting from any $\rho_0$, with $\rho_0 < \delta$, the limiting condition $\lim_{t \to \infty} \rho = 0$ holds uniformly to $x_{i0}$ and $t_0$.

- $x_i(t)$ is "asymptotically stable in the large", if it is asymptotically stable and $\delta$ is very large.

## 6. REMARKS. SPECIAL STABILITIES.

In this section we make a few remarks and, by using the relationships (3), we give a variety of special stability concepts all of which have a practical usefulness.

- Remark 1. Liapunov and Poincaré stabilities.

-10-

If the $\rho$'s in the relationships (3) are interpreted as Liapunov or Poincaré distances, the stability concepts are in Liapunov or Poincaré sense. The Liapunov stability concept is an appropriate one for the discussion of the stability of the motion of a system, while the Poincaré stability concept is appropriate for the stability of the orbit (itself) of the motion.

In periodic motions, the Liapunov stability concept classifies as unstable situations which are practically considered as stable, so this stability concept is a narrow one in periodic motions.
The Poincaré stability concept is the appropriate one for investigations of periodic motions.

- <u>Remark 2</u>. <u>Stability of equilibrium points</u>.

In case the orbit $S$ shrinks to an equilibrium point of the system, the Liapunov and Poincaré stability distances can be considered as identical, when, for the stability of the singular points of dynamical systems, the stabilities in Liapunov and in Poincaré sense are equivalent.

- <u>Remark 3</u>. <u>Uniform, eventual and finite time stabilities</u>.

If $\delta_1$ in 3(a) is independent of the initial time $t_0$, then we speak about "uniform stabilities".

If the solution $x(t)$ is stable for $t \geq t_0$, then the solution will be stable for initial time bigger than $t_0$, but not necessarily for initial time smaller than $t_0$. There are cases of stability in which the initial time has a minimum value $\tau$, which depends on $\varepsilon$, when $\tau(\varepsilon) \leq t_0$. In these cases we speak about "eventual stabilities".[5]

-11-

In the investigation of the stability of a solution $x(t)$ of the model of a system, $x(t)$ may become infinite either when $t \to \infty$ , or when time takes a maximum value which is a finite number $T$ , when $0 \leq t_0 \leq t \leq t_0 + T$. This value $T$ of time is called a "finite escape time".[3(c),10]
The so-called "finite time stability" is characterized by a finite time $T$ , which is not, in general, the "finite escape time". This kind of stability characterizes real systems expressed in general by nonautonomous differential equations, especially if these differential equations contain (structural) parameters. The "practical stability", as we will see, presupposes finite time

- Remark 4. Continuous dependence of a solution on the initial conditions and its stability.

If the solution $x(t, x_0)$ of a system is stable in the special parameter space of the initial conditions, then, for any two sets $x_{10}$ and $x_{20}$ of the initial conditions, and for appropriate numbers $\varepsilon$ and $\delta_1$ , the two inequalities:

$$\rho_0 = \{\Sigma (x_{20} - x_{10})^2\}^{1/2} < \delta_1 \quad , \quad \rho = \{\Sigma [x(t, x_{20}) - x(t, x_{10})]^2\}^{1/2} < \varepsilon$$

are compatible. But these inequalities are the conditions for the solution $x(t, x_0)$ to be continuously dependent on the initial conditions $x_0$ uniformly in $t$ , and this property of the solution is a Hadamard postulate for the solution to have a physical meaning.
Therefore, the continuous dependence of a solution on the initial conditions is equivalent to the stability situation in a special parameter space, which is the initial conditions space.

- Remark 5. Structural stability.

-12-

A real structure differs from the idealized structure designed by an engineer this difference being connected with small imperfections and defects. In order for the real structure to behave approximately as the idealized scheme, the structure must be postulated as stable with respect to small perturbations of the structure during its life.

We call "structural stability" of a system its stability, if it is "invariant" under small perturbations, either perturbations of the system itself or perturbations of its parameters.
The appropriate space of the perturbations of the system or of the parameters, where the system is structurally stable, is the "domain of structural stability" of the system.
The structural stability of a system in its domain implies a variety of important properties of the system.[6] Properties of structural stability suggest methods for investigation of physical problems with well accepted results, as, e.g., in problems of morphological processes or catastrophies in biology, etc.[8]
The structural stability of a system fails at points of "bifurcation" in the space of parameters of the system, where the topological structure of the system changes abruptly.

- Remark 6. The boundedness of a solution and its stability.

The boundedness of a solution $x(t)$ of a system is, by definition, characterized by the boundedness of the distance $\rho = \rho_t = OM = |x(t)|$, Figure 2, of any point $M$ of the orbit $S$ of the solution from the origin $O$ of the coordinate system.

-13-

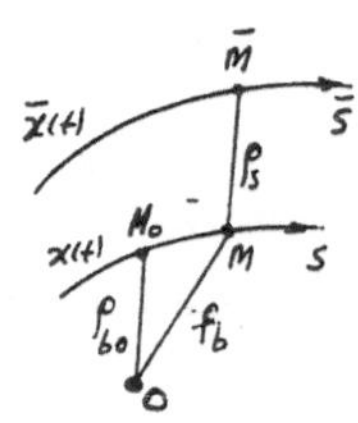


Figure 2

We remark that the relationships (3) contain the concept of boundness as a special case, by interpreting $\rho$ as the distance $\rho_b = OM$, Figure 2.

If, given a point $M_0$ of $S$, corresponding to the distance $\rho_{b0} = OM_0$, such that $\rho_{b0} < \delta_1$, where $\delta_1$ is a positive finite number, and if, for any point $M$ of $S$, there is a positive finite number $\varepsilon$, however large, and $\rho_{b0} < \delta_1$ implies $\rho_b < \varepsilon$, the orbit $S$ is bounded.

To different types of stability there, in general, correspond different types of boundedness.[11]

The concept of boundedness of a solution has been taken, from the time of Lagrange, as a stability concept, especially in celestial mechanics, called "stability in Lagrange sense".

The solar system, for example, is considered as "Lagrange stable" in the sense that none of the members of this system excapes to infinity, and not any two of its members collide. The velocities of the bodies at their collision are unbounded, and the Lagrange stability means that the coordinates and velocities of the bodies of the system are bounded.

The boundedness property of a solution of a system is different from its stability property. In some cases these two properties are identical and one implies the other, and in some other cases methods in studying one of these properties can often be so modified as to be of use also in problems involving the other.[1,3(b),11]

The Lagrange stability classifies as stable any situation of a state in finite distance, and as unstable any situation in a large distance, and such concepts are not, in general, accepted for investigation of stability problems on earth, say.

-14-

We may have unstable situations in bounded distances, and stable situations in unbounded distances. The "boundedness distance" $\rho_b = OM$, Figure 2, may be unbounded as $t \to \infty$, that is it may be: $\lim_{t\to\infty} \rho_b = \lim_{t\to\infty} |x| = \infty$, but the "stability distance" $\rho_s = M\bar{M} = |\bar{x} - x|$ for $t \to \infty$, may be of the form $(\infty - \infty)$, when the stability distance" $\rho_s$ may be either finite or infinite, and the unperturbed situation may be either stable or unstable.

## 7. STABILITY CONCEPTS IN CASE OF PERSISTENT PERTURBATIONS.

In case the perturbations acting on a system are permanent, we have a variation of the second members $X_i$ of (1), when we take the model of the system in the form (2), where $X_i$ and $p_i$ are such that (2) has a unique solution $x_i(t)$. We do not, in general, assume that $p_i(t,0) = 0$, then the origin of the perturbed system is not, in general, a solution of (2).
The stability concepts in the case of persistent perturbations come from the stability relationships (3), where 3(d) is meaningful, by interpreting or restricting appropriately the stability distances $\rho$ and accepting special forms for the norm of the perturbations $p_i$ depending on their magnitude.
We may have the following stability concepts.[9]

- In case the magnitude of $p_i$ is small, the norm is taken in the form: $\|p_i\| = \max_{x_i} |p_i|$, when we have the concept of "total stability".

- If the magnitude of $p_i$ is large in a small interval of time, the norm is taken in the form: $\|p_i\| = \int_0^\infty \max_{x_i} |p_i| dx$, and we have the concept of "integral stability".

-15-

- For both the above restrictions of the magnitude of $p_i$ , the norm is taken as $\|p_i\| = \int_t^{t+T} \max_{x_i} |p_i| dx$ , and we have the "mean stability". All these stability concepts are in Liapunov or in Poincaré sense, if the distances $\rho$ in the relationships (3) are appropriately interpreted.

Some connections between stability concepts can be introduced by the following statements: [9]

- Asymptotic integral stability implies total asymptotic stability in the mean, and conversely

- Asymptotic integral stability implies total asymptotic stability and this implies total stability

- Asymptotic stability in the mean implies stability in the mean, and this implies either integral stability or total stability

- Integral stability does not imply total stability

- Stability in the mean does not imply total asymptotic stability

- Total asymptotic stability does not imply integral, stability.

## 8. STABILITY CONCEPTS IN CASE OF MODELS WITH DEVIATING ARGUMENTS.

If the model of the system is expressed by differential equations with deviating arguments, as, e.g., by differential equations with retarded arguments of retardations $\tau(t)$ , we distinguish two cases, accordingly as the perturbations act either momentarily or permanently.[7(b)]

-16-

(a): In the first case, the model does not contain the perturbations, when the model may be of the form:

$$\left.\begin{aligned} \dot{x}_i(t) &= X_i[t, x_j(t, \tau_{jk}(t))], \quad \tau_{jk} > 0 \\ i,j &= 1,2,\dots,n \; ; \; k = 1,2,\dots,m, \; m \le n \end{aligned}\right\} \tag{4}$$

The initial functions $\varphi(t)$ are defined in the initial interval: $E_{t_0}: t_0 - \tau_{jk} \le t \le t_0$, and the corresponding solution $x_\varphi(t)$ of (4) holds for $t \ge t_0$.

For the Liapunov distances, Figure 3(a), we have $\rho_{10} = M_{01}\bar{M}_{01} = |\bar{\varphi} - \varphi|$ in the interval $E_{t_0}$, while $\rho_\varrho$ in the interval $t \ge t_0$.

For Poincaré stability appropriate remarks hold, Figure 3(b).

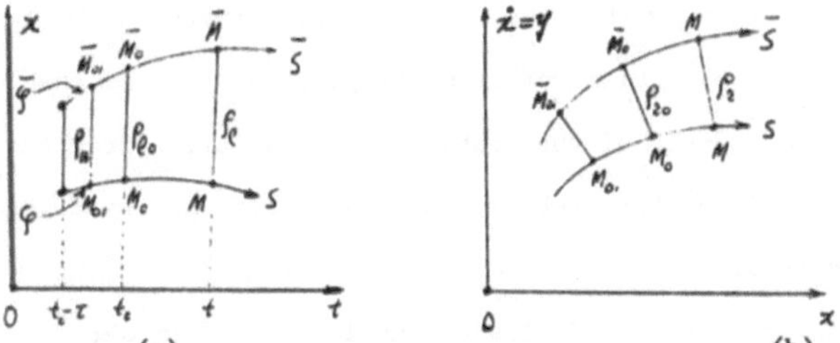

(a) (b)

Figure 3

(b): In the second case, the model contains the perturbations, and the model may be of the form:

$$\dot{x}_i(t) = X_i[t, x_j(t - \tau_{jk}(t))] + p_i[t, x_j(t - \tau_{jk}(t))] \tag{5}$$

The solutions $x_\varphi$ of (5), corresponding to the initial functions $\varphi(t)$, is "stable" with respect to perturbations $p$, if, given $\varepsilon > 0$ and $t_0 \ge 0$, there exist two positive constants $\delta_1 = \delta_1(t_0, \varepsilon)$, $\delta_2 = \delta_2(t_0, \varepsilon)$ such that for $\rho_{10} < \delta_1$ valid in the interval $E_{t_0}$, and $\|p\| < \delta_2$ valid in $t \ge t_0$, the inequality $\rho < \varepsilon$, in $t \ge t_0$, is implied.

(c): In case the retardations $\tau$ are perturbed, we have a new stability concept characterized by the distance $\rho_\tau = |\bar{\tau}(t) - \tau(t)|$, where $\bar{\tau}$ is the perturbed retardation corresponding to the unperturbed $\tau$, which holds in $t \ge t_0$

-17-

## 9. THE STABILITY OF A SYSTEM AND THE COORDINATE SYSTEM.

For the study of a phenomenon or for investigation of a physical system, one selects some of its quantities, called "space variables" of the system, which give the "coordinate system" to which the investigation is referred.[7(d)] The nature of the stability of a physical system is, in general, "not invariant" with the selection of the space variables, or with a general transformation of the coordinate system. In other words, the coordinate system selected and its transformations affect, in general, the nature of the stability.
Also, we may have stability with respect to some coordinates only, that is, the coordinates may not be equivalent from the point of view of stability.
The problem of the dependence of the stability of a solution from the coordinate system is a very important problem, but no specific and systematic investigation upon it exists.
By means of some examples in the following, an idea is given about this problem.

## 10. GEOMETRICAL INTERPRETATION OF STABILITIES.

The stability concepts may be clarified by a geometrical interpretation that follows.

We define as a half "$\varepsilon$-cylinder" around a curve $Q$ in $n$-dimensional space the set of points of which the distance $\rho$ from the curve $Q$ is smaller than $\varepsilon$ . The curve $Q$ is the axis of the cylinder.[3(d),12(b)]

Given the motion $x_i(t, t_0, x_{10}, \dots, x_{n0})$ of orbit $S$ , if we consider $S$ as the axis of "$\delta_1$-cylinder" and "$\varepsilon$-cylinder", where $\delta_1$ and $\varepsilon$ are the constants of the inequalities (3), and we interpret $\rho$ either as a "Liapunov distance", or as a "Poincaré distance", we can have a geometrical interpretation either of "Liapunov stability", or of "orbital stability".

The motion $x_i$ with orbit $S$ starting from an initial point $x_{io}$ in the $(t, x_i)$-coordinates, is "Liapunov stable", if, given an " $\varepsilon$ -cylinder" around $S$ , one can find a " $\delta_1$ -cylinder" such that any perturbed orbit $\bar{S}$ starting from $\bar{x}_{io}$ . in " $\delta_1$ -cylinder"remains in the " $\varepsilon$ -cylinder for all $t \geq t_0$ , Figure 4. $\rho$ is taken as "Liapunov distance".

If, in addition, $S$ and $\bar{S}$ tend to coincide as $t \to \infty$ , $S$ is "asymptotically stable".

For "orbital stability" of $S$ , $\rho$ must be interpreted as a "Poincaré distance", and $S$ and $\bar{S}$ are referred in the phase space $(x_i , \dot{x}_i = y_i)$.

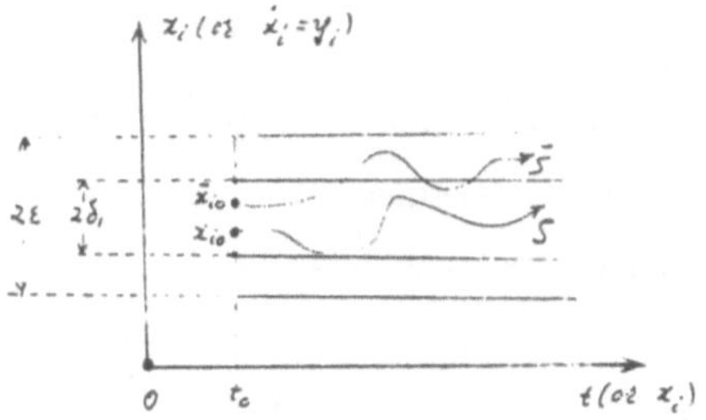


Figure 4

For stability under persistent perturbations, the " $\delta_2$ -cylinder" must be used.

We remark that the widths and lengths of the above cylinders, as well as the nature of the nonlinearities of the model of the system, play, as we will see, an important role for the so-called "practical stability".

-19-

## 11. PRACTICAL STABILITY.

The previous discussion on stability concepts is of theoretical or mathematical type. It represents a possible function of a physical system, but it is, in general, of small practical usefulness.
One can see that a state of a system may be unstable under the previous stability definitions, but the system may oscillate sufficiently near this state and its performance can be accepted practically as a stable one.
Missiles many times have this kind of behavior.
Also, an equilibrium state of a system may be stable mathematically in a very small region, but in practice the perturbations expected may cause the system to go far from the equilibrium state, when the system is practically unstable at the equilibrium state.

In order that the theoretical stability concepts interpret the reality adequately and have a practical usefulness, which must be the ultimate purpose of the investigator, they must be accompanied by appropriate modifications, changes and supplements, when, by all these, we will have the notion of the so-called "practical stability". Practical stability is defined as follows:[4,12(a)]

Given an interval of time $[t_0, t_0+T]$ and two sets: $R_{t_0}$ and $R_t$ in the $n$-dimensional space $(x_1, \dots, x_n)$, if there exists a set $R_a$ in the space of admissible values of the parameters such that the integral curves $x(t)$ of the system, starting from $t_0$ of the set $R_{t_0}$, remain in the set $R_t$ for all time of the interval $[t_0, t_0+T]$, then the solution $x(t)$ is said to be "practically stable".
Practical stability is characterized by the knowledge of:

(a): The size of the region $R_t$ of deviations of the state acceptable for a

-20-

satisfactory operation of the system, that is the "width" of the $\varepsilon$-cylinder. This region is formulated from the definite requirement demanded of the design system.

(b): The size of the region $R_{t_0}$ of the permitted initial conditions that can be controlled, that is the "width" of the $\delta_1$-cylinder. This region is defined by the technical conditions of the functioning of the real system.

(c): The size of the region $R_p$ of the permitted perturbations, that is the "width" of the $\delta_2$-cylinder.

(d): The finite time $T$ over which the stability is investigated, which corresponds to the "lengths" of the cylinders, that is the time interval $[t_0, t_0+T]$ for which the above requirements are satisfied for the solution $x(t)$ of which the stability is investigated.

(e): The size of the region $R_e$ in the space of admissible values of the parameters of the system such that $x(t)$ starting at $t=t_0$ in the $\delta_1$-cylinder fall within $\varepsilon$-cylinder for $t\in[t_0,t_0+T]$ for perturbations in the $\delta_2$-cylinder.

For a perfect practical stability the sizes of the above regions must be large.

For the computation of the above regions the use of the "nonlinearities" of the model of the system is necessary, and, therefore, investigations based on linear approximations of the models must not be taken too seriously.[7(a)(e)]

12. EXAMPLES.

In the following examples we discuss the stability situations by using only the stability definitions, and make appropriate remarks.

-21-

Example 1. Stability of rectilinear motions.

The rectilinear motion of a mass is given in the $t, x$-plane by:

$$x(t) = v(t-t_0) + d \tag{1}$$

where $d$ is distance and $v$ the velocity of the moving mass.
If at the initial time $t_0$ the distance is $d = d_0$ and the velocity $v = v_0$, the orbit of the motion is a straight line $S$, Figure 5(a), starting from the point $M_0(t_0, d_0)$ with velocity $v_0$ (slope of $S$). We consider two cases.

(a): Uniform rectilinear motion.

For this motion the velocity $v$ is considered as a constant, when the perturbation of the motion will affect only the distance $d$ in the equation (1), then the perturbed orbit $\bar{S}$ will have the same slope with $S$, and $S, \bar{S}$ are parallel, Figure 5(a). For any place $M$ of the mass on $S$ the Liapunov distance $\rho_\ell = M\bar{M}$ is constant, also the Poincaré distance $\rho_2 = M\bar{\bar{M}}$ is constant, then the uniform rectilinear motion is stable both in Liapunov and Poincaré sense.

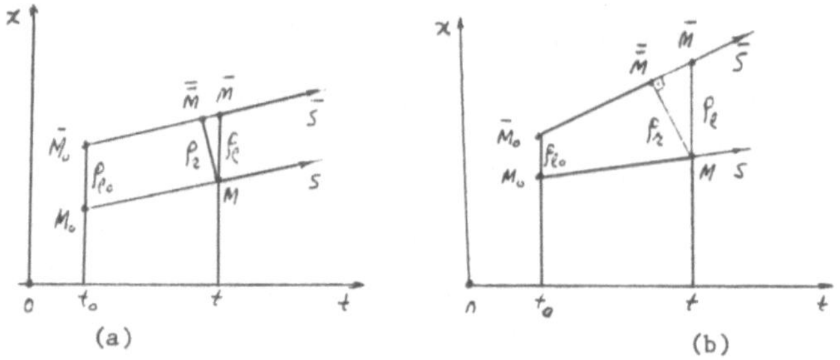


Figure 5

(b): Nonuniform rectilinear motion.

Since during this motion the velocity $v$ is not constant, the perturbation of $S$ will affect both the distance $d$ and the slope $(v)$ of the orbit $S$, and $S$ and its perturbed $\bar{S}$ are not parallel. The stability distances $\rho_\ell$ and $\rho_2$,

-22-

Figure 5(b), increase to infinity as $t \to \infty$ and the motion is unstable both in Liapunov and Poincaré sense.

We remark that the orbits in these motions are unbounded as $t \to \infty$, then these motions are unstable in Lagrange stable.

Example 2. Linear systems with or without forcing terms.

(a):

$$\dot{x}_i(t) = \sum_{j=1}^{n} a_{ij}(t)\, x_j \; ; \quad i = 1, \dots, n \tag{2.a}$$

the coefficients $a_{ij}(t)$ are continuous functions of $t$ in $t \geq t_0$. We see that:[2] "The boundedness of the solutions of the system (2.a) implies the Liapunov stability, and conversely. If $a_{ij}$ are constants and all solutions bounded, we have uniform stability".

(b):

$$\dot{x}_i(t) = \sum_{j=1}^{n} a_{ij}(t)\, x_j + f_i(t) \tag{2.b}$$

$a_{ij}(t)$ and $f_i(t)$ are continuous functions of $t$ in $t \geq t_0$. We see that:

(i): "The boundedness of the solutions of (2.b) implies their Liapunov stability, but the converse is not true". For the converse we have:

(ii): "The Liapunov stability of the solutions of (2.b) implies the boundedness of all solutions, if, in addition, there is at least one solution bounded".

We remark that connections between stability concepts and boundedness of the solutions in nonlinear systems are found in some specific cases.

Example 3. Stability in a problem of astronomy.

We examine: "The stability of the motion of a mass moving in an orbit under the inverse square Newton's law of attraction of an attractive center".[2]

The orbit $S$ of the mass $M$ is in this case an ellipse which can be determined by using initial conditions. The initial conditions are the distance

-23-

of the moving mass from the attractive center $E$ and its velocity at an initial time $t_0$. The motion of $M$ on $S$ is periodic with period $T$. Perturbations acting momentarily affect the initial conditions, so we have a perturbed elliptic orbit $\bar{S}$ with a new period $\bar{T}$.

(a): Stability in Liapunov sense.

The Liapunov distance is $\rho_\ell = M\bar{M}$, Figure 6(a), where $M$ and $\bar{M}$ are places of the mass on the orbits $S$ and $\bar{S}$ at the "same time". The periods $T$ and $\bar{T}$ depend on the length of the major axis of the correspondent ellipse. $M$ and $\bar{M}$ being very close initially and traveling on different ellipses may find themselves at opposition and then at great distance from each other in due course of time. Then, given a small positive number $\varepsilon$, one can not find a $\delta_1$ such that $\rho_0 < \delta_1$ implies $\rho < \varepsilon$, when the motion is "Liapunov unstable".

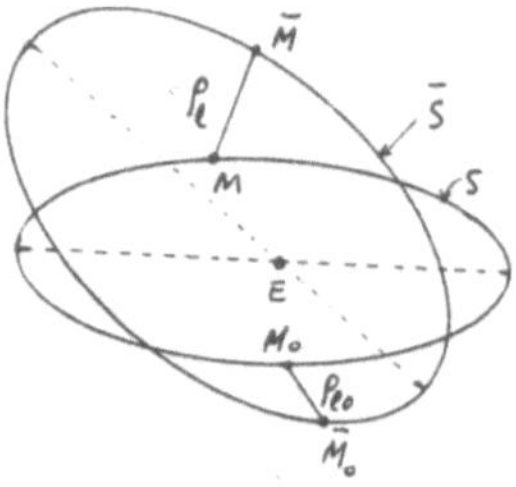


(a)

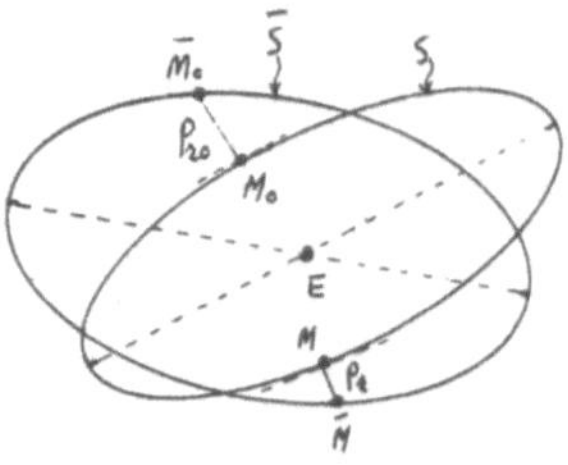


(b)

Figure 6

We can prove that in general a periodic motion of which the corresponding perturbed motion has different period is Liapunov unstable.

The free vibrations of a simple pendulum are Liapunov stable if they are linear, and Liapunov unstable if they are nonlinear. In the case of linearity the isochronous phenomenon exists, while in the case of nonlinearity the isochronous phenomenon is violated.

-24-

(b): Stability in Poincaré sense (orbital).

The Poincare distance $\rho_2 = M\bar{M}$, Figure 6(b), must be the "minimum" distance from $M$ to points of $\bar{S}$. If one wants $\rho_2$ to be smaller than a given positive number $\varepsilon$ for all time, a positive number $\delta_1$ must be found such that, if the Poincaré distance initially is $\rho_{20} < \delta_1$, the inequality $\rho_2 < \varepsilon$ must be implied. This is possible. Given a perturbation $p$, the perturbed orbit $\bar{S}$ is known, when $\rho_2$, corresponding to $M$ on $S$, is known. The deviations for all points of $S$ have a maximum $\rho_M$ and a minimum $\rho_m$, and the smaller the perturbation $p$ the smaller $\rho_M$ and $\rho_m$. Given $\varepsilon$ small, if we want $\rho_2 < \varepsilon$ for all $M$ on $S$, we must use a very small perturbation $p$ such that $\rho_M < \varepsilon$, when $\delta_1$ must be $\delta_1 < \rho_m$, and this means that the motion is orbitally stable. The distance $\rho_2$ never tends to zero as $t$ changes, then the orbital stability is not an asymptotic one.

According to the above the same phenomenon may be stable or unstable depending on the stability definition used. For the stability situation of the phenomenon, the appropriate stability definition must be selected according to practical usefulness.

In the above example practical reasons suggest to accept the stability situation coming from stability concept in Poincaré sense.

Example 4. Stability and the change of coordinates.

(a): The motion:
$$x = e^{-\alpha t} \cdot \cos \omega t \tag{4.a}$$
for $\omega$ in $-\infty < \omega < +\infty$, and $\alpha$ a parameter, is "unstable" in Liapunov sense for $\alpha \leq 0$, but "quasi-asymptotically stable" for $\alpha > 0$.

(b): The nonlinear differential system:
$$\dot{x} = -\mu y \;,\; \dot{y} = \mu x \;,\; \mu \neq 0 \tag{4.b}$$

-25-

has as a solution the motion:

$$x = a\cos(\mu t + b), \quad y = a\sin(\mu t + b) \tag{4.b.1}$$

with orbit

$$x^2 + y^2 = a^2 \tag{4.b.2}$$

$a$ and $b$ are constant parameters, and $\mu$ characterizes the frequency or the period $(\frac{2\pi}{\mu})$ of the motion (4.b.1) on the orbit (4.b.2).

The motion (4.b.1) is "Liapunov stable" if $\mu$ is a constant, but it is "Liapuno unstable" if $\mu$ is changed, say in the interval $\mu_1 \leq \mu \leq \mu_2$.

For $a$ constant, the "orbital stability" is meaningless; and for $a$ changed, the motion is "orbitally stable".

The motions all are "Lagrange stable" for any $a$ and $\mu$.

(c): The nonlinear differential system:

$$\dot{x} = -y(x^2+y^2)^{1/2}, \quad \dot{y} = x(x^2+y^2)^{1/2} \tag{4.c}$$

accepts as a solution the motion:

$$x = a\cos(at + b), \quad y = a\sin(at + b) \tag{4.c.1}$$

with orbit

$$x^2 + y^2 = a^2 \tag{4.c.2}$$

$a$ and $b$ are constant parameters.

For $a$ constant, the motion (4.c.1) is "Liapunov stable", but the orbital stability is meaningless. For $a$ changed, the motion is "Liapunov unstable", but "orbitally stable". The motion (4.c.1) is "Lagrange stable".

– The solution (4.c.1), which is "Liapunov unstable" for $a$ changed, can become "Liapunov stable" by changing the variables $x$ and $y$ in (4.c) according to appropriate relations.[2] If the new variables [redacted] $r$ and $\vartheta$ are related to $x$ and $y$ according to:

$$x = r\cos\vartheta, \quad y = r\sin\vartheta, \quad \vartheta = at + b \tag{4.c.3}$$

-26-

the equations (4.1) are transformed into the new ones:

$$\dot{r} = 0, \quad \dot{b} = 0 \tag{4.c.4}$$

of which the solutions:

$$r = c_1 \quad , \quad b = c_2 \tag{4.c.5}$$

$c_1$ and $c_2$ arbitrary constants, are "Liapunov stable".

(d): The nonlinear equation:

$$\ddot{x} + \sin x = 0 \tag{4.d}$$

accepts as a solution the motion:

$$x = a \sin\{\varphi(a)\, t + b\} \tag{4.d.1}$$

where $a$ and $b$ are arbitrary constants, and the frequency $\varphi(a)$ can be expressed in terms of elliptic functions.

The solutions (4.d.1) are, except the origin, "Liapunov unstable", but by using new coordinates $r$ and $b$, according to the transformations:[2]

$$x = r \sin\{\varphi(r)t + b\}, \; y = r \cos\{\varphi(r)t + b\} \tag{4.d.2}$$

the equation (4.d) leads to the equations

$$\dot{r} = 0 \quad , \quad \dot{b} = 0 \tag{4.d.3}$$

of which the solutions $r = c_1$ , $b = c_2$ are "Liapunov stable".

(e): The nonautonomous normal nonlinear system:[2]

$$\dot{x}_i = X_i(t, x_1, \ldots, x_n); \; i = 1, \ldots, n \tag{4.e}$$

has the functions:

$$\phi_i(t, x_1, \ldots, x_n) = c_i \tag{4.e.1}$$

as solutions in implicit form. $c_1, \ldots, c_n$ are arbitrary constants.

By the transformation:

$$y_i = \phi_i(t, x_1, \ldots, x_n) \tag{4.e.2}$$

(4.e) are reduced to

$$\dot{y}_i = 0 \tag{4.e.3}$$

of which the solutions

$$y_i = c_i \tag{4.e.4}$$

are "Liapunov stable", and this is not, in general, the case for the solutions (4.e.1) of (4.e).

III. <u>REFERENCES</u>


1. H. Antosiewicz: "Boundedness and Stability", in "Nonlinear differential equations and nonlinear mechanics", edited by: J. LaSalle and S. Lefschetz, Academic Press, New York (1963), pg. 259-267.
2. L. Cesari: "Asymptotic behavior and stability problems in ordinary differential equations", Springer-Verlag, Berlin (1959), pg. 6-13.
3. W. Hahn: "Theory and application of Liapunov direct method", Prentice-Hall, Englewood Cliffs, N.J. (1963), (a) pg. 10, (b) pg. 129, (c) pg. 126, (d) pg. 8.
4. J. LaSalle and S. Lefschetz: "Stability by Liapunov's direct method", Academic Press, New York (1961), pg. 121-126.
5. J. LaSalle and R. Ruth: "Eventual stability", Proc., 4th Joint Automatic control conference , Univ. of Minnesota, Minneapolis (1963), pg. 468-470.
6. S. Lefschetz: "Differential equations: Geometric theory", Interscience Publ., New York (1957), pg. 239-245.
7. D. Magiros: (a): "On stability definitions of dynamical systems", proc., Nat. Acad. 5c., Washington, D.C. (U.S.A), Vol. 53, No. 6 (1965), pg. 1288-1294.

-28-

(b): "Stability concepts of solutions of differential equations with deviating arguments", Practica, Athens Acad. SC, Athens (1971).

(c): "Remarks on stability concepts od dynamical systems", Practica, Athens Acad. SC (1974).

(d): "Mathematical models of physical and social systems", General Electric Co., RESD, Phila., PA. (Nov. 1976).

(e): "Characteristic properties of linear and nonlinear systems", Practica, Athens Acad. SC. (1976).

8. R. Thom: "Structural stability and morphogenesis", W. Benjamin, Reading, Mass, U.S.A. (1975).

9. I. Vrkǒv "On some stability problems", Proc. of a conference in Prague, Acad. Press, New York (1963), pg. 217-221.

10. L. Weis and E. Infante: "Finite time stability under perturbations and on product spaces", in "Differential equations and dynamical systems", ed. J. Hale and J. LaSalle, Acad. Press, New York (1967), pg. 341-350.

11. T. Yoshizawa: "Stability theory by Liapunov second method", Math. Soc. of Japan (1966).

12. V. Zubov: "Mathematical methods for the study of automatic control systems", The Mac Millan Co., New York (1963), (a): pg. 13, 123, (b): pg. 9-12, (c): pg. 96-97.

# ON A CLASS OF PRECESSIONAL PHENOMENA AND THEIR STABILITY IN THE SENSE OF LIAPUNOV, POINCARÉ AND LAGRANGE

Demetrios G. MAGIROS *

Abstract

The stability of a class of precessions in the sense of Liapunov, Poincare' and Lagrange is investigated in this paper.

## 1. Introduction

In this paper we discuss the stability of a class of precessional phenomena of the rotational motion of a rigid body around its axis of symmetry. The stability is examined in the three basic stability concepts, that is stability in the sense of Liapunov, Poincare' and Lagrange. In section 2 we specify the class of precessions, of which the stability is discussed in section 4. The discussion of the stability is based on stability remarks and results of section 3. Ref. 1 and 2 are used for sections 2 and 3, respectively.

## 2. A Class of Precessional Phenomena

If a rigid body has an axis of symmetry and rotates around it, precessional phenomena occur depending upon the nature of the resultant external torque $\underline{L}$ acting on the body, the torque taken with respect to its center $o_1$ of the mass of the body. The rotational motion of our symmetric body is governed by the Euler's equations:

$$\dot{\omega}_1 = L_1/I_1,\ \dot{\omega}_2 = L_2/I - \omega_1\omega_3 (I_1 - I)/I,\ \dot{\omega}_3 = L_3/I + \omega_1\omega_2 (I_1 - I)/I \tag{1}$$

where $\omega_1, \omega_2, \omega_3$ are components of the angular velocity $\underline{\omega}$, and $L_1, L_2, L_3$ components of the resultant torque $\underline{L}$ along the axes of the coordinate system $o_1x_1x_2x_3$ which is fixed in the moving body. The constants $I_1, I_2, I_3$ are moments of inertia about the coordinate axes, and if $o_1x_1$ is the axis of rotation, the last two moments of inertia are equal, $I_2=I_3=I$.

Each solution $(\omega_1, \omega_2, \omega_3)$ of (1) characterizes a precessional motion of the body dependent on the nature of the torque components $L_1, L_2, L_3$.

We distinguish here the class of precessional motions which correspond to a time-dependent torque, and especially the class corresponding to the torque with components:

$$L_1 = L_1(t),\ L_2 = L_3 = 0 \tag{2}$$

In this case the solution of (1) is given by:

$$\omega_1 = (1/I_1)\int L_1(t)\,dt + c,\ \omega_2 = A\cos Q(t),\ \omega_3 = A\sin Q(t),\ Q(t) = [(I_1 - I)/I]\int \omega_1 dt \tag{3}$$

c and A are constants determined by specializing the initial conditions.

The locus of the endpoint P of the angular velocity $\underline{\omega} = \underline{o_1P}$ is a curve, the "precessional curve", of which the equations can be found from (3) by eliminating the time t. The precessional curves characterize the precessional phenomena corresponding to the torque component $L_1(t)$.

We specialize $L_1(t)$ as in the following cases:

$$L_1 = 0,\quad L_2 = L_3 = 0 \tag{4.1}$$

$$L_1 = \overline{L}_1 = \text{const},\quad L_2 = L_3 = 0 \tag{4.2}$$

$$L_1 = \sin t,\quad L_2 = L_3 = 0 \tag{4.3}$$

* Consulting Mathematician, General Electric Co., Re-entry and Environmental Systems Division, Philadelphia, Pa., U.S.A.

Reprinted from the *Proceedings of the VIIIth Intl. Symp. Space Tech. Sci.*, Tokyo (1969), 1163–1170.

By using (4), the formulae (3) give:

$$\omega_1 = c = \text{const.}, \ \omega_2 = A\cos Q_1(t), \ \omega_3 = A\sin Q_1(t), \ Q_1(t) = \frac{I_1 - I}{I}\, ct \tag{5.1}$$

$$\omega_1 = \frac{\overline{L}_1}{I_1}\, t + c, \ \omega_2 = A\cos Q_2(t), \ \omega_3 = A\sin Q_2(t), \ Q_2(t) = \frac{I_1 - I}{I}\, ct + \frac{(I_1 - I)\,\overline{L}_1}{2 I_1 I}\, t^2 \tag{5.2}$$

$$\omega_1 = -\frac{1}{I_1}\cos t + c, \ \omega_2 = A\cos Q_3(t), \ \omega_3 = A\sin Q_3(t), \ Q_3(t) = \frac{I_1 - I}{I}\, ct - \frac{I_1 - I}{I_1 I}\sin t \tag{5.3}$$

Equations (5) give three members of the class of precessions (3). (5.1) give the well-known "regular precession", (5.2) the "helicoid precession", and (5.3) a "non-regular periodic precession". Eliminating t in each case of equations (5), one can get the corresponding precessional curves, shown in Fig. 1, 2, 3, respectively.

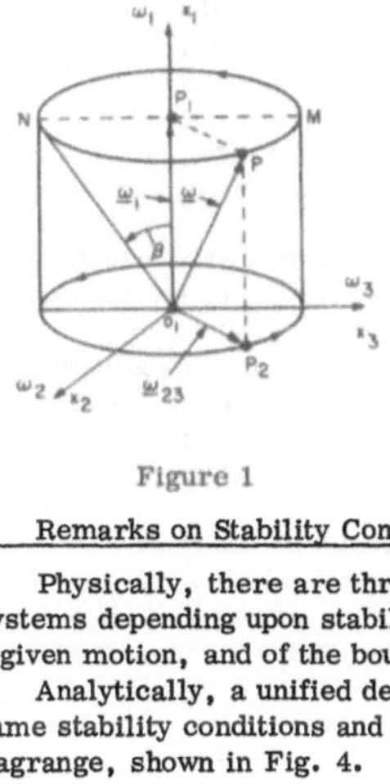


Figure 1

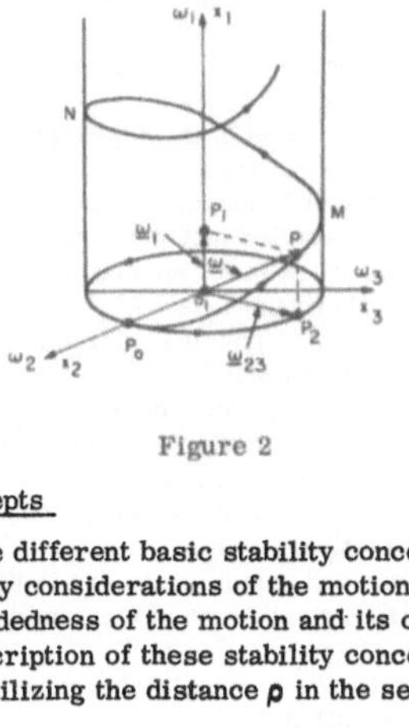


Figure 2

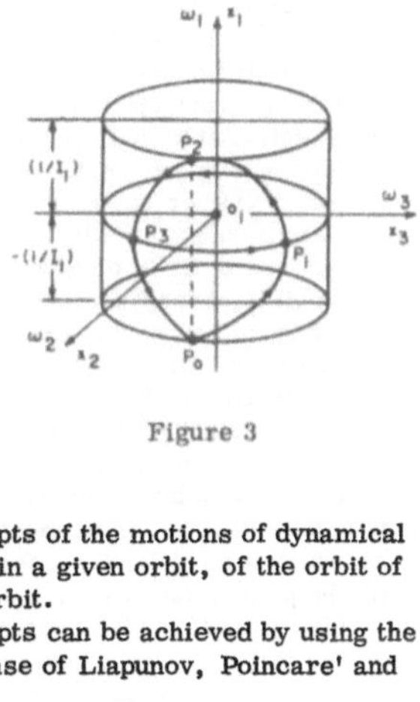


Figure 3

## 3. Remarks on Stability Concepts

Physically, there are three different basic stability concepts of the motions of dynamical systems depending upon stability considerations of the motion in a given orbit, of the orbit of a given motion, and of the boundedness of the motion and its orbit.

Analytically, a unified description of these stability concepts can be achieved by using the same stability conditions and utilizing the distance $\rho$ in the sense of Liapunov, Poincare' and Lagrange, shown in Fig. 4.

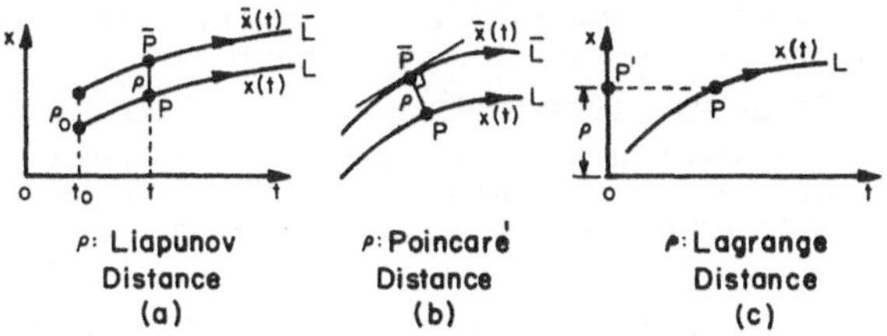


Figure 4

The first two cases of these distances interpret in different ways the correspondence between a point P of the unperturbed orbit L of which the stability is required and a point $\overline{P}$ of the perturbed orbit $\overline{L}$, and these distances measure the effect of a perturbation at P. In Liapunov sense, $\overline{P}$ of $\overline{L}$ corresponds to P of L at the "same-time", Figure 4(a), and in Poincare' sense, $\overline{P}$ of $\overline{L}$ corresponds to P of L in such a way that the distance $\rho = P\overline{P}$ is the minimum of the distances $(P, \overline{L})$ of P from the points of $\overline{L}$, Figure 4(b). In Lagrange sense, $\rho$ is the distance of P from origin in the x-space. The conditions for stability definitions are:

$$\rho_0 < \delta, \quad \rho < \epsilon, \quad \lim_{t \to \infty} \rho = 0, \quad \| p \| < \eta \tag{6}$$

By an appropriate interpretation of the distance $\rho$ and its initial value $\rho_0$ in these conditions, one can get the three stability concepts in the sense of Liapunov, Poincare' (orbital) and Lagrange (boundedness).

In case the perturbations affect only the initial conditions, that is in case of "sudden perturbations", the first two of the conditions (6) define the "stability" of the motion in L, and the first three of these conditions the "asymptotic stability" of the motion in L.

In case the perturbations p(x, t) affect the whole dynamical system, that is in case of "persistent perturbations", the definitions of stability require a restriction of these perturbations, say a restriction of their norm, $\|p\|$, so that for the stability definitions under persistent perturbations the fourth condition of (6), in addition to all the others, is needed.

The above remarks and the unification of the stability concepts in the conditions (6) lead to important results concerning stability, and of these results we mention just three needed in the present paper:

(a) For the equilibrium states of a system, the stability concepts in the sense of Liapunov and Poincare' are equivalent. For periodic states, the Liapunov stability concept is a narrow one compared to the Poincare' stability concept (orbital). The isochronism, that is the constancy of the frequency, characterizes and suggests the use of the Liapunov concept in periodic states.

(b) A state stable in Liapunov sense is stable in Poincare' sense, and a state unstable in Poincare' sense is unstable in Liapunov sense. But a state stable in Poincare' sense may be either stable or unstable in Liapunov sense.

(c) The Lagrange stability concept, applied either to equilibrium states or to periodic motions in finite distance, classifies these special motions as stable, so the Lagrange stability concept is useless in these motions.

According to the above, one and the same physical phenomenon may be, mathematically speaking, stable or unstable depending on the stability concept which is used for the discussion of the phenomenon. It then becomes necessary that for any stability problem one must select beforehand the stability concept on which the stability discussion will be based, and also know which stability concept is the appropriate one for the problem.

The previous stability concepts and definitions are of mathematical type and in order to be physically accepted they must meet practical needs and agree with "practical stability" and for these appropriate modifications, changes and supplements are to be found, as, e.g., to know:

(a) the size of deviations, the "$\epsilon$-region", that is the "acceptable states", for a satisfactory operation of the system;

(b) the size of initial conditions, the "$\delta$-region", that is the "permitted size of the initial conditions", which can be controlled;

(c) the size of the perturbations, the "p-region".

For practical stability of a solution of a system the nonlinearities of the system must be used, and the linearization of the system is not, in general, permitted.

## 4. The Stability of the Precessions (5) in the Sense of Liapunov, Poincare' and Lagrange

Let $\omega_{10}$, $\omega_{20}$, $\omega_{30}$ be initial values of the precessions (5), and the perturbations affecting only these values, when the stability investigation needs only the first three of the stability conditions (6). We discuss the stability of the above precessions separately.

### 4.1 The Regular Precession (Eq. 5.1)

(a) Liapunov stability. By using formulae (5.1), the amplitude and frequency of the regular precession are given by:

$$A = (\omega_{20}^2 + \omega_{30}^2)^{1/2}, \quad f = (I_1 - I)\,\omega_{10}/I$$

If the perturbations affect $\omega_{10}$, the constancy of the frequency, in going from the unperturbed precessional curve to the perturbed one, is violated, and the precession is: "Liapunov - unstable".

If the perturbations do not affect $\omega_{10}$, the frequency remains constant and the precession is "Liapunov - stable", but not "Liapunov - asymptotically stable".

(b) Poincare' (or orbital) stability.

Theorem 1. "The regular precession is "orbitally stable", but not "orbitally asymptotically stable".

Proof. The unperturbed and perturbed precessional curves of the above precession are circumferences on parallel planes perpendicular to the $\omega_1$ - axis, which contains their centers, then the Poincare' distance $\rho$ at any point P of the unperturbed precessional curve is a constant, $\rho = \rho_o$.

According to the first two conditions (6), the unperturbed precession is orbitally stable if, for a given small positive number $\epsilon$, it is possible to find a small positive number $\delta$ such that if the above distance initially is smaller than $\delta$, $\rho_o < \delta$, then the distance $\rho$ at any point P is smaller than $\epsilon$, $\rho < \epsilon$.

Since $\rho = \rho_o$, the above two inequalities are satisfied by selecting $\delta = \epsilon$, when the precession is "orbitally-stable".

The limiting condition $\rho \to 0$ as $t \to \infty$ is not satisfied, and then the precession is not "orbitally-asymptotically stable".

As a result of the above the precessional curve is a "cycle" but not a "limit cycle".

(c) Lagrange stability (boundedness). The precessional curve of the above precession has, for any time, all its points in finite distance from the origin, then the precession is "Lagrange-stable".

4.2 The Helicoid Precession (Eq. 5.2)

(a) Liapunov stability. The vector $\underline{\omega}_{23}$ on the $\omega_2$, $\omega_3$-plane with components $\omega_2$, $\omega_3$, given by (5.2), is periodic with frequency depending on time, when the motion of its endpoint on the circumference with radius A and center at $o_1$, is "Liapunov-unstable", and, as a consequence, the precession is "Liapunov-unstable".

(b) Poincare' stability. The Poincare' (or orbital) stability situation of the helicoil precession can be successfully discussed by using the following definitions and theorems.

A generator of the cylinder surface intersects any two particular helicoid precessional curves L and $\bar{L}$ into two infinite sets of points, s: $P_0, P_1, P_2, \ldots\ldots P_n, \ldots\ldots$ on the curve L, and $\bar{s}$: $\bar{P}_0, \bar{P}_1, \bar{P}_2, \ldots, \bar{P}_n \ldots$ on $\bar{L}$, Fig. 5(a). At any point $P_n$ of L we can define three different distances, shown in Fig. 5(b),

$$D_n = P_n P_{n+1}, \quad d_n = P_n \bar{P}_n, \quad \rho_n = P_n \bar{P}'_n$$

$P_n$ and $P_{n+1}$ are consecutive points of L, $\bar{P}_n$ of $\bar{L}$ is consecutive point of $P_n$ of L, and $\bar{P}'_n$ is the intersection point of $\bar{L}$ and the plane through $P_n$ of L perpendicular to $\bar{L}$ at $\bar{P}'_n$. $D_n$ is the pitch distance of L at $P_n$, $d_n$ the pitch distance of L and $\bar{L}$ at $P_n$, and $\rho_n$ the Poincare' distance of L at $P_n$, if $\bar{L}$ is the perturbed of L.

The distances $D_n$, $d_n$, $\rho_n$ defined as above, have a common limiting property, which is related to the orbited stability of the helicoid precession.

Theorem 2. "The distances $D_n$, $d_n$, $\rho_n$ at a point $P_n$ of any particular helicoid precessional curve L of (5.2) decrease to zero, as the point $P_n$ goes to infinity".

Proof. (i) Let us consider the particular helicoid precessional curve corresponding to initial conditions $t_o = 0$, $\omega_{10} = 0$, $\omega_{20} \neq 0$, $\omega_{30} = 0$, then starting from the point $P_o$ $(0, \omega_{20}, 0)$, Fig. 5(a). From equations (5.2) we get:

$$c = 0, \quad A = \omega_{20}, \quad Q_2(t) = \frac{(I_1 - I)\,\bar{L}_1}{2 I_1 I}\, t^2$$

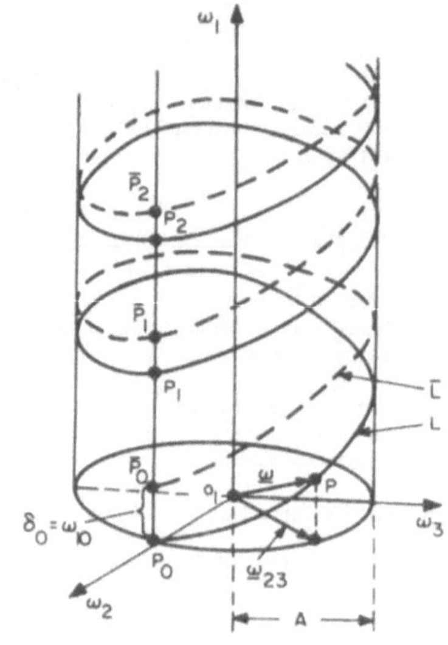


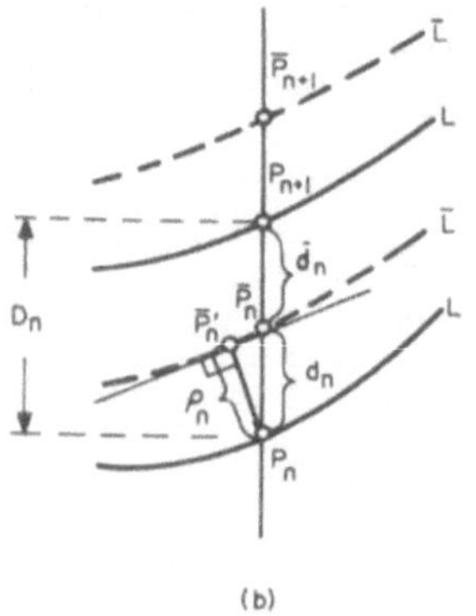


Figure 5

and the equations of L are:

$$\omega_1(t) = \frac{\bar{L}_1}{I_1} t, \ \omega_2(t) = \omega_{20} \cos \left\{\frac{(I_1 - I)\bar{L}_1 t^2}{2 I_1 I}\right\}, \ \omega_3(t) = \omega_{20} \sin \left\{\frac{(I_1 - I)\bar{L}_1 t^2}{2 I_1 I}\right\} \quad (7)$$

The generator through $P_o$ intersects the curve (7) into points of which the projection on the $\omega_2$, $\omega_3$ - plane is the point $P_o$, then for all intersection points we have $\omega_2(t) = \omega_{20}$, $\omega_{30} = 0$, when from equations (7) we have:

$$\frac{(I_1 - I)\bar{L}_1 t^2}{2 I_1 I} = 2\pi n; \ n = 0, \pm 1. \pm 2, \ldots$$

and the time t for the intersection points is:

$$t = \left(\frac{4\pi n I_1 I}{(I_1 - I)\bar{L}_1}\right)^{1/2} \quad (8)$$

We notice that for the reality of t we take positive values of the integer n if $I_1 > I$, and negative if $I_1 < I$.
Inserting (8) into the first of (7) we get the values of $\omega_1$ at the intersection points:

$$(\omega_1)_n = P_o P_n = a\sqrt{n} \quad (9)$$

where the constant a is given by:

$$a = \left( \frac{4 \pi I \bar{L}_1}{I_1 - I) I_1} \right)^{1/2} \tag{9.1}$$

and the integer n is $n \geq 0$.

From (9) we can get the distance $D_n$ at $P_n$:

$$D_n = P_n P_{n+1} = a(\sqrt{n+1} - \sqrt{n}) \tag{10}$$

The sequence $\{\sqrt{n+1} - \sqrt{n}\}$, $n = 0, 1, 2, \ldots$ is monotone decreasing with zero limit as $n \to \infty$, and, since a is a constant, the sequence $D_n$, $n = 0, 1, 2, \ldots$ has the same property, and then, as the point $P_n$ goes to infinity, we have:

$$\lim_{n \to \infty} D_n = 0 \tag{11}$$

We remark that this property of $D_n$ of the curve (7) is property of $D_n$ corresponding to any curve of (5.2) starting from any point.

(ii) From Figure 5(b) we see that: $D_n = d_n + \bar{d}_n$, when, according to (11):

$$\lim_{n \to \infty} d_n + \lim_{n \to \infty} \bar{d}_n = 0$$

and, since $d_n$ and $\bar{d}_n$ are positive constants,

$$\lim_{n \to \infty} d_n = 0 \tag{12}$$

(iii) The Poincare' distance $\rho_n$ at $P_n$, Fig. 5(b), is smaller than the distance $d_n$ at $P_n$, $\rho_n < d_n$, then

$$\lim_{n \to \infty} \rho_n = 0 \tag{13}$$

Theorem 3. "The helicoid precession (5.2) is "orbitally-asymptotically stable".

Proof. Taking any helicoid precessional curve L and its perturbed one $\bar{L}$, we can see that, given an $\epsilon > 0$, we can find a $\delta > 0$ such that $\rho_0 < \delta$ implies $\rho_n < \epsilon$ for any $\rho_n$. Indeed, such a $\delta$ exists, say $\delta = \epsilon$, when if we take initially $d_0 = \delta$, when $\rho_0 < \delta$, we will have $\rho_n < \epsilon$, and the curve L is "orbitally stable"; and since, in addition, $\rho_n \to 0$ as $n \to \infty$, the curve L is "orbitally asymptotically stable".

The $\delta$-region and $\epsilon$-region of an helicoid precessional curve.

If an helicoid precessional curve L starts from the point with coordinates ($\Omega_{10}$, $\Omega_{20}$, $\Omega_{30}$), we can determine the corresponding $\delta$-region and $\epsilon$-region which are needed in order that the above orbital stability of L is of practical importance.

Let $L_0, L_1, \ldots, L_n, \ldots$ be the arcs of L between the points $P_0$ and $P_1$, $P_1$ and $P_2$, ..., $P_n$ and $P_{n+1}$, ...

The coordinates $\omega_1$, $\omega_2$, $\omega_3$ of any point of L are related to the coordinates of the initial point of L according to:

$$\left.\begin{aligned} &\omega_2^2 + \omega_3^2 = \Omega_{20}^2 + \Omega_{30}^2 = \text{const.} \\ &(\omega_1)_n \leq \Omega_{10} \leq (\omega_1)_{n+1} \end{aligned}\right\}$$

By using (9), the last inequality reads:

$$a\sqrt{n} \leq \Omega_{10} \leq a\sqrt{n+1}$$

when

$$n \leq \left(\frac{\Omega_{10}}{a}\right)^2 \leq n+1$$

and the value of n is calculated, and therefore:

$$D_n = a(\sqrt{n+1} - \sqrt{n})$$

is known. This distance $D_n$ is the upper limit of $\delta$ and $\epsilon$.

(c) Lagrange stability. Since the endpoint $P_1$ of $\omega_1 = o_1 P_1$, Fig. 2, according to the first of (5.2), goes to infinity with the time t, the above precession is "Lagrange-unstable".

4.3 The Non-Regular Periodic Precession (Eq. 5.3)

(a) Liapunov stability. From formulae (5.3) we can see that this precession is "Liapunov - unstable".

(b) Poincare' stability.

Theorem 4. "The above precession is "orbitally - stable", but not "orbitally - asymptotically stable".

Proof. For given initial conditions and known sudden perturbations, the above precession and the corresponding perturbed one are known, also is known the Poincare' distance $\rho$, associated with any point P of the unperturbed precessional curve. Let $\rho_M$ and $\rho_m$ of these distances be the maximum and minimum, respectively. The smaller perturbations the smaller $\rho_M$ and $\rho_m$. Now, given a small positive number $\epsilon$ and using the Poincare' distance $\rho$, if one wants $\rho < \epsilon$ for all points P of the unperturbed precessional curve, one must use small perturbations such that $\rho_M < \epsilon$, when an appropriate $\delta$ must be $\delta < \rho_m$, a condition which can be satisfied, when the above precession is "orbitally - stable".

The conditions $\rho \to 0$ as $t \to \infty$ cannot be satisfied, when the precession is not "orbitally asymptotically stable".

(c) Lagrange stability. All the points of the precessional curves of the above precession are in finite distance from the origin, then the precession is "Lagrange - stable".

The above results are summarized in the following table.

TABLE OF RESULTS

| Kinds of Stability \ Kinds of Precession | Regular Precession (Eq. 5.1) | Helicoid Precession (Eq. 5.2) | Non-Regular Periodic Precession (Eq. 5.3) |
|---|---|---|---|
| Liapunov Stability | Stable - if the perturbations do not change $\omega_1$<br>Unstable - if the perturbations do change $\omega_1$ | Unstable | Unstable |
| Poincare' Stability (Orbital) | Stable | Asymptotically Stable | Stable |
| Lagrange Stability (Boundedness) | Stable | Unstable | Stable |

References

1. (a) Magiros, D. G.: The Study of the Orientation of an Earth Satellite, General Electric Company, TIS, 68SD249, RSD, April 19, 1968, Philadelphia, Pa.

(b) Magiros, D. G., and Reehl, G.: Sur quelques sortes de précession dans le mouvement de rotation d'un corps ridige ayant un axe de symétrie, Comptes Rendus, Academie des Sciences, Paris, 266, April 8, 1968.

2. (a) Magiros, D. G.: "On Stability Definitions of Dynamical Systems", National Academy of Sciences (U.S.A.), Proceedings, Vol. 53, No. 6, pp, 1288-1294, June 1965.

(b) Magiros, D. G.: Stability Concepts of Dynamical Systems, Information and Control (U.S.A.) Vol. 9, No. 5, October 1966.

*§2.1. Flight Trajectories*

# ON THE HELICOID PRECESSION: ITS STABILITY AND AN APPLICATION TO A RE-ENTRY PROBLEM

by

D. G. MAGIROS and G. REEHL

General Electric Company, Re-entry System Organization,
Philadelphia, Pennsylvania (U.S.A.)

## Introduction

This investigation deals with problems of helicoid precession, that is its stability and an application to a re-entry problem.

It is found that the helicoid precession is "unstable" in the sense of Liapunov and Lagrange, but "asymptotically stable" in the sense of Poincaré.

The re-entry problem, which is treated by using the concept of the helicoid precession, is the problem of determining the error in the orientation of a spin-stabilized axi-symmetric re-entry vehicle (RV).

In Sec. 1 we give the equations of the helicoid precession, in Sec. 2 the stability of the helicoid precession very briefly, and in Sec. 3 we state a re-entry problem of which the solution is discussed as an application of the concept of the helicoid precession.

## 1. The Helicoid Precession

If a rigid body has an axis of symmetry and rotates around it, and the resultant torque vector, acting on the body, is, in the body axes, a constant vector with magnitude $L$, and direction along the symmetry axis, the components $w_1$, $w_2$, $w_3$ of the angular velocity $w$, in the body axes, are given by

$$\begin{aligned} w_1 &= (L_1/I_1)t+c_1 \\ w_2 &= A\cos Q(t) \\ w_3 &= A\sin Q(t) \\ Q(t) &= \frac{I_1-I}{I}ct+\frac{(I_1-I)L_1}{2II_1}t^2 \end{aligned} \tag{1}$$



Reprinted from the *Proceedings of the XXth Intl. Astronatucial Congress*, Buenos Aires (1969), 491–496.

$c_1$ and $A$ are constants determined by the initial conditions; $I_1$ and $I$ moments of inertia about the symmetry axis and a perpendicular to it, respectively.[1]

Equations (1) give the "helicoid precession", and the curve in space coming from these equations by elimination of time is the "helicoid precessional curve", that is the locus of the endpoint $P$ of the vector $\boldsymbol{w} = \boldsymbol{o}_1 \boldsymbol{P}$, where $o_1$ is the mass center of the body (Fig. 1*a*). An individual helicoid precessional curve corresponds to given initial conditions $w_{10}$, $w_{20}$, $w_{30}$.

## 2. The Stability of the Helicoid Precession

By using sudden perturbations, which affect only the initial conditions, the three basic stability concepts in the sense of Liapunov, Poincaré and Lagrange are unified in the three stability conditions:

$$\varrho_0 < \delta, \quad \varrho < \varepsilon, \quad \lim_{t\to\infty} \varrho = 0 \tag{2}$$

where $\delta$ and $\varepsilon$ are positive constants, and $\varrho_0$ and $\varrho$ can be interpreted as distances in the sense of Liapunov, or Poincaré or Lagrange.[2]

The nature of the stability of the precession (1) can be found by using the conditions (2).[3]

We will have:

(a) *The helicoid precession* (1) *is unstable in Lagrange sense*, since the component $w_1$ is unbounded as $t \to \infty$.

(b) *This precession is unstable in Liapunov sense*, since the vector $w_{23}$ with components $w_2$ and $w_3$ in the $w_2$, $w_3$-plane Fig. 1*a* is periodic with time-dependent frequency, when the motion of the end-point of $w_{23}$ on the circumference with radius $A$ and center at $o_1$ is unstable in Liapunov sense.

(c) *The above precession is asymptotically stable in Pioncaré sense* (*orbitally*). The proof of this statement comes from the property that the distances $D_n$, $d_n$, $\varrho_n'$ defined in the following, decrease to zero as $n \to \infty$.

A generator of the surface of the cylinder, on which the helicoid precessional curves lie, intersects any two such curves $L$ and $\bar{L}$ into two sets of points: $s$: $P_0, P_1, \ldots, P_n, \ldots$ on $L$, and $\bar{s}$: $\bar{P}_0, \bar{P}_1, \ldots, \bar{P}_n, \ldots$ on $\bar{L}$, Fig. 1*a*.

The distances $D_n$, $d_n$, $\varrho_n$ at any point $P_n$ of $L$, Fig. 1*b*, are defined as:

$$D_n = P_n P_{n+1}, \quad d_n = P_n \bar{P}_n, \quad \varrho_n = P_n \bar{P}_n' \tag{3}$$

where $\bar{P}_n'$ is the intersection point of $\bar{L}$ by the plane through the point $P_n$ perpendicular to $\bar{L}$.

$D_n$ and $d_n$ are different kinds of pitch distances, and $\varrho_n$ the Poincaré distance at $P_n$ of $L$.

We can prove that:

$$D_n = \alpha(\sqrt{n+1}-\sqrt{n}), \quad d_n < D_n, \quad \varrho_n < d_n \tag{4}$$

where the constant $a$ is dependent on $L_1$, $I$, $I_1$.

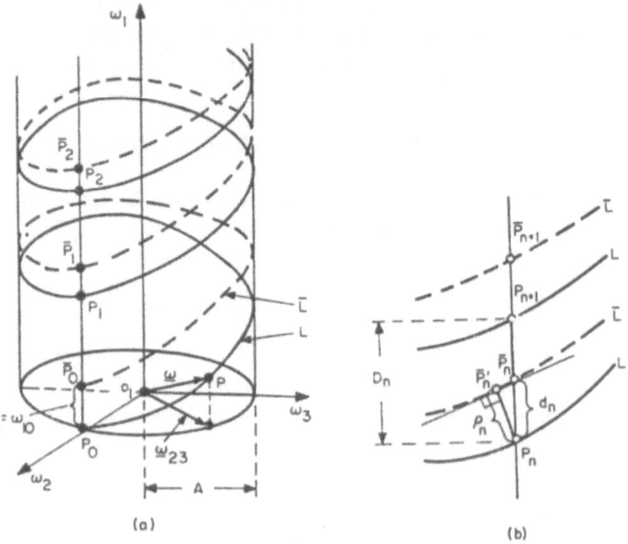


FIG. 1

Since the limit of $\{\sqrt{n+1}-\sqrt{n}\}$, as $n \to \infty$, is zero, we have:

$$\lim_{n\to\infty} D_n = 0, \quad \lim_{n\to\infty} \varrho_n = 0. \tag{5}$$

Combining (2) and (5) we can prove the asymptotic stability of the helicoid precession (1).

## 3. Application of the Concept of the Helicoid Precession to a Re-entry Problem

### 3.1. The problem

Shortly after it separates from its booster, an axi-symmetric RV is spun-up about its roll axis by means of small rockets. The reason for spinning it is to maintain the roll axis in the desired orientation during exospheric flight and to minimize dispersion. During firing of constant thrust rockets the motion is a helicoid precession, but after the spin rockets have fired the RV is spinning with regular precession. "*The problem is to determine the error in the orientation of the RV by finding the values of the coordinate angles at the time of the constant thrust termination* (that is at the time when the helicoid precession of the RV stops and the regular precession starts) *with respect to the initial attitude of the RV*".

The above problem has been studied by numerical methods on a linear basis, but the method presented here provides a quantitative assessment of the general behavior of non-linear type.

### 3.2. The solution

We consider the orthogonal coordinate system $OX_1X_2X_3$ (Fig. 2*a*), fixed in space, the system $o_1 X_1' X_2' X_3'$ parallel to $OX_1X_2X_3$, $o_1$ the mass center of the RV, and the orthogonal system $o_1 x_1 x_2 x_3$, fixed in the body, as a system of principal axes of inertia of the RV. The orientation of the RV moving in space can be specified by the "angular coordinates" $\theta$, $\psi$, $\varphi$ defined as follows (Fig. 2*b*):

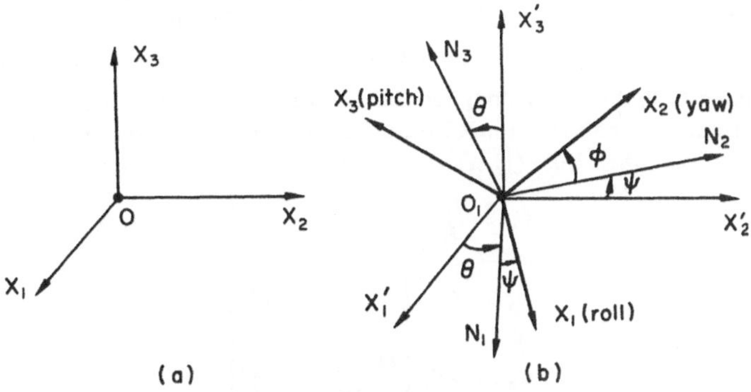


Fig. 2. Angle $\theta$: rotate $X_1'X_3'$-plane about $X_2'$-axis through angle $\theta$ when $X_1'$ comes to $N_1$ and $X_3'$ to $N_3$. Angle $\psi$: rotate $N_1X_2'$-plane about $N_3$-axis through angle $\psi$ when $N_1$ comes to $x_1$-axis and $X_2'$ to $N_2$. Angle $\varphi$: rotate $N_2N_3$-plane about $x_1$-axis through angle $\varphi$ when $N_2$ comes to $x_2$-axis and $N_3$ to $x_3$-axis

If we select the $x_1, x_2, x_3$-axes as the roll, yaw, pitch axes of the RV, and along these axes, respectively, we take the body coordinates $w_1, w_2, w_3$ of the angular velocity vector $w$ of the rotation of the RV, these angular velocity coordinates are given by the following formulae:

$$\left.\begin{aligned} w_1 &= \dot{\varphi}+\dot{\theta}\sin\psi \\ w_2 &= \dot{\psi}\sin\varphi+\dot{\theta}\cos\varphi\cos\psi \\ w_3 &= \dot{\psi}\cos\varphi-\dot{\theta}\sin\varphi\cos\psi \end{aligned}\right\}. \tag{6}$$

By the assumption that the RV has the $x_1$-axis as the symmetry axis, which is a principal axis, the two moments of inertia $I_2, I_3$ at $o_1$ are equal,

$I_2 = I_3 = I \neq I_1$, where $I_1$ is the axial moment of inertia, and $I$ the transverse moment of inertia. The angle $\psi$ will normally be small, and then formulae (6) can be approximated by:

$$\left.\begin{aligned} w_1 &= \dot{\varphi} \\ w_2 &= \dot{\psi}\sin\varphi + \dot{\theta}\cos\varphi \\ w_3 &= \dot{\psi}\cos\varphi - \dot{\theta}\sin\varphi \end{aligned}\right\}. \tag{7}$$

from which we get:

$$\left.\begin{aligned} \dot{\varphi} &= w_1 \\ \dot{\delta} &= w_2\cos\varphi - w_3\sin\varphi \\ \dot{\psi} &= w_2\sin\varphi + w_3\cos\varphi \end{aligned}\right\}. \tag{8}$$

If $t = 0$ is the time of application of the spin-up on the RV, and at that time the coordinate system $X_1'X_2'X_3'$ and $x_1x_2x_3$ are in coincidence, then the initial conditions, associated with equations (8), are:

$$\varphi_0 = \theta_0 = \psi_0 = 0 \tag{8.1}$$

If the spin-up torque is along the roll axis with the constant magnitude $L_1$, the rotational motion of the RV during the action of this torque, is a helicoid precession, when Eqs. (1) read:

$$\left.\begin{aligned} w_1 &= \frac{L_1}{I_1}t \\ w_2 &= A\cos\left\{\frac{(I_1 - I)L_1}{2II_1}t^2\right\} \\ w_3 &= A\sin\left\{\frac{(I_1 - I)L_1}{2II_1}t^2\right\} \end{aligned}\right\}. \tag{9}$$

Inserting the first of (9) into the first of (8) and integrating we have:

$$\varphi = \frac{L_1}{2I_1}t^2 \tag{10}$$

Inserting now the second and third of (9) into the second and third of (8) and using appropriate trigonometric identities and formula (10) we get:

$$\left.\begin{aligned} \dot{\theta} &= A\cos\left(\frac{1}{2}\frac{L_1}{I}t^2\right) \\ \dot{\psi} &= A\sin\left(\frac{1}{2}\frac{L_1}{I}t^2\right) \end{aligned}\right\}. \tag{11}$$

To integrate (11), since the angle is very small compared to unity, we expand the functions "cosine" and "sine" in Taylor Series and take a few terms, when we have the following approximate formulae:

$$\left.\begin{aligned}\dot\theta &= A\left\{1-\frac{1}{2}\left(\frac{L_1}{2I}\right)^2 t^4+\frac{1}{24}\left(\frac{L_1}{2I}\right)^4 t^8+O(t^{12})\right\}\\ \dot\psi &= A\left\{\frac{L_1}{2I}t^2-\frac{1}{6}\left(\frac{L_1}{2I}\right)^3 t^6+O(t^{10})\right.\end{aligned}\right\} \quad (12)$$

The terms omitted in the series expansion are of order $O(t^{12})$ and $O(t^{10})$ in the first and second of Eqs. (12), respectively. By integration of (12) we get:

$$\left.\begin{aligned}\theta &= A\left\{t-\frac{1}{10}\left(\frac{L_1}{2I}\right)^2 t^5+\frac{1}{216}\left(\frac{L_1}{2I}\right)^4 t^9\right\}+O(t^{13})\\ \psi &= A\left\{\frac{L_1}{6I}t^3-\frac{1}{42}\left(\frac{L_1}{2I}\right)^3 t^7\right\}+O(t^{11})\end{aligned}\right\} \quad (13)$$

If $t = t_s$ is the time of constant thrust termination, and $\varphi_s, \theta_s, \psi_s$ the values of the angular coordinates $\varphi$, $\theta$, $\psi$ at that time, we can get:

$$\left.\begin{aligned}\varphi_s &= \frac{1}{2}w_s t_s\\ \theta_s &= At_s\left\{1-\frac{1}{10}\left(\frac{I_1}{2I}w_s t_s\right)^2+\frac{1}{216}\left(\frac{I_1}{2I}w_s t_s\right)^4\right\}\\ \psi_s &= \frac{1}{3}At_s\left\{\frac{I_1}{2I}w_s t_s\right\}\left\{1-\frac{1}{14}\left(\frac{I_1}{2I}w_s t_s\right)^2\right\}\end{aligned}\right\} \quad (14)$$

where:

$$w_s = (w)_{t=t_s} = \frac{L_1}{I_1}t_s. \quad (14.1)$$

After the thrust termination the motion is a regular precession with the roll axis rotating around the angular momentum vector $\boldsymbol{H}_s$ at $t = t_s$ and precession angle $\beta_s$ given by:

$$\tan\beta_s = \frac{A}{w_s} = \frac{AI_1}{L_1 t_s} \quad (15)$$

## References


1 MAGIROS, D. G. and REEHL, G., Sur quelques sortes de précession dans le mouvement de rotation d'un corps ayant un axe de symétrie, Comptes Rendus, Academy of Sciences in Paris, France, t. 266, pp. 770–773 (8 April 1968)

2 MAGIROS, D. G., Stability concepts of dynamical systems, *J. Information and Control, U.S.A.* 9, No. 5 (Oct. 1966)

3 MAGIROS, D. G., La stabilité de la précession hélicoidale dans le sens de Liapunov, Poincaré et Lagrange, Comptes Rendus, Academy of Sciences in Paris, France, t. 268, pp. 652–654 (24 March 1969)

22nd INTERNATIONAL ASTRONAUTICAL

CONGRESS

Brussels, Belgium

September, 1971

---

# ORIENTATION OF THE ANGULAR MOMENTUM VECTOR OF A SPACE VEHICLE AT THE END OF SPIN-UP

ABSTRACT

In this paper, the angular momentum vector of a space vehicle at the end of spin-up is determined.

Use is made of the concept of the helicoid precession, which characterizes the rotational motion of the vehicle during the action of a constant thrust acting on the vehicle.

Reprinted from the *XXIInd Intl. Astronautical Congress*, Brussels (1971), 1–8.

# ORIENTATION OF THE ANGULAR MOMENTUM VECTOR OF A SPACE VEHICLE AT THE END OF SPIN-UP(*)

BY: Demetrios G. Magiros(**)

## 1. INTRODUCTION

This investigation deals with the determination of the angular momentum vector of a space vehicle (S.V.) at the end of spin-up. The concept of the "helicoid precession", introduced and used in the papers of reference 1 and 2, is applied in this investigation.

The procedure in the paper gives a method, which seems to be a unique one, to treat such a subject.

Shortly after it separates from its booster, an axisymmetric re-entry vehicle is spun-up about its roll axis by means of small rockets of constant thrust. During the action of the constant thrust, the "helicoid precession" characterizes the motion of the SV; but, just before and after the action of the constant thrust, the motion of the SV is characterized by different "regular precessions". As a result of this, the magnitude and orientation of the angular momentum vector H of the SV at the time of the termination of the constant thrust have different values than that before the action of the constant thrust.

---

(*)Communicated to the "22nd International Astronautical Congress, Brussels, Belgium, September, 1971.

(**) Consulting mathematician, General Electric Co., Re - Entry and Environmental Division, Philadelphia,Pa.,U.S.A.

The determination of the difference of these values, which gives the "error in the angular momentum vector" is the subject of this paper.

Mr. G. Reehl, an author's collaborator, helped in the calculations.

## 2. THE ANGULAR COORDINATES AND THEIR DERIVATIVES AS TIME FUNCTIONS

Following the reference paper (2), we give here the formulae for the angular coordinates and their derivatives as time functions, which will help the calculation of the angular momentum vector at the time of the constant thrust termination.

We consider the orthogonal coordinate system $OX_1X_2X_3$, Figure 1 (a), fixed in space, the system $oX_1'X_2'X_3'$ parallel to $OX_1X_2X_3$ ; o the mass center of the SV and the orthogonal system $ox_1x_2x_3$, fixed in the body, as a system with principal axes of inertia of the SV.

The orientation of the SV moving in space can be specified by the "angular coordinates" $\theta, \psi, \varphi$ , defined as follows, Figure 1(b).

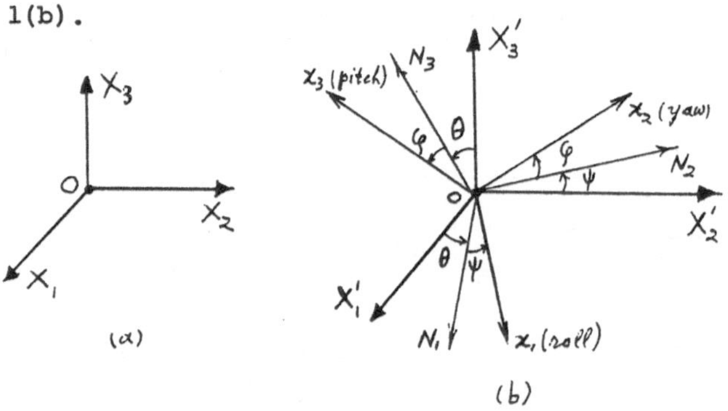


FIGURE 1

Angle $\theta$ : Rotate $X_1' X_3'$ - plane about $X_2'$- axis through angle $\theta$ , when $X_1'$ comes to $N_1$ and $X_3'$ to $N_3$ ;

Angle $\psi$ : Rotate $N_1 X_2'$ - plane about $N_3$ - axis through angle $\psi$ , when $N_1$ comes to $x_1$ - axis and $X_2'$ to $N_2$ ;

Angle $\varphi$ : Rotate $N_2 N_3$ - plane about $x_1$- axis through angle $\varphi$ , when $N_2$ comes to $x_2$- axis and $N_3$ to $x_3$- axis.

If we select the $x_1, x_2, x_3$-axes as the roll, yaw, pitch-axes of the SV, and if $t=0$ is the time of the application of the spin-up on the SV, and if at that time the coordinate systems $X_1' X_2' X_3'$ and $x_1 x_2 x_3$ are in coincidence, when $\theta=\psi=\varphi=0$ , these angles at time $t$ , according to the reference paper 2, are approximately given by:

$$\left.\begin{aligned} \varphi &= \frac{L_1}{2I_1} t^2 \\ \theta &= A\left[t - \frac{1}{10}\left(\frac{L_1}{2I}\right)^2 t^5 + \frac{1}{216}\left(\frac{L_1}{2I}\right)^4 t^9\right] + O(t^{13}) \\ \psi &= A\left[\frac{L_1}{6I} t^3 - \frac{1}{42}\left(\frac{L_1}{2I}\right)^3 t^7\right] + O(t^{11}) \end{aligned}\right\} \quad (1)$$

where $L_1$ is the applied roll torque, $I_1$ the roll moment of inertia, $I$ the moment of inertia about the transverse axis and $A = \left(\omega_{20}^2 + \omega_{30}^2\right)^{1/2}$ ; $\omega_{20}, \omega_{30}$ initial angular velocity components along the $x_2, x_3$ - axes.

The symbol $O(t^m)$ denotes a function $k(t^m)$ such that $\frac{\|k(t^m)\|}{t^m} \leq c$ = constant as $t \to 0$ . From formulae (1) we have:

$$\left.\begin{aligned} \dot\varphi &= \frac{L_1}{I_1} t \\ \dot\theta &= A\left[1 - \frac{1}{2}\left(\frac{L_1}{2I}\right)^2 t^4 + \frac{1}{24}\left(\frac{L_1}{2I}\right)^4 t^8\right] + O(t^{12}) \\ \dot\psi &= A\left[\frac{L_1}{2I} t^2 - \frac{1}{6}\left(\frac{L_1}{2I}\right)^3 t^6\right] + O(t^{10}) \end{aligned}\right\} \quad (2)$$

If $t=t_s$ is the time of the constant thrust termination and $\theta_s, \varphi_s, \psi_s, \dot{\theta}_s, \dot{\varphi}_s, \dot{\psi}_s$ are the values of the coordinate angles and their time rates at $t = t_s$, we have:

$$\left.\begin{aligned}
\varphi_s &= \tfrac{1}{2}\omega_s t_s \\
\theta_s &= A t_s\left[1-\tfrac{1}{10}\left(\tfrac{I_1}{2I}\omega_s t_s\right)^2+\tfrac{1}{216}\left(\tfrac{I_1}{2I}\omega_s t_s\right)^4\right. \\
\psi_s &= \tfrac{1}{3}A t_s\left[\tfrac{I_1}{2I}\omega_s t_s\right]\left[1-\tfrac{1}{14}\left(\tfrac{I_1}{2I}\omega_s t_s\right)^2\right] \\
\dot{\varphi}_s &= \tfrac{L_1}{I_1}t_s = \omega_s \\
\dot{\theta}_s &= A\left[1-\tfrac{1}{2}\left(\tfrac{I_1}{2I}\omega_s t_s\right)^2+\tfrac{1}{24}\left(\tfrac{I_1}{2I}\omega_s t_s\right)^4\right] \\
\dot{\psi}_s &= A\left[\tfrac{I_1}{2I}\omega_s t_s\right]\left[1-\tfrac{1}{6}\left(\tfrac{I_1}{2I}\omega_s t_s\right)^2\right]
\end{aligned}\right\} \quad (3)$$

The first three formulae of (3) give the orientation of the SV at the $t_s$ when the constant thrust stops acting on the SV.

3. FORMULAE FOR THE ANGULAR MOMENTUM

The angular momentum vector $\underline{H}_s$ at $t=t_s$ in vehicle coordinates is given by:

$$\underline{H}_s = \underline{x}_1 I_1 \omega_s + \underline{x}_2 I \omega_{2s} + \underline{x}_3 I \omega_{3s} \quad (4)$$

where $\omega_s, \omega_{2s}, \omega_{3s}$ are the angular velocity components about the $x_1, x_2, x_3$- axes at $t=t_s$. From reference (2) we have:

$$\left.\begin{aligned}
\omega_{2s} &= \dot{\psi}_s \sin\varphi_s + \dot{\theta}_s \cos\varphi_s \\
\omega_{3s} &= \dot{\psi}_s \cos\varphi_s - \dot{\theta}_s \sin\varphi_s
\end{aligned}\right\} \quad (5)$$

The body $x_1, x_2, x_3$- axes are related to the inertial $X_1, X_2, X_3$ - axes by the matrix equation:

$$\begin{bmatrix} X_1 \\ X_2 \\ X_3 \end{bmatrix} = \begin{bmatrix} \cos\theta\cos\psi & \left\{ \begin{array}{l} \sin\theta\sin\varphi - \\ -\cos\theta\cos\varphi\sin\psi \end{array} \right\} & \left\{ \begin{array}{l} \sin\theta\cos\varphi + \\ +\cos\theta\sin\varphi\sin\psi \end{array} \right\} \\ \sin\psi & \cos\varphi\cos\psi & -\sin\varphi\cos\psi \\ -\sin\theta\cos\psi & \left\{ \begin{array}{l} \cos\theta\sin\varphi + \\ +\sin\theta\cos\varphi\sin\psi \end{array} \right\} & \left\{ \begin{array}{l} \cos\theta\cos\varphi - \\ -\sin\theta\sin\varphi\sin\psi \end{array} \right\} \end{bmatrix} \cdot \begin{bmatrix} x_1 \\ x_2 \\ x_3 \end{bmatrix} \tag{6}$$

Since $\theta_s$ and $\psi_s$ are normally small, equation (6) at $t=t_s$ can be approximated by:

$$\begin{bmatrix} X_1 \\ X_2 \\ X_3 \end{bmatrix} = \begin{bmatrix} 1 & (\theta_s\sin\varphi_s - \psi_s\cos\varphi_s) & (\theta_s\cos\varphi_s + \psi_s\sin\varphi_s) \\ \psi_s & \cos\varphi_s & -\sin\varphi_s \\ -\theta_s & \sin\varphi_s & \cos\varphi_s \end{bmatrix} \cdot \begin{bmatrix} x_1 \\ x_2 \\ x_3 \end{bmatrix} \tag{7}$$

Using (4) and (7), the component of $\underline{H}_s$ along $X_1$ is given by:

$$H_{1s} = I_1\omega_s + I\,[\omega_{2s}(\theta_s\sin\varphi_s - \psi_s\cos\varphi_s) + \omega_{3s}(\theta_s\cos\varphi_s + \psi_s\sin\varphi_s)] \tag{8}$$

and substituting from equation (5):

$$H_{1s} = I_1\omega_s + I(\theta_s\dot{\psi}_s - \psi_s\dot{\theta}_s) \tag{9}$$

For values of the parameters normally encountered in practice, the second term is negligible compared to the first. Thus, approximately, we have:

$$H_{1s} = I_1 \omega_s \tag{10}$$

Also, from equations (4) and (7), the components of angular momentum along the inertial axes $X_2$ and $X_3$ are given by:

$$\left.\begin{aligned} H_{2s} &= I_1 \omega_s \psi_s + I\left(\omega_{2s}\cos\varphi_s - \omega_{3s}\sin\varphi_s\right) \\ H_{3s} &= -I\omega_s\theta_s + I\left(\omega_{2s}\sin\varphi_s + \omega_{3s}\cos\varphi_s\right) \end{aligned}\right\} \tag{11}$$

and substituting from equation (5):

$$\begin{aligned} H_{2s} &= I\dot{\theta}_s + I_1\omega_s\psi_s \\ H_{3s} &= I\dot{\psi}_s - I_1\omega_s\theta_s \end{aligned} \tag{12}$$

Since the final orientation of the angular vector will not deviate very much from $X_1'$ in practical cases, it can be specified by two small angles $\Delta\theta$ about $X_2'$ and $\Delta\psi$ about $X_3'$, defined as follows:

$$\Delta\theta = -\frac{H_{3s}}{H_{1s}} = \theta_s - \frac{I\dot{\psi}_s}{I_1\omega_s} \tag{13}$$

$$\Delta\psi = \frac{H_{2s}}{H_{1s}} = \psi_s + \frac{I\dot{\theta}_s}{I_1\omega_s} \tag{14}$$

when substituting from equations (2) and (3), become:

$$\Delta\theta = \frac{At_s}{2}\left[1 - \frac{1}{30}\left(\frac{I_1}{2I}\omega_s t_s\right)^2 + \frac{1}{108}\left(\frac{I_1}{2I}\omega_s t_s\right)^4 \tag{15}$$

$$\Delta\psi = \frac{IA}{I_1\omega_s} + \frac{At_s}{12}\left(\frac{I_1}{2I}\omega_s t_s\right)\left[1 - \frac{1}{108}\left(\frac{I_1}{2I}\omega_s t_s\right)^2\right] \tag{16}$$

The first term of equation (16) is the precession angle of the regular precession, that is, the angle between the vehicle roll axis and the regular momentum vector.

We remark that the procedure and the results of the present paper may be useful to some research area of which here we mention the following two:

a. For manned planetary missions, any spacecraft should be designated to create artificial gravity, which could be accomplished by rotating the vehicle continuously during the flight. The rotation produces precession and special angular momentum vector.

b. If an atom is placed into a magnetic field in a direction, the field exerts a torque on the magmetic dipole moments, trying to align it with the direction. The atom has, in general, an angular momentum, that is an orbital and an intrinsic (spin) one, and the atom in the above situation becomes like any gyroscopic.

REFERENCES


1. D. G. Magiros and G. Reehl: "Sur quelques sortes de précession dans le movement de rotation d'un corps ayant an axe de symétrie," Comptes Rendus, Academie de Sciences, Paris, France, V. 266, 770-776, April 8, 1968.
2. D. G. Magiros and G. Reehl: "On the helicoid precession: Its Stability and an application to a re-entry problem," Proceedings, XX International Astronautical Congress, Buenos Aires, Argentina, October, 1969.

ΠΡΑΚΤΙΚΑ ΤΗΣ ΑΚΑΔΗΜΙΑΣ ΑΘΗΝΩΝ

ΣΥΝΕΔΡΙΑ ΤΗΣ 25ΗΣ ΜΑΪΟΥ 1972

ΠΡΟΕΔΡΙΑ ΓΡΗΓ. ΚΑΣΙΜΑΤΗ

ΑΝΑΚΟΙΝΩΣΕΙΣ ΜΗ ΜΕΛΩΝ

ΕΦΗΡΜΟΣΜΕΝΑ ΜΑΘΗΜΑΤΙΚΑ.— **The stability of a class of helicoid precessions in the sense of Liapunov and Poincaré,** *by Demetrios G. Magiros** Ἀνεκοινώθη ὑπὸ τοῦ Ἀκαδημαϊκοῦ κ. Ἰω. Ξανθάκη.

## Introduction

In a previous paper, Ref. 1, we discussed the stability of a helicoid precession in case of constant torque, whereby, by employing different stability concepts, we found for this precession different stability situations. The concept of this helicoid precession was successfully applied in problems of current interest in Astrodynamics, treated in papers Ref. 2, 3.

In the present paper, we discuss the stability of a «class of helicoid precessions», of which the helicoid precession of the paper Ref. 1 is only a member.

The concepts of stability in the sense of Liapunov and Poincaré, Ref. 4, are employed.

We found that all the members of the class of precessions are

* ΔΗΜΗΤΡΙΟΥ Γ. ΜΑΓΕΙΡΟΥ, Ἐπιστημονικοῦ Συμβούλου τῆς General Electric Co, RESD Philadelphia, PA, U.S.A. : **Ἡ εὐστάθεια μιᾶς κλάσεως ἑλικοειδῶν μεταπτώσεων κατὰ τὴν ἔννοιαν τῶν Liapunov καὶ Poincaré.**

Reprinted from the *Proceedings of the Athens Academy of Sciences* **47** (1972), 102–110.

unstable in Liapunov sense; but in Poincaré sense the stability of a member S of the class is either stable, or asymptotically stable, or unstable, if the limit value of the pitch distance of S is either a constant, or zero, or infinite, respectively.

There are reasons which suggest that the stability situation of the above class of the helicoid precessions in the sense of Poincaré is close to practical stability, then it is preferred.

## 1. The class of the helicoid precessions

The rotational motion of a rigid body around its symmetry axis, is governed by the Euler's equations:

$$\dot{\omega}_1 = \frac{L_1}{I_1}, \quad \dot{\omega}_2 = \frac{L_2}{I} - \frac{I_1 - I}{I} \omega_1 \omega_3, \qquad \dot{\omega}_3 = \frac{L_3}{I} + \frac{I_1 - I}{I} \omega_1 \omega_2 \quad (1)$$

where $\underline{\omega} = (\omega_1, \omega_2, \omega_3)$ is the angular velocity; $\underline{L} = (L_1, L_2, L_3)$ the external resultant torque acting on the body, $I_1$, $I_2 = I_3 = I$ the moments of inertia about the coordinate axis $0_1 \omega_1 \omega_2 \omega_3$, $0_1$ the center of the mass of the boby.

In the case where the torque is:

$$L_1 = L_1(t), \qquad L_2 = 0, \qquad L_3 = 0 \quad (2)$$

the solution of (1) is:

$$\left.\begin{array}{l} \omega_1(t) = \frac{1}{I_1} \int L_1(t) + c, \qquad \omega_2(t) = A \cos Q(t), \qquad \omega_3(t) = A \sin Q(t) \\ Q(t) = \frac{I_1 - I}{I} \int \omega_1(t)\, dt \end{array}\right\} \quad (3)$$

c and A are constants to be determined from the initial angular velocity:

$$\underline{\omega_0} = (\omega_{10}, \omega_{20}, \omega_{30}), \quad \text{and} \quad A = (\omega_{20}^2 + \omega_{30}^2)^{1/2}.$$

In case $\omega_1(t)$ is increasing function of and tends to infinity with time t, the solution (3) is a helicoid curve S on the surface of an orthogonal circular cylinder of radius A and gives a helicoid precession corresponding to the specified function $L_1(t)$, and so (3) gives a «class of helicoid precessions», each member of which is determined by the specification of $L_1(t)$. We remark that if $\omega_1(t)$ does not satisfy the above requirement, the corresponding precession is not helicoid, as, e.g., for $L_1 = 0$, when we have the «regular precession» and its «precessional curve» is circumference on the surface of the cylinder; or for $L_1 = \sin t$, when the precessional curve is closed curve on the surface of the cylinder. But, for $L_1 =$ constant, $\omega_1(t) \rightarrow \infty$ as $t \rightarrow \infty$, and we have a helicoid precession, the simplest one of the class (3).

We discuss here the stability of the class of helicoid precessions (3) in the sense of Liapunov and Poincaré.

## 2. Stability in Liapunov sense

The vector $\underline{\omega}_{23} = (\omega_2, \omega_3)$ on the $\omega_2, \omega_3$ — plane, Fig. 1(a), of which the components are given by (3), is periodic in t, but with period dependent on t, then the motion of the end point of this vector on the circumference with radius A and center $0_1$, is unstable in Liapunov sense, Ref. 4, and, as a consequence, *«the class of helicoid precessions is «unstable» in Liapunov sense»*.

## 3. Stability in Poincaré sense (orbital stability)

The orbital stability of any member of the above class of precessions depends upon the structure of the corresponding function $L_1(t)$. Some auxiliary distances and their properties, shown below, will help to create a criterion for the orbital stability.

### 3.1 Some auxiliary distances and their properties.

Let us take two helicoid precessional curves S and $\overline{S}$ belonging to the same family, that is corresponding to the same function $L_1(t)$, but starting from different points $P_0(0, \omega_{.0}, 0)$ and $\overline{P}_0(\omega_{10}, \omega_{20}, 0)$, respectively, Fig. 1(a).

The generator of the cylinder through $P_0$ intersects S into the points: $P_0, P_1, P_2, \ldots, P_n, \ldots$, and $\overline{S}$ into the points: $\overline{P}_0, \overline{P}_1, \overline{P}_2, \ldots, \overline{P}_n, \ldots, S_0$, at the point $P_n$ of S, we can define, as shown in Fig. 1(b), the pitch distances: $D_n = P_n P_{n+1}$, $d_n = P_n \bar{P}_n$. A third distance at $P_n$ is defined by the plane through $P_n$ perpendicular to $\bar{S}$ at $\bar{P}'_n$, the distance $\varrho_n = P_n \bar{P}'_n$.

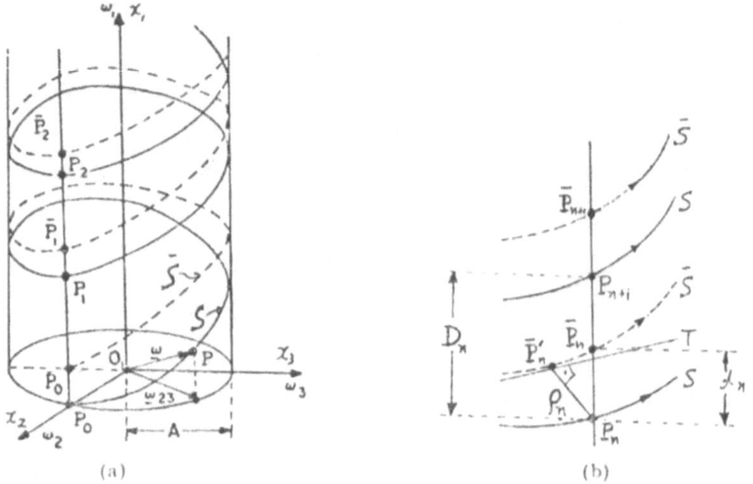


Fig. 1

The above distances have properties very useful for the discussion of the orbital stability of the helicoid precessions. These properties are given by the following theorem.

**Theorem 1.** *The distances* $D_n$, $d_n$, $\varrho_n$ *have the properties*:

. (a): $D_n > d_n > \varrho_n$

. (b): *The limit distance:* $\lim_{n \to \infty} D_n = \bar{D}$ *is either a constant,* or zero, *or infinite, when the* $\lim_{n \to \infty} d_n = \bar{d}$, $\lim_{n \to \infty} \varrho_n = \bar{\varrho}$ *are either constant, or zero, or infinite, respectively.*

**Proof** . (a): The point $P_n$ is a point of the segment $P_n P_{n+1}$, Fig. 1 (b), which means that: $D_n > d_n$.

The plane through $P_n$ and perpendicular to $\bar{S}$ at $\bar{P}'_n$ intersects perpendicularly the tangent $\bar{P}'_n T$ of $\bar{S}$ at $\bar{P}'_n$, so this tangent is perpendicular to the distance $\varrho_n$. The plane through $\bar{P}'_n$ and perpendicular to $\varrho_n$ contains the tangent $\bar{P}'_n T$ and divides the whole space into two parts, one of which contains all the points of $\overline{P'_n}\,\overline{S}$, and the other contains the point $P_n$, so the distance $P_n \overline{P}_n$ is bigger than the distance $P_n \overline{P'_n}$, that is $d_n > \varrho_n$.

(b): For the second part of the theorem the calculation of the limit $\overline{D}$ is needed.

The curve S starts from $P_0\,(0, \omega_{20}, 0)$ at $t_0 = 0$, and corresponds to $c = 0$ in the formulae (3). Its equations are:

$$\left.\begin{array}{c} \omega_1(t) = \dfrac{1}{I_1}\displaystyle\int L_1(t)\,dt, \quad \omega_2(t) = \omega_{20}\cos Q(t), \quad \omega_3(t) = \omega_{20}\sin Q(t) \\ Q(t) = \dfrac{I_1 - I}{I}\displaystyle\int dt \int L_1(t)\,dt \end{array}\right\} \quad (4)$$

For the points $P_0, P_1, P_2, \ldots, P_n$, we have $\omega_2(t) = \omega_{20}$, $\omega_3(t) = 0$, then the quantity $Q(t)$ of (4) for these points must be:

$$Q(t_n) = 2\pi n, \qquad n = 0, \pm 1, \pm 2, \ldots \quad (5)$$

and, if we take into account some restrictions of Q and the nature of n, we can solve (5) for $t_n$, when:

$$t_n = \overline{Q}(n) \ldots \quad (6)$$

Inserting (6) into the first of (4), we can get the value of $\omega_1(t)$ corresponding to the point $P_n$: $\omega_1(t_n) = P_0 P_n = (\omega_1)_n$, when the distance $D_n$ at $P_n$ is:

$$D_n = P_0 P_{n+1} - P_0 P_n = P_n P_{n+1} = (\omega_1)_{n+1} - (\omega_1)_n \quad (7)$$

As $n \to \infty$, the points $P_n$ and $P_{n+1}$ go to infinity, the distances $P_0 P_n$ and $P_0 P_{n+1}$ tend to infinity, and the $\lim\limits_{n\to\infty} D_n = \overline{D}$ tends to get the undetermined form $(\infty - \infty)$, which may be either a constant, or zero or infinite, and, then the limits $\bar{d}$ and $\bar{\varrho}$ may be either constants, or zero, or infinite, respectively.

### 3.2 Orbital stability criterion

Based on the above properties of the distances $D_n$, $d_n$, $\varrho_n$, we can formulate a criterion for the orbital stability of the helicoid precessions (3), expressed by the:

**Theorem 2.** «*Any member of the class of the helicoid precessions (3), corresponding to a given function L(t), is orbitally either stable, or asymptoti-*

*cally stable, or unstable, if, respectively, the limit distance $\bar{D}$ is either a constant, or zero, or infinite».*

**Proof:** We first see that, for sudden perturbation when the initial conditions are only perturbed, the curve $\bar{S}$ can be considered as the perturbed of S and the distance $\varrho_n$, defined above, is the «Poincaré distance» of S at $P_n$, Ref. 4.

If the number $\bar{D}$ is a constant, given $\varepsilon > 0$, we can really find a $\delta > 0$ such that, if the Poincaré distance initially is $\varrho_0 < \delta$, then inequality $\varrho_n < \varepsilon$, for all n, can be implied, since we can select $\delta = \varepsilon$, and $d_0 = \delta$, when $\varrho_0 < \delta$ implies $\varrho_n < \varepsilon$, and S is «orbitally stable».

If $\bar{D} = 0$, then $\bar{\varrho} = 0$, and S is «orbitally asymptotically stable».

If $\bar{D} = \infty$, S is «orbitally unstable».

**3.3 Example.** As an example, we mention the case $L = \bar{L}_1 =$ constant, treated in Ref. 1.

The corresponding helicoid precession is in this case given by:

$$\left.\begin{array}{c} \omega_1(t) = \dfrac{\bar{L}_1}{I_1} t, \quad \omega_2(t) = \omega_{20} \cos Q_1(t), \quad \omega_3(t) = \omega_{20} \sin Q_1(t) \\ Q_1(t) = (I_1 - I)\bar{L}_1 t^2 / 2 I I_1 \end{array}\right\} \tag{8}$$

This precession, due to the form of $Q_1(t)$, is «Liapunov unstable»; but it is «orbitally asymptotically stable», since the distance $D_n$ is given by: $D_n = \alpha(\sqrt{n+1} - \sqrt{n})$, $\alpha =$ constant, and of which the limit, as $t \to \infty$, is $\bar{D} = 0$.

For this example, we can determine the region of the permitted deviations of the precessional curve, the «ε-region», and the corresponding region of the initial points, the «δ-region», for which regions the helicoid precession is orbitally asymptotically stable, when this stability situation of (8) has a practical importance.

Given a point $(\bar{\omega}_{10}, \bar{\omega}_{20}, \bar{\omega}_{30})$ on the surface of the cylinder as a starting point of a helicoid precessional curve (8), the coordinates $\omega_1, \omega_2, \omega_3$ of any point of this curve are related to $\omega_{10}, \omega_{20}, \omega_{30}$ by:

$$\omega_2^2 + \omega_3^2 = \omega_{20}^2 + \omega_{30}^2 = \text{constant} \tag{8.1}$$

$$(\omega_1)_n \leqslant \bar{\omega}_{10} \leqslant (\omega_1)_{n+1} \tag{8.2}$$

The inequality (8. 2), by using $(\omega_1)_n = \alpha\sqrt{n}$, leads to:

$$n \leqslant \bar{\omega}_{10}/\alpha \leqslant n+1 \tag{8.3}$$

from which the integer n can be determined, when, as a result, $D_n = \alpha(\sqrt{n+1} - \sqrt{n})$ is known. This distance $D_n$ is the upper limit of $\delta$ and $\varepsilon$.

We remark that we can calculate the «ε, δ-regions» of any member S of the helicoid precessions (3), if the distance $\bar{D}$ of S is zero or finite, when the orbital stability situation of S, and not its Liapunov stability situation, has a practical meaning.

**3. 4 Remarks.** We saw above that, for the same phenomenon, we have different stability situations, if we apply different stability concepts.

There arises the problem of the selection of the stability concept appropriate to the phenomenon, that is of the selection of the stability situation, which interprets the reality in an adequate way, and it is more close to «practical stability» of the phenomenon.

The possibility of the determination, by using a physical situation, of the region of the permitted deviations of motion and orbit, of the corresponding region of the initial points, and of the appropriate region of the perturbation, in case of persistent perturbations, Ref. 5, makes the stability results practically important and physically accepted.

Stability investigations, which may satisfy mathematical curiosities or needs, will become useful if they are oriented towards «practical usefulness».

REFERENCES

1. Magiros, D. G.: Comptes Rendus, Academy of Sciences, Paris, France 268, No. 12 (1969), Series A, 652-654.
2. » Proceedings XX International Astronautical Congress, Buenos Aires, Argentina (1969).
3. » Proceedings, XXII International Astronautical Congress, Brussels, Belgium (1971).
4. » J. Information and Control, U.S.A. 9, No. 5 (Oct. 1966), 531-548.
5. La Salle, J. and Lefschetz, S.: «Stability by Liapunov's direct method with Applications», Academic Press (1961), New York.

## ΠΕΡΙΛΗΨΙΣ

Εἰς προηγουμένην ἐργασίαν, ἀνακοινωθεῖσαν εἰς τὴν Ἀκαδημίαν τῶν Παρισίων (1969), ἐμελετήθη ἡ εὐστάθεια μιᾶς ἑλικοειδοῦς μεταπτώσεως εἰς τὴν περίπτωσιν σταθερᾶς ἐξωτερικῆς ροπῆς μὲ χρησιμοποίησιν διαφόρων ὁρισμῶν εὐσταθείας, καὶ εὑρέθησαν διαφορετικαὶ καταστάσεις εὐσταθείας, ὑπεδείχθη δὲ ποία ἐκ τῶν καταστάσεων εὐσταθείας τῆς ἑλικοειδοῦς ἔχει πρακτικὴν ἀξίαν.

Εἰς τὴν παροῦσαν ἐργασίαν μελετᾶται ἡ κατάστασις εὐσταθείας μιᾶς κλάσεως ἑλικοειδῶν μεταπτώσεων, ποὺ περιέχει ὡς ἕνα μέλος της τὴν ἑλικοειδῆ μετάπτωσιν τῆς προηγουμένης ἐργασίας.

Χρησιμοποιοῦνται δύο ὁρισμοὶ εὐσταθείας, οἱ κατὰ Liapunov καὶ Poincaré. Τὰ συμπεράσματα τῆς παρούσης ἐργασίας εἶναι :

α. Ὅλα τὰ μέλη τῆς κλάσεως τῶν ἑλικοειδῶν μεταπτώσεων εἶναι εἰς ἀσταθῆ κατὰ Liapunov κατάστασιν.

β. Ἡ κατὰ Poincaré κατάστασις εὐσταθείας οἱουδήποτε μέλους S τῆς κλάσεως ἐξαρτᾶται ἀπὸ τὴν ὁριακὴν τιμὴν τοῦ βήματος τῆς ἑλικοειδοῦς S, καὶ ὅταν ἡ ὁριακὴ τιμὴ εἶναι σταθερὰ ἢ μηδὲν ἢ ἄπειρον, τότε ἡ ἑλικοειδὴς εἶναι εὐσταθής, ἢ ἀσυμπτωτικὰ εὐσταθὴς ἢ ἀσταθής, ἀντιστοίχως.

γ. Εἰς τὴν περίπτωσιν ποὺ ἡ ἑλικοειδὴς S εἶναι ἀσυμπτωτικὰ εὐσταθής, τότε ἡ κατάστασις αὐτὴ καὶ μόνον ἔχει πρακτικὴν ἀξίαν.

*

Ἔχω τὴν τιμὴν νὰ παρουσιάσω εἰς τὴν Ἀκαδημίαν Ἀθηνῶν τὴν ἐργασίαν τοῦ κ. Δημητρίου Μαγείρου, Ἐπιστημονικοῦ Συμβούλου τῆς General Electric τῶν Η.Π.Α. ὑπὸ τὸν τίτλον «Ἡ Εὐστάθεια μιᾶς Κλάσεως Ἑλικοειδῶν Μεταπτώσεων κατὰ Liapunov καί Poincaré».

Ὁ κ. Μάγειρος εἰς προηγουμένην ἐργασίαν του, ἀνακοινωθεῖσαν εἰς τὴν Ἀκαδημίαν Παρισίων, ἐμελέτησε τὴν εὐστάθειαν μιᾶς Ἑλικοειδοῦς Μεταπτώσεως εἰς τὴν περίπτωσιν σταθερᾶς ἐξωτερικῆς ροπῆς.

Εἰς τὴν παροῦσαν ἐργασίαν μελετᾶται ἡ κατάστασις εὐσταθείας μιᾶς κλάσεως ἑλικοειδῶν μεταπτώσεων εἰς τὴν ὁποίαν ἓν ἐκ τῶν μελῶν της εἶναι καὶ ἡ ἑλικοειδὴς μετάπτωσις τῆς ἀναφερθείσης ἤδη προηγουμένης ἐργασίας.

Χρησιμοποιοῦνται πρὸς τοῦτο οἱ ὁρισμοὶ εὐσταθείας κατὰ Liapunov καὶ Poincaré, τὰ δὲ ἀντίστοιχα πορίσματα τῆς ἐρεύνης εἶναι τὰ κάτωθι :

α) Ὅλα τὰ μέλη τῆς κλάσεως τῶν ἑλικοειδῶν μεταπτώσεων εὑρίσκονται εἰς ἀσταθῆ κατὰ Liapunov κατάστασιν.

β) Ἡ κατὰ Poincaré κατάστασις εὐσταθείας οἱουδήποτε μέλους τῆς ἐξεταζομένης κλάσεως ἐξαρτᾶται ἀπὸ τὴν ὁριακὴν τιμὴν τοῦ βήματος τῆς ἑλικοειδοῦς. Οὕτω, ὅταν ἡ ὁριακὴ τιμὴ εἶναι σταθερὰ ἢ μηδὲν ἢ ἄπειρος, τότε ἡ ἑλικοειδὴς εἶναι ἀντιστοίχως εὐσταθής, ἀσυμπτωματικὰ εὐσταθής, ἢ ἀσταθής.

γ) Εἰς τὴν περίπτωσιν ὅπου ἡ ὁριακὴ τιμὴ εἶναι μηδέν, ὁπότε ἡ ἑλικοειδὴς εἶναι ἀσυμπτωματικὰ εὐσταθής, τότε καὶ μόνον τότε ἡ κατάστασις αὕτη ἔχει πρακτικὴν ἀξίαν.

ΕΦΗΡΜΟΣΜΕΝΑ ΜΑΘΗΜΑΤΙΚΑ.— **On the separatrices of dynamical systems,** *by Demetrios G. Magiros**. 'Ανεκοινώθη ὑπὸ τοῦ 'Ακαδημαϊκοῦ κ. Φ. Βασιλείου.

## INTRODUCTION

Separatrices are special trajectories or motions of dynamical systems, and play an important role in the study of problems of the systems of current interest, especially when quantitative aspects enter the problems. But there is no general and systematic discussion on the use of the properties and on the determination of the separatrices on nonlinear dynamical systems, this determination being by itself an important problem.

In this paper we will see remarks and results concerning the properties of separatrices and their use for the study of physical problems.

The definition of separatrices given in topology is supplemented as needed in physical problems, some theorems on separatrices are stated, a list of useful properties of separatrices is given, and by selected examples we see the usefulness of the separatrices in the study of physical problems.

## 1. DEFINITION OF SEPARATRICES

We give the definition of separatrices both from a topological and a dynamical point of view.

We can say that a space W is filled by a collection S of solution curves of a dynamical system, if each solution curve of S lies in W, and each point in W is on exactly one solution curve of S.

The whole space W of the validity of a differential system may be decomposed into subspaces of which the corresponding collection of the solution curves has common properties which characterize each space.

These subspaces are called «canonical regions» of the space W, and

* Δ. Γ. ΜΑΓΕΙΡΟΥ, **'Επὶ τῶν διαχωριστικῶν καμπυλῶν τῶν δυναμικῶν συστημάτων.**

Reprinted from the *Proceedings of the Athens Academy of Sciences* 54 (1979), 264–287.

the paths of the solution curves of the system, which bound these canonical regions, are called «separatrices» of the system [7, 10].

In this topological definition of separatrices the solution curves are considered only as paths, that is as locus of point sets. In the reality, the solution curves of the dynamical systems are time-parametrized curves, that is paths on which the law of the motion of the system is known, when the topological definition of the separatrices, although it helps the investigation in some aspects, is unrealistic.

The time must be included in the concept of separatrices of physical problems. This can be succeeded by accepting the separatrices as «special "limiting" trajectories through special equilibrium states». Supplemented by this property, the topological definition of separatrices satisfies physical requirements and acquires a «physical validity».

By examples which will follow we clarify concepts related to separatrices and emphasize the usefulness of their properties in the investigation of physical problems.

## 2. THEOREMS RELATED TO SEPARATRICES

The separatrices are intimately related to the singular points of the system. It is the nature of the trajectories at the neighborhood of a singular point which guarantees the existence of separatrices through the singular point. We give, without analysis or proof, statements of theorems concerning singular points and corresponding separatrices, and the formulation of these theorems is given as needed in applications.

Theorem 1. Given a «noncritical linear dynamical system» in its normal form, if m is the number of the solution curves through a point of the space W of its validity, we may have the following cases:

a) For $m = 1$, the point is «regular», but for any other value of m the point is «singular»;

b) For $m = 0$ the singular point is a «center»;

c) For m a finite even integer, the singular point is a «saddle» point, and all the solution curves through this point are «branches of a separatrix»;

d) For $m = \infty$, the singular point is either a «node» or a «spiral» point, and among the infinitely many solution curves through the point some of them may be separatrices.

The above singular points are «elementary singular points» and characterize the linear noncritical systems.

Theorem 2. In «noncritical nonlinear systems» it is the order of magnitude of the nonlinearities of the system which decides on the nature of the singular point and of the corresponding separatrices, and we have the following cases:

a) If the order of magnitude of the nonlinearities is appropriately small, the singular point is «elementary» and the situation of separatrices is as in Theorem A.

b) If the smallness of the magnitude of the nonlinearities can not be restricted appropriately, the singular point is «nonelementary» and the situation of separatrices is a complicated matter.

Theorem 3. We distinguish two cases:

a) In «critical linear systems» the singular point may be elementary or nonelementary, and the separatrices will be in a complicated situation, especially if the system has many singular points.

b) In «critical nonlinear systems», or in «nonlinear systems without linear part», the phase portrait near the nonelementary singular point is very complicated. A small neighborhood around such a point may be devided by separatrices into sectors with this point as the apex. These sectors may be of «nodal» (parabolic), or «elliptic», or «saddle» (hyperbolic) type.

We remark that there are cases very complicated, and only a few results are known today for highly nonelementary singular points and the corresponding separatrices. The following theorem is due to Bendixson [**3**].

T h e o r e m 4. If a system is the x, y-phase plane is given by

$$y' = x^{-m}[\alpha y + bx + B(x, y)] \tag{12.1}$$

where $B(x, y)$ is a polynomial of degree at least two, and $\alpha \neq 0$, we have the following four cases:

a) If $\alpha > 0$ and $m =$ even integer, then there is only one branch of integral curves tending to the origin on the left side of y-axis, $x < 0$, while integral curves on the other side, $x > 0$, constitute a nodal distribution; that is, there is a coalescence of «saddle-nodal» points.

In this case there exists a separatrix through the origin.

b) If $\alpha < 0$ and $m =$ even integer, this case can be transformed to the previous case, and we have a coalescence of «nodal-saddle» points (node at $x < 0$, and saddle at $x > 0$).

A separatrix exists through the origin in this case.

c. If $\alpha > 0$ and $m =$ odd integer, the origin is a nodal point, when a separatrix may exist.

d. If $\alpha < 0$ and $m =$ odd integer, the origin is a saddle point and a separatrix exists.

The analysis of the above statements is based on the definition of the separatrices, on the concepts of the «α-limiting» and «ω-limiting» properties of the separatrices, and on other concepts.

## 3. SOME PROPERTIES OF SEPARATRICES

Combining the definition of separatrices and results coming from the theorems, one can find properties of separatrices, which are very useful in applications. In the following we list some of these properties.

— The separatrices may be points, lines, surfaces, depending on the dimensions and the structure of the dynamical system.

— There is no separatrix through a center.

— A separatrix through a singular point may be either a «α-limiting» or a «ω-limiting» trajectory, when, starting from a point of the separatrix, the time to reach the terminating point is «infinite».

— Separatrices starting from a singular point may terminate to the same singular point, when they are «closed» separatrices, and they have a finite length. «Non-closed» or «open» separatrices do not start and terminate at the same singular point. They start from a singular point and they may terminate either to another singular point or to infinity. Some of these open separatrices may have finite length.

— Separatrices through a node have at this point a definite tangent, and separatrices terminating to a spiral point move around it spirally and they do not have a definite tangent at this point.

— An «isolated closed path» of a dynamical system, in case all its points are regular, is a «limit-cycle» of the system, when it corresponds to a periodic phonomenon with a fixed period. But, if this closed path is through a singular point, the periodicity disappears and the closed path is not a limit cycle, but it is a «closed separatrix». The limit cycle is a separatrix according to the topological definition; but it is not a separatrice according to the supplemented definition of separatrices.

— The separatrices have important physical significance. We indicate some of them.

They may determine the whole region of the validity of a dynamical system and separate it from the «empty regions» which are without real solutions of the system.

They may be boundary curves of the regions in each of which the solutions are characterized by different stability situations, when the separatrices are in a «neutral stability situation». This property is of paramount importance in contemporary nonlinear control problems.

They may have some other physical meaning.

## 4. DETERMINATION OF SEPARATRICES

For the determination of the separatrices we see two cases. In case we know the general solution of the mathematical model of the physical problem, the determination of separatrices is identical to the determination of special particular solutions through appropriate singular points of the system.

In case the general solution is not known, approximate methods, either geometrical, or numerical, or analytical, map help the investigation towards the determination of the separatrices.

Remark. We remark that the concept of «index» of singular points, of trajectories in general and of separatrices in particular, introduced by Poincaré, plays an important role for investigation of their nature [**5**, **13**].

## 5. EXAMPLES

In each of the following examples appropriate remarks are given related to properties of separatrices.

Example 1. The separatrices in this example determine exactly the boundary of the canonical regions, which are regions of the validity of the system where real trajectories exist, and empty regions.

The dynamical system [**4** (a)] :

$$\dot{x}^2 = 1 - x^2, \qquad \dot{y}^2 = 1 - y^2 \tag{1.1}$$

has four singular points, the points $(\pm 1, \pm 1)$, which are points of intersection of the lines $x = \pm 1$, $y = \pm 1$.

The system (1. 1) corresponds in the x, y - phase plane to the DE :

$$y'^2 = \frac{1 - y^2}{1 - x^2}. \tag{1.2}$$

For the reality of the solutions of (1. 2), the $(1 - x^2)$ and $(1 - y^2)$ must have the same sign, and this restriction helps to determine the real regions of the validity of (1. 2). By separating the variables and integrating, one can find the general solution of (1. 2):

$$\left.\begin{array}{ll} \text{arc sin}\, y \pm \text{arc sin}\, x = c\,; & |x| < 1, \quad |y| < 1 \\ \text{arc cos}\, hy \pm \text{arc cos}\, hy = c\,; & |x| > 1, \quad |y| > 1 \end{array}\right\} \tag{1.3}$$

c is the arbitrary constant. Figure 1 shows the phase-portrait of (1. 3).

The separatrices are the lines $x = \pm 1$ and $y = \pm 1$, which separate the whole x, y-plane into nine regions five of which have families of real solutions and four are empty regions.

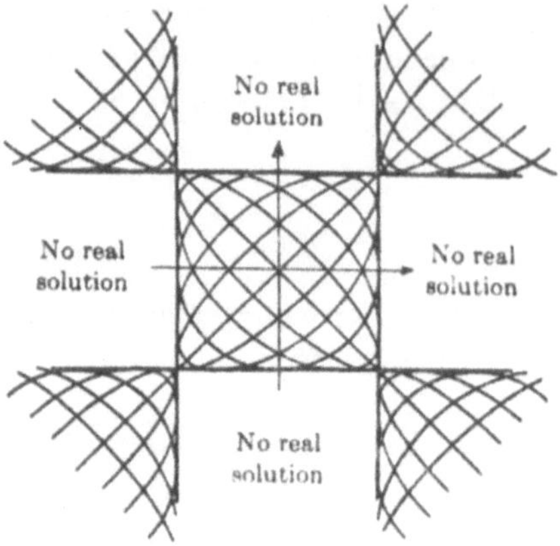


Fig. 1.

Example 2. In this example we see that the separatrices, with the help of some other curves (which are not separatrices), determine the boundaries of the canonical regions, and that the common property of each family of trajectories of the regions is a special stability situation.

The dynamical system:

$$\dot{x} = x(\varepsilon x - 1), \quad x(0) = x_0, \quad t \geqslant 0 \tag{2. 1}$$

has as singular points the points $x = 0$ and $x = \frac{1}{\varepsilon}$. The general solution of (2. 1) is:

$$x(t) = \frac{x_0}{\varepsilon x_0 - (\varepsilon x_0 - 1)\, e^t} \tag{2. 2}$$

of which the portrait is shown in Figure 2.

The separatrices in the t, x-plane are the lines $x = 0$, $x = \frac{1}{\varepsilon}$, $t = 0$. These separatrices, with the help of the line:

$$t = \log \frac{\varepsilon x_0}{\varepsilon x_0 - 1} \tag{2.3}$$

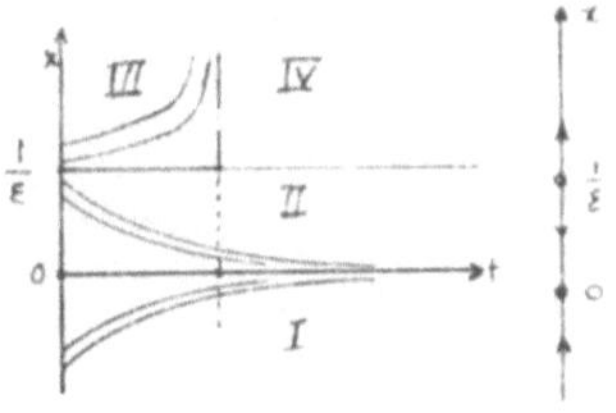


Fig. 2.

separate the half t, x-plane, $t \geqslant 0$, into four regions of which I, II, III are real regions of the validity of (2.1), and IV is the empty region.

Example 3. There exist dynamical systems without singular points, then without separatrices, of which the canonical regions are separated by curves of nature different than the separatrices. In this example these separating curves are «asymptotes».

The dynamical system:

$$\dot{x} = 2, \qquad \dot{y} = y^2 - 1 \tag{3.1}$$

is without singular points, then without separatrices. This system corresponds to the DE:

$$y' = \frac{1}{2}(y^2 - 1) \tag{3.2}$$

of which the general solution is:

$$y = \frac{1 + ce^x}{1 - ce^x}. \tag{3.3}$$

The phase-portrait of (3.3) is shown in Figure 3.

The lines $y = \pm 1$ separate the x, y-plane into three canonical regions, and these lines are «asymptotes» of the families of the solutions of the regions. We remark that in the previous example, Figure 2, the separatrix $x = 0$ is an asymptote for the trajectories of the regions I and II.

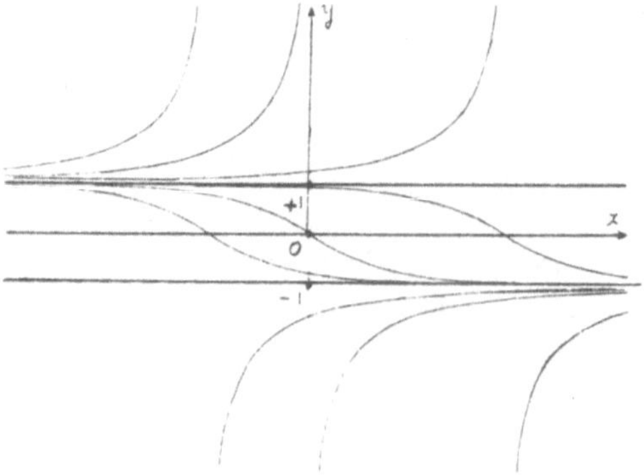


Fig. 3.

Example 4. Here we distinguish the concept of a separatrix from the concept of an envelope of a family of solution curves. Both have the property to separate the region of the validity of the system into canonical regions, but the separatrices are special members of the families of the solutions, while the envelopes have not, in general, this property but they are special singular solutions of the system, tangent to all members of the families of the solutions.

We give appropriate examples.

## a. The motion of a projectile in a vacuo.

1. We imagine all trajectories described by projectiles fired from the same point 0 with the same initial velocity $\bar{v}_0$ on a x, y - plane, each trajectory corresponding to a different direction of firing $\varphi$, Figure 4. All these trajectories belong to a family of parabolas given by [**6**]:

$$y = mx - \frac{g(1+m^2)x}{2v_0} \tag{4.1}$$

where $m = \tan\varphi$ is the parameter, and g serves as a retardation.

For the envelope of the family (4.1), we eliminate the parameter m between (4.1) and its derivative with respect to m, when the result:

$$y = \frac{v_0^2}{g^2} - \frac{gx^2}{2v_0^2} \qquad (4.2)$$

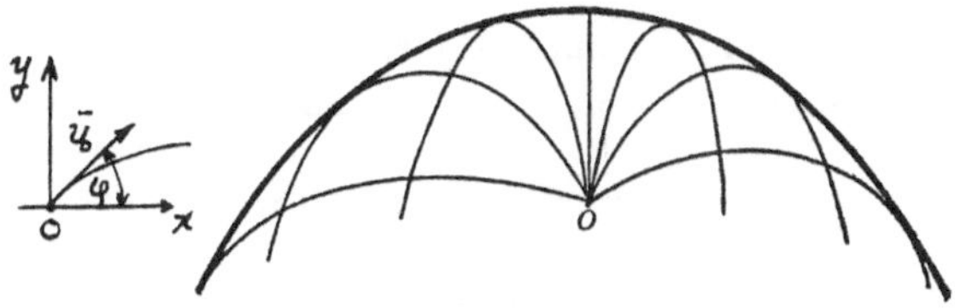


Fig. 4.

a parabola, is the envelope. This envelope separates the points which can be reached from those which can not be reached, and it is not a number of the family (4.1), Figure 4.

The family (4.1) has no separatrix.

2. There are other physical problems of the same nature, e.g., the «caustics» in optics are envelopes of light rays reflected by a mirror.

Let us see another example.

**b. Consider the DE:**

$$y'^2 = \frac{x^2}{1-x^2} \qquad (4.3)$$

which is equivalent to:

$$y' = \pm \frac{x}{\sqrt{1-x^2}} \qquad (4.4)$$

valid in the strip $|x| \leqslant 1$. Its general solution is:

$$f \equiv x^2 + (y+c)^2 - 1 = 0. \qquad (4.5)$$

That is, a family of circumferences with centers on the y-axis and tangent to the lines $x = \pm 1$.

For every «regular» point of the strip two circumferences pass, but this is not the case for the points of the lines $x = 0$, $x = \pm 1$, which are «singular lines». The lines $x = \pm 1$ are boundaries of the strip and they are tangents to every member of the family (4.5), and these lines are «singular solutions» of (4. 4), envelopes of the family (4. 5). There is no separatrix in the family (4. 5).

We remark that in some systems, as in the Example 11, Figure 12, envelope and separatrix exist and are identical.

E x a m p l e 5. This example shows that the boundary of the stability regions of a dynamical system may be not a separatrix. The system:

$$\dot{x} = -y, \qquad \dot{y} = f(x, y) \tag{5. 1}$$

with

$$f(x, y) = \begin{cases} -x + 2x^3y^3, & \text{if: } x^2y^2 < 1 \\ -x, & \text{if: } x^2y^2 > 1 \end{cases} \tag{5. 2}$$

has the origin as the singular point, which is in the region $x^2y^2 < 1$, when the appropriate equation in the x, y - plane is:

$$y' = \frac{-x + 2x^3y^2}{-y}. \tag{5. 3}$$

The eigenvalues of (5. 3) are both real and negative, then the origin is a «node», a «regular attractor», when starting from any point of the

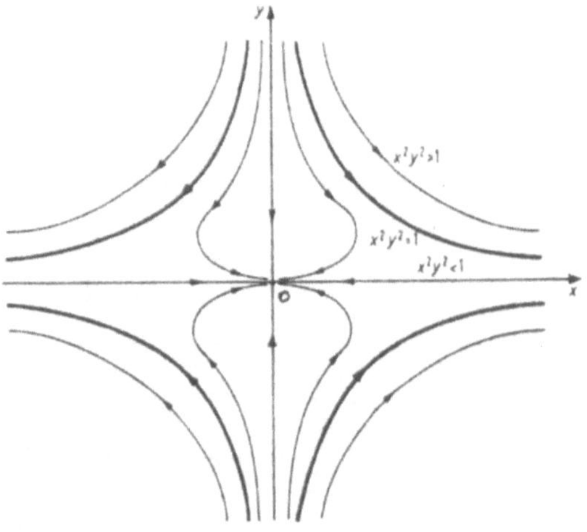


Fig. 5.

region $x^2y^2 < 1$ and following the corresponding trajectory we will terminate to the origin. The order of the magnitude of the nonlinearity of

(5. 3) agrees with this result. The phase-portrait of the solutions of (5. 3) is shown in Figure 5. The curve $x^2y^2=1$, which consists of four branches, determines the region of attractiveness of the origin. Outside of this region the stability situation is different. The curve $x^2y^2=1$, which is the boundary of different stability situations of the regions, is not a separatrix of the system.

Example 6. By this example we see changes in the nature of the separatrices by putting restrictions to the constants of the system [**9**].

Take the system:

$$\ddot{x} = x(\alpha^2 - x^2) + b\dot{x} \tag{6. 1}$$

which has the normal form:

$$\dot{x} = y, \quad \dot{y} = x(\alpha^2 - x^2) + by, \quad \alpha b \neq 0. \tag{6. 2}$$

The singular points are $O(o, 0)$, $A_1(\alpha, 0)$, $A_2(-\alpha, o)$.

For the nature of the origin we find the characteristic equation of (6. 2):

$$\lambda^2 - b\lambda - \alpha^2 = 0$$

when the eigenvalues are:

$$\lambda_{1,2} = \frac{1}{2}\left(b \pm \sqrt{b^2 + 4\alpha^2}\right) \tag{6. 3}$$

and since these eigenvalues are real and of opposite sign, the origin is a saddle point.

For the nature of the points $A_1$ and $A_2$, we use the transformations $x = \bar{x} + \alpha$, $y = \bar{y}$, when (6. 2) is reduced to a perturbed system of which the origin corresponds to $A_1$ and $A_2$, and the characteristic equation of the perturbed system is $\lambda^2 - b\lambda + 2\alpha^2 = 0$, and the eigenvalues are:

$$\lambda_{1,2} = \frac{1}{2}\left(b \pm \sqrt{b^2 - 8\alpha^2}\right). \tag{6. 4}$$

We have two cases.

a. If $b^2 < 8\alpha^2$, $\lambda_1$ and $\lambda_2$ are complex numbers with real part of sign of b, when $A_1$ and $A_2$ are spirals, stable for $b<0$ unstable for $b>0$.

Figure 6(a) has been drawn for $A_1$, $A_2$ stable spirals. For unstable spirals one merely reverses the arrows in this Figure.

The separatrices in this case connect saddle and spiral points and have infinite length.

b. If $b^2 > 8\alpha^2$, $\lambda_1$ and $\lambda_2$ are real and of the same sign as b, when $A_1$, $A_2$ are nodes, stable for $b < 0$, unstable for $b > 0$. Figure 6(b) shows the case of stable nodes $A_1$, $A_2$, and for unstable nodes we reverse the arrows in this Figure. Two of the separatrices are of finite length and two of infinite length.

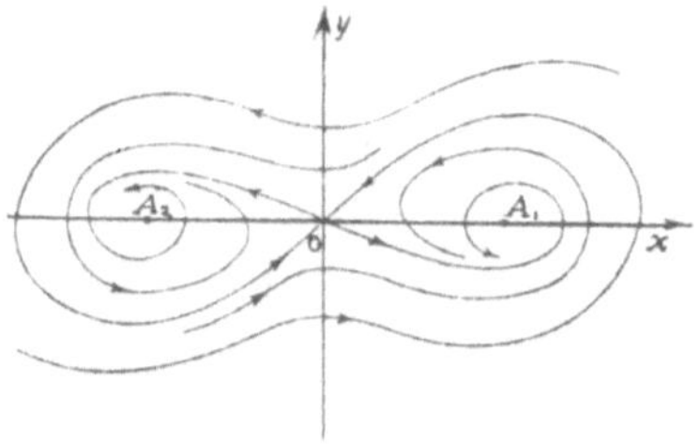


(a)

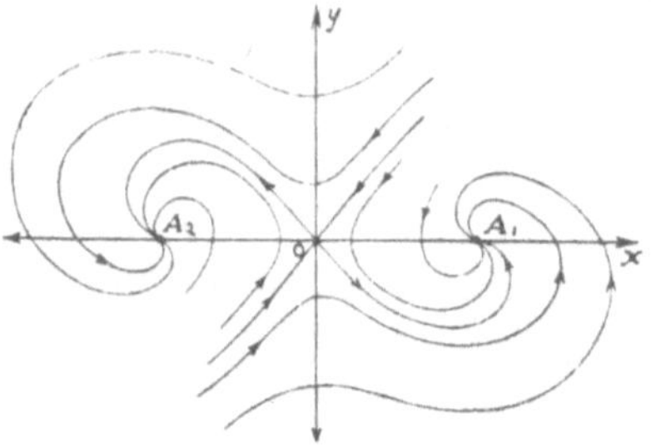


(b)

Fig. 6.

Example 7. Here we have a physical problem of biology or economics in which the separatrices are calculated as special particular solutions of the general solution of the model of the problem. In addition we see a property of separatrices which is very important in interpreting theoretical results.

There are many assemblies around us of which the elements influence each other through competition and cooperation.

The «problem of population growth» is a problem of this nature.

We discuss this problem as a biological problem, but, by appropriate changes in the meaning of the variables and the constants involved, the problem can become a problem in other fields, as, e.g., in economics.

We consider two coexisting species of population numbers x and y at time t, both hunters, that is one species kills members of the other species. By using appropriate assumptions, the correspondent mathematical model is the nonlinear differential system [8]:

$$\dot{x} = \alpha x - cxy, \qquad \dot{y} = by - dxy \tag{7.1}$$

x and y are positive integers, but they can be considered as positive continuous functions of time. The coefficients α, b, c, d have a physical meaning and here are taken as positive integers.

The equilibrium points of (7.1) are the origin and the point $A\left(\frac{b}{c}, \frac{\alpha}{c}\right)$, and we can check that the origin is a «node», and A a «saddle» point.

The system (7.1) corresponds in the x, y - phase plane to the DE:

$$y' = \frac{b - dx}{x} \cdot \frac{y}{a - cy} \tag{7.2}$$

of which the general solution is:

$$y^{\alpha} \cdot e^{-cy} = k \cdot x^{b} \cdot e^{-dx}. \tag{7.3}$$

The constant k in (7.3), which corresponds to the point A, is:

$$k = \left(\frac{\alpha}{c}\right)^{\alpha} \cdot \left(\frac{d}{b}\right)^{b} \cdot e^{b-\alpha}. \tag{7.4}$$

Inserting (7.4) into (7.3) one gets the equation of the separatrix through A. For a specific case, let us take $\alpha = 4$, $b = 3$, $c = 2$, $d = 1$, when the point A and the constant k are: A(3, 2), $k \simeq .218$, and the equation of the separatrix through A is:

$$(y^2/x^2) \cdot e^{x-2y} = .218\,. \tag{7.5}$$

An investigation of (7.5) leads to the Figure 7, in which the four branches of the separatrix are the courves through the point A, and these branches separate the first quadrant into the four regions I, II, III, and IV.

Starting from any point of any of these regions, we see that, as $t \to \infty$, one of the species tends to vanish asymptotically, while the other species tends to become infinite. In addition, we see that the species y eventually dissapears if the corresponding (x, y) - point is in the region

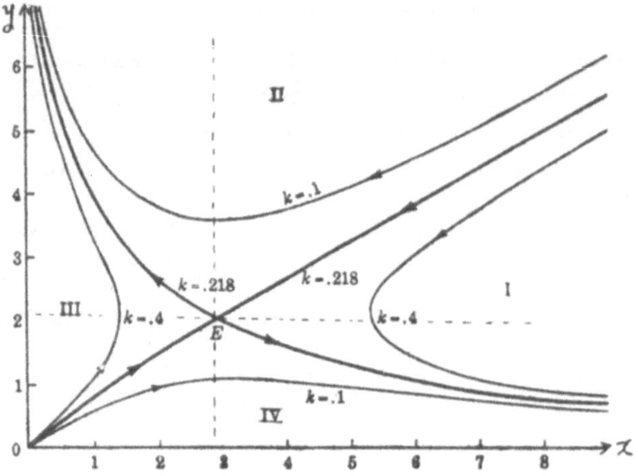


Fig. 7.

I or IV, and we see the opposite situation if the (x, y) - point is in the regions II and III. These results indicate a new property of separatrices, and show how important it is to know the location of the separatrices in the x, y - plane.

The origin O is a repulsor in the regions III and VI, and in III the species $x \to 0$, while in IV $y \to 0$.

Of course, due to the over-simplification of the model (7.1) of the problem, the above results are somehow unrealistic. For better results, the model of the problem must be modified by taking into account other influences for the growth of the species, e.g., the food supply, etc.

We remark that the previous discussion, modified by suitable changes to the problem and appropriate specification on the competitive species and the limiting resources, might be useful for an investigation of a problem of nature different than the above. E. G., one can have a problem in the field of economics if the variables denote the size or extent of commercial enterprizes for a common source and for a common market.

Example 8. By this example we see how the separatrices can be calculated in case of coexistence of many singular points, and also that the spearactrices, either closed or open, may be of finite length.

If the system is expressed by the DE [11]

$$\ddot{x} + 3x - 4x^3 + x^5 = 0 \tag{8.1}$$

by using $\dot{x} = y$, this equation can be reduced to

$$y' = \frac{1}{y}\left\{x\,(x^2 - 1)\,(x^2 - 3)\right\} \tag{8.2}$$

valid in the (x, y) - phase plane. The singular points are the origin and the points $(\pm 1, 0)$ and $(\pm \sqrt{3}, 0)$, and we can check that the origin and $(\pm \sqrt{3}, 0)$ are «centres», while $(\pm 1, 0)$ are «saddle» points.

The general solution of (8.1) in the (x, y) - phase plane is:

$$y^2 = c - 3x^2 + 2x^4 - \frac{1}{3}x^6. \tag{8.3}$$

The value of the arbitrary constant c of (8.3) corresponding to the saddle points $(\pm 1, 0)$ is $c = \frac{4}{3}$, when the separatrix through the points $(\pm 1, 0)$ is:

$$y^2 = \frac{1}{3}\,(4 - 9x^2 + 6x^4 - x^6) \tag{8.4}$$

of which the graph is shown in Figure 8.

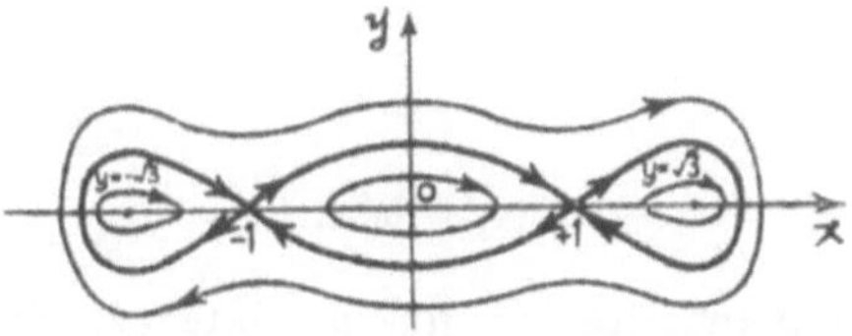


Fig. 8.

The separatrix (8.4) has four branches all of which have a finite lenght. The branches around the points $(\pm\sqrt{3}, 0)$ are «closed», and that around the origin are «open».

From this and the previous example we see a procedure for the determination of separatrices through saddle points of the system, of which the existence guarantees the existence of the separatrices.

Example 9. In these examples we see systems which have infinitely many «open» separatrices with finite or infinite lengths.

a. $$\ddot{x} + \omega^2 \sin x = 0. \quad (9.1)$$

This is the pendulum equation and it is equivalent to the system:

$$\dot{x} = y, \quad \dot{y} = -\omega^2 \sin x, \quad (9.2)$$

$\omega$ is the proper frequency. The singular points are infinitely may and they are the points $x = n\pi$, $n =$ integer, of the x-axis.

For even n the singular points are centers, and for odd n saddle points. There are infinitely many canonical regions and infinitely many separatrices running from a saddle point to the nearest saddle point. Figure 9 gives the corresponding phase-portrait. The separatrices are open and have finite length.

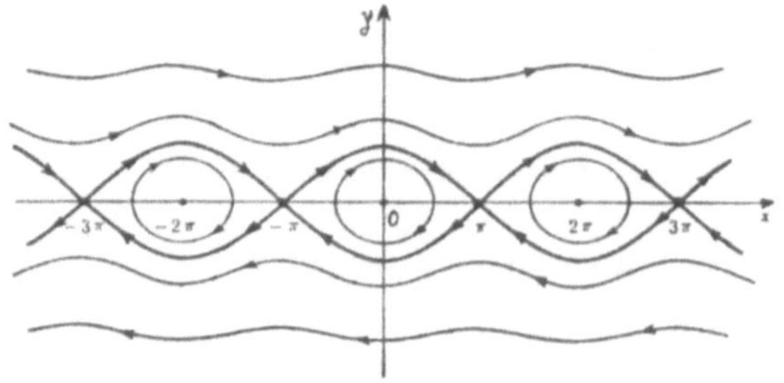


Fig. 9.

b. $$\ddot{x} + k\dot{x}|\dot{x}| + \omega^2 \sin x = 0. \quad (9.3)$$

The singular points are $x = n\pi$, $n =$ integer, on the x-axis. For even n are spirals, and for odd n are saddle points. The infinitely many separatrices are of infinite length and run from a saddle point to the nearest spiral points, or they run from infinity to saddle points. Figure 10 shows the corresponding phase-portrait.

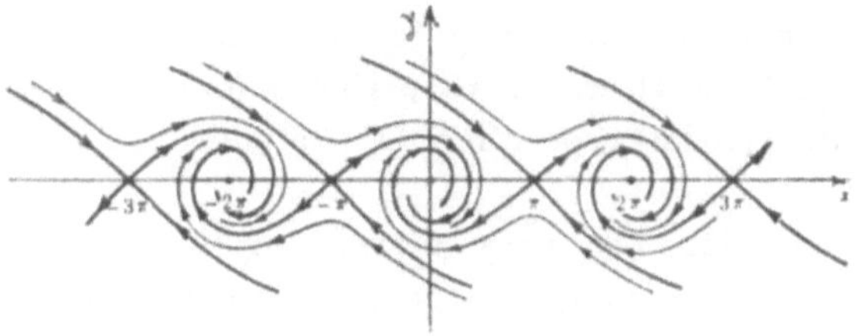

Fig. 10.

E x a m p l e 10. In this example we see an nonelementary singular poit of complicated nature.

We take the system in polar coordinates

$$\dot{r} = r(1-r), \qquad \dot{\vartheta} = \sin^2\left(\frac{\vartheta}{2}\right). \tag{10.1}$$

Its singular points are $O(r=0,\ \vartheta=0)$, $O_1(r=1,\ \vartheta=0)$.

The DE in the phase-plane, corresponding to (10.1) is:

$$\frac{dr}{d\vartheta} = \frac{r(1-r)}{\sin^2\left(\frac{\vartheta}{2}\right)} \tag{10.2}$$

which can be integrated, and the family of the solutions $r = r(\vartheta) + c$ is shown in Figure 11.

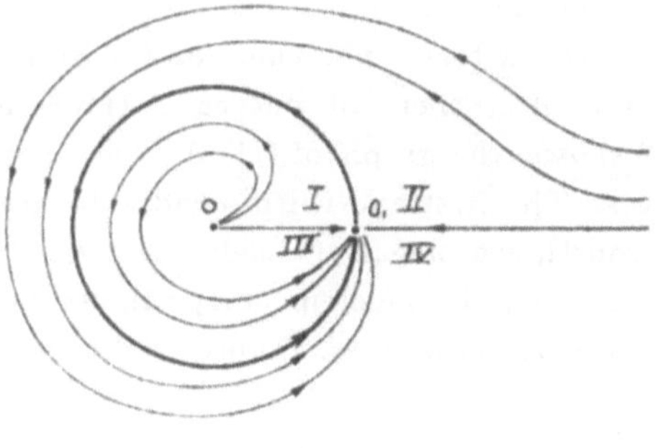


Fig. 11.

The separatrix is the circumference with center O and radius $OO_1 = 1$. The origin is a repulsor or a negative attractor (unstable). The point $O_1$ is a nonelementary singular point of complicated nature. The line $OO_1$ and the separatrix divide the neighborhood of $O_1$ into four sectors of which I and II are of «hyperbolic» (saddle) type, and III and IV of «parabolic» (nodal) type where $O_1$ is a positive attractor.

Example 11. The phase-portrait at the neighborhood of a nonelementary singular point may be complicated, but the separatrices through this point may be very simple; in addition «separatrices» and «envelopes» may be identical.

This is shown by the present example [**4**(b)].

The system:

$$\dot{x} = x(2y^3 - x^3), \qquad \dot{y} = -y(2x^3 - y^3) \tag{11.1}$$

corresponds to the DE:

$$y' = -\frac{y(2x^3 - y^3)}{x(2y^3 - x^3)}. \tag{11.2}$$

The origin is the only singular point and, since the system is without linear part, this point is nonelementary.

The right hand member of (11.2) is a function of the ratio $(y/x)$, when by using the transformation $y = x \cdot u(x)$ one can separate the variables and integrate. The general solution of (11.2) can be found to be

$$x^3 + y^3 - 2cxy = 0, \tag{11.3}$$

c is the arbitrary constant. (11.3) is a one parameter family of curves known as «Folia of Descartes». The equation (11.3) is satisfied at the origin for any value of c, then all curves of (11.3) are through the origin. Figure 12 shows the graph of (11.3). The axes of coordinates are the separatrices. The first and third quadrants are elliptic sectors, the second and fourth are negative nodal sectors. In this example, although the origin is highly nonelementary, the separatrices are very simple lines, the axes of coordinates. Figure 12 shows the phase-portrait of (11.3).

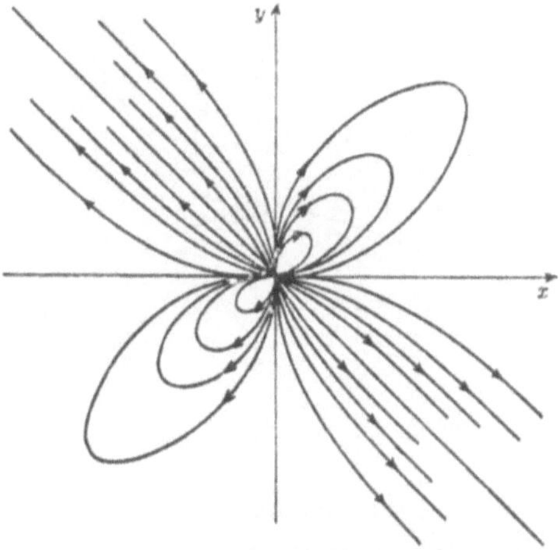


Fig. 12. (Folia of Descartes).

We remark that

The derivative of (11.3) with respect to the arbitrary constant c gives $x = 0$, $y = 0$, then the axes of coordinates of the above system are «envelopes» of the families of the solutions of the system and are identical with the separatrices of the system.

Example 12. In this example the nonelementary singular point is of «nodal-saddle» type, and we have three separatrices.

The system:

$$\dot{x} = -x^6, \qquad \dot{y} = y^3 - yx^4 \tag{12.1}$$

corresponds to:

$$y' = \frac{yx^4 - y^3}{x^6}. \tag{12.2}$$

The origin is the only singular point which is nonelementary.

The phase portrait, shown in Figure 13, can be found approximately by, say, geometrical methods.

There are four sectors I, II, III, IV and three separatrices which are the y-axis and the curve $OO_1$ and $OO_2$. The 180° sector I has negative nodal trajectories.

The origin is a positive attractor in the sector II. The sectors III and IV are of saddle type.

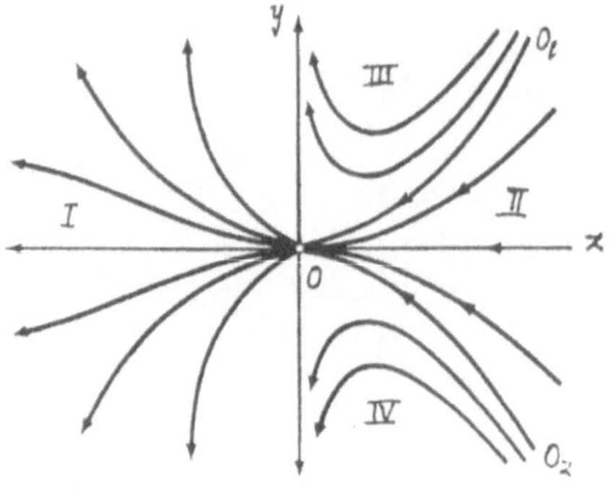


Fig. 13.

Example 13. By this example we give a highly nonelementary singular point, where the separatrices is difficult to be calculated. If the system in its phase-plane is given by [5]:

$$y' = x\left\{\frac{x^2 - xy - xy^2}{x^2 - y^2 - x^2y^2} + \frac{y}{x^2}\right\}. \tag{13. 1}$$

the origin is the nonelementary singular point. Figure 14 shows the graph of the solutions of (13. 1) at the neighborhood of the origin found

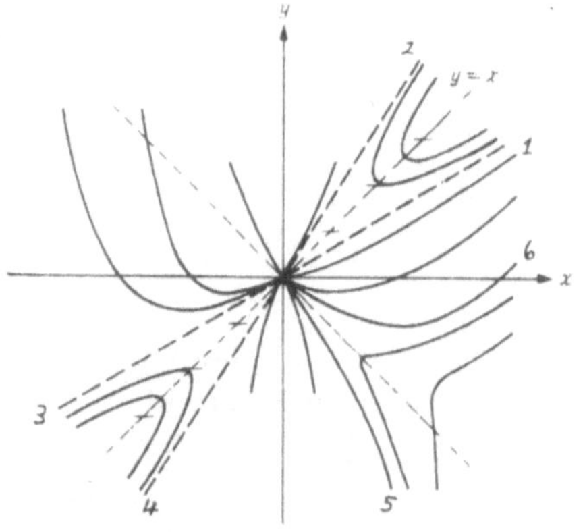


Fig. 14.

by approximate methods. We have six sectors with apex the origin, of which three are of hyperbolic type and three of parabolic type. Two hyperbolic sectors contain the line $y=x$ in the first and third quadrants, and the third hyperbolic sector contains the line $y=-x$ in the fourth quadrant. The six branches of the separatrix are the curves 01, 02, 03, 04, 05, 06 shown in Figure 14.

Example 14. At the neighborhood of the saddle points of the previous examples, either elementary or nonelementary saddle points, the behavior of the trajectories were characterized by the property that these trajectories do not intersect the correspondent separatrices.

Fig. 15.

There are systems with saddle points at the neighborhood of which the behavior of the trajectories is very complicated. The three-body problem and some problems of dynamics show this complexity, Figure 15 [**1**, **2**, **12**].

## ΠΕΡΙΛΗΨΙΣ


Αἱ διαχωριστικαὶ καμπύλαι (ΔΚ) τῶν δυναμικῶν συστημάτων παίζουν σπουδαῖον ρόλον εἰς τὴν ἔρευναν φυσικῶν προβλημάτων τρέχοντος ἐνδιαφέροντος. Ὅμως δὲν ὑπάρχουν σήμερον εἰδικὰ δημοσιεύματα ἀναφερόμενα εἰς τὸν γενικὸν προσδιορισμὸν τῶν ΔΚ, ὅπως καὶ εἰς τὴν χρῆσιν τῶν ἰδιοτήτων των εἰς τὴν ἔρευναν φυσικῶν προβλημάτων ποσοτικοῦ τύπου, ὁ δὲ προσδιορισμός των εἶναι ἀφ᾽ ἑαυτοῦ ἕνα σπουδαῖον πρόβλημα.

Εἰς τὴν παροῦσαν ἐργασίαν δίδονται παρατηρήσεις καὶ εὑρίσκονται συμπεράσματα σχετικὰ μὲ τὴν ὕπαρξιν καὶ τὸν προσδιορισμὸν τῶν ΔΚ, ὅπως καὶ μὲ τὴν χρῆσιν τῶν ἰδιοτήτων των εἰς τὴν ἔρευναν καὶ ἑρμηνείαν φυσικῶν προβλημάτων.

Ὁ τοπολογικὸς ὁρισμὸς τῶν ΔΚ συμπληρώνεται καταλλήλως ὥστε νὰ γίνῃ χρήσιμος εἰς τὴν ἔρευναν πρακτικῶν προβλημάτων, διατυπώνονται θεωρήματα σχετικὰ μὲ τὰς ΔΚ χωρὶς ἀνάλυσιν ἢ ἀπόδειξιν, δίδονται παρατηρήσεις διὰ τῶν ὁποίων ὑποβοηθεῖται ὁ προσδιορισμὸς τῶν ΔΚ, τονίζονται ἰδιότητες τῶν ΔΚ χρήσιμοι διὰ τὴν ἔρευναν.

Διὰ τῶν παραδειγμάτων καταφαίνεται ἡ χρησιμότης τῶν ἰδιοτήτων τῶν ΔΚ εἰς ἐφαρμογὰς κυρίως. Ἐκ τῶν ἰδιοτήτων αὐτῶν τονίζονται δύο κυρίως :

(α) : αἱ ΔΚ εἶναι δυνατόν, μόναι των ἢ καὶ μὲ τὴν βοήθειαν καὶ ἄλλων καμπυλῶν, νὰ προσδιορίζουν τὰ χωρία ὅπου τὰ δυναμικὰ συστήματα ἔχουν λύσεις πραγματικὰς ἀπὸ τὰ χωρία ὅπου δὲν ὑπάρχουν κἂν λύσεις,

(β) : αἱ ΔΚ διαχωρίζουν τὸ χωρίον προσδιορισμοῦ τῶν συστημάτων εἰς χωρία, ὅπου αἱ λύσεις ἔχουν διαφόρους καταστάσεις εὐσταθείας ἕκαστον χωρίον, ὁπότε αἱ ΔΚ εὑρίσκονται ὑπὸ «οὐδετέραν» κατάστασιν εὐσταθείας.

Ἡ ἰδιότης αὕτη τῶν ΔΚ εἶναι μεγάλης σημασίας εἰς προβλήματα εὐσταθείας τῆς νεωτέρας θεωρίας «μὴ γραμμικῶν συστημάτων ἐλέγχου».

Ἡ παροῦσα ἐργασία θὰ συμπληρωθῇ καταλλήλως πολὺ σύντομα.


## REFERENCES


1. V. Arnold, «Small denominators and problems of stability of motion in classical and celestial mechanics», Transl. Division, Foreign Technical Div., WP-AFB, Ohio (May 5, 1964), U. S. A.
2. V. Arnold and A. Avez, «Ergodic problems of classical mechanics», Benjamin Inc., New York (1968), pg. 91, U. S. A.

3. I. Bendixson, «Sur les courbes définies par des équations differentielles», Acta Math. 24 (1899), Sweden.
4. G. Birkhoff and G. Carlo, «Ordinary differential equations», Ginn and Co., New York (1962), (a): pg. 15 - 16, (b): pg. 138 - 142, U. S. A.
5. E. Davis and E. James, «Nonlinear differential equations», Addison-Wesley Publ. Co., New York (1966), pg. 28 - 39, U. S. A.
6. J. Jeans, «Theoretical mechanics», Ginn & Co., New York (1907), pg. 211, U. S. A.
7. W. Kaplan, «Regular curve families filling the plane», Duke Math. J., I. 7 (1940), II. 8 (1941), U. S. A.
8. J. Kemeny and J. Snell, «Mathematical models of the social sciences», MIT - Press, Cambridge, Mass. (1972), pg. 24 - 33, U. S. A.
9. S. Lefschetz, «Differential equations, geometric theory», Interscience Publishers, Inc., New York (1957), pg. 223 - 231, U. S. A.
10. L. Markus, «Global structure of ordinary differential equations in the plane», Trans. Am. Math. Soc. 76 (1954), pg. 127 - 148, U. S. A.
11. N. McLachlan, «Ordinary nonlinear differential equations in engineering and physical sciences», 2nd Ed. (1958), Oxford, Clarendon Press, pg. 196, England.
12. V. Mel'nicov, «On the stability of the center for time-periodic perturbation», Math. Proc., Moscow Math. Soc., U. S. S. R.
13. H. Poincaré, «Sur les courbes definies par des équations differentielles», Journal de Mathematiques, 1881, 1882, . . . , Paris, France.

UDC 517.9

*D. G. Magiros*

**Philadelphia, USA**

## SEPARATRICES OF DYNAMICAL SYSTEMS

Separatrices are special trajectories or motions of dynamical systems, they have a fundamental geometrical and physical significance and play an important role in the study of physical, engineering and social problems of current interest.

There is no systematic discussion on the existence and determination of separatrices, on their properties and their use. Also, their known definitions are somehow defective, they lead to confusion and need a clarification and modification.

In this paper, the known definitions of separatrices are appropriately supplemented, and, based on the supplemented definition, theorems are discussed related to separatrices at the neighborhood of elementary or nonelementary critical points of dynamical analytical systems in the plane. The systems are taken in general as homogeneous, and three cases are distinguished, namely the cases: $\Delta \neq 0$, $\Delta \equiv 0$, $\Delta = 0$, where $\Delta$ is the discriminant of the system.

By an appropriate transformation of the given system, an «auxiliary system» results in the study, which facilitates the investigation. The concepts of «nonlinearities», «eigenvalues», «sectors» are used in this paper.

We intend in another paper to use the concept of the «index» and give a discussion of the properties of separatrices and their use in physical applications.

**On the Definitions of Separatrices.** We can say that a space W is filled by a collection $S$ of solution curves of a system of differential equations, if each solution curve of $S$ lies in $W$ and each point in $W$ is on exactly one solution curve of $S$.

The whole space $W$ of the validity of the differential system may be decomposed into subspaces in each of which the corresponding collection of the solution curves has common properties which characterize each subspace.

These subspaces are called «cononical regions» (or cells) in the space $W$ and the special solution curves of the differential system, which have the property to bound the canonical regions, are called «separatrices» of the system [6, 9].

In this topological definition of separatrices the solution curves are considered only as geometrical entities, as a locus of point sets, as paths which are just carriers of the motion.

By using this definition one cannot distinguish the concept of a separatrix from the concepts of limit cycles, or of asymptotes and envelopes of a family of curves. In the reality, the solution curves of dynamical systems are time-parameterized curves, they are trajectories, that is paths on which the law of the motion is known. The above «topological property» of, separatrices helps the investigation in some aspects; it is necessary for the definition of separatrices, but it is not sufficient. The time must be included in the definition of the concept of separatrices.

Another definition of separatrices is based on stability concepts [1, 2, 7, 10]. But some of the so defined separatrices are stable or unstable in Liapunov or in Poincare' sense, or stable or unstable structurally [1, 2] when this definition is accompanied by confusion. The definition of separatrices must be given independently of the stability concepts.



Reprinted from the *Proceedings of the 9th Intl. Conf. on Nonlinear Oscillations* (Kiev, 1981) v. 2, *Qualitative Methods*

We may define as separatrices of a dynamical system its trajectories which are characterized by the two properties [8]: (a) The «topological property» according to which these trajectories separate the whole region of the validity of the system into canonical subregions, and, (b) The «dynamical property» according to which they are «limiting trajectories», either «ω-limiting», or/and α-limiting» trajectories, through special critical points of the system, with definite tangents there.

By «ω-limiting» trajectory we mean trajectory tending to critical point as $t \to \infty$, and by «α-limiting» as $t \to -\infty$.

Such a definition of separatrices elucidates the concept of separatrices, satisfies physical requirements and acquires a physical validity.

**A relation Between Figenvalues and Slopes at the Origin.** We consider the dynamic system

$$\dot{x} = a_1 x + b_1 y + X_2(x, y), \quad \dot{y} = a_2 x + b_2 y + Y_2(x, y) \tag{1}$$

where $a_1, a_2, b_1, b_2$ are real constants, and $X_2, Y_2$ series convergent in a region of the validity of (1), which have terms of degree at least two in $x, y$. The linearization of (1) is

$$\dot{x} = a_1 x + b_1 y, \quad \dot{y} = a_2 x + b_2 y. \tag{1.1}$$

The origin is a critical point of these systems, and if their discriminant $\Delta = \begin{vmatrix} a_1 b_1 \\ a_2 b_2 \end{vmatrix}$ is different than zero the origin is an «elementary critical point», while if $\Delta \equiv 0$, or $\Delta = 0$, the origin is a «nonelementary critical point». The «eigenvalues» $\lambda$ of the systems are given by

$$\lambda = \frac{1}{2}\{a_1 + b_2 \pm \sqrt{(a_1 + b_2)^2 - 4\Delta}\}. \tag{1.2}$$

For both $\lambda$ real, the elementary critical points are either «nodals» (if both $\lambda$ of the same sign) or «saddles» (if the $\lambda$ are of opposite sign); and when $\lambda$ are complex the critical points are «spirals», while for $\lambda$ purely imaginary they are «centres».

The concept of the «slope» of the trajectories through the origin taken as a critical point gives important help to the investigation of the separatrices through the origin.

For system (1), in which $X_2, Y_3$ are infenitesimals stronger than $x, y$ at the neighborhood of the origin, the slope $K$ of the tangents of the trajectories at the origin can be given by

$$K = \frac{1}{2b_1}\{a_1 - b_2 \pm \sqrt{(a_1 - b_2)^2 + 4a_2 b_1}\}. \tag{1.3}$$

Since the discriminants of (1.2) and (1.3) are equal, $K$ can be given in terms of $\lambda$ by

$$K = (\lambda - b_2)/b_1. \tag{1.4}$$

$K$ must be real, and it is real for real $\lambda$, according to (1.4), then for the investigation of separatrices spirals and centres must be excluded.

**Separatrices of Elementary Critical Points. The Case $\Delta \neq 0$.** We examine the problem of separatrices in case of a nodal or saddle at the origin.

We distinguish three kinds of nodal points: the «ordinary», the «degenerate», and the «dicritical» nodal points. The ordinary nodal is characterized by real distinct eigenvalues both of which are either negative or positive; in the degenerate nodal both eigenvalues are equal, and in the dicritical nodal $a_1 = b_2$, $a_2 = b_1 = 0$ in (1.1). In the system (1), in which the linear part dominates its nonlinear part at the neighborhood of the origin, the separatrices through the origin are as follows.

**Theorem 1.** (a) «All the trajectories in nodal points either stop at or start from these points, and some of them are separatrices.

There are four separatrices in an ordinary nodal point, two in a degenerate nodal, and four in a dicritical nodal.

The separatrices in the ordinary and degerate nodals are all either ω-separatrices» or «α-separatrices». The separatrices in the dicritical nodals are lines of «neutral» type (neutral in the sense of stability) (Fig. 1). (b) In a saddle point there are four separatrices, two «ω-separatrices» and two «α-separatrices», which are the only trajectories of the system through the saddle point.

A «ω-separatrix» and its consecutive «α-separatrix» considered together can be taken as one separatrix, when in a saddle point one can have four separatrices transversing the saddle point».

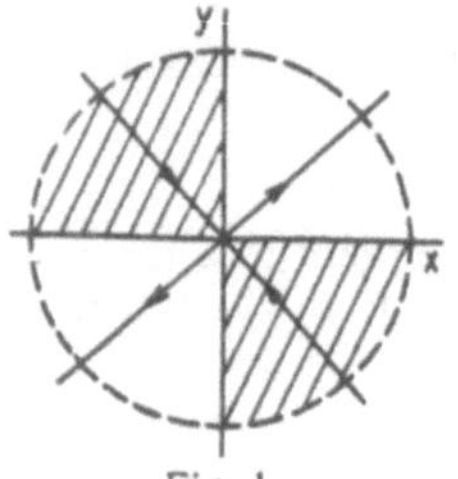

Fig. 1.

**Separatrices of Nonelementary Critical Points. The Case** $\Delta \equiv 0$. We consider a nonlinear system without linear part, ($\Delta = 0$), in the form

$$\dot{x} = X_m(x, y) + f(x, y), \quad \dot{y} = Y_m(x, y) + f_2(x, y) \tag{2}$$

where $X_m$, $Y_m$ are homogeneous polynomials without common factors, and $m > 1$; $f_1$ and $f_2$ are convergent series in $x, y$ with terms of degree bigger than $m$. Omitting $f_1$, $f_2$ in (2) and using the transformation $y = ux$ one gets the «auxiliary, equation»

$$\frac{du}{dx} = \frac{F(u)}{xX_m(1, u)}, \tag{2.1}$$

where the «auxiliary function» $F(u)$ is

$$F(u) = Y_m(1, u) - uX_m(1, u). \tag{2.2}$$

The nature of the critical points of (2.1), which are the real roots of $F(u)$ on $u$ — axis, plays, a decisive role for the investigation of the nature of the critical point at the origin of $x, y$-plane and of the nature of the sector and separatrices through the origin. There exist criteria [3, 4] to check the nature of these critical points in $x, u$-plane which are of «nodal» ($N$), or «saddle» ($S$), or «nodal-saddle» (NS) type. If $F(u)$ is without real roots, the critical points in $x, u$-plane are of «spiral» or «centre» type, and this case is of no interest to us. To each real root $u_i$ of $E(u)$ on $u$-axis a «critical line», perpendicular to $u$-axis, corresponds, and, by the transformation $y = ux$, to each real root $u_i$ the straight line $y = u_i x$ corresponds in $x, y$-plane, which is the carrier of two separatrices through the origin in $x, y$-plane. The number $l$ of the real roots $u_i$, $i = 1, 2, \ldots, l$ depends on the degree of $F(u)$ which the maximum is $m + 1$, and $l \leqslant m + 1$, when the number of the sectors and separatrices through the origin in $x, y$-plane is $2l \leqslant 2(m + 1)$. As for the nature of the sectors and separatrices through the origin of $x, y$-plane, we remark the following[3,4]:

Two consecutive nodals $u_i$ and $u_{i+1}$ on $u$-axis correspond to two elliptic sectors at the origin of $x, y$-plane;

Two consecutive saddles correspond to two hyperbolic sectors;

A saddle and a consecutive nodal correspond to two parabolic sectors;

Two consecutive nodal-saddles give two elliptic sectors, if the sides of nodals are consecutive, or give two hyperbolic sectors if the sides of saddles are consecutive, or give two parabolic sectors, if the side of nodal follows the side of saddle.

The same happens if the consecutive of nodal-saddle is a nodal or a saddle. If the degree of $F(u)$ is its maximum $m + 1$, $F(u)$ has $m + 1$ real distinct finite roots, corresponding to $2(m + 1)$ straight line separatrices through the origin in $x, y$-plane. In case the degree of $F(u)$ is smaller than $m + 1$ the point $u_\infty$ at the infinity of $u$-axis is a critical point in $x, u$ — plane, and the number of the separatrices depends on the multiplicity of $u_\infty$ as a root of $F(u)$ and this number is $< 2(m + 1)$

The nature of the portraits and of the separatrices through the origin of $x$, $y$-plane depend on the nature of the real roots of $F(u)$ on $u$-axis, their number and their multiplicity.

In the following we specialize the above statements in cases of systems of «quadratic» and «cubic» homogeneous forms [3].

**Quadratic Homogeneous Systems.** In systems of quadratic homogeneous forms, one can distinguish four cases depending on the degree of $F(u)$, which may be three (maximum), or two, or one, or $F(u)$ may be a constant, in which cases the number of separatrices is six, four, or two.

**Theorem 2.** (a) **If the degree of F (*u*) is three, we have three cases.**

(i) ***F* (*u*) has three real simple roots.** The number of the separatrices through the origin of $x$, $y$-plane is six. The nature of the sectors and of the separatrices depend on the nature of the three roots, which may be positive, or negative and of N, or S or (NS) type. We distinguish three combinations of the roots: S, S, S, or S, S, N, or S, N, N. Fig. 2 corresponds to N, S, S.

(ii) **One root simple and one double.** The separatrices are four. We have two combinations of the roots: S, (NS), or N, (NS). Fig. 3 (combination N, NS)).

(iii) **One real root either simple or triple,** which will be of N or S type. The separatrices are two. Fig. 4 (N).

(b) ***F* (*u*) of degree two.** $F(u)$ has a simple root $u_\infty$ at infinity of $u$ — axis, and the finite roots may be either two simple or one double. This case can be deduced to the previous cases, and $y$-axis is a trajectory of N or S type.

(c) ***F* (*u*) of degree one.** $u_\infty$ is a double root of $F(u)$, the separatrices are four, and the situation is similar to the above case (a, ii), Fig. 5 (S, (NS)).

(d) ***F* (*u*) is a constant.** $u_\infty$ is a triple root of $Fu$ the separatrices two, and the situation is similar to the case (a, iii), Fig. 6 (S).

**Cubic Homogeneous Systems.** In cubic homogeneous systems the degree of $F(u)$ is $\leqslant 4$, and the separatrices are in an even number $\leqslant 8$. We consider some cases.

**Theorem 3.** (a) **F (u) of degree four.** We have five cases.

(i) ***F* (*u*) with four roots simple.** The number of separatrices through the origin of $x$, $y$- plane is eight. We distinguish five combinations of roots: S, S, S, S, or S, S, S, N, or S, S, N, N, or S, N, S, N, or S, N, N, N. Fig. 7 (combination S, S, S, S).

(ii) ***F* (*u*) with two simple and one double roots.** We have six separatrices, and four root combinations are considered: S, S, (NS), or S, N, (NS), or S, N, (SN), or N, N, (NS), Fig. 8 (S, N, (NS).

(iii) ***F* (*u*) with two real roots and two complex, or one simple and one triple.** There are four separatrices, and the root combinations: S, S, or S, N, or N, N, Fig. 9 (N, N).

(iv) ***F* (*u*) with two double roots.** The separatrices are four, and the root combinations: (NS), (NS), or (NS), (SN). Fig. 10 (SN), (NS)).

(V) ***F* (*u*) with one real double root and two complex, or one real of multiplicity four.** The separatrices are two. Fig. 11 ((SN)).

(b) ***F* (*u*) of degree $<4$.** The $y$-axis in $x$, $y$-plane is a trajectory the nature of which depends on the order of multiplicity of the root $u_\infty$. The separatrices are in a number $<8$.

**The case of a Multiple Critical Point at the Origin of *x*, *u*- plane.** The separatrices in this case are given according to the following [3].

**Theorem 4.** Let the auxiliary system be of the form

$$\frac{du}{dx}=\frac{U_m(x,\ u)}{X_m(x,\ u)}\ , \tag{3}$$

with $X_m$, $U_m$ homogeneous in $x$, $u$ without common factors, and $m>1$. The origin in $x$, $u$-plane is supposed to be an isolated critical multiple point from which

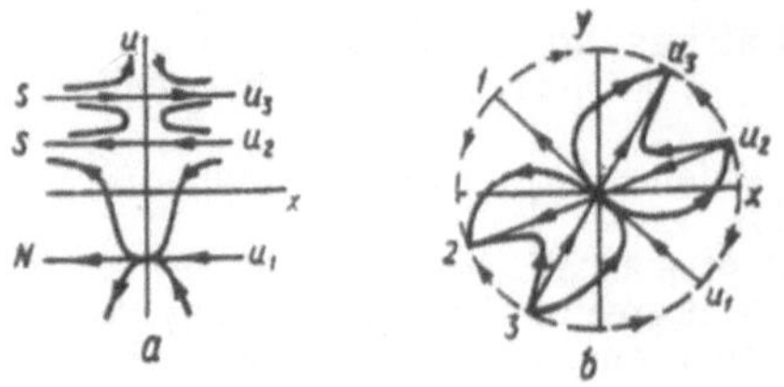


Fig. 2.

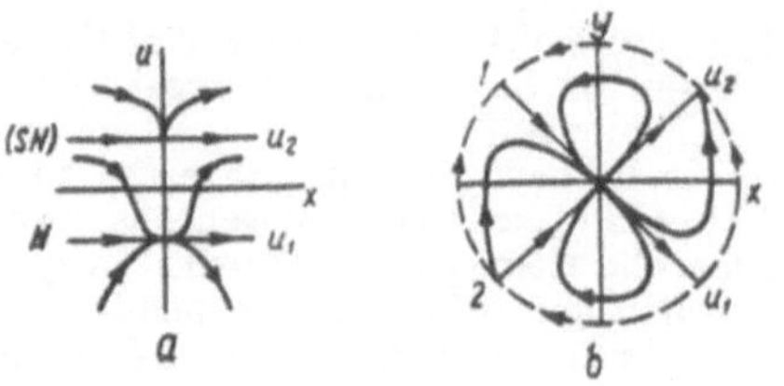


Fig. 3.

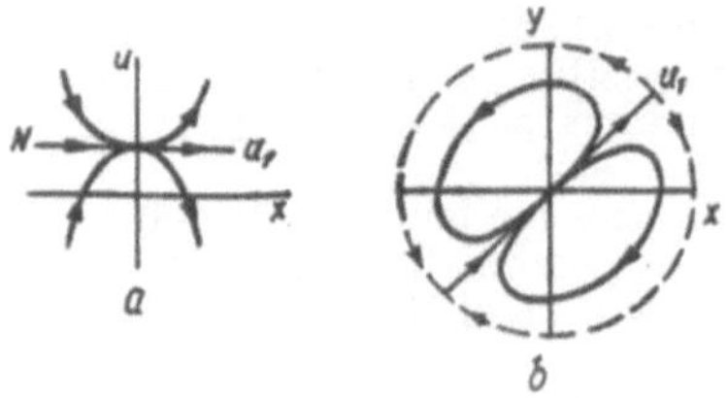


Fig. 4.

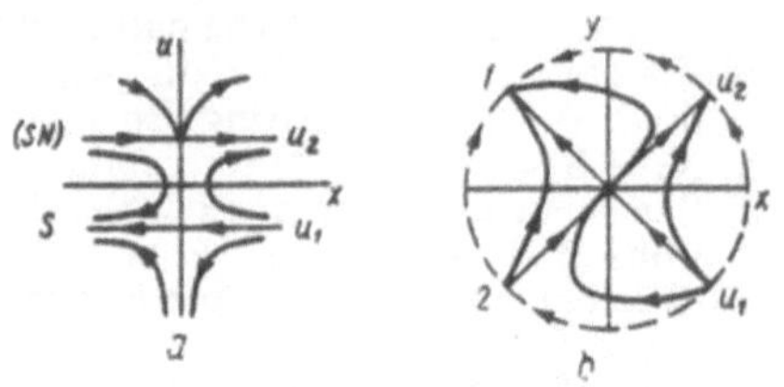


Fig. 5.

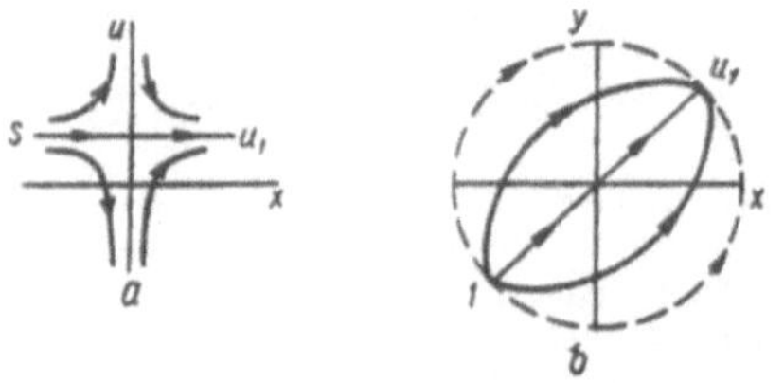


Fig. 6.

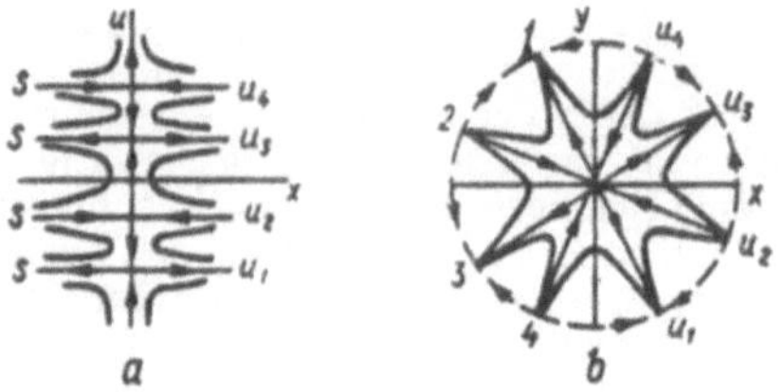


Fig. 7.

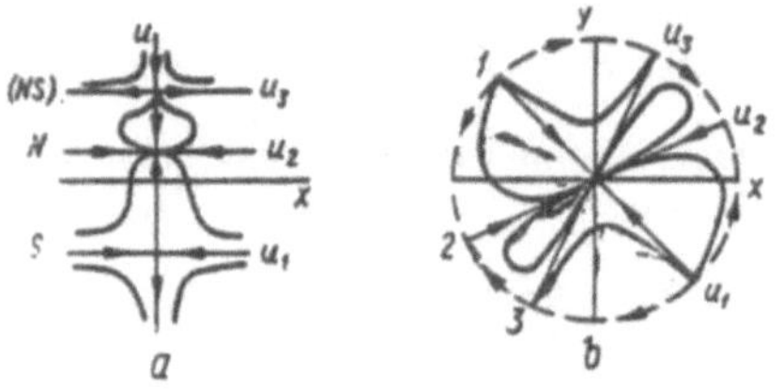


Fig. 8.

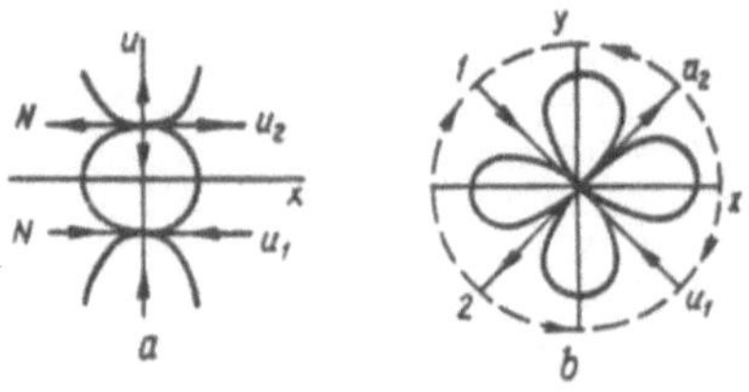


Fig. 9.

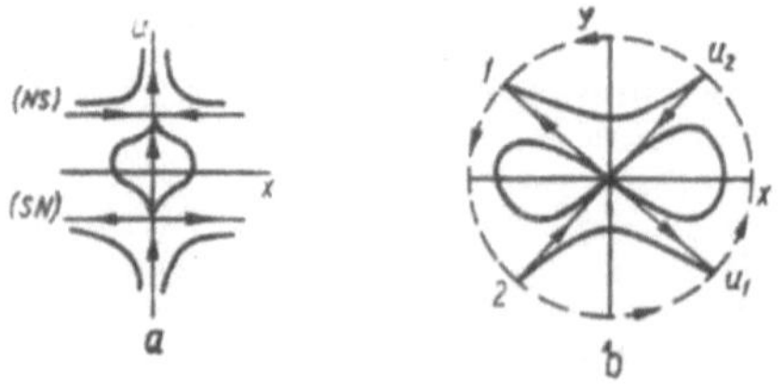


Fig. 10.

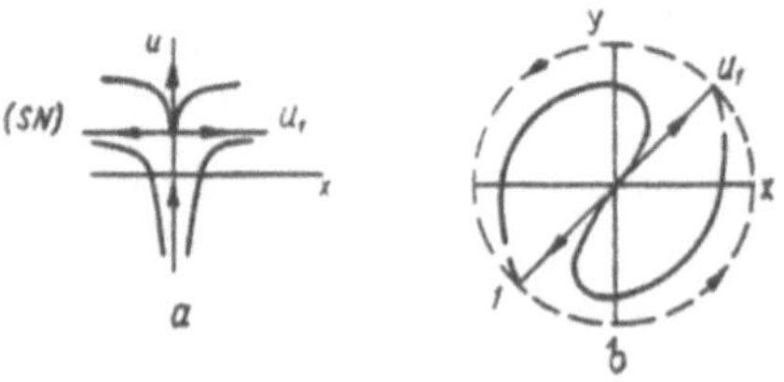


Fig. 11.

rectilinear trajectories of (3) pass with finite slopes. The auxiliary function is

$$F(\alpha) = U_m(1, \alpha) - \alpha X_m(1, \alpha) \tag{3.1}$$

and has finite roots $\alpha_i$. By the transformation $y = xu$ the equation (3) leads to

$$\frac{dy}{dx} = \frac{yX_m(x^2, y) + x^2 U_m(x^2, y)}{xX_m(x^2, y)} \tag{3.2}$$

If $u$-axis is not a trajectory of (3) and $\alpha_i$ are real, then the rectilinear critical trajectories $u = \alpha_i x$ of (3) are transformed in $x, y$-plane into the parabolas $y = \alpha_i x^2$ each of which is the carrier of two separatrices (Fig. 12).

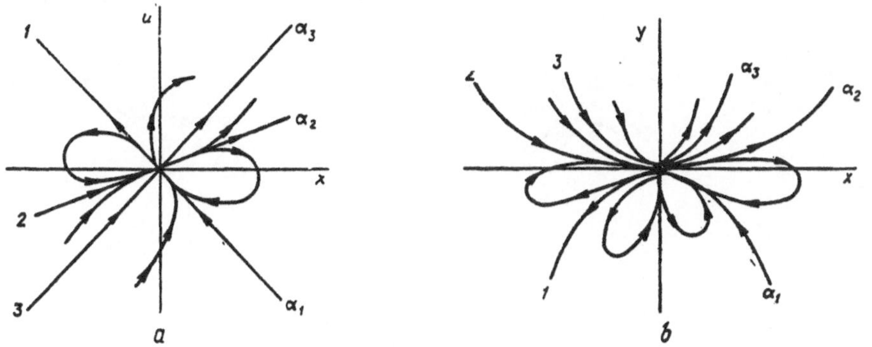


Fig. 12.

**Separatrices Through Nonelementary Critical Points. The case $\Delta = 0$.** We consider the system [3, 11]

$$\dot{x} = y + \sum_{i=2}^{p} X_i(x, y), \quad \dot{y} = \sum_{i=2}^{q} Y_i(x, y), \tag{4}$$

where $X_i$, $Y_i$ are homogeneous polynomials with real coefficients of degree $i$ in $x$, $y$. Transforming by $y = xu$ and taking terms of degree smaller than three the auxiliary system is

$$\dot{x} = xu + ax^2, \quad \dot{y} = bx - u^2 + (c - a)xu + dx^2, \tag{4.1}$$

where $a, b, c, d$ are real coefficients. The nature of the trajectories at the neighborhood of the origin in $x, u$-plane depend on the values of the quantities: $a, b, c, d, E, \alpha_1, \alpha_2$, which may be positive, negative, or zero. Important cases are when $b \geq 0$, $E \gtrless 0$. In case $b = 0$, if $A(x, u)$ and $B(x, u)$ are the members of (4.1) at right, respectively, the auxiliary equation $F(\alpha) = B(1, \alpha) - \alpha A(1, \alpha) = 0$ leads to

$$-2\alpha + (c - 2a)\alpha + d = 0, \quad \alpha = u/x, \tag{4.2}$$

with roots

$$\alpha_\infty, \alpha_i = \frac{1}{4}\{(c - 2a) + (-1)^i \sqrt{E}\}, \quad i = 1, 2, \tag{4.3}$$

$$E = (c - 2a)^2 + 8a. \tag{4.4}$$

The separatrices through the origin of $x, y$-plane in the case of the system (4.1) are given by the following theorems.

**Theorem 5.** (a) If in the auxiliary system (4.1): $b = 0$, $E > 0$, and one of the real and distinct roots $\alpha_1, \alpha_2$ is zero, then the $x$-axis is a trajectory in both

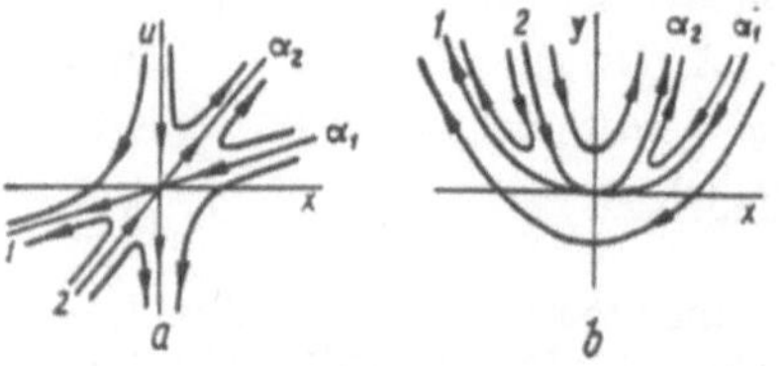


Fig. 13.

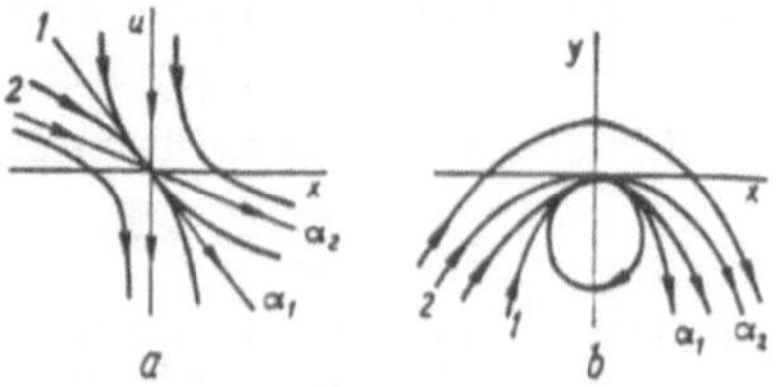


Fig. 14.

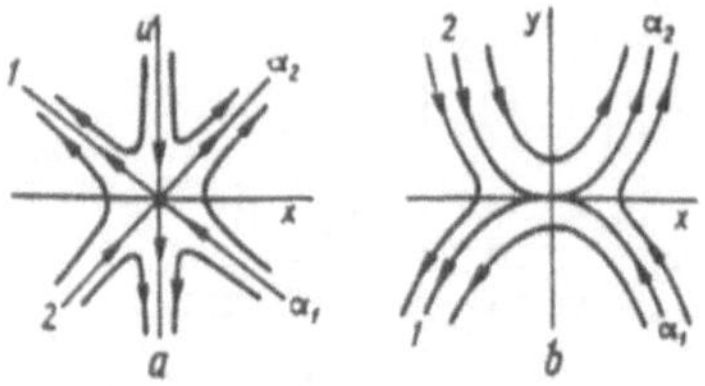


Fig. 15.

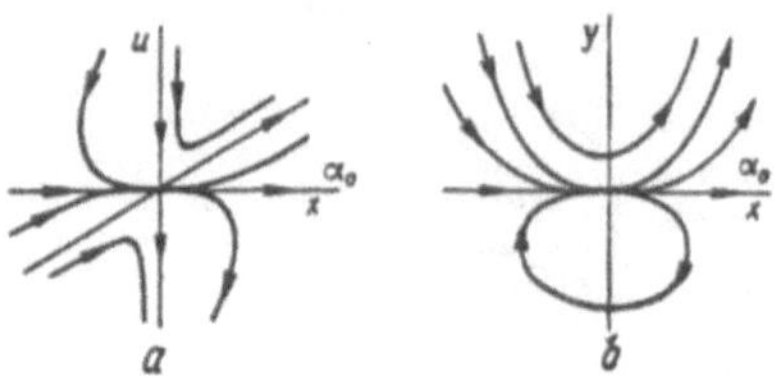


Fig 16.

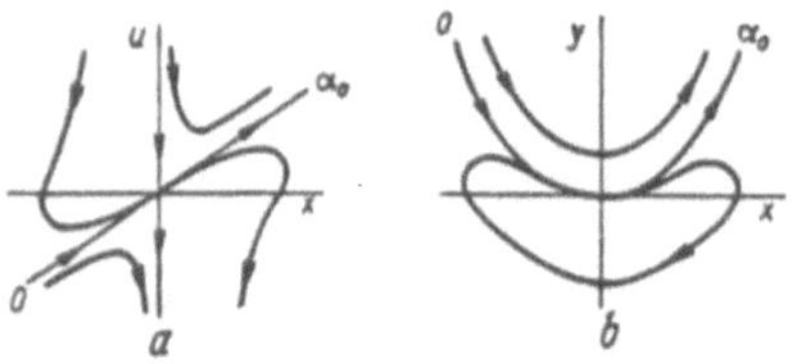


Fig. 17.

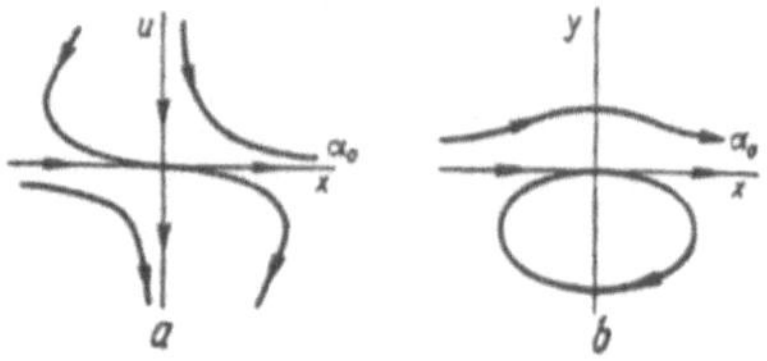


Fig. 18.

Fig. 19.

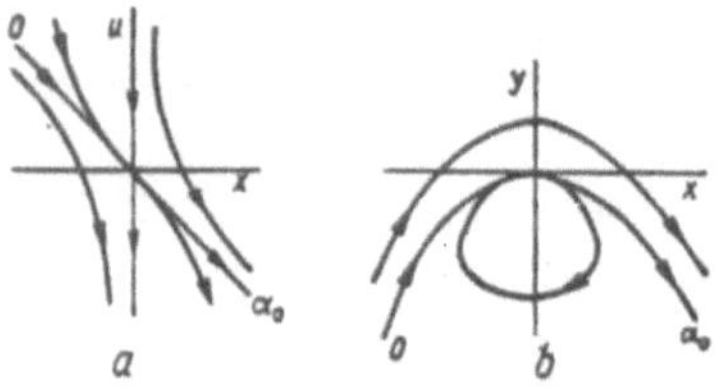


Fig. 20.

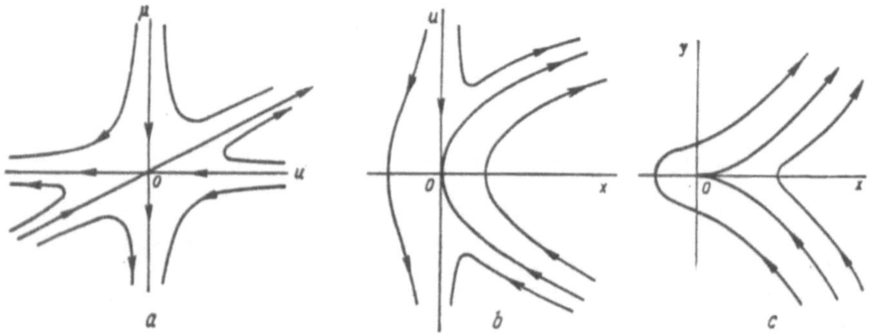

the $x$, $u$-plane and the $x$, $y$-plane, and the straight lines $u = \alpha_i x$, $i = 1, 2$ in $x$, $u$-plane correspond to the parabolas $y = \alpha_i x^2$ in $x$, $y$-plane with $x$-axis tangent to these parabolas at the origin. Each of these two parabolas is the carrier of two separatrices, when the separatrices in $x$, $y$-plane are four. We can distinguish four cases according to the sign of $\alpha_1$ and $\alpha_2$ (Fig. 13—16).

(b) If in the system (4.1) $b = 0$, $E = 0$, $c + 2a > 0$, the finite roots coincide: $\alpha_1$-$\alpha_2$-$\alpha_0$-$\frac{1}{2}(c - 2a)$, and the corresponding parabola $y = \alpha_0 x^2$ in $x$, $y$-plane is the carrier of two separatrices. We can distinguish three cases (Fig. 17—19).

**Theorem 6.** If in (4.1) $b > 0$, by using the transformation $x = u\mu$ and taking the quadratic terms only, one can get

$$\dot{u} = u(b\mu - u) = C(u, \mu), \quad \dot{\mu} = \mu(2u - b\mu) = D(u, \mu) \tag{4.5}$$

The critical points of (4.5) are given by

$$K(\alpha) = D(1, \alpha) - \alpha C(1, \alpha), \quad \alpha = \mu/u \tag{4.6}$$

and they are

$$\alpha_\infty, \ \alpha_0 = 0, \quad \alpha_1 = 3/2b. \tag{4.7}$$

The portraits are symbolically represented by

$$(u, \mu) \xrightarrow[x=u\mu]{} (x, u) \xrightarrow[y=ux]{} (x, y)$$

the origin in $x$, $y$-plane is of «cusp» type, and the separatrices are two (Fig. 20).

**General remark.** In the preceding the separatrices of some dynamical systems in the plane were examined locally at the neighborhood of a critical point (the origin). The global investigation of separatrices is also very important. Various physical problems have solution — curves connecting two critical points, and there are conditions which assume the existence of such solutions, which are physically realistic if they are locally unique and (in some sense) persistent to small perturbations.

The separatrices connecting two critical points can be thought of as being formed by intersection of two manifolds each of which is associated with a critical point [5]. The separatrices may be closed or open. In the closed separatrices the starting and terminal critical points are identical (the separatrices are loops). In the open separatrices the terminal point is different then the starting one they may have finite or infinite length and if they have infinite length either go to infinity or go to a spiral point, when they become regular spiral curves and they lose the property of being separatrices.


1. *Андронов А. А., Леонтович Е. А., Гордон И. И. и др.* Качественная теория динамических систем второго порядка.— М. : Наука, 1966.— 568 с.
2. *Андронов А. А., Леонтович Е. А., Гордон И. И. и др.* Теория бифуркаций динамических систем на плоскости.— М. : Наука, 1967.— 488 с.
3. *Argemi J.* Sur les points singuliers multiples de systèmes dunamiques dans $R^2$.—Theses, Université d'Aix—Marseille, France, 1967.
4. *Bendixson I.* Sur les courbes définies par les équation différentielles.—Acta Math., 1899, 24, Sweden.
5. *Gordon P.* Paths connecting elementary critical points of dynamical systems.— SIAM, J. Appl. Math., 1974, **26**, N 1.
6. *Kaplan W.* Regular curve families filling the plane.— Duke Math. J., 1940, **1**, 7; 1941, **11**, N 8, U.S.A. p. 155—185, p. 11—46.
7. *Lefschetz S.* Differential equations : Geometric theory.— Interscience Publishers, Inc., New York, 1957, p. 212.
8. *Magiros D.* Separatrices of dynamical systems.— Practica. Athens Academy June 1979, Greece.
9. *Markus L.* Global structure of ordinary differential equations in the plane.— Trans. Amer. Math. Soc., 1954, **76**, p. 127—149.
10. *McLachlan N.* Ordinary non linear differential equations in engineering and physical sciences, 2nd Ed. Oxford, the Clarencon Press, 1950, p. 188.
11. *Sansone G., Conti R.* Nonlinear differential equations. The McMillan Co., New York, 1964.

# APPENDIX: PAPERS IN RUSSIAN

Ukrainian Academy of sciences,
Ukrainian Mathematical journal
Vol. 30, No. 2, 1978

D. G. Magiros: NONLINEAR DIFFERENTIAL EQUATIONS WITH SEVERAL GENERAL SOLUTIONS.

A nonlinear differential equation may have several general solutions. The investigation of the existence and construction of such solutions may be succeeded by "restricting quantities of the equation", or by "factorizing the equation".

УДК 517.93

*Д. Г. Мажирос*

## Нелинейные дифференциальные уравнения с несколькими общими решениями

Нелинейное обыкновенное дифференциальное уравнение может иметь несколько общих решений. Их существование и определение может быть рассмотрено путем «ограничения решений» дифференциального уравнения или путем «факторизации дифференциального уравнения».

1. Введение. Общие решения нелинейных обыкновенных дифференциальных уравнений — наиболее желательные решения, особенно для приложений, однако только для некоторых классов уравнений можно найти их в замкнутом виде.

В то время как линейное обыкновенное дифференциальное уравнение имеет одно общее решение, нелинейное обыкновенное дифференциальное уравнение может иметь одно или несколько общих решений.

В этой статье рассматриваются классы нелинейных обыкновенных дифференциальных уравнений, имеющие несколько общих решений. Для этого класса уравнений можно получить хорошие результаты либо путем ограничения величин дифференциального уравнения, либо путем его факторизации.

Уточним некоторые понятия, связанные с понятием общих решений дифференциальных уравнений.

Рассмотрим нелинейное дифференциальное уравнение $(F)$, справедливое в области $(R)$ пространства своих переменных, и рассмотрим функцию $\Phi$, зависящую от переменных дифференциального уравнения $(F)$, содержащую ряд произвольных постоянных и принимающую значения в области $(C)$



Reprinted from the *Ukrainian Mathematical Journal*, Ukrainian Academy of Sciences **30** (1978), 238–241.

пространства этих постоянных. Функция, возникающая из ($\Phi$) при определении всех ее произвольных постоянных, и такая, что удовлетворяет ($F$), является частным решением уравнения ($F$).

Функция ($\Phi$), рассматриваемая как совокупность всех частных решений уравнения ($F$), возникающих из ($\Phi$), является общим решением этого уравнения ($F$).

Любая функция, возникающая из общего решения ($\Phi$) нелинейного дифференциального уравнения ($F$) при определении некоторых произвольных постоянных в ($\Phi$), которая содержит оставшиеся неопределенными постоянные в качестве произвольных параметров, является «частью» общего решения уравнения ($F$).

Если имеются функции ($\Phi_i$), $i = 1, 2, \ldots, n$, обладающие свойствами, аналогичными свойствам ($\Phi$) относительно ($F$), и, кроме того, частные решения любых двух функций ($\Phi_i$) не тождественны, то их следует рассматривать как «разные общие решения» дифференциального уравнения ($F$); тогда говорим, что ($F$) имеет «несколько общих решений». Другими словами, рассматривая ($\Phi_i$) как множество решений дифференциального уравнения ($F$) таких, что ни одна из любых двух функций ($\Phi_i$) не включает другую, но они могут иметь пересечение, мы считаем их разными общими решениями нелинейного дифференциального уравнения ($F$).

**2.** Метод ограничения величин дифференциального уравнения. Ограничение решений или других величин дифференциального уравнения может дать несколько общих решений дифференциального уравнения. Приведем примеры.

Пример 1.

$$y'y''' - y''^2 = 0. \qquad (1)$$

Ограничим решение $y$ и получим 3 случая:

а) если $y$ ограничено посредством $y' = 0$, откуда следует $y'' = 0$, тогда (1) удовлетворяется посредством $y' = 0$, $y'' = 0$, так что $y = c$ удовлетворяет (1); $c$ — произвольная постоянная;

б) если $y$ ограничено посредством $y'' = 0$, откуда следует $y''' = 0$, тогда (1) удовлетворяется посредством $y'' = 0$, $y''' = 0$, так что

$$y = a_1x + a_2, \qquad (2)$$

с произвольными постоянными $a_1$, $a_2$, является общим решением уравнения (1);

в) если $y$ ограничено посредством $y' \neq 0$ и $y'' \neq 0$, дифференциальное уравнение (1) можно записать в виде

$$\frac{y'''}{y''} = \frac{y''}{y'},$$

откуда посредством интегрирования можно получить

$$\frac{y''}{y'} = c_1, \qquad (3)$$

$c_1$ — произвольная постоянная. Интегрирование (3) дает

$$y' = c_2e^{c_1x}, \qquad (4)$$

$c_2$ — вторая произвольная постоянная. Интегрирование (4) дает

$$y = \frac{c_2}{c_1}e^{c_1x} + c_3, \qquad (5)$$

которое содержит три произвольных постоянных.

Функции (2) и (5) — два различных общих решения дифференциального уравнения (1), содержащие прямые линии, параллельные оси $x$-ов, как общую часть.

Пример 2.

$$y'''(1+y'^2)-2y'y''=0. \tag{6}$$

Ограничим решение $y$ двумя способами:

а) если $y$ ограничить посредством $y''=0$, откуда следует $y'''=0$, тогда дифференциальное уравнение (6) будет одновременно удовлетворяться посредством $y''=0$, $y'''=0$, и функция

$$y=a_1x+a_2 \tag{7}$$

будет являться общим решением уравнения (6);

б) если $y$ ограничивается посредством $y''\neq 0$, тогда (6) может быть точно проинтегрировано [1], и функция

$$x^2+y^2+c_1x+c_2y+c_3=0, \tag{8}$$

т. е. семейство окружностей в плоскости $(x, y)$ будет общим решением уравнения (6).

Так как определение произвольной постоянной (7) и (8), которое отождествляет частное решение (7) с частным решением (8) не осуществлено, то в таком случае функции (7) и (8) являются различными общими решениями уравнения (6).

Пример 3.

$$y''+\frac{2}{x}y'+y^n=0. \tag{9}$$

В этом примере ограничим скаляр $n$. Это уравнение хорошо известно в астродинамике как дифференциальное уравнение «Емдена» [2]. Его решение в замкнутом виде известно в случаях $n=0$, $n=1$, когда (9) линейно; в случае $n=5$, функция, содержащая произвольную постоянную, удовлетворяет этому дифференциальному уравнению, и рассматривается как «часть» «неизвестного» общего решения. Ни в каких других случаях скаляра $n$ дифференциальное уравнение (9) не имеет общего решения.

Хотя общие решения и наиболее желательны, но встречаются случаи, когда они не в состоянии отразить всей сути рассматриваемых физических явлений и любая попытка найти общие решения должна побуждаться только теоретическим любопытством. Емденовское дифференциальное уравнение и является таким примером.

В своих исследованиях Емден нашел, что решения уравнения (9) имеют физический смысл в случае, когда $n$ находится между 0 и 5. Он также нашел, что в физических ситуациях исследуются граничные условия: $x=0$, $y=1$, $y'=0$, и при этих условиях частные решения уравнения (9), известные в случаях $n=0, 1, 5$, не приемлемы. Емден и его последователи нашли решение, удовлетворяющее физическим требованиям, и этим решением было разложение Тейлора относительно $x=0$ с последующим применением аналитического продолжения ряда.

Заметим, что, используя ограничения величин дифференциального уравнения, можно получить несколько общих решений нелинейного дифференциального уравнения, отличных от приведенных примеров.

3. Метод факторизации дифференциального уравнения. Факторизация дифференциального уравнения может привести к нескольким общим решениям дифференциального уравнения. Проиллюстрируем этот метод на нескольких примерах.

Пример 4.

$$x^3y''y^{IV}+x^2y''^2-2xy'y''+2yy''=0, \tag{10}$$

$y''$ — общий множитель всех членов уравнения (10), так что либо

$$\text{а)}\ y''=0, \quad \text{либо б)}\ x^3y^{IV}+x^2y''-2xy'+2y=0,$$

откуда имеем

$$\text{а) } y = a_1x + a_2; \quad \text{б) } y = c_1x + c_2x^2 + c_3x^{-1}. \tag{11}$$

Функции (11)— два общих решения уравнения (10). Видим, что семейство прямых линий, проходящих через начало координат,— общая часть этих общих решений.

Пример 5. $F(x, y, y') = 0$, когда это выражение есть полином по $y'$ степени $m$.

В этом случае имеем

$$F(x, y, y') \equiv y'^m + P_1(x, y)\, y'^{m-1} + \ldots + P_{m-1}(x, y)\, y' + P_m(x, y) = 0. \tag{12}$$

Мы можем решить (12) относительно $y'$, если $F_{y'}(x, y, y') \neq 0$; если $y'_1, y'_2, \ldots, y'_m$ являются $m$ простыми корнями, то можно записать

$$F(x, y, y') \equiv [y'_1 - P_1(x, y)]\,[y'_2 - P_2(x, y)] \ldots [y'_m - P_m(x, y)] = 0. \tag{13}$$

Приравнивая к нулю каждый множитель, из соответствующих дифференциальных уравнений получаем $m$ функций:

$$\varphi_1(x, y, c_1) = 0,\ \varphi_2(x, y, c_2) = 0, \ldots, \varphi_m(x, y, c_m) = 0, \tag{14}$$

являющихся $m$ разными общими решениями выражения (12). Из них выбирают такие, которые являются действительными функциями от действительных переменных. Например, в дифференциальном уравнении

$$y'^2 y''' + y^2 y''' + y''' = 0 \tag{15}$$

факторизация дает: $y'''(y'^2 + y^2 + 1) = 0$ и, так как $y'^2 + y^2 + 1 \neq 0$, это дифференциальное уравнение (15) эквивалентно $y''' = 0$, так что функция

$$y = c_1x^2 + c_2x + c_3 \tag{16}$$

является общим решением дифференциального уравнения (15). Это общее решение состоит из двух частей разного характера — семейства «парабол» ($c_1 \neq 0$) и семейства «прямых линий» ($c_1 = 0$).

## ЛИТЕРАТУРА

1. Davis H. Introduction to Nonlinear Differential and Integral Equations. Dover Publications Inc., New York, 1962, p 371—377.
2. Goursat E. Differential Equations. Ginn and Company, Boston, 1917, pg. 5.

США

Поступила в редакцию
19.V. 1977 г.

Magiros D.G., UNIFICATION OF STABILITY CONCEPTS. -Mat. Fizika, V. 33 (1983): 16-21.

A great number of concepts of stability are illustrated and unified in the present paper by means of the same mathematical relations, which give the possibility to obtain concrete definitions of stability and some correlations between them.

The author considers the systems under the conditions of instantly and constantly effecting perturbations, and unifies the concepts of stability on the basis of metrics. Relations between stability according to Lyapunov, Poincaré and Lagrange are discussed.

УДК 517.91

Д. Г. МАЖИРОС

# УНИФИКАЦИЯ ПОНЯТИЙ УСТОЙЧИВОСТИ

Исследования основных физических, технологических и социальных проблем приводят к формулировке проблем устойчивости динамических систем. Эти исследования наталкиваются на трудности различной природы — в понимании концепции устойчивости, приемлемых с физической точки зрения и совместных с математической; в разработке эффективных критериев, с помощью которых можно было бы сделать заключение об устойчивости или неустойчивости состояния системы. Разнообразие понятий устойчивости возникает по различным причинам, обусловленным формой математической модели системы; природой выбора системы переменных; способами описания состояния системы; способами приближения или отталкивания от другого состояния системы; характером возмущений, действующих на систему; способом измерения нормы возмущений и их действия на систему. В данной работе множество понятий устойчивости выражено и унифицировано с помощью одних и тех же математических соотношений, которые посредством конкретизации параметров дают возможность получить конкретные понятия устойчивости и некоторые соотношения между ними. Эта унификация приносит естественное упрощение в понимании проблем, касающихся устойчивости, и является источником для получения новых результатов.

**Системы при мгновенно или постоянно действующих возмущениях.** Опишем математические модели физических систем, устойчивость решений которых будем исследовать. Нарушения состояния системы зависят от возмущений, которые можно рассматривать, как возмущающие силы, действующие на систему



Reprinted from *Matematich. Fizika*. Ukrainian Academy of Sciences **33** (1983), 16–21.

мгновенно или постоянно. Мгновенно действующие возмущения нарушают начальные состояния, но не появляются в математической модели, т. е. они не проявляются в уравнениях, описывающих движение системы. В случае мгновенных возмущений модель можно взять в векторной форме

$$\dot{x}(t) = X(t, x), \quad x(t_0) = x_0, \quad X(t, 0) = 0, \quad t \geqslant t_0, \tag{1}$$

где $X = (X_1, \ldots, X_n)$, $x = (x_1, \ldots, x_n)$ — переменные состояния; $x_0 = (x_{10}, \ldots, x_{n0})$ — начальные условия. Эти уравнения рассматриваются в $(t, x)$-пространстве. Постоянно действующие возмущения изменяют состояние системы, $X$ изменяется в зависимости от возмущений, а математическая модель имеет вид

$$\dot{x}(t) = X(t, x) + p(t, x). \tag{2}$$

Здесь $p$ — постоянно действующие возмущения. Для существования единственного решения уравнений (1), (2) величины $X$ и $p$ должны удовлетворять определенным условиям.

Когда модель выражается дифференциальными уравнениями с отклоняющимися аргументами, например дифференциальными уравнениями с запаздывающим аргументом $\tau(t)$, различаем два типа моделей:

1) при мгновенно действующих возмущениях модель имеет вид

$$x(t) = X[t, x(t - \tau(t))], \quad \tau > 0, \tag{3}$$

при этом начальная функция $\varphi(t)$ определена на начальном интервале $E_{t_0} : t_0 - \tau \leqslant t \leqslant t_0$, а соответствующее решение $x_\varphi(t)$ (3) при $t \geqslant t_0$;

2) при постоянно действующих возмущениях —

$$x(t) = X[t, x(t - \tau(t))] + p[t, x(t - \tau(t))]. \tag{4}$$

**Действие возмущений. Некоторые определения.** Следующие соображения и определения дадут возможность унифицировать понятия устойчивости.

Результатом действия возмущений на систему является изменение определенных величин, принадлежащих системе, и это воздействие может быть наглядно выражено путем изменения траектории $S$ начального (невозмущенного) решения системы в траекторию $\bar{S}$ нового (возмущенного).

Состояние системы обозначается точкой траектории, и если точка $M$ невозмущенной траектории $S$ соответствует точке $\bar{M}$ возмущенной траектории $\bar{S}$, расстояние $\rho = M\bar{M}$ (рис. 1) может рассматриваться как мера величины действия возмущений на $M$.

Это расстояние играет важную роль в понимании устойчивости и формулирования ее соотношений. Назовем его расстоянием устойчивости.

Данной точке $M$ траектории $S$ могут соответствовать разные точки $\bar{M}$ траектории $\bar{S}$. Такое соответствие допускает предположение о физической пригодности и характеризует конкретное понятие устойчивости траектории $S$. Если имеется два основных понятия устойчивости, то можно выделить два наиболее физически осмысленных соответствия между $M$ и $\bar{M}$:

1) одно соответствие между $M$ и $\bar{M}$ на $S$ и $\bar{S}$ справедливо тогда, когда эти точки взяты в один и тот же момент времени (рис. 2); в этом случае расстояние $\rho = \rho_l = M\bar{M}$ называется ляпуновским и характеризует одно из основных понятий устойчивости — устойчивость в смысле Ляпунова, а математическая модель относится к пространству $(t, x)$;

2) при другом соответствии расстояние $\rho = \rho_r = M\bar{M}$ (рис. 3) берется как минимальное расстояние от точки $M$ до точек траектории $\bar{S}$ и называется расстоянием Пуанкаре, характеризующим другое основное понятие устойчивости — устойчивость в смысле Пуанкаре, или орбитальную устойчивость, в этом случае модель относится к пространству $(x, y = \dot{x})$, где $t$ является параметром.

**Унификация понятий устойчивости.** Разнообразие понятий устойчивости может быть количественно описано следующими соотношениями:

$$\rho_0 < \delta_1, \tag{5}$$

$$\rho < \varepsilon, \qquad (6)$$

$$\lim_{t\to\infty} \rho = 0, \qquad (7)$$

$$\|p\| < \delta_2, \qquad (8)$$

$$\rho_\tau < \delta_3, \qquad (9)$$

которые унифицируют понятия устойчивости. Здесь $\varepsilon$, $\delta_1$, $\delta_2$, $\delta_3$ — положительные постоянные; $\delta_1$, $\delta_2$ зависят от $\varepsilon$ и начального момента $t_0 \geqslant 0$; $\rho_0$, $\rho$ — расстояния устойчивости; $\| p \|$ обозначает норму возмущений; $\rho_\tau$ является расстоянием между невозмущенным и возмущенным запаздываниями.

Укажем некоторые практически полезные понятия устойчивости, вытекающие из соотношений (5) — (9) путем их комбинаций, конкретизации $\rho$ и из ограничений других величин.

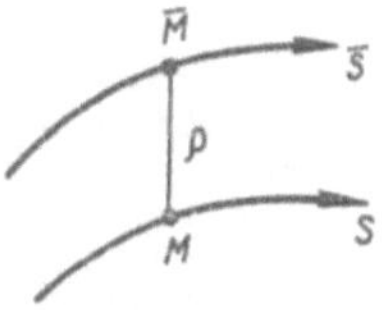

Рис. 1

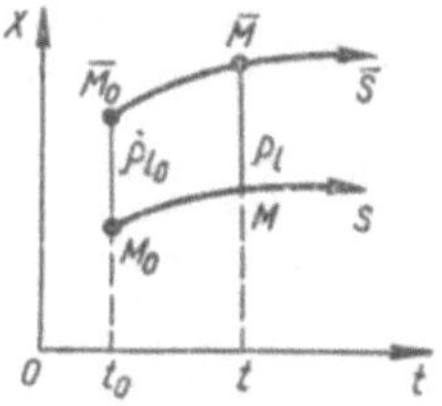

Рис. 2

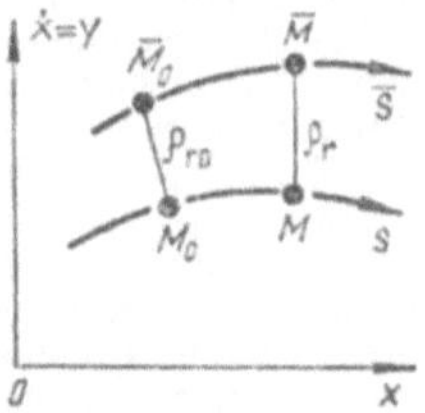

Рис. 3

**Понятия устойчивости, вытекающие из соотношения (5).** Для понятий устойчивости в случае модели (1) используем соотношения (5) — (7). Неравенство (8) используется в случае модели (2), а неравенство (9) — если в моделях (3), (4) возмущается запаздывание.

1. Решение $x(t)$ системы (1) называется устойчивым, если для данного сколь угодно малого $\varepsilon$ можно найти $\delta_1 = \delta_1(t_0, \varepsilon)$ такое, что из неравенства $\rho_0 < \delta_1$ следует, что $\rho < \varepsilon$ для $t \geqslant t_0$. Расстояние $\rho = |\bar{x} - x|$, где $\bar{x}$ — возмущение $x$; $\rho_0$ — начальное расстояние. Если $\rho$ является ляпуновским расстоянием, то $x(t)$ устойчиво в смысле Ляпунова, а если $\rho$ — расстояние Пуанкаре, то имеет место устойчивость в смысле Пуанкаре (орбитальная устойчивость). Если к тому же выполняется условие (7), то имеем асимптотическую устойчивость.

2. Для устойчивости решения уравнения (2) является важным неравенство (8). Различным нормам $p$ соответствуют различные типы устойчивости. Укажем три из них [10]: траектория $S$ является устойчивой в целом, если $p$ мало и его норма определяется как $\| p \| = \max_x | p |$; $S$ интегрально устойчива, если $p$ может быть большим на малых интервалах времени, а его норма определяется как $\| p \| = \int_0^\infty \max_x | p | \, dx$; $S$ устойчива в среднем, если $\| p \| = \int_t^{t+T} \max_x | p | \, dx$.

Существуют утверждения, дающие соотношения между этими понятиями устойчивости. Указанная выше устойчивость может пониматься как в смысле Ляпунова, так и в смысле Пуанкаре.

**Некоторые замечания по понятиям устойчивости.** 1. Понятие устойчивости по Ляпунову соответствует устойчивости движения по траектории, а понятие устойчивости Пуанкаре — устойчивости самой траектории. В то время как понятие устойчивости Ляпунова является слишком ограничительным для периодических движений, понятие устойчивости Пуанкаре является именно тем, которое нужно для периодических движений.

2. При исследовании устойчивости точек равновесия системы расстояния по Ляпунову и Пуанкаре идентичны, так как устойчивость точек равновесия в смысле Ляпунова и Пуанкаре эквивалентна.

3. Если $\delta_1$ не зависит от $t_0$, $\delta_1 = \delta_1\varepsilon$, обе устойчивости являются равномерными.

4. Непрерывная зависимость решения от начальных условий эквивалентна устойчивости решения по отношению к начальным условиям, рассматриваемым как параметры решений.

5. Понятие ограниченности решений, которое может быть дано как частный случай соотношений (5) — (9), есть частный случай понятия устойчивости, называемый устойчивостью по Лагранжу. Но свойство ограниченности решения является отличным от свойства устойчивости, хотя в некоторых случаях одно вытекает из другого [1; 3, с. 129; 11].

6. Природа устойчивости физических систем является неинвариантной по отношению к выбору пространственных переменных, используемых как координаты системы, или же по отношению к преобразованию этой системы координат [6]. Можно также иметь устойчивость только по отношению к части координат и неустойчивость по отношению к другим координатам [3, с. 10; 12, с. 96—97]. Зависимость устойчивости решения от системы координат и ее преобразований является весьма важной проблемой, но по ней нет никаких систематических исследований.

7. Система определяется как структурно устойчивая, если ее устойчивость инвариантна при малых возмущениях параметров либо формы правых частей самой системы. Из структурной устойчивости системы вытекает много важных свойств [5]. Свойства структурной устойчивости подсказывают методы исследования физических проблем с хорошо понимаемыми результатами, как в задачах, связанных с морфологическими процессами, катастрофами в биологии и т. д. [9].

8. Обсуждаемые выше понятия устойчивости носят теоретический или математический характер, и чтобы адекватно интерпретировать реальные соотношения, необходимы соответствующие модификации, изменения, дополнения [3, с. 126; 4; 12, с. 13; 123].

Устойчивость, имеющая практическую пригодность, называется практической устойчивостью. Для нее требуется знание следующих параметров:

(*i*) размера области отклонений системы, пригодной для удовлетворительной работы;

(*ii*) размера области допустимых начальных условий, которые могут быть управляемы;

(*iii*) размера области допустимых возмущений;

(*iv*) размера области допустимых значений параметров;

(*v*) конечного времени, в течение которого исследуется устойчивость.

Для безупречной практической устойчивости размеры указанных выше областей должны быть большими. Вычисляются эти области с использованием нелинейностей в математической модели системы, и в этом случае метод линеаризации модели не допускается.

Рассмотрим примеры, являющиеся приложением определений устойчивости.

1. Прямолинейное движение: $x(t) = v(t - t_0) + a$ в плоскости $(t, x)$, где $a$ — расстояние, $v$ — скорость движущейся массы (наклон прямолинейной траектории движения):

*i* является устойчивым в смысле Ляпунова и Пуанкаре, если движение равномерное;

*ii* является неустойчивым в смысле Ляпунова и Пуанкаре, если движение неравномерное.

Упомянутое выше движение в общих случаях неустойчиво по Лагранжу.

2. Движение массы, притягиваемой к центру по обратно квадратическому закону Ньютона, является периодическим с эллиптической орбитой. Это движение по Ляпунову неустойчиво, а по Пуанкаре и Лагранжу устойчиво.

3. Система нелинейных дифференциальных уравнений

$$\dot{x} = -y(x^2 + y^2)^{\frac{1}{2}}, \quad \dot{y} = x(x^2 + y^2)^{\frac{1}{2}} \tag{10}$$

в качестве решения имеет

$$x = a\cos(at + b), \quad y = a\sin(at + b) \tag{11}$$

с орбитой

$$x^2 + y^2 = a^2, \tag{12}$$

где $a$, $b$ — параметры. Имеем ($i$), если $a$ — постоянная, решение (11) устойчиво по Ляпунову, но не имеет смысла по Пуанкаре;

($ii$) если $a$ — переменная, решение (11) неустойчиво по Ляпунову, но устойчиво по Пуанкаре. В обоих случаях движение устойчиво по Лагранжу.

Отметим, что ляпуновская неустойчивость в случае переменного $a$ может быть устранена посредством замены переменных $x$, $y$ новыми $r$, $b$ по формулам перехода [2]

$$x = r\cos v, \quad y = r\sin v, \quad v = at + b, \tag{13}$$

когда неравенство (6) переходит в уравнения $\dot{r} = 0$, $\dot{b} = 0$, решением которых является $r = c_1$, $b = c_2$, где $c_1$, $c_2$ — произвольные постоянные. Это решение устойчиво по Ляпунову.

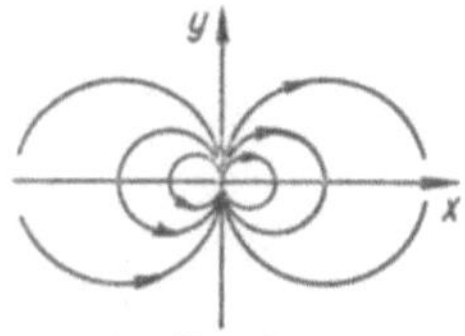

Рис. 4

4. Система нелинейных дифференциальных уравнений

$$\dot{x} = 2xy, \quad \dot{y} = x^2 - x^2 \tag{14}$$

в качестве решения имеет однопараметрическое семейство кривых (циклов)

$$(x - r)^2 + y^2 = r^2, \tag{15}$$

единственной точкой равновесия которого является начало координат (рис. 4). Этот пример напоминает картину магнитного силового поля вокруг электрона. Решение, начинающееся в любой точке на любой интегральной кривой, стремится при $t \to \infty$ к предельной точке — началу координат, т. е. $\lim\limits_{t\to 0} \rho = 0$. Видно, что для данного $\varepsilon > 0$ не существует $\delta_1$ такого, что для любого начального $\rho < \delta_1$ все точки соответствующей кривой решения будут находиться внутри окружности радиуса $\varepsilon$, т. е. соотношение $\rho < \varepsilon$ не может выполняться. Так как выполняется только условие (7), то начало координат является «аттрактором» специального типа и псевдоасимптотически устойчиво.

5. Если твердое тело вращается вокруг своей оси симметрии, то возникает явление процессии, которое зависит от природы внешнего момента вращения, действующего на тело. В частном случае момент вращения процессии соответствует некоторой кривой на поверхности ортогонального кругового цилиндра и эта кривая стремится к бесконечности при $t \to \infty$, когда процессия относится к геликоидному типу. Этот вид процессии неустойчив по Ляпунову, но асимптотически устойчив по Пуанкаре. Из практических соображений в случае геликоидной процессии для описания устойчивости следует принять асимптотическую устойчивость по Пуанкаре [6].

6. Исследование численных процессов с использованием быстродействующих ЭВМ дало начало математическим моделям с возмущающими членами, устойчивыми в смысле Ляпунова и Пуанкаре.

1. *Antosiewicz H.* Boundedness and stability.— In: Nonlinear differential equations and nonlinear mechanics. New York : Acad. press, 1963.— 270 p.
2. *Cesari L.* Asymptotic behavior and stability problems in ordinary differential equations. — Berlin : Springer, 1959.— 271 p.
3. *Hahn W.* Theory and application of Liapunov direct method.— Englewood Cliffs : Prentice-Hall, 1963.
4. *LaSalle J., Lefschetz S.* Stability by Liapunov's direct method.— New York : Acad. press, 1961.
5. *Lefschetz S.* Differential equations : geometric theory.— New York : Intersci. publ., 1957.— 364 p.
6. *Magiros D.* Mathematical models of physical and social systems.— Practica, Athens Academy of sciences, 1976.
7. *Magiros D.* Characteristic properties of linear and nonlinear systems.— Ibid.
8. *Magiros D.* On the stability of a special class of precessions.— In: Yolume in memorium D. Eginitis. Athens, s. a., p. 189—198.
9. *Thom R.* Structural stability and morphogenesis.— Reading. (Mass.) : Benjamin, 1975.— 362 p.
10. *Vrkoc I.* On some stability problems.— In: Proc. conf. Prague. New York : Acad. press, 1963, p. 217—221.

11. *Yoshizawa T*. Stability theory by Liapunov second method.— Tokyo : Math. Soc. Jap., 1966.— 223 p.
12. *Zubov V*. Mathematical methods for the study of automatic control systems.— New York : MacMillan, 1963.— 324 p.

Корпорация «Дженерэл Электрик»

Поступила в редколлегию 24.10.80

УДК 62.502

# BIOGRAPHICAL NOTE OF D.G. MAGIROS

Demetrios G. Magiros was born in Euboia Island, Greece, December 29,1912. He died in Philadelphia, Pennsylvania, January 19, 1982.

Magiros studied pure mathematics at the University of Athens, Greece, where he prepared his doctor's thesis in 1940. He begun his teaching career at the National Metsoveion Polytechneion (National Technical University) of Athens as a lecturer of mechanics and geodesy, and later as a professor of mathematics of the same institute. In 1949 he left for the United States with the purpose of carrying out further studies and advanced research in applied mathematics. Magiros showed early in his career his enormous interest in the applications of mathematics and his conviction that pure mathematics had important applications that mathematicians had a duty to seek to facilitate. In this attitude he differed sharply from many mathematicians who praised pure mathematics for its beauty and logical rigor alone, while disdaining any thought of application.

In the United States he continued his studies in applied mathematics at Brown University, the Courant Institute of Mathematical Sciences of New York University, and the Massachusetts Institute of Technology. He was first engaged in research at the IBM Watson Laboratory of Columbia University, and continued at the Republic Aviation Corporation and the Courant Institute (NYU). At Hofstra University he was appointed professor of mathematics and mechanics, while at the same time he accepted the duties of scientific consultant at the Space and Missile Center of General Electric Company in Philadelphia.

By 1960, when Magiros assumes full-time responsibilities as a researcher and mathematical consultant of the G.E. Space and Missile Center (presently, Re-Entry and Environmental Systems Division), he has developed a remarkable multi-dimensional scientific activity. He keeps in constant communication with distinguished scientists and academicians of various countries, such as, Leon Brillouin, Henry Villat, S. Lefschetz, C. Friedricks, L.H. Thomas, Y. Mitropolsky, G. Duboshin, C. Kodradiyev, who influenced to a degree his research orientation.

It was in this scientific center, one of the most advanced and dynamic space centers of the world where, with incomparable enthusiasm and dedication, he prepared most of his important work at a time when the knowledge on space vehicles and their orbits was limited while the world competition was highly intense.

His contribution to mathematical formulation and solution of high research space problems, such as, long-range ballistic missiles, ballistic re-entry, hypersonic aerothermodynamics with high speed re-entry, planetary entry, absorb energies in certain spectra, theories for improvement of the system effectiveness, etc., is of fundamental significance. At the same time his influence as a consultant expanded in many G.E. scientific centers in U.S. and his seminars and lectures were attended intensively by numerous scientists from all over the world.

The Space Center of General Electric Co. and the Athens Academy of Sciences repeatedly honored Magiros' distinguished service to mathematical advancement.

Magiros' remarkable scope of interests and his important contribution to mathematical research and technical progress rendered him an internationally recognized scientist. An intense, driving yet sensitive and modest man, a prolific mathematician and an inspiring teacher Magiros will continue for a long time to be an influence on mathematical research through his own works and through those he will be inspiring.

## COMPLETE CHRONOLOGICAL LIST OF MAGIROS' PUBLICATIONS


[1] 1946 Ο τόπος κορυφής μεταβλητής αλυσοειδούς {On the locus of the vertex of a variable catenary}. Tech. Chron., Sc.J.Tech.Chamber,Greece,V.23,no. 265-266 (1946):99-102.

[2] 1946 Περί της τάσεως εις μεταβλητήν αλυσοειδή {The study of the tension of a variable catenary} .Proc. Athens Acad.Sci.,V.21 (1946):41-46;V.23,no.273-274 (1947): 3-7.

[3] 1946 Περί τασικών διαγραμμάτων των διατομών πακτώσεως εις ανηρτημένα σύρματα. {Diagrams of the tension of the cross-section in the suspension of hanging wires} . Proc.Athens Acad.Sci.,V.21 (1946):46-50; Tech. Chamber, Greece, V.23,no.273-274 (1947): 8-10.

[4] 1956 Ελεύθεραι πλευρικαί ταλαντώσεις απλής ράβδου υπό πλαστικήν εξαίτησιν, {I. Lateral free vibrations of simple prismatic bar in plasticity} . Proc. Athens Acad.Sci.,V.31 (1956):16-21.

[5] 1956 Ελεύθεραι πλευρικαί ταλαντώσεις απλής ράβδου υπο πλαστικήν εξαίτησιν, II. {Lateral free vibrations of simple prismatic bar in plasticity,II} .Proc. Athens Acad.Sci.,V.31 (1956):168-171.

[6] 1957 Subharmonics of any order in case of nonlinear restoring force, pt.I.Proc.Athens Acad.Sci.,V.32 (1957):77-85.

[7] 1957 Subharmonics of order one third in the case of cubic restoring force. Proc.Athens Acad.Sci.,V.32 (1957):101-108.

[8] 1957 Remarks on a problem of subharmonics. Proc.Athens Acad.Sci.,V.32. (1957):143-146.

[9] 1957 On the singularities of differential equations, where the time figures explicity. Proc.Athens Acad. Sci.,V.32 (1957):448-451.

[10] 1958 Subharmonics of any order in nonlinear systems of one degree of freedom:application to subharmonics of order 1/3. Inf.and Control,V.1,no.3 (1958): 198-227.

[11] 1959 On a problem of nonlinear mechanics. Inf.and Control,V.2,no.3 (1959):297-309;Proc.Athens Acad.Sci., V.34 (1959):238-242.

[12] 1960 The motion of a projectile around the earth under the influence of the earth's gravitational attraction and a thrust. Proc.Athens Acad.Sci.,V.35 (1960):96-103.

[13] 1960 The Keplerian orbit of a projectile around the earth, after the thrust is suddenly removed. Proc. Athens Acad.Sci.,V.35 (1960):191-202.

[14] 1960 A method for defining principal modes of nonlinear systems utilizing infinite determinants, I . Proc. Natl. Acad. Sci.,U.S.,V. 46,no.12 (1960):1608-1611.

[15] 1961 A method for defining principal modes of nonlinear systems utilizing infinite determinants, II . Proc. Natl.Acad.Sci.,U.S.,V.47, no.6 (1961):883-887.

[16] 1961 Diffraction by a semi-infinite screen with a rounded end (with Joseph B. Keller). Comm.Pure Appl. Math.,v.14,no.3 (1961):457-471.

[17] 1961 Method for defining principal modes of nonlinear systems utilizing infinite determinants. J.Math. Phys.,V.2,no.6 (1961):869-875.

[18] 1961 Remarks on Rosenberg's paper "The normal modes of nonlinear n-degree-of-freedom systems". J.Appl. Mech.,Trans.ASME,V.30,ser.E,no.1 (1963):151.

[19] 1963 On the convergence of series related to principal modes of nonlinear systems. Proc.Athens Acad.Sci., V.38 (1963):33-36.

[20] 1963 On the convergence of the solution of special two-body problem. Proc.Athens Acad.Sci.,V.38 (1963): 36-39.

[21] 1963 The impulsive force required to effectuate a new orbit through a given point in space. J.Frank.Inst., V.276, no.6 (1963):475-489;Proc.XIVth Intl.Astron. Congress, Paris, 1963.

[22] 1964 Motion in a Newtonian forced field modified by a general force, {I.}. J. Frank.Inst.,V.278,no.6 (1964):407-416;Proc.XVth Intl.Astron.Congress, Warsaw, 1964.

[23] 1965 Motion in a Newtonian force field modified by a general force,{II}. XVIth Intl.Astron.Congress, Athens, Greece (1965):349-355.

[24] 1965 On the stability definitions of dynamical systems, Proc.Natl.Acad.Sci. (U.S.),V.53,no.6 (1965):1288-1294.

[25] 1965 Physical problems discussed mathematically.Bull. Soc.Math.Greece,nouv.ser.,t.6II,fasc.I (1965): 143-156.

[26] 1966 Motion in a Newtonian force field modified by a general force, III.Application:the entry problem (with G. Reehl). XVIIth Intl.Astron.Congress,Madrid (1966):149-154.

[27] 1966 The entry problem (with George Reehl) Proc.Athens Acad.Sci.,V.41 (1966):246-251.

[28] 1966 Stability concepts of dynamical systems,Inf.and Control,V.9,no.5 (1966):531-548.

[29] 1967 Unification and classification of stability concepts of dynamical systems. Proc.VIIth Intl.Symp. Space,Tech. and Sci.,Tokyo, 1967.

[30] 1967 Attitude stability of a spherical satellite. Remarks on stability.Proc.XVIIIth Intl.Astron.Congress,Belgrade, Yugoslavia, 1967.

[31] 1967 A class of nonautonomous differential equations reducible to autonomous ones by an exact method. Proc.Natl.Acad.Sci.,V.58,no.2 (1967):412-419.

[32] 1968 Sur quelques sortes de précession dans le mouvement de rotation d'un corps rigide ayant un axe de symetrie. Comptes Rendus, Acad. Sci.,Paris, ser.A, t. 266 (1968):770-773.

[33] 1968 Attitude stability of a spherical satellite (with A.J.Dennison).J.Frank.Inst.,V.286,no.3 (1968):193-203;Bull.Amer.Phys.Soc.,ser.2,V.12,no.3 (1967): p.288 (Abstract).

[34] 1969 La stabilité de la précession hélicoidale dans le sens de Liapunov,Poincaré et Lagrange. Comptes Rendus, Acad.Sci.,Paris, sér.A, t. 268 (1969): 652-654.

[35] 1969 On a class of precessional phenomena and their stability in the sense of Liapunov, Poincare and Lagrange. Proc.VIIIth Intl.Symp.on Space Tech. Sci., Tokyo (1969): 1163-1170.

[36] 1969 The stability in the sense of Liapunov, Poincaré and Lagrange of some precessional phenomena. Proc. Vth Intl.Conf.Nonl.Oscill.,Kiev (1969):347-357

[37] 1969 On the helicoid precession:its stability and an application to a re-entry problem (with G. Reehl). Proc. XXth Intl.Astron.Congress,Buenos Aires, Argentina (1969):491-496.

[38] 1969 Stability concepts of dynamical systems. Applications to flight dynamics and numerical analysis. Proc.Athens Acad.Sci.,V.44 (1971):270-284.

[39] 1970 Actual mathematical solutions of problems posed by reality, I:a classical procedure. Proc.Athens Acad.Sci.,V.45 (1971):179-187.

[40] 1971 Actual mathematical solutions of problems posed by reality, II:applications. Proc.Athens Acad. Sci.,V.46 (1971):21-31.

[41] 1971 Orientation of the angular momentum vector of a space vehicle at the end of spin-up. Proc.XXIIth Intl.Astron.Congress, Brussels, Belgium, 1971.

[42] 1971 Stability concepts of solutions of differential equations with deviating arguments. Proc. Athens Acad.Sci.,V.46 (1971):273-278.

[43] 1972 The stability of a class of helicoid precessions in the sense of Liapunov and Poincare. Proc.Athens Acad.,Sci.,V.47 (1972):102-110.

[44] 1974 Remarks on stability concepts of solutions of dynamical systems. Proc.Athens Acad.Sci.,V.49 (1975): 408-416.

[45] 1974 On the stability of a special class of precessions. In memoriam of "Demetrios Eginitis". Athens Acad. Sci.(1975);Proc.XXVth Intl.Astron.Congress,

[46] 1975 (a) Μαθηματικά μοντέλα φυσικών και κοινωνικών συστημάτων (Mathematical models of physical and social systems). Math.Epith.,Greek Math.Soc., V.3 (1975):81-124.

(b) Mathematical models of physical and social systems. Genl.Electric Co.,R.S.D.,Philadelphia, 1980.

[47] 1977 On the linearization of nonlinear models of the phenomena,pt.I:linearization by exact methods. Proc.Athens Acad.Sci.,V.51 (1977):659-668.

[48] 1977 On the linearization of nonlinear models of the phenomena,pt.II:linearization by approximate methods. Proc.Athens Acad.Sci.,V.51 (1977): 669-683.

[49] 1977 Characteristic properties of linear and nonlinear systems. Proc.Athens Acad.Sci.,V.51 (1977): 907-935.

[50] 1977 (a) Nonlinear differential equations with several general solutions. Proc.Athens Acad.Sci.,V. 52 (1977):221-229.

(b) Nonlinear differential equations with several general solutions (In Russian). Ukrainian Math. J.,Ukrainian Acad.Sci.,V.30,no.2 (1978):238-241.

[51] 1978 The general solutions of nonlinear differential equations as functions of their arbitrary constants. Proc.Athens Acad.Sci.,V.52 (1978):524-532.

[52] 1979 On the separatrices of dynamical systems.Proc. Athens Acad.Sci.,V.54 (1980):264-287;Proc.IXth.Intl. Congress on Nonl.Oscill.,Kiev, 1981.

[53] 1980 Unification of stability concepts. Philadelphia: Genl.Electric.Co.,R.S.D.,1980. Mathematich. Fizika, Ukrainian Acad.Sci. (April) 1983.

[54] 1980 Stability concepts of dynamical systems. Philadelphia:Genl.Electric Co.,R.S.D.,1980.

## MAGIROS' UNPUBLISHED WORKS

### LECTURE NOTES

1963 Special functions and applications.
1963 Integral transforms.
1964 On some topics of nonlinear mechanics.
1965 Selected topics in applied mathematics.
1966 Stability concepts and criteria of dynamical systems.
1967 Nonlinear ordinary differential equations with closed form solutions.
1968 Methods for solutions of nonlinear differential equations.
1972 Linear, nonlinear and linearized phenomena.
1973 Exact solutions on nonlinear ordinary differential equations.
1974 Geometrical methods for solution of nonlinear ordinary differential equations.
1974 Numerical methods for solution of Nonlinear ordinary differential equations.
1975 Calculus of variations and applications, I.
1975 Optimal control systems. Applications.
1975 Matrices and transformations.

### MONOGRAPHS

1976 Differential equations with one argument.
1976 Linearization of nonlinear models of the phenomena.
1977 Mathematics and real world.
1977 General and singular solution of nonlinear differential equations.
1977 Approximate methods for solutions of nonlinear differential equations: an introduction with some remarks.
1978 Approximate analytical methods for solution of nonlinear differential equations.
1979 Differential equations with one argument. Applications.
1979 Mathematics in economics.

### REGULAR REPORTS

1961 Problems related to the motion of a projectile under the influence of a Newtonian center and a thrust (Genl. Electric Co., TIS no. 61SD36).

1961 Motion of a projectile under the influence of the attractive force and a Newtonian center and a thrust (Genl. Electric Co., TIS no. 61SD142).
1962 An actual solution of a problem of nonlinear mechanics (Genl. Electric.Co., TIS no. 62SD144).
1963 Problems of classical celestial mechanics and astrodynamics (Genl. Electric Co., TIS no. 63SD957) {Rev. ed. 1978}
1964 Linear and nonlinear phenomena (Genl. Electric Co., TIS no. 64SD291).
1968 The study of orientation of an earth satellite (Genl. Electric Co., TIS no. 68SD249).
1968 Groups of problems of practical importance in numerical analysis, flight dynamics and nonlinear mechanics (Genl.Electric Co., TIS no. 68SD273).
1971 Remarks on linearization of models of nonlinear phenomena (Genl. Electric Co., TIS no. FM#71-1).
1972 Linear, nonlinear and linearized phenomena (Genl. Electric Co., TIS no. 72SD 225).
1972 The study of mathematical models of physical and real life phenomena.
1972 The role of mathematics in the investigation of physical and real life phenomena (Genl. Electric Co., TIS no. 72SD228).
1973 Exact solutions of nonlinear ordinary differential equations (Genl. Electric Co., TIS no. 73SD240).
1974 Geometrical methods for solution of nonlinear ordinary differential equations (Genl. Electric Co., TIS no. 74SD204).
1974 Numerical methods for solution of nonlinear ordinary differential equations (Genl. Electric Co., TIS no. 74SD20).
1976 Optimal control systems: Theory and applications, pt. I. (Genl. Electric Co., TIS no. 76SDR007).
1977 Matrices and transformations. Applications (Genl. Electric Co., TIS no. 76SDR042).

## INTERIM REPORTS

1961 Periodic solutions of the general linear inhomogeneous system with periodic coefficients (NAC Analysis and synthesis, no. 131-096).
1961 On a system of differential equations of which the coefficients are functions of many parameters (OAO Stability contract).
1963 The nature of the stability of nonlinear control systems.
1963 Criteria of stability of nonlinear control systems.
1965 On the blast problem.
1969 Linearization of nonlinear models of physical and social phenomena.
1970 A problem of evolution.
1970 Competition between two species, I: Mathematical remarks.
1970 Competition between two species, II: The stability of equilibrium.
1970 Competition between two species, III: Equilibrium point of the model with depletion terms.

1970 Competition between two species, IV-V: The stability of the equilibrium state of the model with completion terms (2 v.)